BIOLOGY

PRINCIPLES AND PERSPECTIVES

John E. Silvius

Professor of Biology

Cedarville College

KENDALL/HUNT PUBLISHING COMPANY
4050 Westmark Drive   Dubuque, Iowa 52002

Copyright © 1985, 1994 by John E. Silvius

ISBN 0-8403-9031-9

All rights reserved. No part of this publication may be reproduced, stored in a retrieval system, or transmitted, in any form or by any means, electronic, mechanical, photocopying, recording, or otherwise, without the prior written permission of the copyright owner.

Printed in the United States of America
10 9 8 7 6 5 4 3 2 1

Contents

Preface		v
Acknowledgments		vii

PART 1	LIFE AND LIFE SCIENCE	1
1	A Scriptural Perspective of Life	3
2	Unifying Characteristics of Living Systems	23
3	Biological Classifications: Patterns Amid Diversity	33
4	Atoms and Molecules of Life	49

PART 2	LIFE AND ENVIRONMENT	69
5	Energy Flow in Ecosystems	71
6	Nutrient Cycles	93
7	Population Ecology	113
8	Global Ecology and Stewardship	137

PART 3	LIFE WITHIN ORGANISMS	159
9	Diversity of Life	161
10	Nutrition	185
11	Reproduction	213
12	Genetics	241
13	Nervous Systems and Hormonal Systems	267
14	Homeostasis and Coordination of Life Processes	291
15	Genetic Variation and Selection	319

PART 4	CELLS — FUNCTIONAL UNITS OF LIFE	359
16	Cell Structure and Function	361
17	Genetic Control of Cellular Metabolism	381
18	Energy Capture and Conversion in Cells	407

APPENDIX A	A Review of Basic Chemistry	427
APPENDIX B	An Abbreviated Taxonomic Classification	435
INDEX		437

Preface

This textbook is intended for use in one-quarter or one-semester, general education course in introductory college biology. *Biology: Principles and Perspectives* offers a unique approach to the study of life science. First, like all biology texts, it reflects the particular "worldview" of the author. However, this text is written from a biblical, theistic worldview. Chapter 1 explains how such an approach is not only consistent with "good science," but also fundamental to the integrity and future progress of the natural sciences.

Second, *Biology: Principles and Perspectives* has a rather unique organization. After a brief introduction to the science of life and the nature of living matter in Part 1, the text introduces the ecosystem level of organization (Part 2), and treats organismic biology in Part 3. Part 4 concludes the sequence with a presentation of cellular and molecular biology. Dr. Alan W. Haney introduced me to this approach when I assisted him in teaching introductory botany at the University of Illinois. I have continued to use this "ecosystem-to-molecule" approach for the past fourteen years at Cedarville College. Many biological concepts involving organismic and cellular metabolism can be comprehended more easily and meaningfully when students have prior experience with the "context of life" through ecology.

A third feature of *Biology: Principles and Perspectives* is the emphasis upon principles *of biology* as models for understanding life and life processes. Individual concepts and facts are introduced only as a means of assisting students in understanding principles that attempt to explain biological processes. Although factual learning is still essential, this text strives to introduce biological science as a relevant discipline in today's world. As students are challenged anew by the *processes, people,* and *principles* of biology, perhaps you as professor will see some students choose a career in biology.

In keeping with the aim of emphasizing principles and preparing students to be lifelong learners, *Biology: Principles and Perspectives* provides several end-of-the-chapter aids. The QUESTIONS AND DISCUSSION TOPICS provide useful questions that both students and professor can use to spark discussions. Section 3-I provides a suggested approach to studying biology that incorporates these questions.

The TOPICS FOR FURTHER STUDY at the end of each chapter provide relevant topics for more detailed reading, whether or not research papers are assigned. No text can serve as a complete "library," but the text should encourage the use of the library. The REFERENCES listed at the end of the chapters should be helpful in this regard.

Like any text, *Biology: Principles and Perspectives* is intended to be a tool for use along with lecture, discussion, laboratory, and multi-media. May it be an effective aid to inquiry into life processes of body, community, and world; and may it challenge its readers to make a difference for eternity by being faithful stewards.

J.E.S.
Cedarville, Ohio

Acknowledgments

This book is the result of the inspiration and support of more persons than can be named here. However, some deserve special acknowledgment for their direct contributions. First, I am thankful to the Person of Jesus Christ through Whom, by God's grace, I received "new life" and a new "worldview" through faith in His Word. I am thankful for God's call to be a teacher, and for those mentors that He has provided to prepare me and to affirm that call.

I am also thankful to my wife, Alvadell, for her support, patience, and confidence in me during this project. My son, Brad, and daughter, Melinda, have also been patient as I have had to restrict my time with them.

I am indebted to my colleagues and students at Cedarville College for academic, financial, and spiritual support. Dr. Daniel Wetzel, Chair of the Science Department, and the college administration and trustees deserve special thanks for their provision of academic leave time for my research and writing. I have appreciated the support and helpful suggestions of Dr. Terry Phipps and Dr. Joseph Francis, both colleagues at Cedarville College, and of Dr. Jerry Wenger (Covenant College) and Prof. Jerry Johnson (Western Baptist College). These colleagues have used the 1985 edition of *Biology: Principles and Perspectives*, "the fledgling" which was published locally and printed by Aldine Printing Co., Xenia, OH.

My son, Brad Silvius, prepared many of the illustrations of food webs, human anatomy, and plant anatomy. The cover layout and Figures 6-3 through 6-7 were prepared by artist, Jamey Gregory of Tiffin, Ohio. Beverly Check scanned the illustrations into computer files and made helpful modifications. Lisa Lightly provided six months of encouraging assistance with word processing. Connie Winch and Donna Brock assisted with word processing and correspondence.

This text would not have reached its present stage without the helpful support of the staff at Kendall/Hunt. Special thanks to Matthew Krise, editor, and his predecessor, Theodore Hansen for their timely help and encouragement.

I am especially thankful to Dr. Alan W. Haney with whom I was privileged to serve as Visiting Lecturer at the University of Illinois. He taught me, by example, how to love my students and to love teaching. The "ecological-conceptual approach" used by Dr. Haney in his introductory botany course has been incorporated into my own teaching, as reflected in this text.

Finally, I thank my families, both biological and spiritual, my pastors, and my church. Many have been such an encouragement through their interest and prayers.

John Silvius
June 23, 1993

Part 1

Life and Life Science

Biology, of all the sciences, has the greatest potential for making major advances in aid of humanity; at the same time, most problems facing society today are either biological or have large biological components[1].

A *course* of study...

Welcome to a *course* of study in biology, the science of life. Embarking upon a course of study is like an orientation during the first day at a new work place. This textbook, along with lecture and laboratory, will serve as an important tool while you are 'employed' in this course of study. Hopefully, your 'income' will be a more mature understanding of your life and your responsibility as a biological, social, and spiritual being on this earth. Section 3-I includes additional suggestions on how to use this text.

with important benefits...

and a necessary context.

PART 1 of this text will emphasize that your 'employment' in studying biology cannot be isolated from other ways of understanding. The *thesis* of PART 1 is that even though biology is a natural science, the *inquiry process and information provided by biology cannot be understood or applied apart from a moral, ethical, and philosophical context.*

Science and faith.

In Chapter 1, you will be confronted with the wonder of human curiosity and ingenuity that have propelled the natural sciences to great accomplishments and understanding of the universe. There would seem to be no limits. Yet science is limited. Both the direction it takes and the way in which data are interpreted and applied are all influenced by the value and belief systems of people of science. In Chapter 1, you will explore the relationship between the natural sciences and faith.

Characteristics of life

In Chapter 2, you will examine the common characteristics of living systems—common features which unify the living world in spite of the obvious diversity among these creatures. These characteristics will serve as major themes around which our study of living systems will be organized.

Organization of *biology*

Chapter 3 introduces some of the ways in which knowledge of living things is organized. Here, you will begin to develop a framework upon which to arrange the more specific concepts of biology. Like the first day on the job, this "tour" may be a bit bewildering. However, you will soon "know your away around the building" as we begin to study each "room" and its function in the larger context of life.

Composition of life

Chapter 4 concludes PART 1 by providing answers to several intriguing questions: What are living organisms made of? Are living organisms simply the result of their atoms and molecules reacting according to natural laws? Can a bionic human be considered *alive*? Once again we will see the reliance of biology upon a moral and philosophical context.

[1] American Institute of Biological Sciences, Mission, Goals, and Objectives—1991. *Bioscience* 41 (Dec.): 791-791.

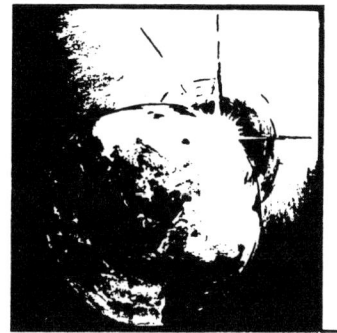

Chapter 1

A Scriptural Perspective of Life

Science extends limits...

of distance...

time...

and size.

Science is an expression of the human spirit of curiosity and of determination to penetrate, explore, describe, and understand the world of the unknown. In recent centuries, men and women of science have used telescopes, satellites, and manned spacecraft to penetrate worlds that were formerly *too distant in space to explore*. Space exploration, radioactive isotope dating, studies of fossils, and mapping of the ocean depths have allowed scientists to probe into the prehistoric era for clues as to the time and manner of Earth's origin, an era formerly *too early in time to investigate*. Electron microscopy and the studies of isolated cells, organelles, and molecules have provided new knowledge of cell and molecular biology, a dimension of living creatures formerly *too small in size to understand*.

How far can we go?

There would seem to be no limits to what humans can accomplish through science. If there are limits, who can say what they are? It was once said, "Surely, humans will never walk on the moon!" Now, history books record this feat! Will genetic engineering techniques, now used to improve crop plants and livestock, be used to repair human genetic disorders? Today's science journals are recording these accomplishments. Will we be able to live in orbiting space laboratories and occupy colonies in space? Who knows how far human scientific endeavor will go?

Science impacts our lives...

and is guided by values.

The impact of science upon our lives is increasingly evident through medical technology, computers, communications, and public policy. Many ask whether these influences are really contributing to the betterment of mankind. For example, <u>should</u> a bone marrow transplant or an *in vitro* fertilization be performed just because it <u>can</u> be done? These considerations remind us that science cannot function apart from the *values* of the culture. Science is not value-free. This principle is evident in three aspects of modern scientific endeavor.

values of individuals...

...and groups.

First, science is the process and product of humans. Human beings are emotional, social, political, and sometimes illogical beings. Therefore, the particular *research projects developed* by scientists are often influenced by social and political priorities, and sometimes by the moral and ethical standards of society. For example, political influences of special interest groups concerned about such issues as cancer, AIDS, or environmental deterioration have a major influence in determining federal budget allocations for scientific research among these areas.

Interpretation of data...

Second, the *interpretation of scientific data* is also influenced by the values and beliefs of the scientist. For example, there is an incredible *biodiversity* of life on Earth. Yet, in spite of this diversity in size, form, and living habit, there is a surprising similarity, or *unity*, among these forms when one considers certain basic features. How do we explain this unity within great diversity? Is it simply the result of natural laws operating upon inanimate matter to produce

depends upon personal beliefs.

cellular forms which then gradually changed over time through the process of *evolution*? Or, were living things *created* through a process that operated beyond the natural laws as we observe them today? Regardless of your personal interpretation, it depends upon your own *belief system* which is rooted in a domain beyond that of the natural sciences, and is not subject to scientific verification.

Applied science is value-directed.

Finally, the *applications of scientific discoveries* are influenced by cultural values and standards. Scientific knowledge in itself is neither good nor bad. It is the application of the information or technique that has moral implications. For instance, the discovery of atomic energy opened the door to many applications, some of which were constructive, others destructive.

Another way of knowing:

In summary, these three aspects of science—project development, interpretation of data, and application of discoveries—all suggest a more fundamental question: *Is there a source of truth that exists beyond what humans may acquire through scientific investigations?*

supernatural revelation

The author of this text has come to believe that the Bible is a supernatural revelation of truth from an omniscient, omnipotent God. Your own personal values and conduct may be based upon biblical revelation, or upon another authority. In this chapter, we will examine the way in which one's belief system influences ones views of science and scientific discoveries?

1-A *HOW* DO YOU KNOW *WHAT* YOU KNOW?

Allow me to begin this discussion by asking you to respond to two personal questions:

QUESTIONS: Are you *alive*? How can you be sure?

Use your five senses.

Did you say you can see, hear, smell, feel, and taste? That's good! Now, pause a few moments and *list* some of the objects or phenomena within the range of your sensory receptors.

analyze

Have you made a list? If so, study Table 1-1. Then, on a separate sheet of paper, analyze several objects or phenomena on your list, and write what you *know* to be *true* of each. When you are finished, consider the following questions:

a. For every detail you have written, can you say, "I *know* it is true?" Explain.

b. Which details do you accept as *truth* that would not be accepted as truth by a non-Christian friend?

c. Should these details be separated into two categories: *real truth* and *relative truth?* Explain.

Table 1-1. Some Sensory Perceptions and Related Knowledge.

Phenomenon	What I "Know" about Each
Human Voice	> sound waves are transmitted through colliding air molecules > from Sally Smith's vocal cords > my friend; an education major > creature of God; Christian
Apple Tree	> green plant; growing; shading > source of O_2 and food > creature of God; beautiful

Science requires faith.

Several principles relating to truth and knowledge are evident from this brief exercise. First, humans can use sensory perception to gather knowledge of their surroundings. This activity is a fundamental role of the natural sciences. The scientist must have faith in his or her sensory perception of the natural world.

framework of truth

Second, humans can use their rational capacity to process sensory information to construct a framework of truth about themselves and the universe. Thus, when you come to know Sally Smith (Table 1-1), you will recognize the sound of her voice. Sounds are being explained by theories of physics which are constructed on the basis of experimental studies of energy imparted to matter. This energy causes vibrations that are transmitted through a given medium to sensory receptors of the ear. Humans use sounds to communicate, share interests, and exchange mutual affections. Because of such communication, Sally may become a true friend (Table 1-1).

epistemology

objective truth

There is a third principle relating to truth and knowledge. This principle emphasizes that every individual possesses an epistemological framework, or "filter", which governs one's acquisition and use of knowledge. *Epistemology* is the study of theories of knowing—of "how we know what we know". For instance, how can Sally *know* that she is a "creature of God" and "a Christian". This is possible because she acknowledges that God's Word, the Bible, is *objective truth*. Objective truth is that which originates outside of mankind, and is independent of mankind for its validity (Romans 1:17; Hebrews 11:1-3).

Through faith one 'sees' beyond the senses...

and forms a worldview.

On the basis of *faith* in God's objective revelation, Christians can conceive a reality beyond the senses, and make rational judgements that shape their view of a sovereign God, His creation, and His sustaining power and purposes. This epistemology is the basis for a Christian world and life view, or *worldview*. According to Walsh and Middleton (14), a worldview is "a way of seeing", "a vision *for* a way of life". Our worldviews are influenced by our culture, are reflected in our language, and help us "interpret the world around us". Worldviews are based upon a belief system which answers "four basic questions facing everyone" (14): *Who am I? Where am I? What's wrong? What's the remedy?*

Each person has a worldview.

All humans, whether they are Christians or not, have a worldview, and they all exercise faith in some object or objects. Returning to our exercise, Sally Smith has many acquaintances besides you, and each one recognizes her visually in a similar way when they spot her. However, if the worldviews of some of her acquaintances are different, they will *see* her origin, purpose, and destiny *in a different light*.

We have attempted to present this text from a Christian worldview. Our thesis is that *the activities of the natural sciences* (i.e. project selection; interpretation and application of data) *are influenced by the worldviews of the people of science*. This point will become evident as we now examine the biblical view of scientific endeavor, the nature of scientific truth, and the conflicts between natural science and theology.

1-B THEOLOGY AND SCIENCE: THE ESSENTIAL HARMONY

Natural Science *versus* Theology

Natural science has been defined as "that body of knowledge dealing with the structural and causal or functional relationships of the physical and time-space aspects of the universe (7)." **Theology**, on the other hand, presents and verifies our claims of our knowledge of God as revealed through the study of the Scriptures by Christian scholars. Thus, science deals with the external, causal, formal, and predictable; while theology deals with the internal, spontaneous, personal, and supernatural.

Fact *versus* Faith

Given the above comparison of science and theology, one might conclude that science deals with **fact**, whereas theology is based upon **faith.** On the contrary, both science and theology rely upon faith. All scientific research rests upon certain *unprovable presuppositions* that are accepted by faith.

faith in objective reality

First, the scientist must accept by faith that the universe is a **reality** which he can examine because his sensory perceptions are reliable. This reality is **objective reality**, existing in some sense independent of human observers. [How does a worldview based upon this faith assertion contrast with one based upon existentialism?]

unity of the universe

Second, science presupposes that the universe is one whole and operates under one set of laws. Therefore, its behavior can be predicted. Indeed, scientific endeavor was hindered until the Reformation. Previously, the predominant Greek worldview, influenced by Plato's and Aristotle's "sacred-secular dichotomy", saw the universe as unyielding and unworthy of investigation.

cause and effect

Finally, science accepts by faith that all observable phenomena are the **effects** of previous measurable, physical **causes**—the "cause and effect" relationship. When a given cause and effect relationship is understood, one should be able to *predict* outcomes (effects) of causes.

Max Planck (1857-1947), the author of the quantum theory, saw faith as a common link between religion and the natural sciences. He stated (6):

A scientist testifies

> Anyone who has taken part in the building up of a branch of science is well aware from personal experience that every endeavor in this direction is guided by an unpretentious but essential principle. This principle is faith—faith which looks ahead. It is said that science has no preconceived ideas: there is no saying that has been more thoroughly or more disastrously misunderstood. It is true that every branch of science must have an empirical foundation: but it is equally true that the essence of science does not consist of this raw material but in the manner in which it is used. The material always is incomplete.. ...[it] must therefore be completed, and this must be done by filling the gaps; and this in turn is

done by means of associations of ideas. And associations of ideas are not the work of the understanding but the offspring of the investigator's imagination—an activity which may be described as faith, or, more cautiously, as a working hypothesis.

Planck's claim that science is rooted in faith is consistent with a biblical worldview. For example, note how the three presuppositions of science are addressed from the perspective of biblical revelation as shown in Table 1-2.

Table 1-2. Scriptural Support for Scientific Presuppositions.

Presupposition	Logical Question	Answers in Divine Revelation
1) I can perceive reality.	Does this reality have any meaning?	Psalm 19:1-3; Rom. 1:19-20 Prov. 25:2; John 1:1,14; 8:32
2) The universe is orderly.	What is the source of the order?	Genesis 1; Colossians 1:16-17
3) Cause and effect	What was first cause?	Gen. 1:1; John 1:1; Heb. 11:3

Distinguish two ways of knowing...

Biblical revelation undergirds and orients scientific mankind who is ultimately in search of meaning and purpose in life. However, unless man by faith acknowledges the validity of divine revelation as *another kind of knowing*, he is limited to what he may know (Figure 1-1). Ramm (7) has discussed this relationship as follows:

about one universe...

> The emphasis of science is on the visible universe, and in theology the emphasis is on the invisible universe, but *it is one universe*. If it is one universe then the visible and the invisible inter-penetrate epistemologically and metaphysically. Through theology we discover the origin, purpose, and end of the created universe. Through science we find clues, analogies, and reinforcements of the existence and nature of the invisible universe.

means* and *ends

As Figure 1-1 illustrates, human reason which rejects faith in the objective truth of God can still experiment, measure and devise means of accomplishing tasks (e.g., medical and bioengineering technology). However, even very elaborate **means** cannot have **meaning** without a world and life view acquired by faith in God's Word.

creation magnifies God

Now that we have established that theology provides an essential context for scientific endeavor, let us consider whether science contributes anything of value to theology. According to Psalm 19, God uses both His Word and His world to reveal His personal nature. Henry Morris (5) states that "...the physical processes in nature continually speak of the power and nature of God, and the biological processes continually bear witnesses of His grace and redeeming love".

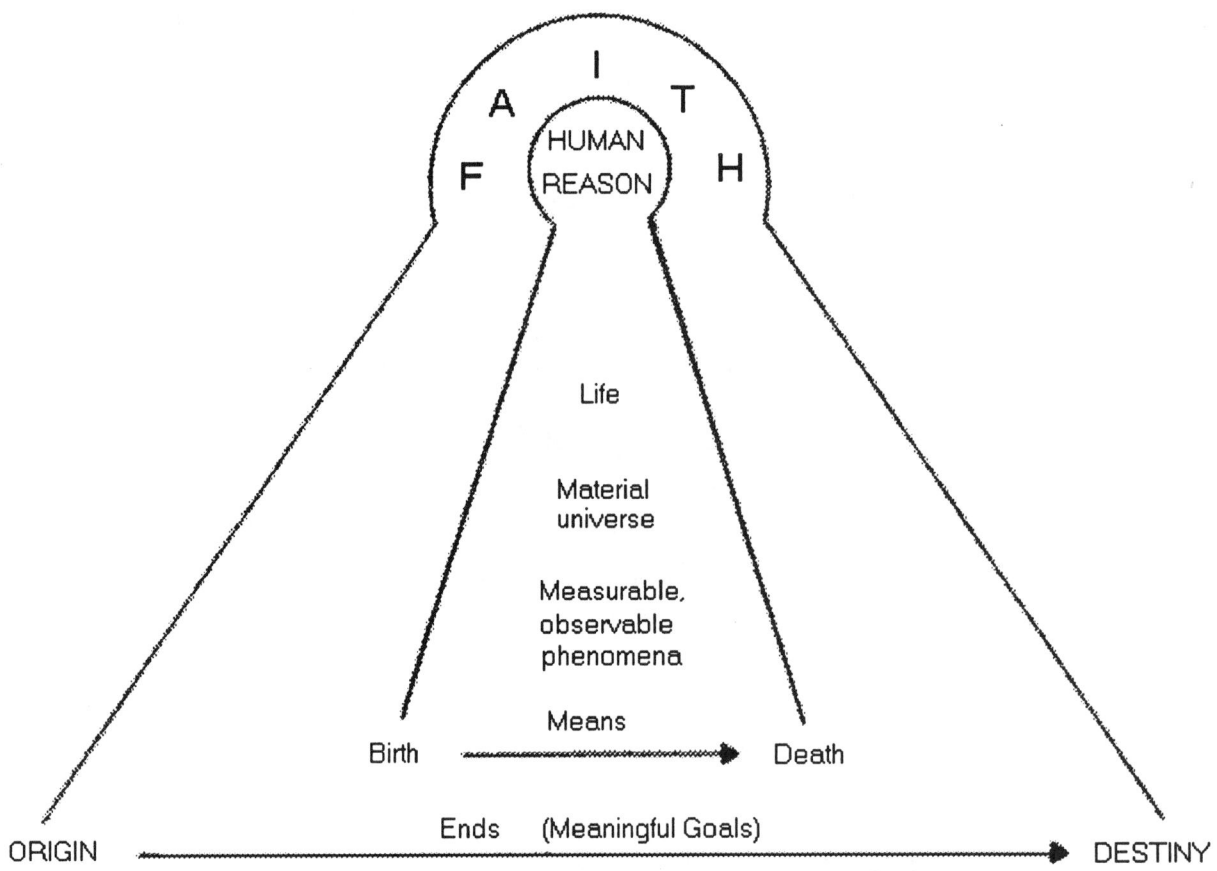

Figure 1-1. Faith in divine revelation is essential to provide a context that orients and undergirds human reason and inquiry.

Table 1-3 summarizes a few of the many ways in which knowledge of the creation can enlighten our understanding of its glorious Creator. Creation indeed cries out in joyous testimony that there is a Creator (Rom. 1:20).

Walter Thorson, in discussing the philosopher Michael Polanyi's notion of "Personal Knowledge", made the following claim:

> Both [religious and scientific knowing] involve faith (or "responsible commitment") as an essential and dynamic element, and both affirm a reality which is **objective**, that is, not of our making and therefore in some sense "out There," not "in Here."

Summary: In summary, a proper interpretation of Scripture is essential to any scientific quest for truth in the natural realm. Furthermore, scientific discovery of truth about the universe expands our idea of who God is, and His immensity, eternity, omnipotence, immutability, and omniscience. God reveals objective truth through the Scriptures and through the creation. However, both theologians and scientists are fallible, and if they are to know truth, they must approach both of God's revelations without pride in human reason, but with humble faith, under the direction of the Holy Spirit. Unfortunately, as we shall see next, this has been a difficult lesson to learn.

Table 1-3. Creational Characteristics That Reflect the Nature of the Creator.

Attribute of Creator	Attribute of Creation	Scripture Reference
Creative wisdom and power	Through what He has made	Romans 1:20 Job 12:7-10
Sustaining power over creation	Laws of Thermodynamics	Genesis 1:1; 2:1 Colossians 1:16-17
Intelligent designer	Evidence of intelligent design in living systems	Psalm 139:13-16 Job 12:7-10
Provider	Provision for the creatures	Psalm 104 Matthew 6:25-33
Character qualities	Demonstrated in the creatures	Proverbs 6:6-8 Job 39:19-25
Power over life and death; 'new birth'	Birth of a new generation	John 3:3-8

Summarized from Morris (5), pp. 130-133.

1-C THEOLOGY AND SCIENCE — THE UNFORTUNATE CONFLICT

Greek science

Greek and medieval "science" *produced much philosophical and theological speculation, but no empirical science* (i.e. hypothesis testing by observation and experimentation), according to Walsh and Middleton (14). They conclude:

Reformation's mandate

Renaissance's machine

What was missing in the Greek and medieval worldviews, and what was new in the sixteenth and seventeenth centuries, was a positive appreciation of this world and of the human task in it.... Due to the Reformation's emphasis on the Scriptures, the biblical vision of God's good creation and of the cultural mandate to develop that creation was recovered. On the other hand, the Renaissance proposed a *secularized* (italics mine) version of this affirmation. God's good creation was reduced to a machine, an object humans could manipulate.

Science: means of "redemption"

scientism

Seventeenth century *biblical* and *secular* worldviews had similar answers to the question, "Who are we?" — i.e. fallen creatures of God. But the two views parted company in their answers to "What's the remedy?". The secular worldview, based upon *dualism*, saw the fall of man in *two parts*, a fall from "innocence", and a fall from dominion over the Earth. Therefore, as Francis Bacon claimed, restoration required two agents: God's redemptive role was to restore the *spiritual* part of mankind, while human dominion was to restore the *physical* world. Natural science would provide the means by which this dominion over nature would be exercised by "human lords". The idea of "restoration" which implied a once-perfect state was soon dropped in favor of the idea of a future utopia, the goal of Bacon's "empire of man over things" (14). Science and the arts would be the tools by which humans would bring on redemption. Walsh and Middleton state: "This is **scientism**: the absolutization of science...reason considered a law to itself and therefore subject to no other law."

Rene Decarte's philosophy of nature as a "machine" which can be understood by the human mind was a ready-made match with Bacon's logical, scientific method. Autonomous mankind could now discover the "chain of cause and effect" by which the machine of the universe operates. The linkage between a secular worldview and the theoretical or natural sciences was completed through the humanistic philosophy of scientism. Table 1-4 outlines several worldviews, described in detail by J.W. Sire (10) and by R.T. Wright (15).

Table 1-4. A Sketch of Major Worldviews and Related Philosophies

Theism	God is the infinite, personal Creator and Sustainer of the whole universe
Deism	God created the universe to operate by natural laws, then departed; reduces God to one who is impersonal and aloof
Teleology	Philosophy that acknowledges existence of providential "design" and "purpose" in nature
Naturalism	God does not exist; "all we see is all there is"; denial of the supernatural
Mechanism	All parts of the universe, "living" and nonliving, can be understood as a "machine" operating by natural laws
Reductionism	A system (or "machine") at any level can be explained entirely by applying laws that govern its parts
Determinism	Phenomena in nature, including human acts of the will, are predetermined by natural laws and cause/effect

determinism

Scientism that would make mankind "god", instead began to remove his freedom. After all, humans, like other biological creatures, are simply part of the "great machine" operating by laws of cause and effect. **Theology** and the **human sciences** (psychology, sociology, etc.) can be dismissed. Instead, in this view, human behavior can be *determined* by the laws of physics and chemistry. Modern *sociobiology* which seeks to explain and predict the totality of human behavior is based on the philosophy of determinism (Table 1-4). [Is there a distinction in your view between biology and psychology? Explain.]

The "scientific revolution"...

has alluring benefits.

Historically, the misunderstandings were not limited to the scientific community. Many theologians tended to view science improperly. The "scientific revolution" introduced both new scientific theory and practical application. How could anyone object to a logic that proposed not only new theories of matter and the universe, but delivered such practical products as steam engines, electricity, and anaesthetic surgery? It seemed to many that science was offering a new world and life for mankind and many began to pursue the 'gospel of science' in place of the seemingly antiquated Scriptures.

Theologians lost crediblity.

Orthodoxy was forced to react to the wave of change out of ignorance of even basic scientific facts, and without a well developed philosophy of science. Many such ill-prepared clergy and theologians lost credibility and undermined the witness of the Church and the Scriptures. The situation was worsened when the religious orthodoxy retaliated with verbal attacks on scientists in a spirit of sarcasm and denunciation. The hyper-orthodox view went so far as to teach that nothing of value could come from science because it is under the influence of unsaved men in rebellion against God. Scientific knowledge is, at best, 'relative truth', especially when in conflict with the Bible. The real explanation must come from theology rather than by empirical studies.

Bernard Ramm (7) summarizes the unfortunate impact which is still in evidence today by saying:

atheistic education

> The prestige of science went to the scientists and to their philosophical and religious views. Science was developed on non-Christian premises. The thousands of students passing through science courses were influenced for naturalism and against religion by the anti-Christian or naturalistic convictions of their professors of science.

Speaking of the present century, Ramm states:

> It is the popular belief that the Bible and science are at odds, that intelligence is on the side of unbelief, and that only childish or sentimental or uneducated people still trust the contents of the Bible.

In assessing the conflict, Ramm suggests how it might have been minimized:

> ...if the theologian and the scientist had been careful to stick to their respective duties, and to learn carefully the other side when they spoke of it there would have been no disharmony between them save that of the non-Christian heart in rebellion against God.

But what are the 'duties' of scientist and theologian? Is there common ground between the mechanistic and ateleological philosophies of scientism, and the scriptural revelation?

1-D THEOLOGY AND SCIENCE — CATEGORICAL COMPLEMENTARITY

We began our discussion of the relationship between the biblical and the natural revelations by emphasizing the potential harmony that should exist between theology and natural science. As man is enlightened concerning the written and creational revelations, God the Revealer is glorified. However as we have seen, the historical record has been marked by apparent contradictions rather than harmony between science and Scripture.

avoiding contradictions

If there is *one* cosmos and *one* Creator who reveals truth through the Scriptures and through science, then there should be no contradiction between these two views. Instead, two other possibilities exist—science and Scripture may be **concordant**, or **complementary**.

concordant view

Those who see the two as concordant believe that both the biblical and natural revelations are important sources of information. Concordists may be of two persuasions. Some attempt to *shape scripture to agree with scientific data*.

A caution to creationists

For example, the **day-age theory** of creation attempts to bring the six-day creation of Genesis 1 into agreement with geological and biological data which suggest that creation occurred over a much longer period. Others who hold the concordant view attempt to *conform scientific data to interpretations of Scripture*. Thus, **scientific creationists** attempt to explain the same scientific data by fitting it into a literal six-day creation period. Concordists face the risk of creating a "God of the gaps" by assigning phenomena to God that are not explainable at present, only to find that God "becomes smaller" when the phenomena is attributed to "natural" causes. Or, when theology is tied incorrectly to an interpretation of the natural world, the credibility of that belief may be diminished along with the refuted theory. A case in point from early church history is given by Hummel (2). Claiming that it was inconsistent with the Scriptures, church leaders opposed Galileo's scientific support of a heliocentric view of the solar system. Instead, they hitched their theology to Aristotle's geocentric view which, when finally rejected, took the credibility of the church and Scriptures down with it (Section 1-C).

Science and scripture are complementary.

Van Till (13) believes that the scientific and the scriptural views of the natural world are complementary. However, he believes that the two views are from different standpoints and each provides answers to different categories of questions. He terms the relationship **categorical complementarity**. This approach is based upon three principles. To illustrate, suppose the student or investigator wishes to answer the questions listed in Table 1-5. Try to arrange these questions into two categories, first those that deal with the *internal affairs* of the natural, observable world; and second, those that concern *external relationships* of the natural world. Table 1-5 outlines the distinctions between internal affairs and external relationships.

Table 1-5. Typical Questions About Life and the Natural Realm

1. What is the age of the earth?

2. Did life on earth originate from inanimate matter as governed by physical and chemical laws?

3. What is the chemical composition of the human brain?

4. How is information stored in the human brain?

5. Will man be any different from an "organic computer" equipped with artificial intelligence and human characteristics?

6. What is it that provides the continual energetic drive to particles of matter as they react, change state, collide, absorb, and radiate to other particles according to natural laws?

7. How does one explain the sudden disappearance of a cancerous growth from a persons body prior to medical treatment?

8. What causes a person to move his/her hand from a hot object?

Distinguish the questions . . .

Van Till's first principle calls for a careful distinction of the questions into the two categories outlined in Table 1-6, part A. Note the general types of questions suited to each category, and attempt to match each question in Table 1-5 with a specific subcategory under category #1 or #2. [Can the unregenerate mind even distinguish these two categories?]

and answers.

A second principle of categorical complementarity requires that the investigator recognize two distinct sources of answers (Table 1-6). Thirdly, the questions must be directed to the appropriate source. In this way, confusion and apparent contradiction can be reduced. Each source of truth should recognize its own distinct realm, and its limitations when attempting to address matters in the other category. Each source can provide *complete answers* to a *partial list* of questions. Therefore, both are needed to provide answers to the range of questions that arise from the human experience. As noted earlier (Figure 1-1), human reason alone is inadequate for such a task. Man must exercise humble faith when addressing both categories of questions, and especially when addressing questions involving external relationships of the natural world where the Scriptures provide the only source of truth.

Table 1-6. Principles of Categorical Complementarity

A. CATEGORIZE QUESTIONS ABOUT THE NATURAL WORLD AS FOLLOWS:

    1. Questions concerning **internal affairs**:

        a. Physical properties—e.g. chemical, physical, biological

        b. Material behavior—e.g. how particles interact

        c. Cosmic history—time component of processes overlaps (2.b.)

    2. Questions concerning **external relationships**:

        a. Status—position of the natural world in relation to deity

        b. Origin—of material world, life, etc.

        c. Governance—What agent **causes** material behavior?

        d. Valuation—What is basis for value (if any) of material entities in relation to an external being?

        e. Purpose—To what end/purpose is the natural realm moving?

B. RECOGNIZE TWO DISTINCT SOURCES OF ANSWERS

    1. Scientific method

    2. Scriptural exegesis

C. DIRECT THE QUESTIONS TO THE APPROPRIATE SOURCE

---

science uses mechanistic approach . . .

often rejects teleology

Several of the questions that challenge biological scientists today have been included in Table 1-5. You may have noticed in attempting to categorize the questions that some require facts which relate to both realms. For example, a complete answer to question #8 requires that it be separated into two parts, one for category #1 (Table 1-6) and one for category #2. The two questions are given in Table 1-7 along with an explanation. Note that the biologist whose worldview accommodates naturalism and denies teleology, is restricted to the mechanistic explanation (category #1). Granted, it is important that we know the **how** of muscular reflexes. Indeed much of our discussion of this text will be presented from a mechanistic viewpoint simply because the "mechanistic how" questions lend themselves more readily to biological investigation and description. However, the window of scriptural truth reveals to us the "teleological why" of this question (category #2).

Table 1-7. Categorical Complementarity: An Example

THE QUESTION: What causes a person to move the hand from a hot object?

ANALYSIS—Two Questions:

 1. Question 1—How does a person move the hand from a hot object?
 2. Question 2—Why does a person move the hand from a hot object?

Categorize the Questions...Turn to Appropriate Source

 1. Question concerning internal affairs of the natural world—physical properties, material behavior

    > Question 1 belongs here; realm of scientific inquiry

 2. Question concerning external affairs of the natural world—man's status, value, purpose in relation to God

    > Question 2 belongs here; realm of scriptural exegesis

ANSWERS TO QUESTIONS:

Question 1 ("mechanistic how"):

 The hot object acts as a stimulus which causes an impulse to pass along sensory neurons to the spinal cord, and via connecting neurons to motor neurons which cause muscle contraction and withdrawal of the hand away from the hot object.

Question 2 ("teleological why"):

 God has equipped many of the motile creatures including humans with stimulus-response mechanisms for the purpose of protecting them from pain, danger, and death. This is in keeping with the character and purpose of God as revealed throughout Scripture.

---

analysis

Our example question requires an **analysis** [Gr. *ana* = up + *lysis* = loosening] resulting in two very different questions. We direct the questions to the appropriate sources of truth which we accept as two entirely different "ways of knowing." Yet, Scriptural revelation provides the objective truth to support a theistic worldview from which we originate both "mechanistic how" and "teleological why" questions, and then direct these questions to the appropriate sources of inquiry. Likewise, we are able to **synthesize** [Gr."syn" meaning together + "tithenai" meaning to place] information that is available from the two sources to arrive at a measure of understanding, and perhaps new questions that await further scientific investigation and/or scriptural enlightenment.

synthesis

As noted earlier (Section 1-B), the natural sciences deal with observable phenomena of the natural world. Science cannot verify or refute moral or religious beliefs by its experimental, or empirical approach. Yet, as we have seen, faith and morality are an inseparable part of the natural sciences (Section 1-B). We are now ready to consider specifically the method of the natural scientist, the scientific method.

1-E SCIENTIFIC METHOD

Scientific method begins with alertness

How observant are you? Did you ever try to visually survey familiar or unfamiliar places in an effort to find something new—a new object or organism, a new process or relationship? The scientific method is not a magical formula. Rather it is initiated whenever a person exercises alertness and concern for details in the surrounding world. We should cultivate alertness to our environment as one of the blessings of simply being alive.

hypothesis: by inductive reasoning

A scientist will make observations and measurements in a regular and systematic manner until sufficient information has been gathered. Then, relevant information is examined, sorted, and summarized. This phase of the study prepares the investigator to propose a **hypothesis**, a tentative explanation for the relationships, patterns, or phenomena observed. The process of moving logically from specific observations to a generalization is called **inductive reasoning**. It is this phase of scientific inquiry that involves creativity and insight which seems to spring from the investigator's imagination and intuition. Recall the earlier quotation from Max Planck who claimed that the process of associating ideas is guided by a "faith which looks ahead". While this is a most intriguing phase of science, even this "inductive jump" must have a solid springboard of careful observations of reality.

experiment: by deductive reasoning

The hypothesis, often referred to as an "educated guess", must then be tested experimentally. Scientists employ **deductive reasoning** to examine the logical consequences of a hypotheses and to determine whether the predictions made by the hypothesis are consistent with specific observations. Thus, deductive reasoning proceeds from the generalization in the form of a hypothesis to a specific conclusion implied by the generalization. This conclusion is always an "if—then" statement. For example, "If a certain facial ointment reduces "zits" (hypothesis), then the "zit density" (number per square inch of skin) should be reduced on the "right" side of your face when the ointment is applied there (deduction). The untreated, "left" side of your face is the **control**." [Why is a *control* necessary?]

When valid experiments provide data which are inconsistent with the predicted outcome, the hypothesis must be altered or rejected. You may wish to repeat the "zit experiment" several times before either accepting or rejecting the hypothesis. A hypothesis is retained as long as experimental results are consistent with its predictions. Figure 1-2 shows the logical progression of hypothesis formulation and testing.

Hypotheses cannot be "proven"!

Hypotheses never reach a 'proven' status because it is impossible to perform all of the possible tests of the predictions of a given hypothesis. Thus, a hypothesis is retained until it is disproved or falsified. When a hypothesis withstands repeated and rigorous attempts at testing and falsification, and demonstrates its validity in the integration of a range of natural phenomena, it is referred to as a **theory**. It is incorrect to use the word "theory" to mean "an idea with little or no support". To the scientist, a theory is a generalization which accounts for a wide variety of observations and offers a frame of reference for further investigations. Long-established theories that come to occupy a fundamental part of science are called **laws**.

sources of error

The scientific method is a very effective means of acquiring knowledge of the natural world. However, the method is also prone to error because humans are not infallible. Table 1-8 lists several of the common sources of error in the scientific process. The scientist must be continually alert to these potential pitfalls. Furthermore, the scientist must carefully scrutinize the work of his or her colleagues. Scientific journal articles and regular meetings of professional societies and symposia serve an important role in communication of experimental results so that one scientist may analyze, empirically verify, refute, or extend the work of another scientist. On the other hand, professional societies can foster a dogmatism which, according to Kuhn (3), prolongs periods of *normal science* in which scientific inquiry is dominated by one reigning **paradigm** — philosophical framework through which a worldview influences scientific inquiry or other disciplines. Bauman (1) notes that "there is a remarkably comprehensive scientific orthodoxy to which scientists must subscribe if they want to get a job, get a promotion, get a research grant, get tenured, or get published".

dogmatism in science

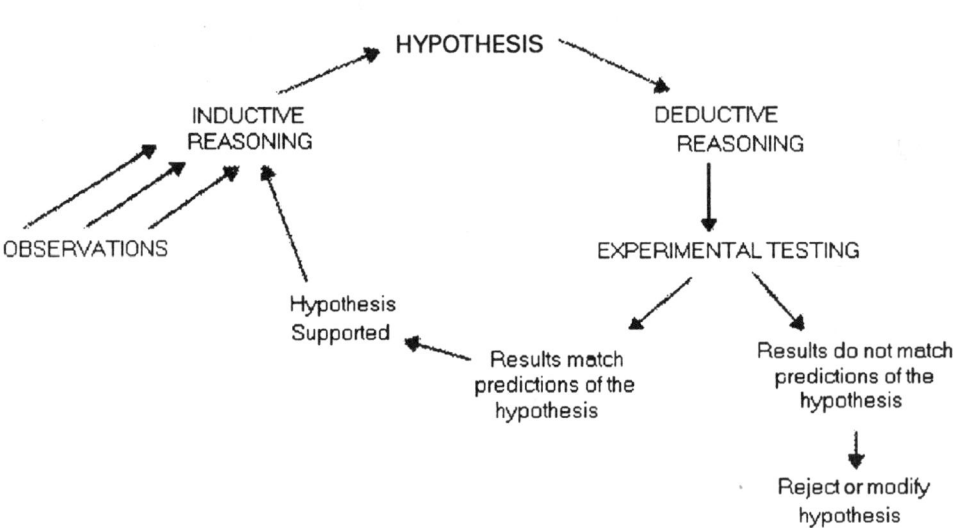

Figure 1-2. Hypothesis formulation and testing in the scientific method.

Table 1-8. Sources of Error in the Scientific Method

Philosophical Bias—stemming from ones worldview, which may influence any of the following:

    a. Hypothesis formulation (inductive logic)

    b. Hypothesis testing—tendency to design experiments to verify the hypothesis rather than to falsify it.

    c. Data interpretation

    d. Development of conceptual and/or visual models to represent reality

Sampling Error—sample too small, biased, lacks replication

Measurement Error—instrument and/or technique lacks precision

Data Manipulation—mathematical or statistical errors

Data Misinterpretation

"scientific correctness"?

Today, the naturalistic worldview which is coupled to an *evolution* paradigm is being propagated in schools and colleges. According to Morris (5), "students are encouraged to 'inquire' and to 'discover', but it is really only important that they become adjusted to the current consensus". The excitement of inquiry is being stifled. Thorsen (12) associates the growing rejection of science with the increasing refusal of mankind to acknowledge the existence of objective truth, or to put forth the rigorous discipline of mind necessary to comprehend it. Francis Schaeffer (9) emphasized that man cannot deny the objective reality of God and the possibility of knowing God without progressively emptying the natural world, and ultimately his own existence, of objective meaning. The apostle Paul lamented that "men will be always learning and never able to come to the knowledge of the truth" (II Timothy 3: 1,2,7). But Christ said,

If you abide in My word, then you are truly disciples [learners] of Mine; and you shall know the truth, and the truth shall make you free (John 8:31b-32).

We can indeed know truth, but note the condition stated above by the Lord: our hearts and minds must be submissive to the authority of Scripture. Let us look more closely at what the Scriptures teach concerning the origin and divine purpose of science.

1-F A SCRIPTURAL MODEL OF SCIENCE

What caused the universe?

As noted earlier, science rests on certain unprovable presuppositions, one of which is the existence of "cause-and-effect" relationships. Theologians and scientists agree that the cosmos is a grand **effect** which demands that there be a **cause** (8). Indeed, the first statement of the divine revelation proposes a divine, supernatural cause—God (Genesis 1:1). Genesis 1:31 tells us that "God saw all that He had made, and behold, it was very good". The created world was "good" not only because of its content, splendor, and perfect function, but because God **saw** that it would be a glorious testimony of His wisdom and power (Psalm 19:1).

Who are we?

What is our responsibility?

Unlike the other biological creatures, man was made in the image and likeness of God, and given the responsibility to exercise dominion over the created realm (Genesis 1:26-28). From the very beginning, dominion required an exercise of responsible management, or stewardship, of the creation. The Greek word *oikonomia*, from which we derive the word **economy**, emphasizes the management role of man. However, if man was to manage and steward the creation, a knowledge of the natural world was essential. Science was instituted by God so that man the steward could formulate an understanding of the natural laws and mechanisms employed by the Creator to sustain the universe.

Science—instituted by God...

When Adam was given the responsibility and privilege of naming the creatures (Genesis 2:19), we see what might be considered the beginnings of **biology**, the science of life. Apparently a great diversity existed among the various kinds of creatures. Therefore, a naming system would be fundamental to the development of a body of knowledge of these creatures. Thus, we have the beginning of taxonomy which is an essential part of the biological sciences today. However, the Hebrew concept of "name" (*qara*) extends beyond simply labeling. Each name was meant to reflect the significance and relationships surrounding the creature that bore the name. Adam was challenged to name each kind of creature as he perceived it in relationship to himself, to other kinds, and to the natural environment. This endeavor represents the beginnings of **ecology** Gr. (*oikologos*), the study of interrelationships among organisms and between organisms and their physical environment.

with guidelines.

God not only instituted science but prescribed certain guidelines for the relationship between Himself and the man of science (i.e. *man* in a generic sense). First, the man of science must actively employ his rational capacities in order to be fruitful in scientific studies of the creation. Adam's assignment to name the creatures would have required use of his reasoning skills as implied from our earlier discussion. It is noteworthy that King Solomon, known for his great wisdom, was also active in scientific pursuits and discoveries (I Kings 4:33).

Science—freedom to discover...

Second, Genesis 1:19 indicates that the Creator displayed His creatures and waited "...to see what [Adam] would name them". King Solomon viewed the opportunity to discover and invent as a privilege given by God (Proverbs 25:2). These are scriptural endorsements of the importance of cognitive freedom and a spirit of thankfulness for the privilege of conducting scientific investigations.

but it is influenced by...

personal interests...

Along with the freedom to discover and invent comes the responsibility for evaluation. The scientist must decide what areas of the vast creation merit his attention. During much of early human history, topics of scientific investigation were decided by individuals primarily out of their personal interest and curiosity, and often as an avocational activity. For example, in the mid-seventeenth century, Robert Hooke, "Curator of Instruments" for the Royal Society of England, designed primitive microscopes that enabled him to view the cellular structure of bark from the cork oak tree. Such **pure science** was of no apparent benefit to mankind at that time. However, Hooke's discovery stimulated additional studies which eventually led to the **cell theory** which is fundamental to the understanding of all biological processes.

sources of funds...

and human values.

Today, scientists often work in groups and utilize sophisticated instrumentation. For this reason, the choice of their research problem is often heavily influenced by funding institutions whose responsibility it is to direct funds to scientific projects of greatest benefit to society. The result has been a shift toward **applied science**. Man's capacity to understand natural processes and thereby manipulate them for his benefit is an expression of his dominion. However, it is evident that human values determine where research will be concentrated; for example, space research versus cancer research. Furthermore, certain applications of scientific discoveries require decisions that have moral and ethical implications. Recent discoveries in genetic engineering are a case in point. Science can perfect the technique for gene transfer from one species to another, but cannot determine the rightness or wrongness of the process. This apparent dilemma brings us to the third guideline for the Godly man of science.

Science— submissive stewardship and...

The Godly scientist must not only be able to reason and make responsible decisions, but thirdly, he must be a submissive steward. The submissive steward of science sees himself as a servant of the omniscient Creator of the universe, having been entrusted with science as a tool to probe into that universe. The outcome should be knowledge which not only benefits his fellowman but also brings glory to the Creator. In matters having moral or ethical implications, the submissive scientist looks to Scripture as the ultimate authority and recognizes that only a Spirit-controlled mind can properly evaluate and apply scientific knowledge.

...a tool to be used wisely

The importance of viewing science as a tool in the hand of the submissive steward can be seen by contrast before and after the fall of man. Adam, the submissive steward for a time, rebelled against God and the authority of His Word. Given superior rational powers, freedom to probe, and God's Word as a guide, Adam chose instead not to submit, and conducted mankind's most unfortunate "experiment". As a result of sin, the history of science is blemished with vain philosophies of those who proudly rely entirely on human reason. Freedom is replaced with fear, and scientific knowledge is often misapplied.

But God through His grace offers redemption to fallen man through faith in Jesus Christ. He has given us the Scriptures to enlighten our minds. Through reliance upon these gifts we will proceed to examine the principles of biological science.

QUESTIONS AND DISCUSSION TOPICS

Questions and Discussion Topics are included at the end of each chapter as an aid to your study. These are intended to serve as one phase of your study of each chapter. Before reading each chapter, skim over the chapter and notice the section headings, captions, and illustrations. Then read the assignment and list the major terms. For each term, write an accurate definition in your own words, and include an illustration or example where appropriate. See Chapter 3, Section 3-I for more detailed study suggestions. When you have mastered the terminology of a given section or chapter, you should be ready to apply your understanding to the "Questions and Discussion Topics".

1. Analyze the following statement: "The *Bible* is to the *theologian* as the *natural world* is to the *scientist.*"

2. Humans have observed the moon on starlit nights for years. What faith presuppositions concerning the natural world had to be accepted as a launching pad for the sciences that eventually put humans on the moon?

3. Do you believe that natural science will eventually explain all there is to know about the moon as suggested by the view of scientism? Explain. Is there legitimate knowledge of the moon that one might acquire from an aesthetic or spiritual perspective? Please explain and illustrate.

4. Was Doubting Thomas acting totally apart from faith when he insisted upon seeing first-hand the wounds of the risen Savior? Can anyone act totally apart from faith?

5. With reference to Figure 1-1, list and discuss several modern scientific developments that represent improved *means* but have unimproved or uncertain *ends*. What is the role of your faith in addressing questions of *means* and of *ends* (or purpose) in your life?

6. The doctors of the Roman church argued that Galileo was wrong to believe in the Copernican theory of planetary motion because they supposed that it contradicted the Scriptures e.g. (Ecclesiastes 1:4-5). Analyze this argument in light of general principles presented in this chapter.

7. In light of the instance cited in question 6, what cautions are in order for both scientists and theologians with regard to the creation-evolution issue?

8. Three principle characteristics of the 'godly man of science' were given in this chapter. How were these evident in God's instruction to Adam? Should the same three characteristics be evident in the 'godly student of science'? Prayerfully examine your personal attitudes and values, and determine to approach your study of biology in a God-pleasing manner.

9. If you were to ask a mechanistic biologist to explain why mammals hibernate in winter how would he or she approach the question? How would one approach the question from a theistic epistemology?

10. Isolate the major principles discussed in this chapter and incorporate them into a statement of your own 'philosophy of science' as it exists at this time in your educational experience.

TOPICS FOR FURTHER STUDY

At the end of most chapters, you will find a selection of topics that you may wish to pursue on your own, perhaps for a special research paper. Many of these are topics of major significance. However, we were unable to present a discussion of them in detail in the text of the chapter. One or two references have been provided to start your library research, in hopes that you will uncover additional references during the course of your study.

1. What is the role of human imagination in science?

    REFERENCES:
    A.M. Taylor. 1970. *Imagination and the Growth of Science.* Schocken Books, New York.
    A.C. Leopold. 1978. The act of creation: creative processes in science. *Bioscience* 28(7): 436-440.

2. What do the Scriptures and certain secular writers say is necessary for a person to move from a naturalistic worldview to one which recognizes the existence of a supernatural God and objective truth?

    REFERENCES: See REFERENCES CITED, and the following:
    T.S. Kuhn. 1970 *The Structure of the Scientific Revolutions,* 2nd ed. Univ. of Chicago Press, Chicago, IL.

3. Who were the real opponents of Galileo's position that the Earth moves, and it is *not* the center of the universe?
    See REFERENCES CITED, especially, Hummel (1986).

REFERENCES CITED

1. Bauman, M. 1992. From beyond the laboratory: A theologian looks at science. *Faculty Dialogue,* #17 (Spring): 131-145.

2. Hummel, C.E. 1986. *The Galileo Connection: Resolving Conflicts between Science and the Bible.* InterVarsity Press, Downers Grove, IL.

3. Kuhn, T.S. 1970. *The Structure of Scientific Revolutions,* 2nd ed., University of Chicago Press, Chicago.

4. Little, P.E. 1968. *Know Why You Believe.* InterVarsity Press, Downers Grove, IL.

5. Morris, H.M. 1983. *Education for the Real World.* Master Books, San Diego, CA.

6. Planck, M. 1936. *The Philosophy of Physics.* W.W. Norton Publ., New York, NY.

7. Ramm, B. 1954. *The Christian View of Science and Scripture.* W.B. Eerdmans, Grand Rapids, MI.

8. Ryrie, C.C. 1978. A synopsis of Bible doctrine. In *The Ryrie Study Bible,* pp. 1933-1954. Moody Press, Chicago.

9. Schaeffer, F.A. 1968. *The God Who Is There.* InterVarsity Press, Chicago.

10. Sire, J.W. 1976. *The Universe Next Door: A Guide to WorldViews.* InterVarsity Press, Downers Grove, IL.

11. Thorson, W.E. 1978. The spiritual dimensions of science. In C.F. Henry, ed. *Horizons of Science: Christian Scholars Speak Out.* pp. 217-257. Harper and Row, New York.

12. Thorson, W.E. 1984. Scientific objectivity and the Word of God. *Journal of the American Scientific Affiliation* 36 (2): 88-97.

13. Van Till, H.J. 1984. Categorical Complementarity and the Creationomic Perspective. Paper presented at the 39th Annual Meeting of the American Scientific Affiliation, Miami University, Oxford, OH.

14. Walsh, B.J. and J.R. Middleton. 1984. *The Transforming Vision.* InterVarsity Press, Downers Grove, IL

15. Wright, R.T. 1989. *Biology through the Eyes of Faith.* Harper and Row, San Francisco, CA.

Chapter 2

Unifying Characteristics of Living Systems

"biological consciousness"

A "course" of study in biology is actually an extension of the "course of your life" which began at the time of development of consciousness of the world around you. [When would this have been?] The "course" continues through such experiences as your first view of a colorful flower bathed in bright rays of sunlight; the first opportunity to glance into a bird nest or to hold a living bird or insect in your hand; the first exposure to a warm spring rain or to a gentle winter snowfall. All the while, you were forming a perception of the living world. Opportunities to travel have given first hand exposure to the varied landscapes of our nation and world. The media of books, television, CD's, and magazines have provided a wealth of vicarious experience in addition to one or more previous academic "courses" in the life sciences. You may have "travelled" further or perhaps not so far as your peers who now study biology with you.

Relate "new knowledge" to previous experience.

Thanks to your past experience, your task will now be to relate new knowledge to previous learning. Perhaps some areas of shallow understanding will give away to more complete perceptions; areas of misunderstanding, confusion, and error will surrender to truth and a deeper appreciation of the living world. These changes in your understanding will come as we journey into the **science** of **biology**, a growing, systematized body of knowledge about life.

Like any journey, a "course of study" must have a destination in view as specified in instructional objectives to facilitate effective learning. No doubt, your instructor has informed you of these objectives his or her approach to guide you in the accomplishment of these objectives. You will need to assume an active role in the process as learner, disciple, and seeker of knowledge.

Be an active "learner".

This text is written so as to actively involve you in thinking and reasoning. Questions will be raised throughout each chapter to stimulate your thinking. Some questions have rather straightforward answers; others will require careful thought and discussion. You will need to master some of terminology and concepts of biology. But, rather than stopping there, you will need to apply this information in addressing questions and discussions that require analytical reasoning. When you have "finished the course", it will be the products of your careful thought and reasoning that will accompany you as an essential part of your preparation to participate in our scientific and technological society.

Biology is . . .

Biological science, like a living organism, is dynamic. Like a tree with spreading branches, biology is growing and expanding as new knowledge is acquired. Yet, certain parts like the roots and trunk of a tree, are foundational and provide a theoretical core of support and anchorage to the superstructure of the science. Our analogy of a growing tree suggests two aspects that are true of the growing science of biology—a process of growth, and a resultant

**process . . .
(inquiry)**

**and product
(knowledge).**

structure. Our study of biology will focus upon these two aspects. First, we will consider the 'growth' aspect as we study the acquisition of new knowledge via the **scientific method**. [In this context, are there also occasions for "pruning" branches? See Section 1-E.] Second, we will consider the 'structure' or body of knowledge itself which is interpreted through a framework of **principles of biology**.

To begin our study of the science of life, please join me for an imaginary hike into a meadow and forest area. I have hiked and studied this area many times and am honored to share some of my observations with you. Your imagination will be aided by previous hikes you may have taken during the "course" of your life. Others have studied and described all of the phenomena we will be observing. We could read about these studies in records contained in the various "branches" of the "tree" of biology. However, we will first make fresh observations and then relate them to the foundational principles of biology that serve as an interpretive framework for our study of life.

2-A OBSERVATIONS FROM A HIKE IN A MEADOW

We enter the meadow on this sunny summer morning by carefully stepping over the single wire of an "electric fence". One or two encounters with this energized little wire is enough to remind grazing cattle to stay within the territory defined by it. Evidently, there are benefits to learning, even for animals.

The landscape of pastures is rather unique. In the eastern United States, as in many parts of the world, land which is inappropriate for cultivation due to steep topography or poor soil conditions, is often used for pasture. This particular pasture covers the steep hillsides surrounding a beautiful valley. As long as the farmer avoids overgrazing the pasture, which destroys vegetative cover, the sod-forming grasses will hold the steep slopes against erosion.

As we hike down the slope toward the valley, you may notice the rather unique community of pasture plants. Various grasses including Kentucky bluegrass, timothy, and bromegrass dominate the hillside. However, interspersed among the grasses are dandelion, plantain, and hawkweed. All of these are among the notoriously successful members of pasture and lawn because like the grasses their growing points are too close to the ground for grazers, or rotary mowers, to reach. Other plants that prosper in the pasture include thistles, brambles, hawthorn, and horse nettle. Grazing animals have learned to avoid these plants because of their nasty thorns and sharp spines.

Already we have seen evidence that the pasture is a living community, full of dynamic relationships between sun, soil, plants, and animals. How do we explain these relationships? How did they originate and how long have they existed? What can we learn about the nature of life from these observations? Think about these questions and note others that come to mind as we travel on.

As we descend into the valley, we enter a "cowpath" which leads to the bottom of the slope at the head of the valley. Here at the convergence of cowpaths from other directions is a source of spring water emerging from the steep hillside. As the frequently travelled paths testify, this water source is important for cattle today. Water was even more important in earlier centuries before the advent of wells and "running water" in houses and before concerns about chemical contamination of water.

The hillside spring offers a refreshing drink. Nearby, oak trees provide shade as their leaves intercept the sunlight and photosynthesize organic substances from carbon dioxide and water. Seated quietly in the cool shade, we begin to see various creatures emerge from dens and nests. A woodchuck, I'll call "Herbie", climbs a stump to survey his domain before beginning to feed among the plants near his burrow. Ground squirrels scamper up and down the trunk of a nearby tree. The trills of songbirds and the sound of rustling oak leaves overhead blend with the trickling sound of the spring water to form a serenade. This is a "serenade of energy". Energy, that "capacity to do work" is all around us in many forms, flowing through the "bloodstream" of this pasture community; sunlight to green plant, to woodchuck; sunlight to evaporating water, to misty rainfall, to flowing spring water, to thirsty animals; all making sounds. Life goes on.

Following the spring-fed stream down the valley, we enter a marsh community decorated with an array of wildflowers including swamp milkweed, blue lobelia, boneset, and Joe-pye weed. Each flower is humming with honeybees and other nectar-seeking insects. Many plant species have a strong dependency upon specific insect species for pollination as part of their reproductive process. The pollinators, in turn, obtain energy so essential for life.

Much more could be observed in this meadow community. No doubt your memory of similar past experiences has been awakened with much that could be added to these observations. Let us now attempt to relate our observations to the foundational principles of biology as a basis for understanding living systems.

2-B OBSERVATIONS AND CONCEPTUALIZATION

organizing observations

Before continuing your reading, return to the beginning of the previous section and reread the account of the meadow hike. As you read, list in the margin each word or phrase that refers to an object, phenomenon, or relationship that would seem relevant to biology. Read slowly and be analytical. Remember, the aim of biological science is to present a systematized body of knowledge about life—a **synthesis** [Gr. *syn* = together) + (*tithenai* = to place)] of knowledge into a complete whole. Certainly man is far from completing his understanding of life. However, before you can understand even the present extent of the synthesis, you must do some **analysis** [Gr. *lysis* = loosening) + (*ana* = up)] or loosening up of the complex to examine the parts. As you complete the above-suggested learning activity, you will be analyzing a series of observations. We will then attempt to combine these components into a series of biological principles describing life.

Have you completed your list? If so, compare your list to Table 2-1.

Table 2-1. Meadow Observations

meadow	soil	oak tree
sunlight	erosion	woodchuck
seasonal cycles	naming organisms	energy intake
daily cycles	adaptations	marsh
electric fence	cowpath	flowers
territory	spring water	nectar
animal behavior	photosynthesis	insects
topography	chemical constituents (water, carbon dioxide)	aesthetic beauty

If your list is similar to Table 2-1, you are demonstrating an ability to relate various random observations into a conceptual framework that characterizes biological systems. Study your list and Table 2-1 and attempt to describe the biological significance of each.

characteristics of life:

In moving from the specific observations of the meadow community (Table 2-1) to a more general framework, we can identify at least five characteristics of living systems within which our observations may be grouped (Table 2-2). These in turn will serve as major themes upon which our study of biology will be organized. For introductory purposes, we shall briefly consider each characteristic as exemplified in the "meadow experience".

Table 2-2. Characteristics of Life and Examples from Meadow Observations.

1. **Living systems are organized hierarchically.**
   * seasonal and daily cycles (BIOSPHERE)
   * meadow community and physical environment (ECOSYSTEM)
   * meadow community (COMMUNITY)
   * woodchuck (POPULATION)
   * woodchuck ("Herbie") (ORGANISM)
   * "Herbie's heart (ORGAN)
   * "Herbie's heart muscles (TISSUE)
   * heart muscle cells (CELL)
   * myosin, contractile protein of muscle (MOLECULE)

2. **Living systems use energy from the environment**
   * e.g. sunlight -> meadow plant -> woodchuck -> hawk
   * photosynthesis by plants
   * ingestion of plants by woodchuck

3. **Living systems reproduce**
   * flower structures, pollination, etc.
   * classification (naming and grouping) based upon distinctions among species

Table 2-2. Characteristics of Life and Examples from Meadow Observations. *(continued)*

4. **Living systems respond to stimuli from the environment and maintain internal homeostasis**
   * daily and seasonal cycles
   * electric fence for behavior modification
   * insect attraction by flowers
   * value judged by man as aesthetically valuable

5. **Living systems are adapted to their environments**
   * low growing point (e.g. grass, dandelion)
   * thorns and spines
   * nests and dens in trees and underground

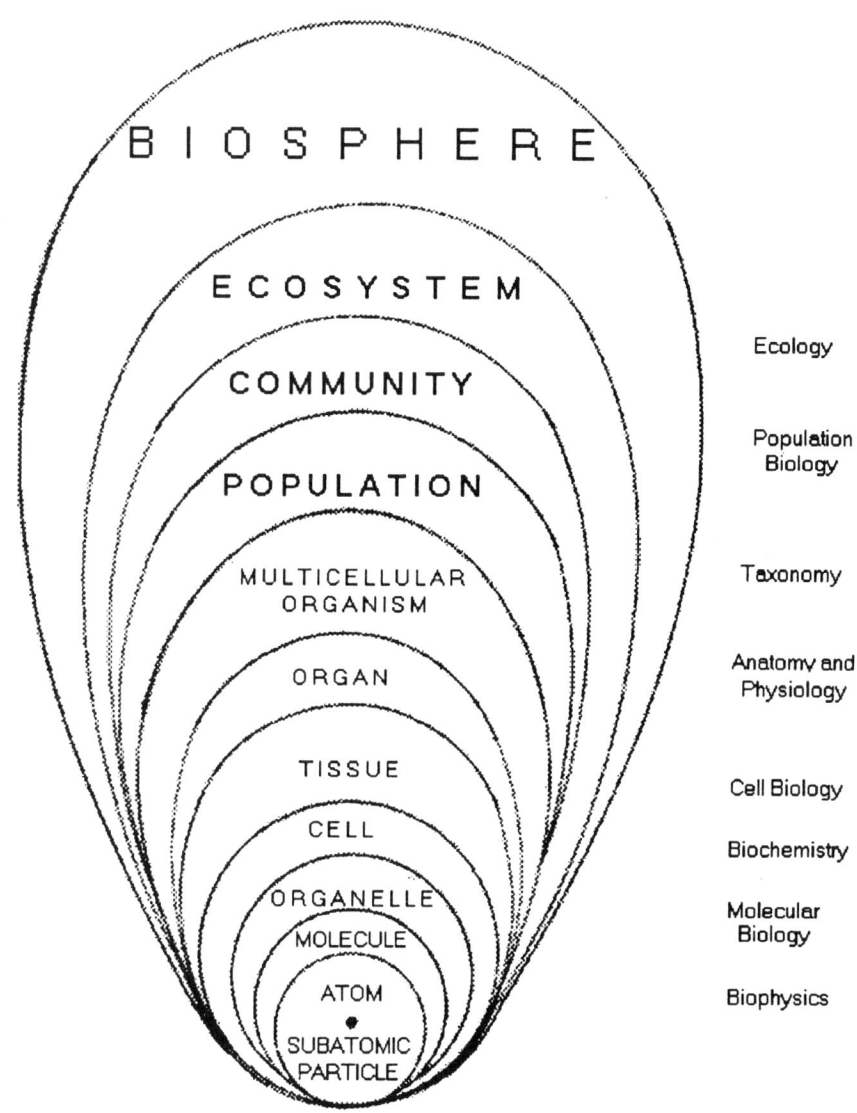

Figure 2-1. Levels of organization of biological systems.

2-C COMMON CHARACTERISTICS OF LIFE

levels of organization...

from atoms...

1. **Living systems are organized hierarchically.** That is, each component is an arrangement of smaller components, each of which, in turn, consist of yet smaller components. This hierarchical series forms eleven or more **levels of organization** as represented in Figure 2-1. At the submicroscopic level, living organisms are composed of **atoms** of different chemical elements (Figure 2-1). Many of these atoms are bonded together to form **molecules** (Chapter 4). The molecules of living matter are arranged in precise fashion to form *membranes*. **Cells**, the smallest (usually microscopic) units of living matter, are distinguished by a membranous boundary, the **cell membrane** (Figure 2-2).

Within the cell membrane are smaller, membranous compartments called **organelles** (Fr. "little organs"). Organelles are a part of the *cytoplasm*, an aqueous medium containing suspended molecules and dissolved salts. Certain types of organisms are composed of single cells and are referred to as *unicellular* organisms. *Multicellular* organisms are composed of many cells, often assembled into more complex arrangements known as **tissues** and **organs.**

...to ecosystems.

Each tissue or organ structure is suited to the performance of a specific life function or process within the multicellular organism. For example, the leaf of the swamp milkweed plant (Table 2-2) is a photosynthetic organ in which cells containing chlorophyll **molecules** are arranged within membranes of organelles called *chloroplasts*. Chloroplasts function to absorb light energy necessary to synthesize sugars and other organic molecules necessary for life. The swamp milkweed, like most multicellular organisms, consists of a various components arranged in hierarchical fashion. Each swamp milkweed plant is an **organism**, an individual member of the swamp milkweed **population**. This population is an important part of the marsh **community** because it provides nectar as food for various insect populations that are also part of this community. The marsh community depends upon sunlight, air, water, and soil, which, together with the community, make up an **ecosystem**. Ecosystems may be of various sizes —e.g. a pond ecosystem, a forest ecosystem, or a desert ecosystem. Taken together, the earth's ecosystems form a circumglobal zone of atmosphere, land, and water that support life, called the **biosphere**.

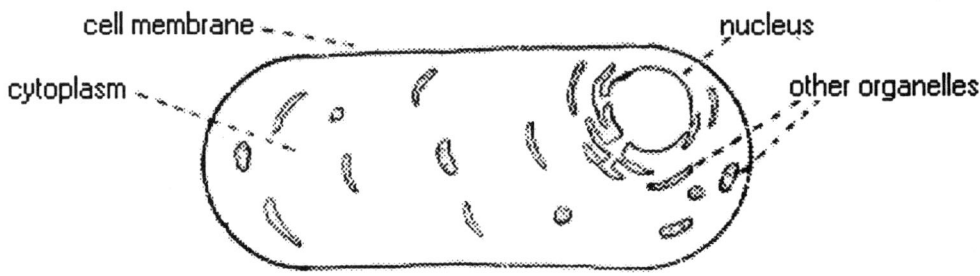

Figure 2-2. Diagrammatic view of the basic structural components of a cell. All lines in the sketch represent membranes composed of precise arrangements of molecules.

Many biologists are specialized.

The science of biology spans the entire hierarchy of levels from subatomic particles to biosphere. Our lives and those of every other creature are influenced by processes represented at each level. Such biological, chemical, and physical processes are the focus of research biologists who specialize in one or more levels of organization. Figure 2-1 lists just a few of the branches of biology along with the level of organization at which their efforts are concentrated.

Models aid research and our learning.

Note the immense range of size represented in Figure 2-1; from subatomic particle to biosphere. For this reason, visual and conceptual **models** (L. *modulus* = small measure) will be used to convey the essence of such abstractions as molecules, cells, and ecosystems. Your challenge will be to realize the fact that we must depend upon diagrams and models, and utilize them to build a conceptual understanding of biological systems. Each of the models we will use in this study can be related back to the model of *levels of organization* (Figure 2-1). This model will serve as a "master conceptual framework" upon which to arrange and relate various specific models of the subcomponents and relationships of the natural world.

Energy drives "life processes"...

2. Living systems use energy from the environment. Energy is a fundamental part of processes at each level of organization (Figure 2-1), and of the assembly of each level into successively higher levels. The energy of atomic particles and molecules propels them in constant motion, enabling them to collide, chemically react, diffuse, and exert pressure to maintain the physical structure of cells. At higher levels of organization, tissues and organs can contract and produce bodily motion, or pump body fluids. For example, insects fly to a swamp milkweed flower and acquire energy in the form of nectar. The nectar was manufactured within the plant leaves by the process of photosynthesis. In this way, energy even links two or more populations together within a biotic community.

...collectively known as metabolism.

Characteristic #2 is the basis for a major principle of biology: **Living systems require a continuous supply of energy and matter in usable form.** Each organism has a means of acquiring energy and utilizing it to transport raw materials into its cells, synthesize complex constituents, degrade other constituents, excrete waste products, or produce movement. These processes, collectively known as **metabolism**, are both unique and essential to the living condition. All living systems perform metabolism. They take in energy from their environment and use it to conduct the basic life processes of *reproduction*, *growth*, *maintenance*, and *responsiveness* to environmental conditions. We now direct our attention to these processes as additional characteristics of life.

Reproduction of new cells...

3. Living organisms reproduce. This characteristic of life provides perhaps the clearest distinction between the living and the non-living world. The metabolism of individuals of one living generation is directed so as to produce a cell or cells which can participate in the production of a new generation. The appearance of the mature offspring is similar to the previous generation — i.e. each *kind* of organism reproduces only more of its own kind.

...is controlled by genetic material...

How is it that offspring of a given generation will bear the main characteristics of their kind, yet, will not be identical to each other. Genetic theory suggests that a **genetic control system** permits *variation in appearance within kinds* but restricts variation so as to *maintain distinctions between kinds*? Each generation *inherits* structural and behavioral characteristics through the transfer of *hereditary material* from a previous generation. This hereditary

called DNA. material serves as a "blueprint" for the offspring as coded on **DNA, deoxyribonucleic acid** molecules. The initiation of growth and development from seed to plant, from infant to adult, or from fertilized egg to larva, pupa, and adult butterfly are all metabolically controlled along a precise timetable or **life cycle** for each kind of creature. These processes are not fully understood but what is already known is quite fascinating.

Organisms must interact with environment...

4. Living organisms respond to stimuli from the environment, and maintain internal homeostasis. Living organisms are intimately linked to their environment. Your survival at this moment depends upon a continual and regular exchange of matter and energy form your environment. [How many essentials can you list?] Animal pollinators have sensory receptors to locate flowers by color, shape, and odor as a means of procuring their nourishment. Plants, growing in the unidirectional sunlight near a window, can orient their leaves and stems toward the light. An animal that fails to respond to your approach or a gentle nudge is considered "dead".

...which fluctuates.

From the above examples, it is evident that living organisms must be responsive to their environment. This requirement for life is made more challenging by the fact that environmental conditions fluctuate daily and seasonally. Water, foods, temperature, light, pressure, oxygen, carbon dioxide, toxic substances, and dissolved salts all determine the chances of survival. Meanwhile, organisms must regulate their metabolism within narrow limits. Most organisms possess mechanisms which permit them to maintain internal metabolic conditions within tolerable limits in the midst of environmental fluctuations. When such conditions are maintained, the processes in question are said to be at **homeostasis** [Gr. (*homios* = similar) + (*stasis* = standing)]. Any disruption of homeostasis beyond the limits of tolerance may become evident in **disease**. Unless the diseased organism can restore homeostatic conditions, **death** may occur.

Homeostasis is essential.

5. Living organisms are adapted to their environment. The earth offers a seemingly endless array of habitats — high mountains and dark ocean depths, marine water and fresh water, tropical forests and arid deserts. Even more incredible is the great diversity of living creatures native to these environments. The intimate association between each plant and animal population and their environment was noted above. This association becomes very evident when man transfers so-called exotic creatures from their native habitat into urban zones, museums, and botanical gardens. Careful provision must be made to match environmental conditions with the structural (or *morphological*), functional (or *physiological*), and behavioral features of each exotic creature. These morphological, physiological, and behavioral features that enable an organism to carry on the basic life processes within its native environment are called **adaptations**.

Organisms "fit" into habitats.

2-D UNITY AMID DIVERSITY

We cannot help but be amazed at the great **diversity** within and among the kinds of life forms throughout the biosphere, each adapting to its environment. However, it is equally amazing that a significant **unity** exists amid this diversity. For example, "molecular unity" exists because DNA molecules function as the hereditary material for all known forms of life. With few exceptions such as the viruses, all living organisms are composed of cells with cytoplasmic and membranous components. Furthermore, all organisms must perform the basic life processes noted above.

exciting questions of biology.

The discovery of unity amid diversity has been a major source of the excitement of biology. The great diversity of life became more evident with the advent of explorations of the New World in the 15th and 16th centuries. Imagine the excitement of discovering and studying creatures never known to exist before. On the other hand, the unity of life became evident through the exploration of the microscopic world of the cell beginning in the 17th century. Then, in our century, investigations of the submicroscopic world of molecular biology revealed that all known organisms have nucleic acids as their genetic blueprint. Out of these exciting discoveries, came even more questions. Why is there such an incredible diversity of life? How is it that each environment on earth has such a unique array of adapted life forms? What sustains and propels all of these creatures generation after generation in their endeavor to survive and reproduce? Perhaps the ultimate question is, what is the origin of life as we know it? These are just a few of the challenging questions we shall encounter in our study of biology.

In Chapter 1 we studied the importance of the integration of science and Scripture as a prerequisite to fruitful inquiry into the natural realm and man's role in it. Chapter 2 has introduced some biological perspectives and basic life processes of living systems. In the next chapter, we will consider the manner in which knowledge of living systems is organized or systemized as a result of scientific endeavor.

QUESTIONS AND DISCUSSION TOPICS

1. For each term in this chapter that is printed in **boldface** type, write a meaningful definition, and illustrate by example where appropriate. *See Chapter 3, Section 3-I.*

2. What is **life**? Are living systems the only systems that can grow, reproduce, and respond to stimuli? Illustrate and explain how this influences your definition of *life*.

3. What are the so-called "basic life processes"? Are they common to all known forms of life?

4. Explain how each term in Table 2-1 relates to one or more levels of organization.

5. How would you determine the vertical dimensions or "thickness" of the biosphere at a given location on earth? Estimate the range of thickness of the biosphere (maximum and minimum) at different global locations, considering such locations as the Pacific Ocean and polar regions.

6. Why is the biosphere represented by the largest concentric circle in Figure 2-1? As one moves from the subatomic level of organization to the biosphere (Figure 2-1) what is the trend of change with respect to size? Complexity? Lifelikeness? Scope of ones knowledge in order to comprehend a given level?

7. List the environmental components necessary to sustain life? How would a list for plants differ from that of animals?

Chapter 3

Biological Classifications: Patterns Amid Diversity

Biological sciences have...

boundaries...

and are systematically organized.

We have defined **biology** as a systematized body of knowledge of living organisms. The body of knowledge has been forming and growing as a product of scientific investigations. Use of the word *body* as an analogy implies that there are limits or boundaries to what we may rightly consider biological knowledge. Scientifically speaking, these limits include only that knowledge acquired by man's proper use of the scientific method. As noted in Chapter 1, such information must originate from observable or measurable phenomena, testable hypotheses, and logical generalizations that arise from experimental results.

The body of biological knowledge not only has limits around it, but also organization within it. We have already examined the *foundations* relating to the common features of living creatures and the basic life functions (Chapter 2). However, as we venture from commonality to confront the great diversity of the living world, we must emphasize the *superstructure* of the body of knowledge and the manner in which new information fits into this structure. Recall for a moment the great diversity of kinds of organisms which you have encountered during the course of your life. Add to this thought the variety of habitats, relationships, and lifestyles represented by this sample of life. Biological scientists have sought to detect patterns within the diversity of the living world and to formulate generalizations that describe these patterns. Generalizations exist in the form of definitions, rules, and other broad statements intended to highlight common characteristics amid the diversity of the living world. You may recall that we used this process in the discussion of our meadow excursion (Table 2-1). Now let's examine the more detailed biological organization that rests upon the general foundation considered in the previous chapter.

3-A LEVELS OF ORGANIZATION OF MATTER

Do you understand the hierarchy?

The hierarchy of levels of organization (Figure 2-1) is a comprehensive and useful model by which we may perceive the order within the body of biological knowledge. As we learned in Chapter 2, any living system or component thereof can be linked back to one or more levels of organization. To illustrate, suppose you wish to learn something about the following items:

>   white oak tree
>   white oak leaf
>   oak forest
>   carbon dioxide

From experience you may be able to picture an oak tree, its leaf characteristics, and the appearance of an oak forest. If asked to distinguish among these items, you would probably classify carbon dioxide separately from the others,

Can you match structures to it?

based upon the criterion of living versus nonliving. However, any additional biological discussion of this list will require an understanding of how each item fits into the hierarchy of levels of organization. For example, your understanding of the biological role of the oak leaf will require that you view the leaf as an integral part of a larger system, the oak tree. Both the leaf and the tree are alive and yet they represent different degrees of biological dependence. In general, this is the relationship of *organ* to *organism*. Carbon dioxide is found within and around the oak leaf and is related physiologically to the life processes of the leaf. Identification of carbon dioxide with the *molecule* level of organization is essential to your understanding of its properties and role in biology.

Table 3-1 outlines what we have been discussing. Note that these conceptual arrangements are only the starting point for a person seeking to comprehend the meaning and significance of a biological system. Nevertheless, it is a fundamental step that you will need to make each time you encounter a new term or concept.

one *body* with many levels

We must integrate among the levels.

Each level of organization serves as a major informational component within the body of biological knowledge. As noted in Chapter 2 (Figure 2-1), there are many branches within the biological sciences. Each of these represent the body of knowledge that is developing through scientific research at specific levels of organization. For example, cell biology focuses primarily upon the cell and organelle levels. Overlap of knowledge does exist between the various branches of biology, thus reflecting the interrelationships perceived to exist between levels of organization. Acquisition of a complete knowledge of any living system will require the contributions of many biologists, each specialized in research at one or more levels of organization, and each being able to integrate their findings with those at other levels.

Classifications for each level

The body of knowledge at each level of organization has its own degree of orderliness because of certain **classification systems**. These systems are theoretical constructs which enable biologists to deal conceptually with the diverse structures, processes and relationships of the living world. Therefore it seems expedient for us to identify these biological classification systems and to use them as a guide to our study of biology. We shall begin with classification at the most all-inclusive level, the biosphere.

3-B BIOSPHERE AND GLOBAL ECOSYSTEMS

global-level classification

The physical structure of the earth can be classified according to the three states of matter—solid, liquid, and gas. The solid phase of Earth, the soil and underlying bedrock, is referred to as the **lithosphere** (Figure 3-1). The oceans, lakes, and rivers represent the liquid phase, or **hydrosphere**. The gaseous phase, the **atmosphere**, consists mainly of nitrogen (79%), oxygen (21%), carbon dioxide (.03%), water vapor, particulate matter, and lesser quantities of other gases, many of which are reactive in biological processes.

Table 3-1. Levels of Organization—A Model for Understanding Biological Systems

Level of Organization	Example	Relationships
Biosphere		
Ecosystem	white oak forest	A whole land environment including white oak and other plant and animal populations existing interdependently as they share the physical resources of that area.
Community		
Population		
Organism	white oak tree	An individual living creature composed of organs, tissues, cells
Organ	white oak leaf	One of many organs each of which contribute to the life of the whole organism and which are dependent upon other organs (e.,g. roots, stems) for survival.
Tissue		
Cell		
Organelle		
Molecule	carbon dioxide	Atoms of carbon and oxygen bonded together into a stable unit; nonliving substance that is both a reactant and a product of metabolism of the oak leaf.
Atom		

FIG 3-1

ATMOSPHERE

BIOSPHERE

LITHOSPHERE

HYDROSPHERE

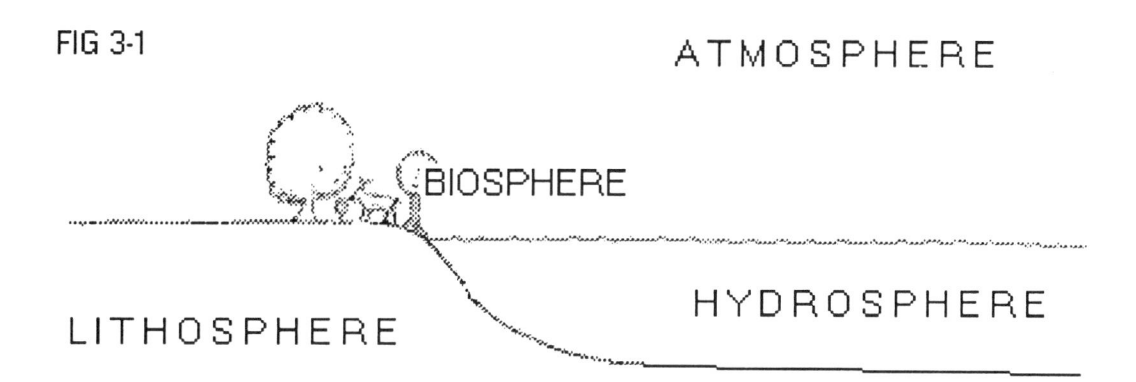

Figure 3-1. The physical environment of the earth may be divided into three components, the atmosphere, the lithosphere, and the hydrosphere. The regions of each component which living organisms occupy is known as the biosphere.

Living organisms inhabit essentially every major area of the surface of the lithosphere, the lower atmosphere, and the hydrosphere. The portion of the earth occupied by living organisms is called the **biosphere**. As any global traveller will testify, the Earth has a tremendous variation in landscapes. These differences are associated with climatic differences which influence the kinds of plant and animal inhabitants of these areas. Ecologists have drawn this diversity into a general classification of global ecosystems according to their overall similarities in climate and resident organisms. These major groupings of ecosystems are called **biomes**. The major classes of biomes are listed in Table 3-7.

3-C POPULATIONS AND SPECIES — Taxonomic Classification

Planet Earth contains a biological diversity of plants, animals and microbes which, according to current estimates, number from 5 to 30 million species! The need for a system of classification of species is obvious. Suppose a biochemist and oncologist were to publish a scientific article describing the anti-cancer properties of an extract from a Peruvian plant. They called the plant species "maqua moqua", or "maku moko", names used by two neighboring villages of nationals. Without some more universally accepted name of this plant, how could other researchers be sure of obtaining samples of the same species? Does the scientific literature already report such findings, but referenced to a different name for the same species? How could one determine which, if any, similar species have been reported? Are there similar species in Peru which might provide a better supply of the drug? These and other questions could not be answered with certainty without a classification system.

Accepted names are important.

Universal naming...

Carl von Linne' (in Latin, Linnaeus) (1707-1778) is credited with developing the first widely used taxonomic classification, called the **binomial nomenclature**. [*bi* = two)+(*nomen* = name)], or "two-name naming". Linnaeus classified all organisms having the same anatomical features into the same group called a **species**. He gave each species a double name, roughly analogous to our first and last names. Thus, the white oak is *Quercus alba* which is internationally recognized by taxonomists as the name for this species. *Quercus*, the first part of the name (analogous to a surname like Smith), indicates the *genus*. The second part of the name, *alba,* indicates this particular *species* as distinct from all other species of oak trees. Note that the generic name is always capitalized and the specific name is left uncapitalized; both are underlined or italicized. Having one universally accepted name for each species avoids the confusion often associated with multiple common names (Table 3-3).

A **population** is a group of individuals of the same species occupying a given area. The **species**, in turn, includes all of the natural populations that interbreed, actually or potentially, to produce fertile offspring, and have one or more anatomical features that distinguish them from other species.

Table 3-3. Confusion Among Common Names of *Brassica nigra* L.

Country	Common Name	Other English Names
United States	black mustard	brown mustard
France	moutarde noire	kerlock
Germany	senfkohl	cadlock
Italy	mostarda	warlock
Saudi Arabia	Khardal aswad	scurvy
Turkey	Kara hardal	senvil

is not easy . . .

The species concept is not an easy one to apply. First, the definition requires that breeding experiments be conducted for two generations to determine whether two given individuals are of the same species. This procedure is difficult and time-consuming. Second, sometimes individuals of two "different species" may produce sterile or fertile offspring, called *hybrids*. In other cases, the anatomical distinctions between species become blurred as, for example, among certain oak species which hybridize. In spite of these difficulties, the species concept remains useful as a basic unit of taxonomic classification. The difficulties should remind us of the dynamic nature of biology, the uncertainties encountered in attempting to develop a theoretical framework to describe the real world, and the need for caution and humility.

but, is the basis for taxonomic classification.

The taxonomic classification system arranges species into successively larger, comprehensive groups, or *taxa*. Thus, closely related *species* are included in the same *genus*, and similar genera occupy the same *family*. Likewise, families are grouped into *orders*, orders into *classes*, classes into *phyla* (or *divisions* in the plant Kingdom), and phyla into *kingdoms* (Table 3-4). Any taxon may be divided into subgroups in certain instances (e.g. subspecies (race), subgenus, etc.).

Table 3-4. Taxonomic Classification of Common Organisms

TAXON	WHITE OAK	LEOPARD FROG	CHIMPANZEE	HUMAN
KINGDOM	Plantae	Animalia	Animalia	Animalia
PHYLUM or DIVISION	Tracheophyta	Chordata	Chordata	Chordata
CLASS	Angiospermae	Amphibia	Mammalia	Mammalia
ORDER	Fagales	Salientia	Primates	Primates
FAMILY	Fagaceae	Ranidae	Pongidae	Hominidae
GENUS	Quercus	Rana	Pan	Homo
SPECIES	*alba*	*pipiens*	*troglodytes*	*sapiens*

As one would expect, kingdoms are the largest groups and are distinguished by major differences in anatomy and physiology. The most widely accepted classification system at present is the one proposed in 1969 by R. H. Whittaker of Cornell University. His five-kingdom system is outlined in Table 3-5. We will survey the taxonomic kingdoms in Chapter 9.

Table 3-5. The Five-Kingdom System of Classification

KINGDOM	COMMON REPRESENTATIVES	DISTINGUISHING FEATURES
Monera	Bacteria; and blue-green bacteria (Cyanobacteria) (4,760 species)*	Single-celled organisms; cells *prokaryotic*, lacking membranous compartments, or *organelles;* autotrophic or heterotrophic
Protista	Golden algae, diatoms, flagellates, ciliates, amoeboids (45,200 species)	Single-celled; cells are *eukaryotic*, having membrane-enclosed organelles; may be autotrophic or heterotrophic
Fungi	Slime molds; true fungi (46,983 species)	Multicellular; eukaryotic; heterotrophic; external digestion absorption of digestion products
Plantae	Green, brown, & red algae; mosses, ferns, and seed plants (265,000 species)	Multicellular; eukaryotic; mostly autotrophic
Animalia	Sponges, jellyfish, worms, arthropods, mollusks, amphibians, reptiles, birds, and mammals (1,033,614 species)	Multicellular; eukaryotic; heterotrophic; digestion occurs after intake of food material

*Number of described species; from E.O. Wilson. 1988. *Biodiversity*. Nat. Acad. Sci., Washington, DC

Taxonomy orders and unifies biology.

Taxonomic classification provides us with a powerful model by which to perceive order and patterns in the midst of diversity. First, there is greater precision in identifying and naming species. Biologists representing different languages and cultures can communicate research findings without confusion. Second, because similar species are grouped according to anatomical and physiological similarities, knowledge of one species within a taxon can often be applied to the understanding of other species within that group.

origin questions—

one taxonomy, but two interpretations

The taxonomic classification system has also contributed to the development of a new interpretation regarding the origin of life. Linnaeus was a creationist and seems to have believed, near the end of his life, that the Genesis concept of "kinds" corresponded to the "order" in taxonomic classification. Linnaeus and other 18th and 19th century taxonomists based their classifications upon observed anatomical and physiological characteristics which were assumed to be distinct and unchanging. However, in the 1800's, Jean-Baptiste Lamarck (1744-1829), Charles Darwin (1809-1882), and others, introduced a new interpretation of the same taxonomic model. This interpretation, embodied within the theory of evolution, proposes that anatomical similarities among species are evidence for evolutionarily similar lines of descent. We shall examine this important part of biology in more detail later. However, we should note that a single theoretical system can often be interpreted in entirely different ways depending upon the worldview of the interpreter.

3-D POPULATIONS — Ecological Classification

Taxonomic classification continues today.

How would you like to spend a winter in the tropical rain forest working with a team of taxonomists collecting and identifying plant and animal species. As biologists seek to understand the workings of the various ecosystems on earth, they must develop a list of species inhabiting each ecosystem. This task has become formidable for taxonomists, especially those studying the complex tropical forests where, each day, creatures are encountered that have never been catalogued before. So far, approximately 1.5 million species have been named and described world-wide. Yet biologists estimate that there may be as many as 30 million species on Earth.

Suppose we were able to identify every population of plants, animals, monerans, protists, and fungi (Table 3-5) that inhabit a given biotic community. Having obtained this taxonomic "Who's who?" list, we would still be far short of understanding the *ecological* relationships within the community. Clearly, we need another classification system to characterize the manner in which these species interact with one another as they acquire their energy or food, water, nutrients, and shelter. Such an **ecological classification** is more relevant to the ecologist who is concerned with *interrelationships among species* and what makes the community "tick".

Ecological classification is based upon energy needs.

trophic models

Ecologists classify organisms according to the manner in which they acquire food energy from their environment. Recall the foundational principle that every living organism requires a continuous supply of usable energy. Although this is true for all species, it does not mean that every species has the same mechanisms for acquiring its energy. In fact, many of the visible differences in form and behavior among plants and animals are related to their differing mechanisms of acquiring nutrition. Ecological classification is helpful in pointing out patterns in this diversity. The result is a more complete understanding of the **trophic structure** (Gr. *tropho* = nourish; food) of biotic communities. The term **food chain** is often used to describe the flow of energy (food) through the trophic structure of communities.

The simplest model portraying the trophic structure of a community has four major components: producers, consumers, detritivores, and decomposers (Figure 3-2). **Producers** are **autotrophic** [Gr. (*auto* = self) + (*tropho* = nourish)], or self-feeding in the sense that they can use energy (usually solar) to synthesize complex molecules from simple inorganic substances such as carbon dioxide, water, and minerals. Green plants that perform photosynthesis are the most common producers.

Consumers are organisms that occupy the second component of the trophic structure and are **heterotrophic** (Gr. *hetero* = other). Heterotrophs rely upon food energy that originates directly or indirectly from autotrophs. Heterotrophs that ingest and digest autotrophs are called primary consumers, or **herbivores**; those that feed on herbivores are secondary consumers, or **carnivores** (L. *carni* = flesh). **Detritivores** are scavengers such as the vultures which feed on **detritus**, representing the remains of dead animals and plants. **Parasites**, which obtain nutrition from a living host, are another type of consumer. **Decomposers** form the final group of heterotrophs and are represented by fungi and bacteria. However, decomposers are distinguished from consumers because they *absorb* their food after performing extracellular digestion (L. *extra* = outside of). External digestion releases some inorganic substances that are not absorbed by the decomposers. These substances become available

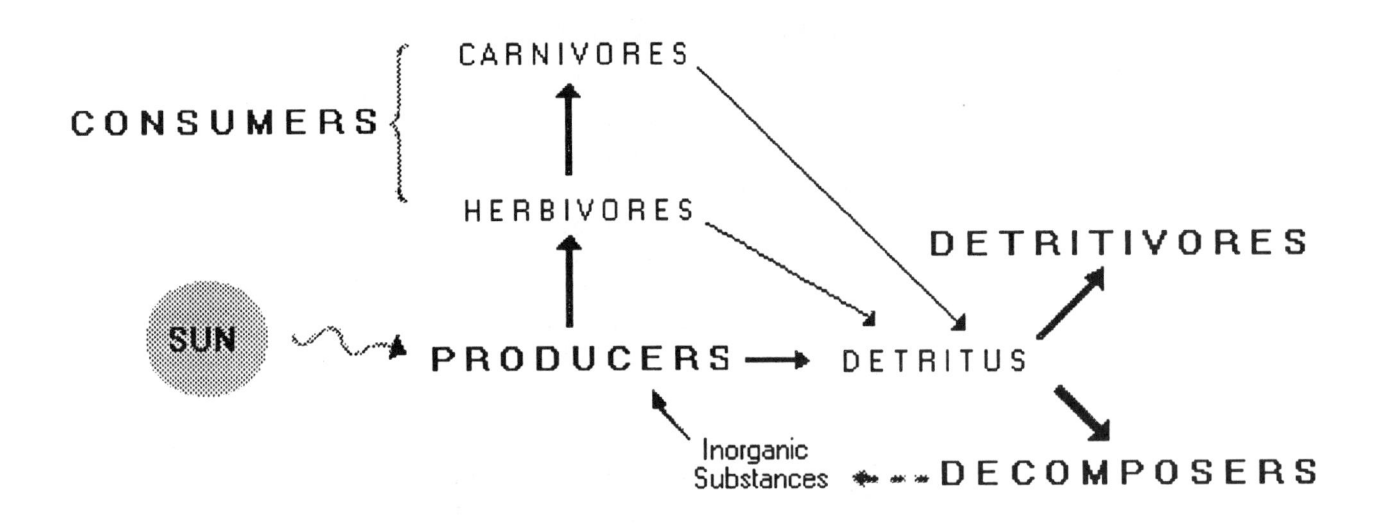

Figure 3-2. A simplified model of the trophic structure of a community. Populations are classified according to their mode of energy or food acquisition.

for producers to reuse (Figure 3-2). [Could heterotrophs live on Earth without autotrophs? Do autotrophs need heterotrophs to survive?]

Obviously, ecological classification of organisms provides a valuable pattern or model for study of relationships among populations in a given community. As one might expect, the groupings differ from taxonomic classification because the criteria for classification differ. Note from Table 3-5 that most taxonomic kingdoms have both autotrophic and heterotrophic members.

3-E INDIVIDUAL ORGANISMS

So far we have surveyed the means by which biologists organize knowledge of the living world at the ecosystem and population levels of organization. Biologists also find it useful to make distinctions among individual organisms within a population.

classification by sexual roles

Suppose we were to perform a survey of animal populations in a forest community. Within any single population, we will notice minor differences in anatomical features attributable to genetic variation. However, we would need to be aware of two other major distinctions that exist within many populations. The first is a **sexual** distinction, involving differences in form and behavior of *male* and *female* members of most populations. Male and female sex organs differ in such a way as to facilitate transfer of sex cells, or **gametes**. The male is more basically defined as an individual that produces motile gametes, or **sperm**. The male is responsible for transfer of sperm to the often nonmotile female gamete, or **ovum** (pl. ova). After fusion of male and female gametes, a process known as **fertilization**, the female generally assumes the role of nurturing the new generation. Although many variations exist, this is the basic pattern of **sexual reproduction** so widespread in the living world. The events involved in reproduction are pictured diagrammatically in a **life cycle**.

life cycle models

Distinctions also extend to what are called *secondary sex characteristics*. For example, the males and females of many bird species differ in color of plumage. Males of the hoofed mammal species usually develop horns or antlers which are employed in establishing dominance. Knowledge of sexual roles *within* a population is as important as making taxonomic distinctions *among* populations noted earlier. Can you imagine the embarrassment of a taxonomist upon discovering that he had listed as two separate populations what turned out to be male and female of the same population? In fact, Linnaeus made this error when he classified male and female mallard ducks as separate species! So much for the continuing challenge of biologists to develop conceptual patterns in the midst of diversity.

classification by stage of development

A second major biological distinction must be made during our survey among individuals of a population. This distinction becomes evident as we consider more closely the concept of the life cycle. Stated simply, we will notice differences in anatomy and behavior that relate to the *stage of development* of each individual. Immature individuals of certain species appear as miniature versions of the adult body form. For example, the immature grasshopper or white oak seedling are fairly easily recognized as members of their species. However, others such as butterflies, moths, and beetles have larval (caterpillar) stages which appear entirely different from the winged adult stage. The stage of development of any organism including humans is an important variable to consider in collecting and reporting biological data.

3-F ORGANS AND ORGAN SYSTEMS

Systems perform basic life functions.

In Chapter 2, we considered the general characteristics that are evident among all forms of life. Several of these characteristics involve what we have called *basic life functions*. These life-sustaining activities include energy or food intake and associated metabolism (e.g. digestion, and excretion of wastes), reproduction, maintenance of bodily homeostasis, and response to external stimuli. Most multicellular species have organs and organ systems designed to perform these functions. However, unicellular species of kingdoms Monera and Protista perform the same basic life functions within a single cell. Thus, whereas species are very diverse in form, they share the same life processes. We shall use an outline of basic life functions as a guide to deal systematically with the diversity of anatomical and physiological features found among the creatures. In many cases the taxonomic classification of a group of organisms includes species with similar features.

3-G CELLS AND TISSUES

Microscopes revealed cells.

When we study the cell level of organization, we will find that cell biologists have also been busy at systematizing their findings. This process began in the 17th century with the pioneering work of Robert Hooke who was the first to publish a description of cells. He used a primitive microscope to observe thin slices taken from the bark of a cork oak tree. These cells lacked cytoplasm and appeared to Hooke as little rooms (L. *cellulae*). Cells of other plant material were observed to contain "juices". Hooke did not realize that he was observing a distinction between living and nonliving matter. However, he was able to explain why cork floats on water: the little rooms contained air which provided buoyancy.

Robert Hooke published *Micrographia* in 1665, and this stimulated many others to fashion similar microscopes and to examine all manner of materials, both living and nonliving. One such investigator was Antony van Leeuwenhoek, a Dutch shopkeeper and amateur microscopist who described blood cells, sperm cells, bacteria, and protista. Imagine the thrill of observing a whole new world of tiny creatures never known to exist before. We trust that you will respond with enthusiasm and curiosity as you experience the privilege of using a much refined microscope, superior slides, and accessories.

Curiosity . . .

inductive thinking, . . .

and scientific publications . . .

promote theory development.

Curiosity and wonderment are essential motivations for scientific progress. However, as we have been emphasizing in this chapter, there must be those who can envision patterns and orderliness in the midst of apparent randomness and disorder. One of the first to see a general pattern amid the multiplicity of microscopic observations was the German naturalist Lorenza Oken. In 1805, he stated, "All organic beings *originate from* and *consist of* vesicles or cells." After several more decades and the contributions of many notable investigators, this statement came to be known as the **cell theory**. Matthias Schleiden and Theodor Schwann, a German botanist and zoologist, respectively, were the first to use the term "cell theory", in 1838. Subsequent studies of dividing cells led another German scientist, Rudolf Virchow, to conclude in 1859 that all cells originate from division of preexisting cells. Stated in another way, Virchow's theory proposes that living cells do not originate by spontaneous generation from nonliving matter. This theory of **biogenesis** [(*bio* = life) + (*genesis* = origin)] was confirmed in 1862 by Louis Pasteur whose experiments demonstrated that sterilized liquid media could not give rise to living microbes.

theories in conflict

The cell theory and the related biogenesis theory have not been refuted in over a century. However, the modern theory of evolution still maintains that life originated spontaneously from nonliving matter by strictly natural mechanisms. How could spontaneous generation (abiogenesis) of life have occurred at one time (or several times?) in earth's history if the phenomena has never been observed during recent recorded history? Science has provided empirical support for biogenesis, but cannot validate a proposed abiogenesis in the past.

Having established that living matter is cellular in structure, we should now point out criteria for distinctions among cells. There are two common criteria for distinctions among cells. First, cells may exist independently as do many **unicellular** monerans and protists (Table 3-5), or as one among many cells of **multicellular** organisms. Secondly, cells may be **prokaryotic**, lacking membranous compartments (*organelles*), or **eukaryotic**, possessing such organelles. This distinction separates Kingdom Monera from the other four kingdoms as shown in Table 3-5. The organelles of eukaryotic cells provide a compartmentation of different metabolic processes. Much of our knowledge of cell biology is based upon studies of organelles isolated from living cells.

Many cells are specialized.

The cells of multicellular organisms differ in size, structure, number, and kind of organelles. These differences reflect their particular function or **specialization** within the tissue and organ structure. Each specialized cell depends upon cells of other tissues and organs to meet its complete array of biological requirements. Thus, multicellular organisms are said to employ a **division of labor** among their many body cells.

3-H ATOMS AND MOLECULES

The membranes and organelles of every living cell are composed of precise arrangements and concentrations of atoms and molecules. Even at this basic level of organization, certain general patterns are evident. Cell and molecular biologists, and biochemists draw from the domains of chemistry and physics the theoretical basis for understanding the chemical reactions that occur in living cells. We shall examine the chemical basis of life in more detail in the next chapter.

3-I A SUGGESTED APPROACH TOWARD LEARNING BIOLOGY

Learn to ask questions.

Aristotle believed that the person who would succeed must ask the right questions. If this is true, and I believe it is, this text should help you to ask good questions. The first three chapters of this text have emphasized the broad themes and principles of biology, and major concepts and relationships we will encounter later. We have completed the "tour of the workplace" which we began in the introduction to Part 1. Take courage if you feel like the first day at a new job. By developing a conceptual framework, you will be able to assimilate information in a more orderly and meaningful way. First, here are a few study suggestions.

Table 3-6 presents one approach which the author has found effective in understanding and teaching biological concepts. It is presented here in hopes that your learning efforts will be more fruitful and rewarding as you use this or a similar systematic approach to the subject matter.

Table 3-6. An Approach to Learning Biology

QUESTIONS TO POSE	COGNITIVE LEVEL	COGNITIVE ACTIVITY
1. Definition?	KNOWLEDGE	define, recall,
2. Identification?	COMPREHENSION	identify, describe, illustrate
3. Structure?	APPLICATION	analyze, organize, apply
4. Function?	and ANALYSIS	
5. Relationships?	SYNTHESIS	synthesize, generalize
6. Significance?	EVALUATION	evaluate, integrate

*Bloom, B.S. (ed.) 1956. *Taxonomy of Educational Objectives: Handbook I: Cognitive Domain.* David McKay Co., Inc. New York.

Definition is crucial.

Some of the terminology presented in this text is unique to the biological sciences. When you encounter new terminology, begin at a basic learning level by asking, "What is the *definition* or meaning of each term within the context of the sentence?" A definition is usually given in a sentence as part of the flow of your reading. Often the definition will contain other biological terms introduced earlier in the discussion. Mastery of each definition means that you can verbally and mentally "draw a line around it" to show the perimeter of that term.

To demonstrate the proposed study approach, let's apply the progression in Table 3-6 to a common biological term, **consumer**.

> Consumers are organisms that occupy the second component of the trophic structure, and are heterotrophic [Gr. *hetero* = other) + (*tropho* = nourish)].

This DEFINITION illustrates the prerequisite of being able to define *trophic structure* and *heterotrophic*.

Illustrations aid learning.

Having mastered the definition and the helpful word roots of *consumer*, can you **illustrate** by an example (Table 3-6, Question 2)? Many times there is confusion here in distinguishing whether the term refers to a structure, a process, a condition, a living or nonliving system, etc. Therefore, identifying and remembering one or more examples along with your definition will be helpful. Perhaps you are familiar with deer and rabbit as consumers.

Distinguish *structures* from *processes*.

The question regarding **structure** and **function** (Table 3-6) requires that you have some understanding of the characteristics and requirements of consumers. The fact that consumers are defined as heterotrophic suggests that food energy is obtained from autotrophs. This implies that structural and behavioral features for procurement of food, digestion, and waste excretion are present in consumers.

... and *relationships*.

The question of what these consumers feed on brings you to question #5, **relationships**. By now you are hopefully able to fit the consumer into the general model of trophic structure (Figure 3-2). Prior to such a synthesis, you would need to define, identify, and analyze each of the other components of the model.

Strive to determine *significance*.

When you understand the general model of a community and how consumers fit into the model, there still remains the question of **significance**. Both questions of relationship and significance are more open ended and often require mastery of related material in a given study unit. For example, having understood the relationship of white-tailed deer to other populations including humans in a forest community, one may be able to develop a model of the trophic structure of that community. From this model, recommendations could be made as a basis for hunting laws (e.g. length of season, limit per hunter, etc.). However, there remains the question of significance and value of a deer population. Should men be allowed to stalk and kill deer? Is land more valuable as deer habitat or when it is cleared for shopping centers and housing developments? Thus, your personal value structure will be drawn into the learning process.

We have come a long way from simply defining "consumer"; and, rather hurriedly at that. Nevertheless, it should be evident that learning begins with mastery of biological terminology. But it must not stop there. Terms and concepts are the basis for understanding principles that together form the core of the discipline. The product of your learning experience in biology should by a working knowledge of these key principles. You will find them useful as you integrate biology with other disciplines of study in your effort to learn of new developments in our scientific and technological society.

3-J SUMMARY

This chapter has provided an overview of the structure of the science of biology. Your learning progression from definition and comprehension to synthesis and application will be enhanced as you consciously attempt to relate each new portion of the subject matter to this "skeletal structure" of biology. Table 3-7 provides a summary outline of this structure. Note that the levels of organization of matter, from biosphere to atom, forms the "backbone" of the skeleton. This same hierarchy is also used as the outline of our text in order to guide you in a systematic learning experience. In Chapter 4, we will consider the **atomic** and **molecular** level because these are the most basic units of living matter. Chapters 5 through 8 will guide our study of **ecosystems** and **communities**. Here we shall consider the interrelationships among **populations** as they each acquire energy and nutrients from the environment. From there it will be appropriate to examine the life processes of the individual **organism** —intake processing and excretion; reproduction of kind; maintenance of internal homeostasis; and, response to environmental stimuli. Finally, we will examine **cells** as the basic units of living matter.

Table 3-7. Some Commonly Used Classifications of Biological Information

LEVEL OF ORGANIZATION	CLASSIFICATION	EXAMPLES
Ecosystems	Biomes—global ecosystems	Tundra Coniferous Forest Deciduous Forest Tropical Forest Grassland Desert

Populations	Taxonomic
	(Binomial Nomenclature)
		Kingdom:
		    Monera
		    Protista
		    Fungi
		    Animalia
		    Plantae
		Phylum:
		    etc.

	Ecological (Trophic Structure)	Producers Consumers Decomposers Detritivores
Organisms	Sexual	Male Female
	Stage of Development	Life Cycle Stages (vary with species)
Organs/Organ Systems	Basic Life Functions	Energy/Food Intake Circulatory System Excretory System Reproductive System Hormonal System Nervous System Immune System
Cells	Structural Features	Prokaryotic Eukaryotic
	Degree of Specialization	
Molecules	Molecular Structure	Carbohydrates Proteins, etc.

QUESTIONS AND DISCUSSION TOPICS

1. Apply the learning approach outlined in Section 3-I to at least one biological term of your choice in each section of this chapter.

2. Explain why biologists use so many different classification systems. How does the tendency to develop these theoretical systems relate to the scientific method?

3. Relate the following terms and concepts to specific categories within the structural organization of biological knowledge: plant, producer, oak leaf, female, gamete, red blood cell.

4. Why is the word "potentially" inserted into the definition of a *species*?

5. How is taxonomic classification different from ecological classification? How is each system useful?

6. Use the historical development of the cell theory to illustrate various stages of the scientific method. How did Leeuwenhoek's contribution differ from that of Oken?

7. Do you believe life originated by *abiogenesis*? How does your concept of life's origin differ from that of naturalistic evolution?

8. How did Linnaeus' interpretation of the taxonomic classification system differ from that of Darwin? How do you account for this difference?

Chapter 4

Atoms and Molecules of Life

What is *life*?

As scientists attempt to understand a living system, they move down from dimension to dimension, from one level of complexity to the next lower level. I followed this course in my own studies. I moved from anatomy to the study of tissues, then to electron microscopy and chemistry, and finally to quantum mechanics. This downward journey through the scale of dimensions has its irony, for in my search for the secret of life, I ended up with atoms and electrons, which have no life at all. Somewhere along the line, life has run out through my fingers. So, in my old age, I am now retracing my steps, trying to fight my way back toward the cell.

— Albert von Szent-Gyorgyi
Nobel Prize Laureate, 1937

Life as "machine" or...

"vital force"

The function of complex machines can often be understood by disassembling them to study their parts. Does this approach apply to biological systems? Can life be explained by the physical and chemical behavior of the atomic and molecular parts? The **mechanistic** philosophy (Section 1-C) would answer "yes". To the mechanist, life processes are expressions of matter behaving according to natural laws of physics and chemistry. A different view, known as **vitalism**, holds that there is more to life than can be explained simply by natural laws. Rather, it is the "vital forces" operating over and above the natural laws that make a living organism different from nonliving matter and machines. [How do you view life on this point?]

Mechanistic hypotheses are useful...

Biologists cannot deny the existence of "vital forces" or any other metaphysical component of life. However, "vital force hypotheses" are not testable because they raise questions that involve relationships that are external to the natural world (Chapter 1). At the same time, vitalism discourages efforts to understand how natural laws influence life processes.

but there is more to life than "machine".

This chapter is presented with a mechanistic emphasis because much of the current theory and understanding of living systems has come from a mechanistic approach. To illustrate, consider one mechanistic hypothesis based upon a "life-as-a-machine" metaphor: "The heart *pumps* blood". This view of the heart provided the scientific momentum that has led to the development of an artificial heart. Who can deny the validity of the mechanistic approach? But, can life be understood entirely as "machine"? Keep this question in mind as you read.

Szent-Gyorgyi stated, in the epigraph above, that the life is gone by the time we reduce a living organism down to its atoms and electrons. [Did he forsake mechanism for vitalism? Reread his statement.] Assuming the mechanistic approach is valid, let us consider some basic findings of biochemistry and molecular biology regarding the processes that contribute to life at the sub-microscopic levels of organization. As you read, think about what it is that makes atoms and molecules together a part of "living matter"?

4-A WHO NEEDS CHEMISTRY?

We need it!

As you probably know by now, the title of this section is a rhetorical question. We all "need chemistry". We live in a world that increasingly requires that we understand some basic chemistry, whether it is in regard to products we purchase or the pollutants released in product manufacture. Furthermore, the mechanistic view suggests that we need an understanding of atoms and molecules to understand living organisms.

This chapter has been written with the assumption that you have had one course in secondary level chemistry. If your chemistry background is weak or needs refreshment, please study Appendix A before proceeding to Section 4-C.

4-B ATOMIC COMPOSITION OF LIFE

God created the *elements* ...

According to Genesis 2:7, "God formed man out of the dust of the earth". For those who believe that science and scripture are complementary, the development of the *atomic theory* added wonderful insight into the meaning of the Scriptures. By incomprehensible wisdom and power, God created the Earth which is composed of 92 known, naturally occurring, chemical elements. From these elements of the earth (Heb. *adamah*), He then formed the living creatures and the human being (Heb. *adam*), "and breathed into his nostrils the breath of life". Thus, atomic theory helps us with the *composition* of the earth

and used them to form *life*.

and the human, but there is also this *enlivening* force placed within the animals (Genesis 1: 21,24) and humans which escapes the comprehension of the natural sciences. [Is there a scriptural support for "vitalism"?]

major elements of living cells

Table 4-1 lists the most common elements of living matter. Note that, with the exception of oxygen, the "big four" elements (carbon, hydrogen, oxygen, and nitrogen; or C-H-O-N) are present in much greater abundance in living matter than in the lithosphere. This difference in composition indicates that living cells are active in accumulating some elements and excluding others.

Each element has one or more important roles in life processes as summarized in Table 4-1. A listing of elements known to function in animal and/or plant life and/or microbial life may be summarized by the following sentence:

C. HOPKINS CaFe Mg(ood) B(y) Mn CuZn Mo CuB

Additional elements known to be biologically essential are being added to the list. For example, in 1992, the essentiality of nickel (Ni) in plant growth and reproduction was confirmed.

Table 4-1. Global Distribution and Roles of Common Elements of Living Matter

ELEMENT	Atomic Composition (Atoms per 100) LITHOSPHERE	LIVING MATTER	COMMON BIOLOGICAL ROLES
Hydrogen	2.92	49.8	Component of water and all organic molecules; H+ ions influence biological reactions and processes.
Oxygen	60.4	24.9	Component of water and organic molecules; needed for aerobic respiration.
Carbon	0.16	24.9	"Backbone" of all organic molecules
Nitrogen	*	0.27	Component of organic molecules *e.g.* proteins and nucleic acids
Calcium	1.88	0.07	Component of bones, teeth; involved in membranes and cellular fluids; muscle contraction
Potassium	1.37	0.05	Nerve impulse transmission; enzyme activator; body water balance
Silicon	0.04	0.03	Cell wall component in grasses, diatoms, and walls of arteries
Magnesium	1.77	0.03	Component of chlorophyll and other molecules; enzyme activator
Phosphorus	0.08	0.03	Component of nucleic acids, lipids, energy-carrying molecules (e.g. ATP)
Sulfur	0.04	0.02	Component of amino acids and proteins
Sodium	2.49	*	Ionic balance in fluids; nerve impulse transmission
Iron	1.90	*	Component of molecules of energy-yielding reactions, and of blood oxygen carriers
Chlorine	*	*	Formation of gastric juice; ionic balances

* Indicates element found only in trace amounts. Not all essential elements are listed; some (e.g. silicon) are not essential for all forms of life.

4-C BONDING OF ATOMS IN LIVING MATTER

Chemical bonds may be...

covalent...

Atoms that compose living cells are commonly attracted to other atoms to form **chemical bonds**. Three types of chemical bonds commonly encountered are covalent bonds, ionic bonds, and hydrogen bonds. **Covalent bonds** are formed when two or more atoms share pairs of electrons. **Molecules** consist of one or more atoms bonded covalently. Molecules of living cells will be discussed in Sections 4-D through 4-J. See APPENDIX A for a more basic discussion of chemical bonding.

or ionic.

Ionic bonds form between **ions**, atoms that have gained an e⁻ (negatively charged like Cl⁻) or lost an e⁻ (positively charged like Na⁺). Ionic substances that are common in the cytoplasm and body fluids include Ca^{++}, K^+, Mg^{++}, Mn^{++}, Cu^{++}, Zn^{++}, and Fe^{++}, or Fe^{+++}. Table 4-2 lists common **polyatomic ions**, consisting of two or more atoms covalently bonded to one another and remaining together through most reactions because of their stability. How many covalently bonded atoms are in NO_3^-? See APPENDIX A for a discussion how ions dissolve in H_2O.

Hydrogen bonds form between polar molecules.

In addition to covalent and ionic bonding, a third type of bonding which influences life processes is **hydrogen bonding**. A hydrogen bond forms between a hydrogen atom covalently bound to one molecule and an atom such as oxygen or nitrogen covalently bound within the same molecule or a nearby molecule. Covalently bonded oxygen and nitrogen atoms usually have a greater affinity for the shared electron pair than hydrogen atoms to which they are bonded. The result is a *polar covalent bond* within the molecule (Appendix A). That is, regions of a molecule occupied by oxygen or nitrogen tend to be negatively charged, whereas regions with hydrogen atoms are slightly positive. The resultant polarity between regions within molecules or between molecules is the basis for hydrogen bonding.

The simplest illustration of hydrogen bonding is shown in Figure 4-1 where hydrogen bonds (represented by dotted lines) are formed between the polar ends of water molecules. The polar nature of water molecules is crucial to living cells. Without it, water would boil and vaporize at temperatures too low for living cells to exist.

Table 4-2. Common Polyatomic Ions of Living Organisms and the Environment.

ION	FORMULA	MAJOR BIOLOGICAL SIGNIFICANCE
Ammonium	NH_4^+	Constituent of proteins; released during decay
Carbonate/ bicarbonate	$CO_3^=/HCO_3^-$	Formed from CO_2 gas dissolved in H_2O; source of CO_2 for aquatic plants; cellular buffer
Nitrate	NO_3^-	Usable form of nitrogen for plant growth; toxic to humans when ingested
Phosphate	PO_4^{3-}	Usable form of phosphorus for all life forms
Sulfate	SO_4^{2-}	Usable form of sulfur for autotrophs

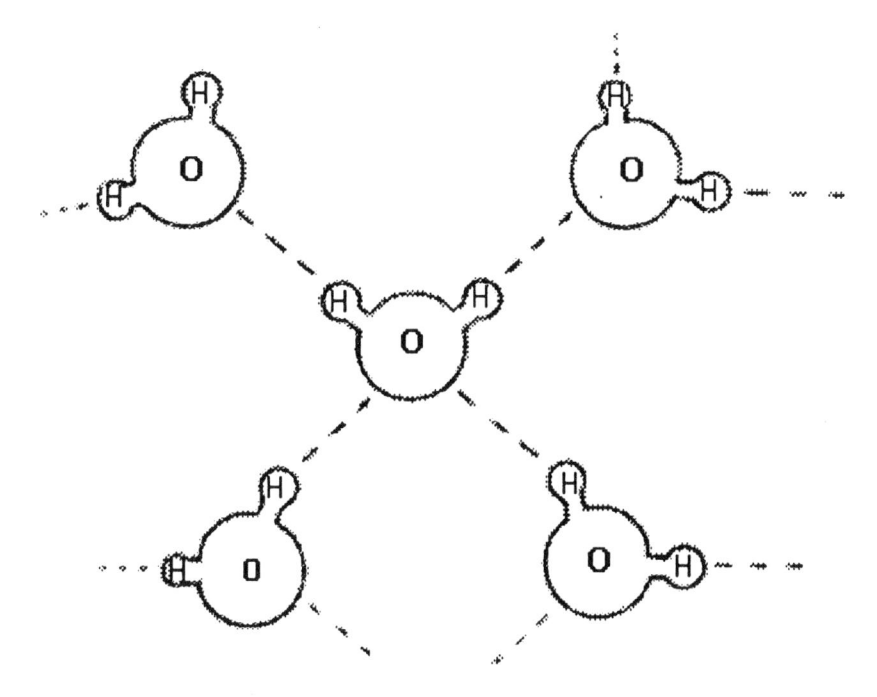

Figure 4-1. Hydrogen bonds (dotted lines) between polar ends of water molecules. These bonds stabilize water in liquid form over a wide temperature range.

In large, chain like molecules such as proteins and nucleic acids, hydrogen bonds form between hydrogen atoms and covalently bonded oxygen or nitrogen located *in the same molecule*. These bonds are quite weak compared to the covalent bonds that hold the atoms into the molecular structure. However, often a great number of hydrogen bonds are formed and collectively they exert a major influence on the *three-dimensional configuration* of proteins and nucleic acids.

Summary: In summary, living matter is composed of atoms held together by chemical bonds of various types between atoms. Bonds are formed in predictable manner under the precise influence of the electronic configurations within the electron shells of each atom. Let us now survey the major classes of molecules that make up living matter.

4-D MAJOR TYPES OF MOLECULES IN LIVING MATTER

Molecules that compose living matter may be divided into two broad groupings. **Organic molecules** are those which contain carbon atoms usually covalently bound to hydrogen and oxygen, and sometimes nitrogen. Organic molecules may be grouped into four major categories based upon their atomic composition, chemical properties, and biological role. The four groups are *carbohydrates, lipids, proteins, and nucleic acids*.

Inorganic molecules, like organic molecules, may contain C, H, O, or N atoms. However, the latter do not display covalent bonding between carbon and hydrogen or nitrogen atoms. Water (H_2O) as well as O_2, CO_2, N_2 (nitrogen gas), and NH_3 (ammonia gas) are important inorganic molecules of living cells. A significant portion of the mass of living organisms consists of water which is the topic of the next section.

4-E WATER AND OTHER INORGANIC MOLECULES

Water works in cells...

Water is the most abundant molecule in living matter, and may compose up to 90% of the mass of a living organism! Because of its abundance and its chemical and physical properties, water has a major influence upon the metabolic reactions of living cells. Here, we will give special attention to water as a dissolver of substances, its role in metabolic reactions, and its relationship to acids, bases, and buffers.

because of its properties.

Water exists as a liquid over a wide range of temperatures suitable for life because of the hydrogen bonds formed between water molecules Section 4-C). The resultant **cohesion** between water molecules permits the existence of surface tension which allows certain insects and spiders to "walk on water". Liquid water also provides the pressure necessary for structural support of cells and tissues, and for the movement of body fluids. [Consider why plants "wilt"; and, feel your pulse.]

a good solvent

Water has an amazing capacity to dissolve a large variety of organic and inorganic substances. That is, water is a good **solvent**. For example, water can dissolve salts such as sodium chloride (NaCl) with apparent ease, in contrast to the fact that large quantities of heat are required to melt dry salt crystals. Crystals of NaCl are composed of individual Na^+ and Cl^- ions arranged in orderly fashion to form a crystal lattice (Figure 4-2a). The electrostatic attraction of the ionic bonds are so strong that NaCl and other salts such a magnesium chloride ($MgCl_2$) and calcium nitrate [$Ca(NO_3)_2$] all have very high melting points. At these high melting temperatures enough heat energy has been supplied to free the ions from the lattice. On the other hand, simply sprinkling NaCl from the salt shaker into water and stirring will quickly **dissolve** the crystalline lattice and release the resultant ions into water. Here, water molecules are able to dislodge the Na^+ and Cl^- ions from the lattice. The ions are then held in solution because they are surrounded by layers of H_2O molecules as shown in Figure 4-2b. The *slight polarity of water* molecules (see APPENDIX A, Section A-3) determines the orientation of the water molecules around the Na^+ and Cl^- ions. The attraction of unlike-charged regions of ion and molecule holds the water molecules in these three dimensional *spheres of hydration*. It is this capacity of water among others that makes it so ideal as the "solvent (dissolver) of life".

slightly polar

Many of the metabolic reactions of living cells involve water as a direct chemical reactant or product. These reactions include the digestion of large molecules into smaller constituents by **hydrolysis**. [(*hydro* = water) + (*lysis* = splitting)]; and the synthesis of larger molecules from simpler constituents by **dehydration** (water removal). These reactions are illustrated in Figure 4-3.

sodium ion
chloride ion

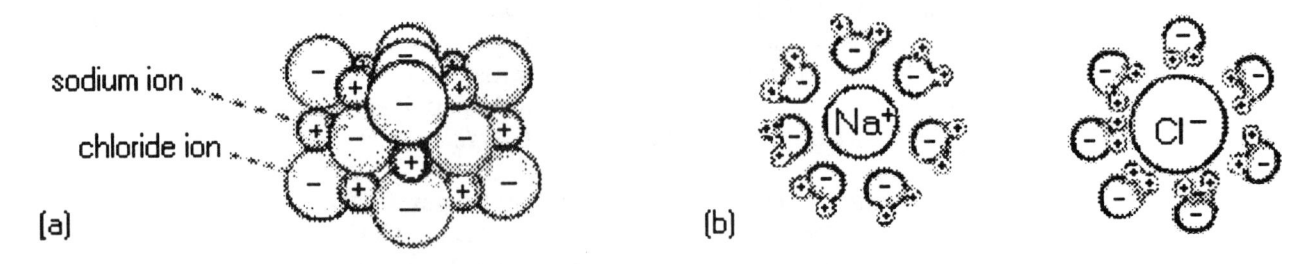

[a]

[b]

Figure 4-2. (a) Diagrammatic view of part of a sodium chloride crystal. (b) Water molecules form "spheres of hydration" around sodium and chloride ions by attractions between unlike charges of ion and the oxygen or hydrogen ends of the polar water molecules.

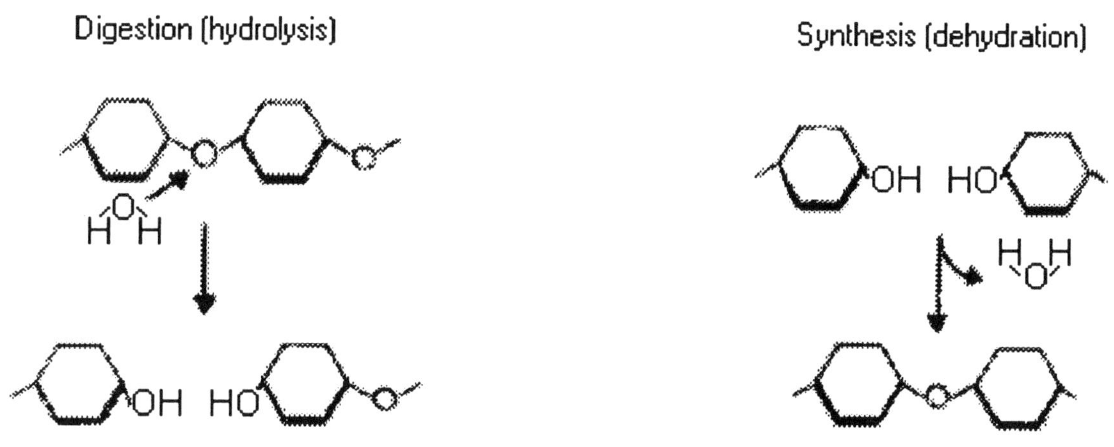

Digestion (hydrolysis)   Synthesis (dehydration)

Figure 4-3. Participation of water in digestion and synthesis reactions.

Water molecules have a very slight tendency to *dissociate* into H^+ and OH^- (hydroxyl) ions as follows:

$$HOH \Longleftrightarrow H^+ + OH^-$$

acids and bases

Other substances that contain hydrogen atoms held with polar covalent bonds or ionic bonds also have a tendency to dissociate, releasing H^+ and some other negatively charged ion. **Acids** are substances that increase the concentration of H^+ ions in water. **Bases** decrease the concentration of H^+ ions in water and produce more *alkaline* conditions. Bases such as sodium hydroxide (NaOH) decrease the H^+ ion concentration by providing OH^- ions which combine with H^+ ions and hence neutralize the acidity. For example, excess stomach acid which originates, in part, from hydrochloric acid (HCl) may be neutralized by taking a dose of milk of magnesia which contains the base, magnesium hydroxide [$Mg(OH)_2$]. The neutralization reaction occurs as follows:

$$\underset{\text{acid}}{2H^+\ 2Cl^-} + \underset{\text{base}}{Mg^{++}\ 2OH^-} \longrightarrow \underset{\text{salt}}{MgCl_2} + \underset{\text{water}}{2H_2O}$$

pH scale

The degree of acidity or basicity of a solution depends upon the concentration of H^+ ions relative to OH^- ions (or other ions that can "scavenge" H^+ ions). The **pH scale** which ranges from 0 to 14 is useful to portray this relationship. Solutions that have equal concentrations of H^+ and OH^- ions are said to be *neutral* and have a pH of 7. Solutions with a pH below 7 have a higher H^+ concentration than OH^-, and are *acidic*; *alkaline* solutions have a higher OH^- concentration than H^+, and a pH greater than 7. Because pH is a logarithmic function of the H^+ or OH^- concentration, each change in pH of 1 pH unit equals a 10-fold change in ion concentration. Life processes are very sensitive to changes in pH of the cellular fluids as we shall see. Figure 4-4 illustrates the pH of common fluids that one may encounter in daily life.

more important inorganic ions

Other inorganic molecules that participate in biological processes include O_2, CO_2, N_2, and NH_3. All of these are present in varying amounts in the atmosphere. Both O_2 and CO_2 are involved in respiration and photosynthesis.

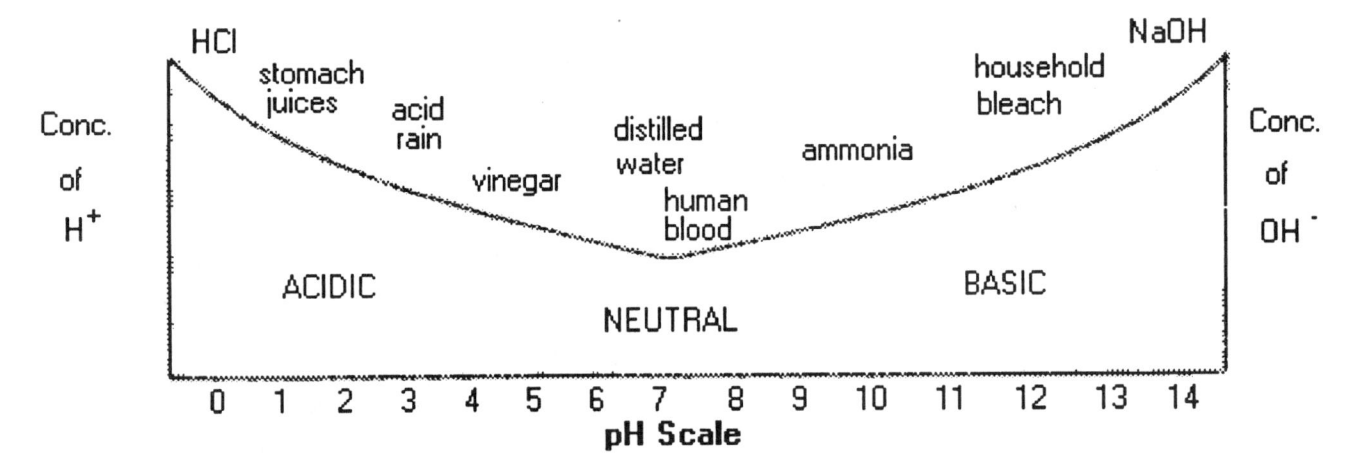

Figure 4-4. The pH scale and H⁺ concentration of familiar solutions.

Carbon dioxide, when dissolved in aqueous solution, causes ionization of water leading to the formation of carbonic acid (H_2CO_3):

$$CO_2 + H_2O \rightleftharpoons H_2CO_3 \rightleftharpoons H^+ + HCO_3^-$$

Carbonic acid in the cytoplasm or cell surroundings has little tendency to dissociate into bicarbonate (HCO_3^-) and release H⁺ ions. Therefore, it is considered a *weak acid*. The carbonic acid-bicarbonate ion system provides an effective mechanism for minimizing fluctuations in cellular pH. Such systems are called **buffers**. If cells come in contact with excess alkalinity (e.g. OH⁻), the following reaction tends to prevent an increase in pH:

Buffers can stabilize pH.

$$OH^- + H_2CO_3 \rightleftharpoons H_2O + HCO_3^-$$

Or, if excess acidity is present, bicarbonate ions scavenge the H⁺ ions and prevent a lowering of pH as follows:

$$H^+ + HCO_3^- \rightleftharpoons H_2CO_3$$

Phosphate and various organic acids also function in cellular buffer systems.

Ammonia is often associated with the decay of dead organisms and excretory waste products. When dissolved in aqueous solution, ammonia gas causes ionization, resulting in the formation of ammonium ion (NH_4^+):

$$NH_3 + H_2O \longrightarrow NH_4^+ + OH^-$$

Plants can use NH_4^+ and (NO_3^-).

Plants and microbes absorb NH_4^+ and nitrates (NO_3^-) as important sources of nitrogen for synthesis of proteins and nucleic acids. Certain groups of microorganisms have the capacity to convert atmospheric N_2 gas into ammonia. This process, known as *nitrogen fixation*, provides a key route of entry for the otherwise unusable N_2 gas into the biological world.

4-F ORGANIC MOLECULES

Living organisms and the undecayed remains of dead organisms (e.g. coal and oil) are the major sources of organic molecules on earth today. Prior to the nineteenth century, organic molecules were distinguished as those *containing*

In 1845, we did what only cells could do.

carbon atoms and synthesized by living organisms. The above definition of organic molecules was accurate and provided a uniqueness to living matter as well. Many believed that man would never be able to match the wondrous capacity of living cells to synthesize organic molecules from inorganic substances such as CO_2 and H_2O. However, the first synthesis of an organic compound from inorganic substances was finally achieved by Adolph Kolbe, in 1845. Since then, the definition and uniqueness of organic molecules has shifted more to the special properties of the carbon atom than upon whether or not the molecule was produced in living cells.

the versatile carbon atom

Carbon is the most versatile of all elements found in living organisms. Carbon atoms can combine with one another and with atoms of other elements such as hydrogen, oxygen, nitrogen, and sulfur to form a seemingly endless array of organic molecules. However, science aims at systematizing knowledge, and the disciplines of **organic chemistry** and **biochemistry** have developed around systematic efforts to understand carbon chemistry.

The versatility of carbon is due to the electron configuration of its outer shell. The carbon atom has four electrons in its second electron shell and needs eight to achieve stability (APPENDIX A, Figure A-2). Therefore, carbon is able to form four covalent bonds. For example, the methane gas molecule is formed when a carbon atom bonds with four hydrogen atoms (Figure 4-5). Each hydrogen atom is oriented so as to achieve maximum separation from each of the other three. The resultant molecule has a three-dimensional structure represented in Figure 4-5a.

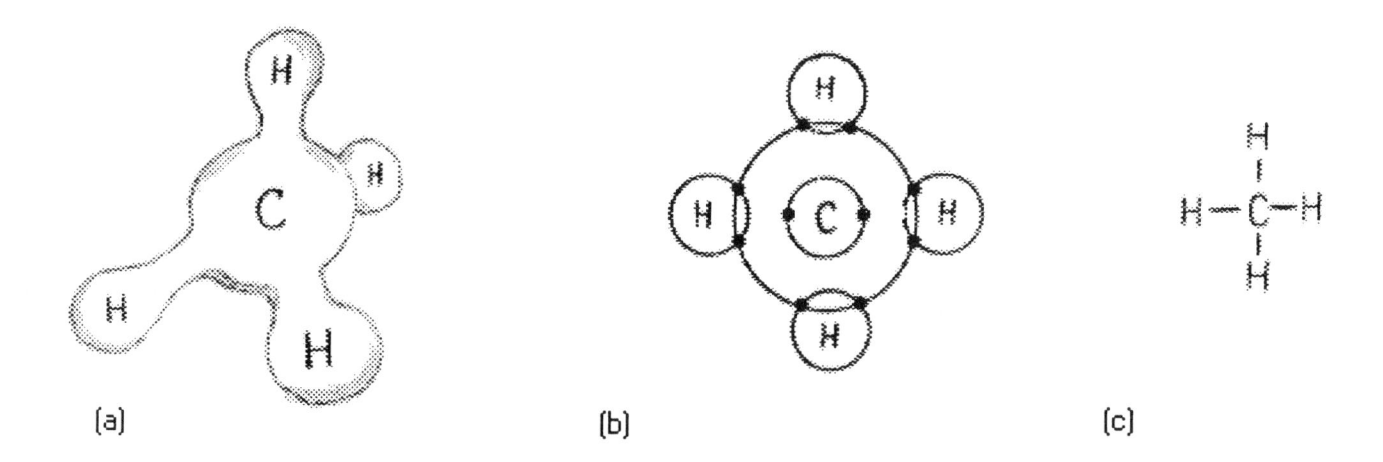

Figure 4-5. The methane (CH_4) molecule depicted in three different ways.

carbon "skeletons"

Larger organic molecules have a much more complex three-dimensional configuration which is crucial to their biological role. The configuration is also difficult to represent on paper; therefore, it is convenient to use flat diagrams wherein dashes represent covalent bonds. Following the basic pattern of methane, and adding the feature of carbon-carbon bonds, makes possible longer chains with carbon as the "skeleton". These chains may occupy any one of a number of configurations as follows:

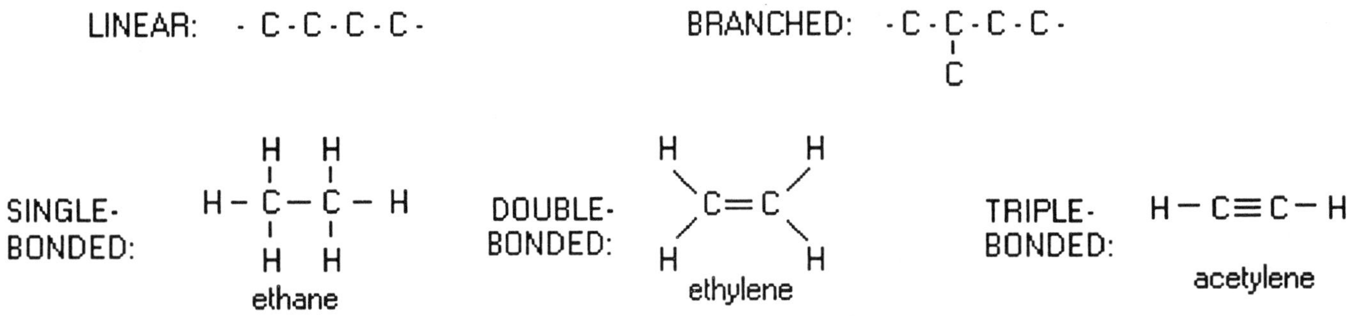

LINEAR: - C - C - C - C -   BRANCHED: - C - C - C - C -
                                                       |
                                                       C

SINGLE-BONDED:

H H
| |
H - C - C - H
| |
H H
ethane

DOUBLE-BONDED:

H   H
 \\   /
  C = C
 /   \\
H   H
ethylene

TRIPLE-BONDED: H - C ≡ C - H

acetylene

Carbon chains may also close to form a ring structure, diagrammed as follows:

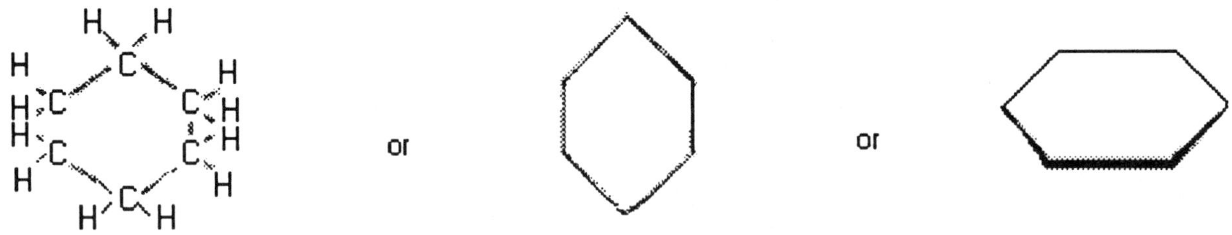

predictable bonding patterns

In spite of the almost bewildering variety of molecules, there are two principles that will allow us to predict the possible structures and chemical behavior of these molecules. First, the number of covalent bonds formed by carbon and atoms of other elements are predictable. Study the diagrams in the previous paragraph and in Table 4-3. Make a table of your own, listing the elements C, H, O, N, and S, found in these organic molecules and the number of covalent bonds formed by each molecule. How does this number compare to the electronic configurations of their outer shells shown in APPENDIX A, Table 4-1? Amazingly, only a few, predictable bonding patterns compose such a great variety of organic molecules. These molecules, in turn, are responsible for the great diversity of life forms.

reactive groups

Covalent bonding patterns allow us to predict molecular structure, whereas, the type of *reactive group* allows us to predict the chemical behavior. When two molecules collide, bonding or other chemical changes occur between reactive groups. Common reactive groups of organic molecules are shown in Table 4-3 along with a brief description of their chemical behavior, or reactivity.

Molecules are classified by reactive groups.

Reactive groups influence the chemical behavior of organic molecules so much that organic chemists use these groups as a basis for classifying molecules. This classification carries over into biochemistry. We shall consider four major classes of organic molecules: carbohydrates, lipids, proteins, and nucleic acids.

Table 4-3. Common Reactive Groups of Organic Molecules

GROUP NAME	STRUCTURE	BIOLOGICAL SIGNIFICANCE
ALCOHOL	-C-OH	Part of the structure of carbohydrates; reacts with carboxylic acids and amino acids
ALDEHYDES and KETONES	-C(H)=O ; -C(C)=O with C	Reactive group of carbohydrates and related molecules
CARBOXYLIC ACID GROUPS	-C(=O)(OH)	Ionizes to release H^+ ions into solution (weak acids); part of fatty acids, amino acids, and other organic acids; provides buffer capacity
AMINO GROUPS	-C-N(H)(H)	Reactive group of amino acids; forms peptide bonds in reaction with carboxyl group of another amino acid; acts as a base in scavenging H^+ to form NH_4^+
PHOSPHATE	-O-P(=O)(O$^-$)-O$^-$	Acts as acid to release H^+ ions; serves as a link between organic molecules such as nucleic acids; participates in energy reactions as an energy carrier
SULFHYDRYL	-C-S-H	Part of certain amino acids; reacts with another sulfhydryl group to form disulfide linkage (-S-S-) which stabilizes protein molecular structure

4-G CARBOHYDRATES

Can you recognize a carbohydrate?

Carbohydrates are composed of carbon, hydrogen, and oxygen atoms in an approximate ratio of 1:2:1, or $(CH_2O)_n$, where "n" represents the number of subunits in the carbon chain of a given molecule. The basic building blocks of the carbohydrates are called **monosaccharides** [(*mono* = one) + (*saccharo* = sugar)]. Monosaccharides may contain from three to seven carbon atoms. Triose (three-carbon), pentose (five-carbon), and hexose (six-carbon) monosaccharides are the most common biologically. Specific representatives of these groups are shown in Figure 4-6. Note that molecules with five or more carbon atoms can form a ring structure. Lines between atoms represent covalent bonds. For convenience, certain bonds are not shown.

simple sugars

isomers are intriguing

Monosaccharides such as ribose, glucose, and fructose are often referred to as simple sugars. Interestingly, both glucose and fructose have the same molecular formula ($C_6H_{12}O_6$), and yet they have different properties. For example, your taste buds can detect the greater sweetness of fructose as compared to glucose. Molecules such as these with the same molecular formula but different bonding patterns are called **isomers**. Isomeric differences among molecules are common in the living world and pose a number of intriguing questions regarding the design of creation at the molecule level. [What questions about taste can you think of?]

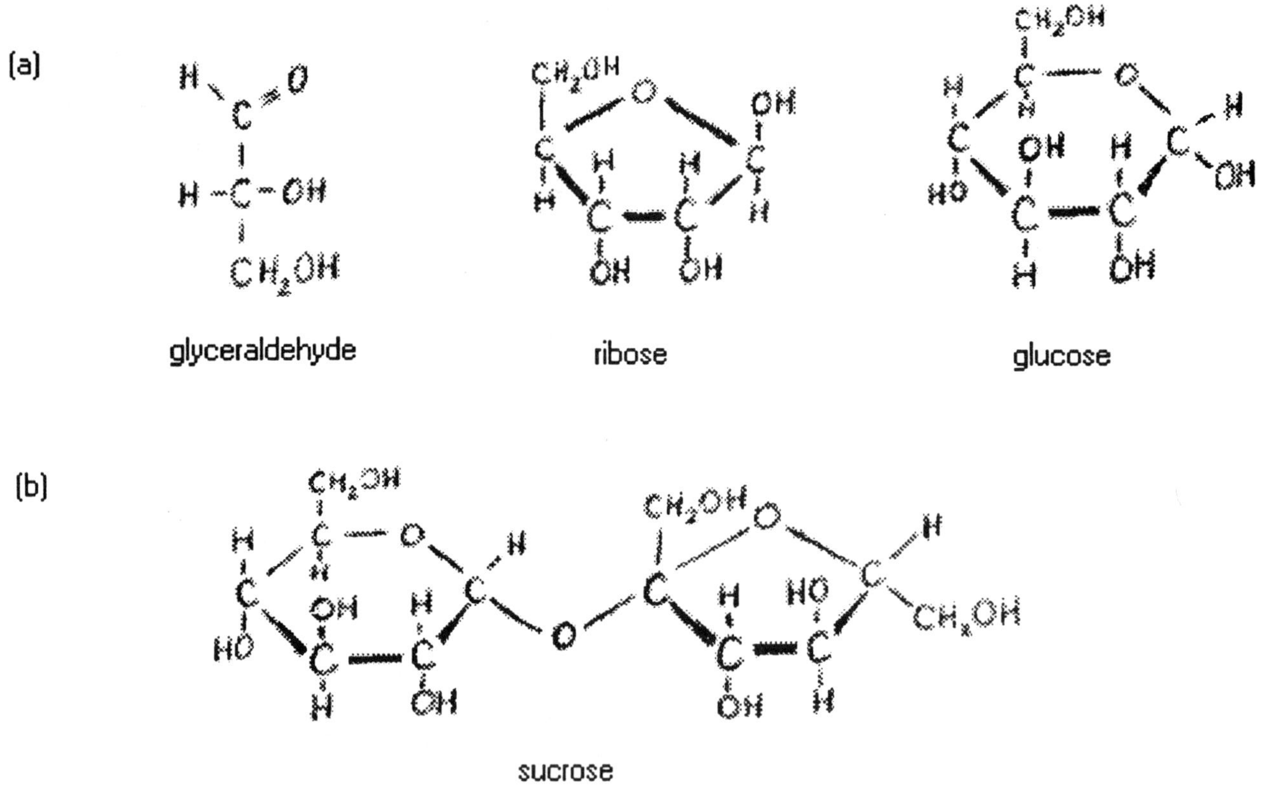

(a) glyceraldehyde   ribose   glucose

(b) sucrose

Figure 4-6. Representative monosaccharides (a) and a disaccharide, sucrose (b), formed when glucose is bonded to fructose.

disaccharides

A great variety of more complex carbohydrates are formed by dehydration synthesis reactions (Figure 4-3) in which monosaccharides are linked together. Sucrose, common table sugar, is formed when the isomers glucose and fructose bond to each other to from a "double sugar" or **disaccharide** (Figure 4-6). Sucrose is the most common of all sugars, and functions in the transport of chemical bond energy in green plants. Some plants such as sugarcane and sugarbeets store great quantities of sucrose and are grown commercially as sources of table sugar. Other common disaccharides include maltose (glucose + glucose), common in germinating seeds, and lactose (glucose + galactose), known as milk sugar.

polysaccharides

Bonding makes a difference

Carbohydrates that form when more than two simple sugars are bonded together are called **polysaccharides** (*poly* = many). Polysaccharides are straight-chain or branched **polymers**, often synthesized by repeated addition of the same monosaccharide to the growing chain. Two common polysaccharides are shown diagrammatically in Figure 4-7. Amylose and cellulose both consist of repeating glucose **monomers** (subunits). However, two different isomeric forms of glucose are involved, *alpha* and *beta* glucose. Consequently, the two differ in physical and chemical properties. For example, amylose functions as an energy reserve in plant leaves and storage organs and is readily digestible (Figure 4-3). On the other hand, cellulose, a major component of plant cell walls, resists digestion and is a major component of the "fiber" that passes through the human digestive tract. [How then is cellulose from dead plants, or that ingested by ruminant animals, digested?]

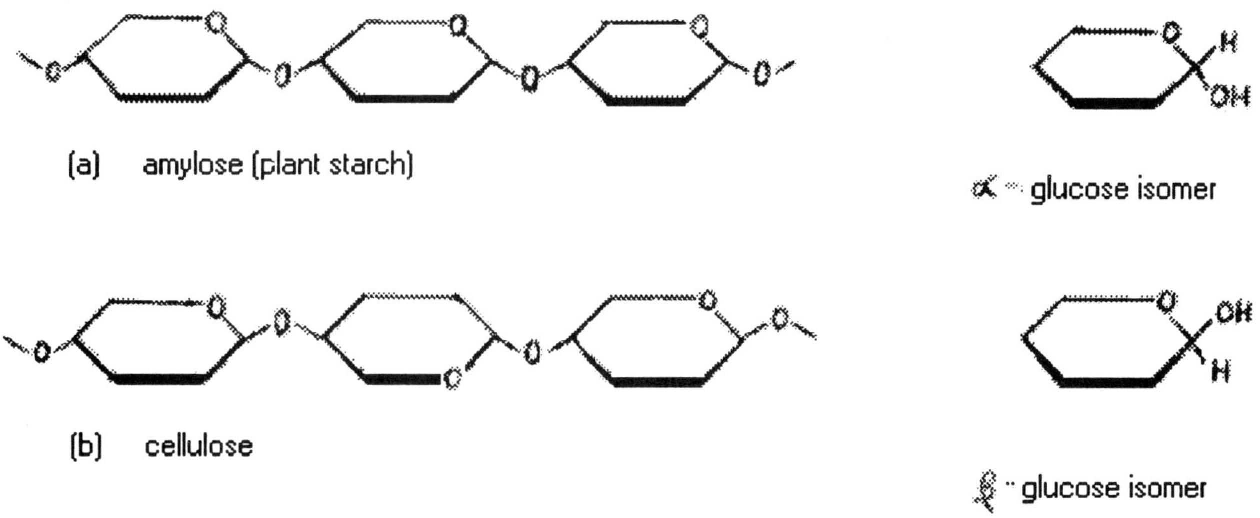

(a) amylose (plant starch)

α - glucose isomer

(b) cellulose

β - glucose isomer

Figure 4-7. A small portion of the chains of two common polysaccharides, (a) amylose, formed from *alpha*-glucose monomers; and (b) cellulose, composed of *beta*-glucose. See text for discussion of implications.

A third polysaccharide, glycogen, functions as an energy reserve in animals. Glycogen is similar to the structure of plant starch but features a branched-chain structure.

4-H LIPIDS

Lipids have mostly carbon and hydrogen.

Lipids are a second group of organic molecules which include fats, phospholipids, waxes, and steroids. The common characteristics of this diverse group include the predominance of carbon and hydrogen atoms, and very few oxygen atoms. This is evident in many of the **fatty acid** components of fats and phospholipids. For example, linoleic acid

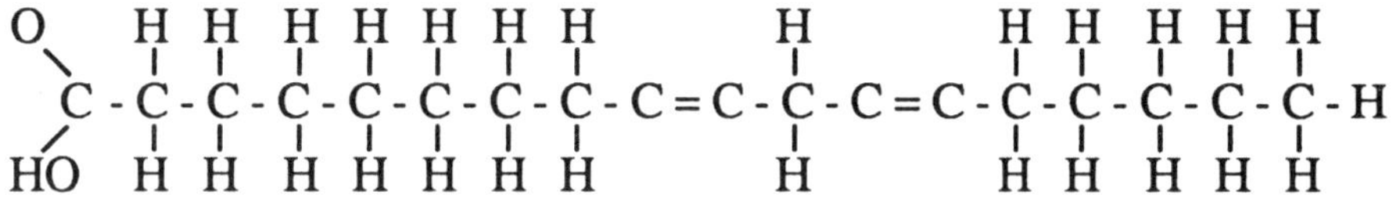

$$\begin{array}{c} O \\ \diagdown \\ C \\ \diagup \\ HO \end{array} - \overset{H}{\underset{H}{C}} - \overset{H}{\underset{H}{C}} - \overset{H}{\underset{H}{C}} - \overset{H}{\underset{H}{C}} - \overset{H}{\underset{H}{C}} - \overset{H}{\underset{H}{C}} - \overset{H}{\underset{H}{C}} - \overset{H}{C} = \overset{H}{C} - \overset{H}{\underset{H}{C}} - \overset{H}{C} = \overset{H}{C} - \overset{H}{\underset{H}{C}} - \overset{H}{\underset{H}{C}} - \overset{H}{\underset{H}{C}} - \overset{H}{\underset{H}{C}} - \overset{H}{\underset{H}{C}} - H$$

is composed of a long *hydrocarbon* chain containing nonpolar covalent bonds. This nonpolarity makes fatty acids and the lipids of which they are a part very insoluble in water which is a polar solvent. The insolubility of lipid molecules in water is a key property in their function as membrane components (e.g. phospholipids) and energy-storage molecules (e.g. fats). In both cases, lipids tend to exclude water and the polar substances dissolved in water. **Fat** molecules contain at least one fatty acid bonded by dehydration synthesis to one molecule of **glycerol** (Figure 4-8. Glycerol is a type of alcohol (see Table 4-3):

```
                    H
                    |
              H - C - OH
                    |
glycerol      H - C - OH
                    |
              H - C - OH
                    |
                    H
```

chemical bonding and lipid properties

Butter and other animal fats are usually solid or semisolid at room temperature because of the tight packing of fatty acid chains. This tight alignment is made possible by the fact that the carbon chains are saturated with hydrogen atoms, hence the term **saturated fats** (Figure 4-8). In contrast, vegetable oils are called **unsaturated fats** because some of the carbon atoms have double bonds instead of bonds formed with hydrogen atoms (see linoleic acid, shown above). At each double bond, the three-dimensional configuration is different. Consequently, unsaturated chains cannot pack as tightly together and the resultant unsaturated fats tend to be more fluid (oily) at room temperature.

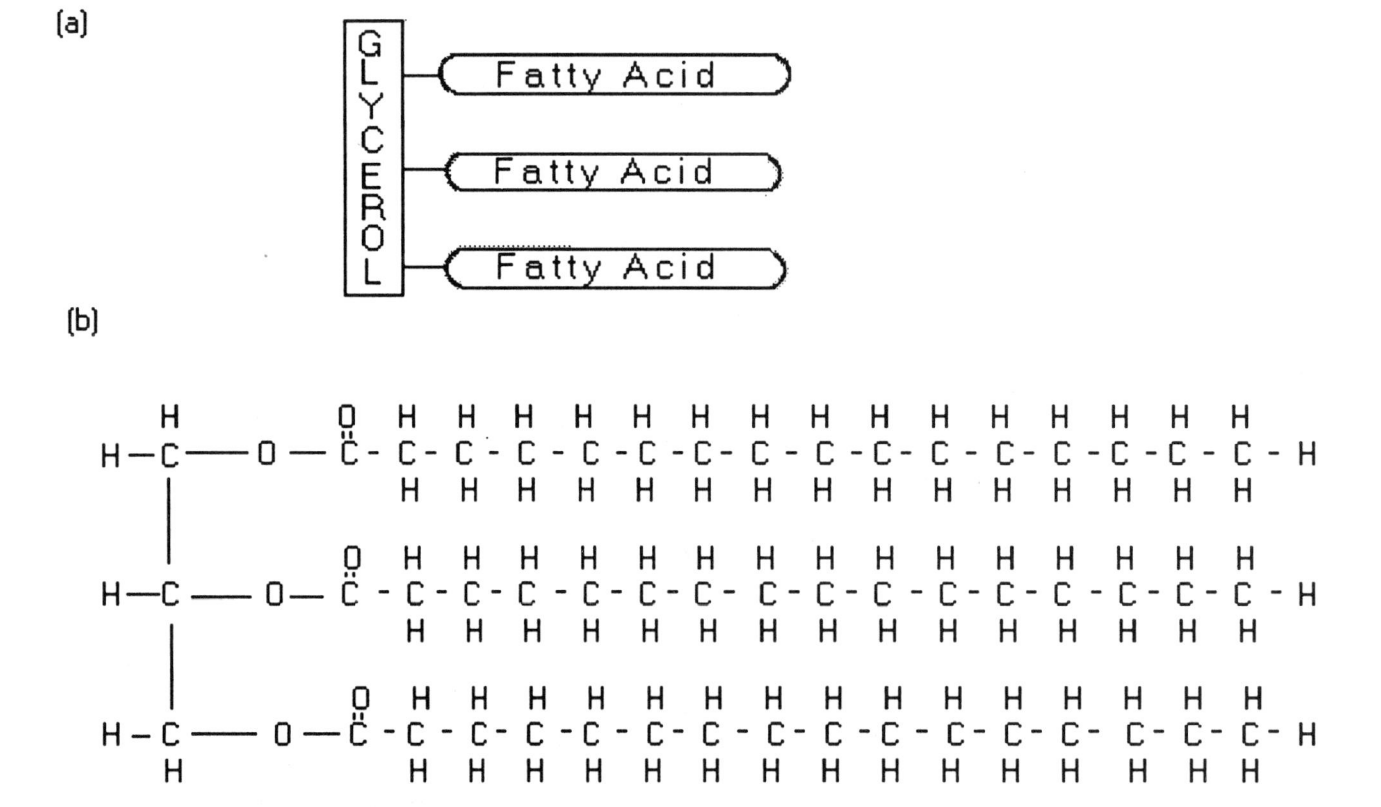

(a)

```
┌─────────┐   ┌────────────┐
│ G       │───│ Fatty Acid │
│ L       │   └────────────┘
│ Y       │   ┌────────────┐
│ C       │───│ Fatty Acid │
│ E       │   └────────────┘
│ R       │   ┌────────────┐
│ O       │───│ Fatty Acid │
│ L       │   └────────────┘
└─────────┘
```

(b)

```
      H        O  H H H H H H H H H H H H H H H
H — C —— O — C - C-C- C - C - C -C- C - C -C- C - C- C -C- C - C - H
      |        H H H H H H H H H H H H H H H H
      |        O  H H H H H H H H H H H H H H H
H — C —— O — C - C- C- C - C - C- C- C - C - C - C - C- C - C- C - C - H
      |        H H H H H H H H H H H H H H H H
      |        O  H H H H H H H H H H H H H H H
H — C —— O — C - C - C- C -C - C- C - C - C -C - C - C- C - C- C -C- H
      H        H H H H H H H H H H H H H H H H
```

Figure 4-8. (a) The general structure of a fat molecule, composed of a glycerol "head" and three fatty acid "tails". (b) Diagrammatic view of the structure of glyceryl tripalmitate, a saturated fat molecule containing glycerol bonded to three fatty acids (palmitic acids).

Lipids function in membranes...

in protection...

and as hormones.

Phospholipids are a major component of cell membranes. Their molecular structure is similar to fats except that one fatty acid chain is replaced by a phosphate group to which is attached a nitrogen-containing alcohol (Figure 16-5). **Waxes** are composed of long-chain fatty acids and alcohols. This class of lipids serve a protective role on the surface of leaves, on the bark of trees, and on the skin and fur of animals. The **steroids** are molecules with a complex carbon ring structure to which a variety of atoms may be attached. Cholesterol and a number of hormonal substances are steroids.

4-I PROTEINS

Proteins, like polysaccharides, are polymeric in structure. However, proteins are much more diverse in molecular structure and function. Whereas polysaccharides are usually composed of chains of the same monomer repeated throughout (Figure 4-7), proteins are chains of **amino acids** of which there are 20 different forms. Four common amino acids are shown in Figure 4-9, differing only in the structure of the side group.

Amino acids are monomers of proteins.

awesome variety of proteins

Such a variety of monomers makes an awesome variety of possible sequences of amino acids, even among protein molecules of the same length. For example, for a protein molecule containing 50 amino acid monomers, there are 20^{50} different possible sequences, each molecule having different structure and properties. You will need an electronic calculator to compute this awesome number, which someone has estimated to represent one unique protein for every gram of matter in the universe! Even more impressive is the fact that large proteins may have over 50,000 amino acid monomers. Imagine $20^{50,000}$ different sequences!

Proteins may be enzymes...

membrane components...

cause movement.

and function as hormones.

The diversity of protein structures is accompanied by an impressive diversity of biological functions. First of all, the **enzymes**, agents that accelerate and control metabolic reactions, are proteins. Second, proteins serve as structural components of cell membranes, bones, and cartilage. Third, proteins function to promote the movement of living matter at the subcellular and cellular levels of organization. Furthermore, proteins are responsible for the motility of whole organisms, as small as the protozoa and as large as the mammals. Fourth, like the steroids, some proteins such as **insulin** are hormones (Figure 4-10), while others such as **hemoglobin** function in transport processes. Hemoglobin is a red iron-containing protein that facilitates oxygen transport in the blood.

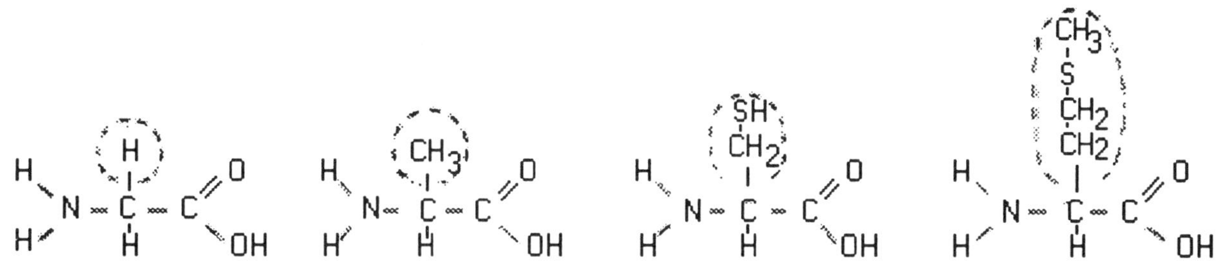

glycine (gly)   alanine (ala)   cysteine (cys)   methionine (met)

Figure 4-9. Four of the twenty amino acids commonly found in protein molecules. Each has an amino group (-NH$_2$) and a carboxylic acid group (-COOH) on the left and right ends, respectively. Each of the different amino acids are distinguished by a different side group shown within the dotted circle.

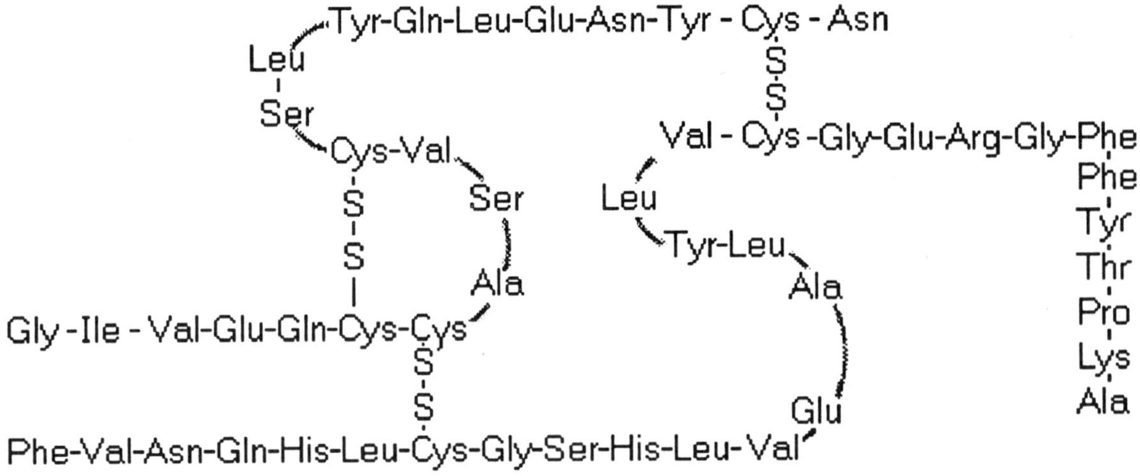

```
          ┌Tyr-Gln-Leu-Glu-Asn-Tyr - Cys - Asn
      Leu                            S
       │                             ·
      Ser                            S
        ╲Cys-Val╲             Val - Cys -Gly-Glu-Arg-Gly-Phe
         S       ╲     Leu   ╱                           Phe
         ·        Ser    ╲Tyr-Leu╲                       Tyr
         S         ╲              Ala                    Thr
         │         Ala                ╲                  Pro
Gly -Ile - Val-Glu-Gln-Cys-Cys╱                          Lys
                          S                              Ala
                          ·
                          S                  Glu
Phe-Val-Asn-Gln-His-Leu-Cys-Gly-Ser-His-Leu-Val╱
```

Figure 4-10. The hormone insulin consists of two amino acid chains totalling 51 amino acid monomers (names abbreviated). The three-dimensional structure of the molecule is formed by the precise sequence of amino acids which allows the sulfhydryl groups (-SH) of cysteine (Cys) (Figure 4-9) to form disulfide linkages (-S-S-). Hydrogen bonding is also an important stabilizing force.

Peptide bonds link amino acids . . .

into precise sequences.

Disulfide bonds influence 3-D structure.

Three types of bonding are essential to the structure of protein molecules, peptide bonds, disulfide bonds, and hydrogen bonds. **Peptide bonds** are covalent bonds between carbon and nitrogen. These bonds are formed by dehydration synthesis which links the carboxyl group of one amino acid to the amino group of another amino acid (Figure 4-11). The sequence of amino acids, linked end to end, ultimately determines the three dimensional structure of the molecule. Insulin, shown diagrammatically in Figure 4-10, cannot function properly as a hormone that regulates blood glucose levels unless the amino acid sequence is formed exactly as shown. When the synthesis of the protein such as insulin is complete, it has assumed a three dimensional structure which is made possible by proper positioning of cysteine (cys) portions of the chain so as to form **disulfide linkages** (Figure 4-10). Finally, **hydrogen bonding** between hydrogen atoms and atoms of oxygen or nitrogen also aids in stabilizing the molecule through attractive forces between different portions of the coiled polymer.

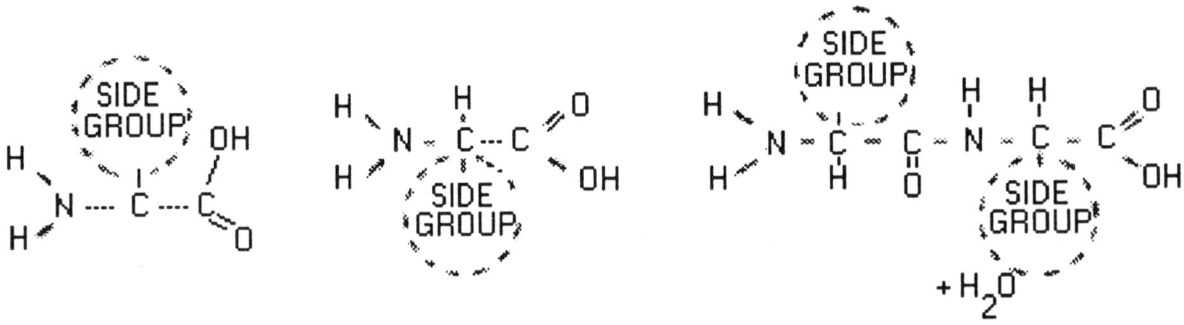

Figure 4-11. Formation of a peptide bond between two amino acids by dehydration synthesis.

4-J NUCLEIC ACIDS

Nucleic acids carry a code...

in nucleotide sequences.

Nucleic acids function as the hereditary material for all known forms of life. Deoxyribonucleic acid, or DNA, is commonly referred to as the "genetic blueprint" of life. Like a blueprint, DNA molecules convey information which influences the structure of growing, developing organisms. Genetic information is coded in a sequence of **nucleotides** which are linked together to form a polymeric molecule. Each nucleotide consists of a pentose sugar to which is attached a phosphate group and a nitrogen-containing molecule called a **nitrogen base**:

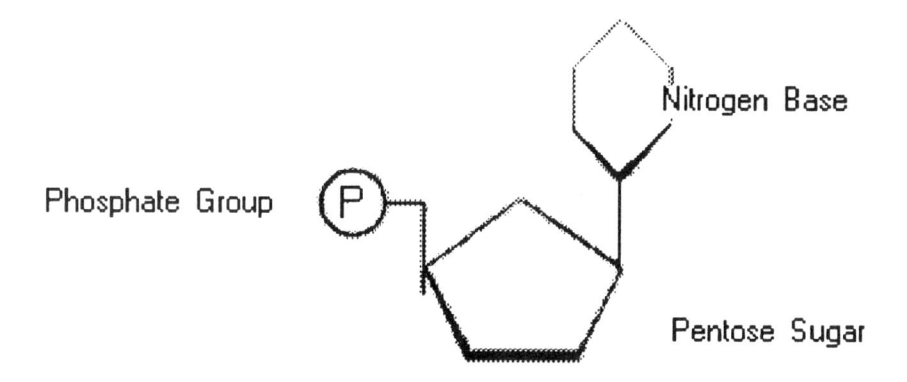

Five different nucleotides are commonly found in nucleic acid polymers, differing in the particular nitrogen base they contain. The pentose sugar is also a point of variation. Deoxyribonucleic acid (DNA) contains deoxyribose sugar, and ribonucleic acid (RNA) contains ribose sugar (Figure 4-6). Nucleic acid polymers usually contain hundreds of nucleotides in precise sequence. A short section of a single chain of DNA is shown in Figure 4-12. Note that the nitrogen bases are a distinguishing feature of each nucleotide and hence create the potential for a coded sequence. We shall explore the structure and function of the nucleic acids in more detail in Chapter 11.

4-K REFLECTIONS ON ATOMS AND MOLECULES OF LIFE

Confused? Reread, and see APPENDIX A.

As we conclude our survey of the atoms and molecules of life, you may be tempted to retreat in discouragement. Perhaps you have been overwhelmed by the complexities of chemistry. If so, refer to APPENDIX A, and reread this chapter. Try to identify problem areas, and strive to master the material. A tutor or other references may help. Remember, you are studying the diagrams and models that explain systems that are quite abstract, and yet fundamental to our understanding of biological processes.

Summary:

Perhaps you are one who has mastered the chemistry, but you came looking for some keys to what makes living creatures unique and "came up empty". In one sense, like Szent-Gyorgyi, we have "ended up with atoms and electrons, which have no life at all". Yet mastery of this chapter will give a clear idea of the chemical makeup of the amazing substance of living cells.

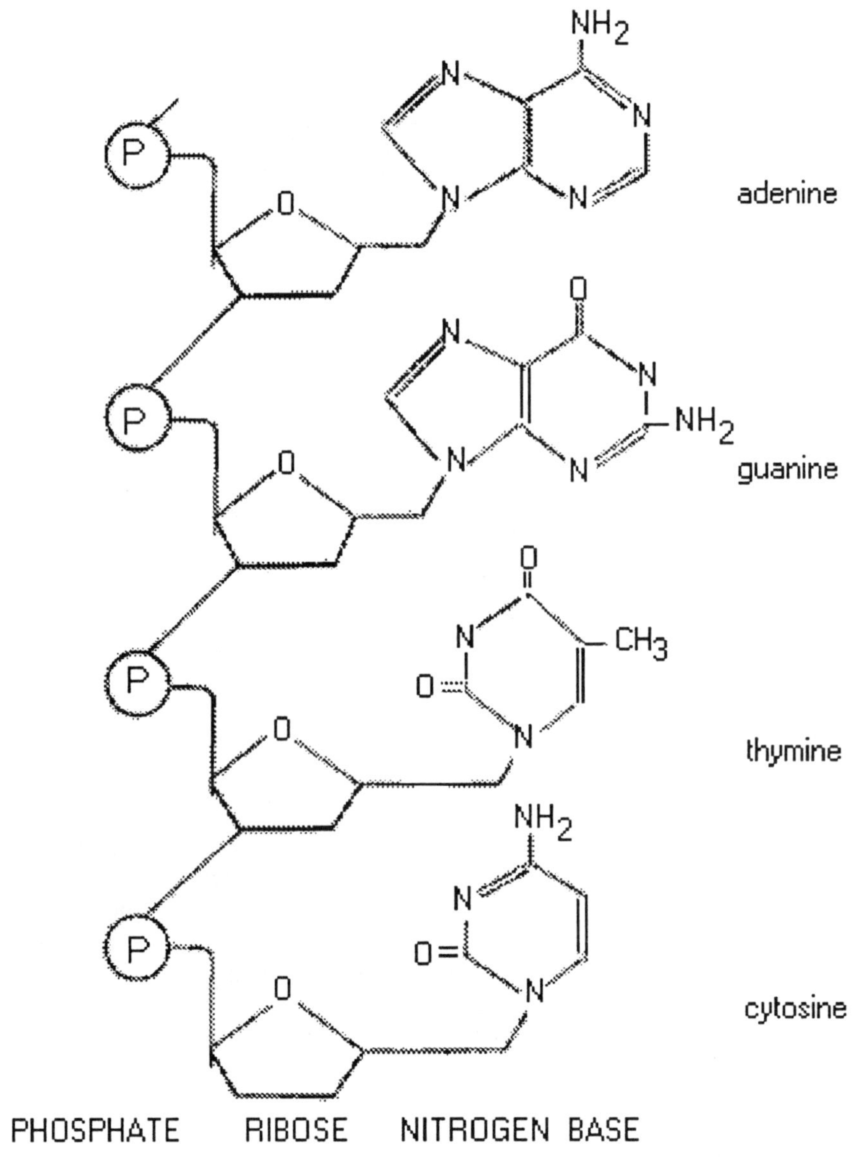

PHOSPHATE   RIBOSE   NITROGEN BASE

Figure 4-12. A portion of a single chain of a DNA polymer. Carbon and certain hydrogen atoms of the ring structures are omitted for clarity. Each nucleotide differs from the adjacent one only with regard to the nitrogen base portion. Nucleotides may occur in different sequences and often the same nucleotide will be repeated two or more times.

We have considered four groups of organic molecules, and a host of inorganic molecules and ions, each known to participate in life processes. These substances must be sought out and taken in from the environment of each organism. What are the sources of these substances in the physical environment? How are they taken in by organisms? How does one population influence another as each seeks to meet its needs? How are humans affecting the life of other creatures on planet Earth? We shall address these and other questions as we depart (not retreat!) from the molecule level and escalate to the levels of biosphere and ecosystem. The emphasis of Unit II will be upon relationships among organisms, and between organisms and the environment.

Next: A return to ecosystems

QUESTIONS AND DISCUSSION TOPICS

1. You are given a small vial containing a green gelatinous mass and asked to determine whether it is a living organism or not. What criteria and experiments would you use to solve the problem?

2. From Appendix Table A-1 what chemical similarities can you see between oxygen and sulfur atoms? In bonding with hydrogen atoms, oxygen forms H_2O, while sulfur forms H_2S, hydrogen sulfide gas. The H_2S molecules, unlike H_2O do not form a hydrogen-bond lattice. How does this difference in chemical properties account for the difference in physical properties between the two substances?

3. How do buffers illustrate the concept of homeostasis?

4. Can you propose an explanation for the existence of fat droplets on chicken broth? Why might some brands of margarine "melt" more readily at room temperature than others?

5. If you lived in the 1840's when Adolph Kolbe synthesized the first organic substance, what concerns might you and others have raised regarding the uniqueness of life and man's "tampering" with it?

6. Are you a vitalist or a mechanist? How does ones philosophy of science influence his view of life processes and life in general?

7. What do proteins have in common with carbohydrates and nucleic acids? How do they differ? How can there be so many different kinds of proteins?

REFERENCES

1. Baker, J.J.W. and G.E. Allen. 1982. *Matter, Energy, and Life*, 4th ed., Addison-Wesley Publ. Co., Reading, Mass.

2. Baum, S.J. and C.W. Scaife. 1975. *Chemistry: A Life Science Approach.* Macmillan, New York.

3. Bloomfield, M.M. 1992. *Chemistry and the Living Organism*, 5th ed., John Wiley and Sons, Inc., New York.

4. Frieden, E. 1972. The chemical elements of life. *Scient. Amer.* 227(1): 52-64.

5. Holum, J.R. 1987. *Elements of General and Biological Chemistry*, 7th ed., John Wiley and Sons, Inc., New York.

6. Snyder, C.H. 1992. *The Extraordinary Chemistry of Living Things.* John Wiley and Sons, Inc., New York.

Part 2

Life and Environment

But now ask the beasts, and let them teach you;
And the birds of the heavens, and let them tell you.
Or speak to the earth, and let it teach you;
And let the fish of the sea declare to you.
Who among all these does not know
That the hand of the Lord has done this,
In whose hand is the life of every living thing,
And the breath of all mankind?
—Job (Job 12:7-10)

How do you interpret the world?

By faith, we understand that God created the physical environment in which we live. *And God saw all that He had made, and behold, it was very good* (Genesis 1:31). The creation, in spite of being marred by sin, is a perfect testimony of the wisdom and greatness of its Creator, and continues to be sustained by the One *in whose hand is the life of every living thing, and the breath of all mankind* (Job 7:10).

Studying parts to understand the whole

The Scripture above was written thousands of years before the modern scientific era, yet it gives clear testimony to the existence of an array of many different creatures which are interdependent. In a feeble attempt to understand this intricacy, we have first broken down the whole and named each conceptual part—*biosphere, ecosystem, community, population, organism*, etc. Then, we arrange these parts into what we understand to be an organizational hierarchy from biosphere "down" to subatomic particle (Section 2-C). Further, we hope that, by understanding some of the individual parts of the hierarchy, we will gain understanding of the whole through "summing up" these individual understandings. But, Job reminds us that no organism is an "island"; each is an integral part like a thread woven into the 'tapestry of life'.

Preview: Part 2

Part 2 will emphasize the relationships between organism and environment. In Chapter 5, we will study the flow of energy in the biosphere; and, in Chapter 6, the cycling of atoms and molecules in the biosphere. Chapter 7 will direct you to explore principles of population growth and interactions of populations. Chapter 8 will challenge you to apply principles of ecology learned in Chapters 5 through 7 to the human population and its predicament on Earth.

Chapter 5

Energy Flow in Ecosystems

5-A INTERRELATIONSHIPS AMONG LIVING SYSTEMS

Ecology is the branch of biology dealing with the interrelationships among organisms and between organisms and their environment. The derivation of the term ecology [Gr.(*oiko* = house) + (*logos* = study of)] emphasizes that this branch of biology is a study of organisms in the "place where they live"—i.e. their environment.

a definable system to study

The ecologist, employing the scientific method, must define (set boundaries around) the system under consideration before controlled experiments can be performed. In theory, an **ecosystem** is considered to be a defined environment plus the **populations** that compose the **biotic community** of that environment (Section 3-C). Figure 5-1 is a simple diagrammatic representation of an ecosystem. For simplicity, the model includes a community of only two populations. But the tapestry of life in the biosphere is complex and tightly woven. Therefore, it is difficult to isolate a forest ecosystem, or lake ecosystem with assurance that we have truly isolated it from "outside factors". [Can you explain and illustrate this difficulty with an example of a forest or lake in your community?]

Organisms are interdependent.

The interdependence among living organisms and the linkages between organism and environment were prominent themes in each of the **characteristics of life** noted in Chapter 2, Section 2-C. [Take a few minutes to review these "life characteristics" and notice the emphasis upon the concepts of *interdependence* and *environment.* Also, try to identify the nature of the various interactions implied. For example, the interactions may involve energy (food), shelter, mating, etc.]

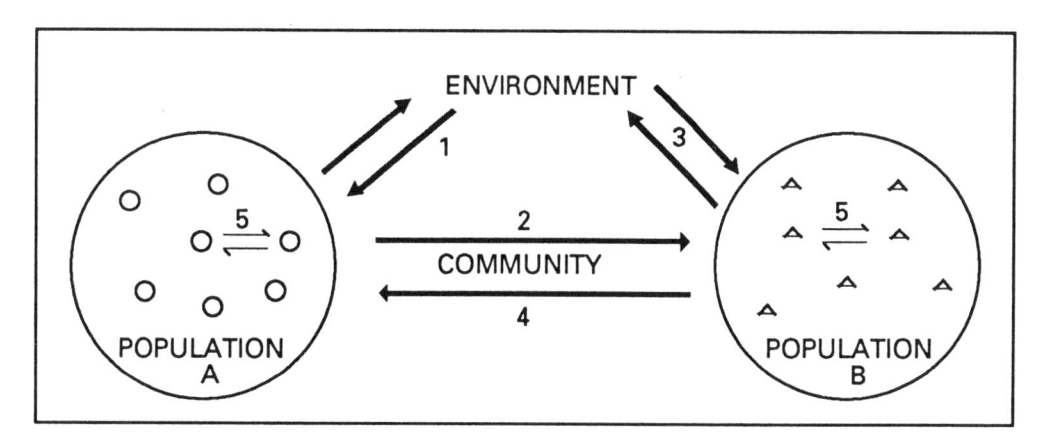

Figure 5-1. Diagrammatic representation of an ecosystem. The rectangle represents the theoretical boundary of an ecosystem. Arrows represent interrelationships among organisms and their environment at different levels of organization. See text for explanation.

levels of organization

Figure 5-1 illustrates the most common types of interrelationships among living systems and their environment. As noted in *life characteristic #1* (Section 2-C), living systems exist along a hierarchy of levels of organization. Cells, organs, organisms, and ecosystems are all living systems with the ability to use energy to grow and maintain homeostasis. Each population of organisms is linked to other populations of the biotic community housed within the defined environment of an ecosystem. In our study of ecology, we will focus primarily upon interrelationships at the population, community, and ecosystem levels.

Continuous need for energy . . .

Life Characteristic #2 states that *living systems use energy from the environment*. The requirement for a continuous supply of energy is a thread that draws organisms together and links them to their environment. Thus, a deer population severed from its deciduous forest community during urban expansion is as much in danger as a human heart severed from the human body. Energy relationships help us to understand how the tapestry is woven.

draws creatures together . . .

Let's suppose that Population-A in Figure 5-1 is a red clover population within the meadow community into which we hiked in Chapter 2. Then arrow #1 represents the flow of energy from sunlight to clover plants which use this energy to grow and produce leaves and flowers. If Population-B represents a hive of honeybees, then arrow #2 portrays the flow of energy from clover flower to the nectar-feeding honeybees. Thus, the energy shared between clover and bee "binds" them into a community relationship. The typical biotic community has many populations, interrelated through their demand for energy. Each occupies one or more positions in the trophic structure of the community. The human population is the largest user of energy in the biosphere. For this reason, it is important that we understand our position in the ecosystems of Earth.

into *community*.

reproductive relationships

According to *life characteristic #3*, *living organisms reproduce themselves*. Sexual reproduction, involving mating and gamete transfer (Section 3-E) occurs, by definition, only within populations (Section 3-C). Therefore, interrelationships involving mating and nurture of offspring are illustrated by arrows numbered "5" within each population.

Sexual reproduction in flowering plants such as our red clover (Population-A, Fig. 5-1) requires transfer of pollen from male to female sex organs. The honeybee (Population-B) plays an important role as pollinator. This relationship involves mutual benefit between the two populations as illustrated by arrows #2 and #4—i.e. energy supply for honeybee, and gamete transfer for red clover. This type of mutual-benefit relationship between two species is called **mutualism**.

mutual benefit

responsiveness and homeostasis

Life Characteristic #4 emphasizes that *living organisms respond to stimuli from their environment and maintain internal homeostasis*. These are essential, moment-by-moment functions of the living organism. Arrows #1 and #3 represent the various environmental components of light, moisture, temperature, oxygen, etc. that impinge upon life. Each individual organism must be able to regulate its metabolism within the levels of each of these environmental influences.

Some organisms of each population may not survive to produce offspring for a new generation. Thus, the **natural selection** of individuals within Populations-A and -B involves the removal of individuals with genetic traits unsuited to the environment of that ecosystem, and the survival and reproduction of

Populations adapt through natural selection.

those that have suitable traits. This differential survival and reproduction over the short term (generation-to-generation) has the apparent result in the fine-tuning of structural, functional, and behavioral traits (adaptations) among the survivors. In this way, as stated in *Life Characteristic #5, living organisms are adapted to their environment*. Thus, homeostasis and responsiveness to environment are concepts that apply to individual organisms and to populations. The long-term welfare of the population depends upon the short-term survival of individuals.

The two arrows that are unnumbered represent effects of populations on their environment. As populations occupy a given environment and use its resources, the environmental conditions may be altered. This, in turn, can affect other populations. [How would light and soil temperature change during the growth of a forest over a 100-year period after a field was abandoned? Relate to the principle above.]

By this time, if you are at least partly overwhelmed by the apparent complexities of ecology, be encouraged. Experts in the field of ecology also tend to be humbled by the immensities of size and complexity of ecosystems. However, with a determined effort to follow our systematic approach, you will come away with a better understanding of this vital discipline.

5-B ENERGY

LIFE

LIVING   LIVELY

MOVING   GROWING   REPRODUCING

SOAKING IN THE RAYS OF SUNSHINE

EATING   BEING EATEN

DEATH

Energy exchange is always occurring.

Living organisms are intimately linked to their environment because they require a continuous supply of energy and raw materials. Each activity and condition described in the composition above is driven by energy, or involves energy exchanges. Even energy contained in the lifeless body after death is usable by scavengers and decomposers; and, life goes on. What is energy, and why is it so essential to life processes?

energy in motion . . .

Energy is defined as the capacity to do work, resulting in the displacement of matter from one location to another. Under natural conditions, the atoms and molecules that compose matter are in constant motion. **Kinetic energy** is energy of motion, the energy of stirring winds, flowing water, and tumbling boulders.

or stored

Molecules such as glucose and amylose also contain chemical bond energy locked in covalent bonds (Section 4-G). This is stored energy, or **potential energy**, having the potential to do work when the bonds are broken to release the energy. When you ingest foodstuffs containing starch, your digestive system hydrolyzes the polysaccharides (Section 4-E, 4-G) to release energy-rich glucose molecules. The glucose is transported to body cells where

"tired molecules"

chemical *versus* nuclear energy

chemical potential energy is released and utilized to do work in sustaining your life processes, including muscle contraction for locomotion and heartbeat, and the maintenance of body temperature through heat production. Heat is a form of kinetic energy.

Water, CO_2, and other inorganic molecules (Section 4-E) also contain covalent bonds. However, these bonds have no potential energy that is usable by living cells. These substances function in ways other than as energy sources. Matter may contain not only chemical bond energy, but also nuclear energy. Nuclear energy is released as kinetic energy when atomic nuclei undergo fusion or fission. The sun is a giant thermonuclear reaction believed to involve the fusion of hydrogen nuclei to form helium as follows: $2 H \longrightarrow He + Energy$. The energy from this reaction is released as a spectrum of **electromagnetic radiation** (Figure 5-2). This spectrum ranges from very short-wavelength radiation such as gamma rays and ultraviolet rays, through a narrow range of visible radiation (light), to long-wavelength radiation, including infrared and radio waves. Ultraviolet rays are intercepted in large part by ozone (O_3) gas and water vapor in the upper atmosphere. The high energy content of the **ionizing radiation** that is not trapped by the atmospheric filter can damage living cells through the breakage of covalent bonds of organic molecules. Thus, the atmosphere serves as a protective envelope to hold in vital gases, while screening out harmful radiation.

Ozone filters harmful rays.

Infrared rays provide heat.

The infrared portion of the solar spectrum represents radiant energy of longer wavelengths that penetrate the earth's atmosphere and provide heat to maintain a temperature range suitable for life. The earth orbits the sun at just the right distance to avoid either scorching or freezing temperatures. Moreover, the rate of the earth's rotation on its axis aids in the distribution of solar energy via wind and ocean current systems. The four seasons are caused primarily by the combined effect of the earth's revolution around the sun and the tilt of the earth's axis, about 23.5° from the perpendicular.

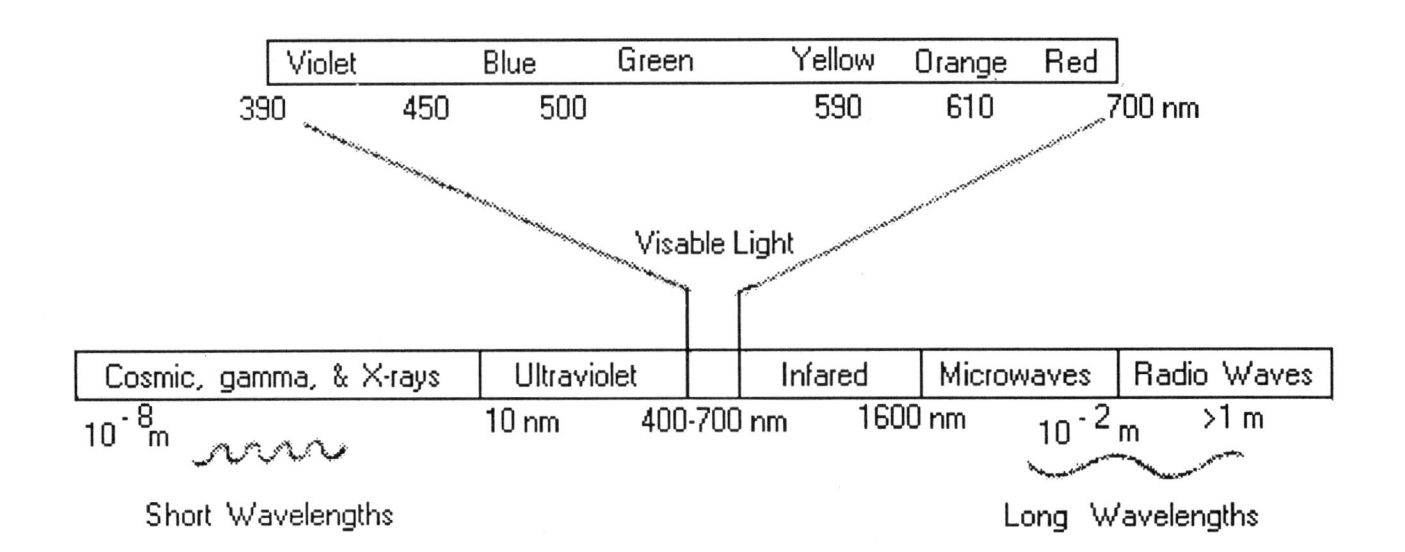

Figure 5-2. The electromagnetic spectrum. Wavelengths are given in meters (m) and nanometers (nm). (1 nm = 10^{-9} meter)

Visible rays are *light*.

You are no doubt most familiar with the visible portion of the electromagnetic spectrum (Figure 5-2). Visible light includes wavelengths ranging from about 400 to 700 nanometers (nm) (1 nm = 10^{-9} meter) which are detected by the visual receptors of many forms of life. The visible portion of the solar spectrum is also utilized in photosynthesis.

5-C THERMODYNAMICS AND ENERGY CONVERSIONS

From our discussion thus far, it is evident that energy can exist as stored energy (potential energy), or be released when matter is set in motion (kinetic energy). The kinetic energy of solar radiation is essential for life on earth. Autotrophs have the capacity, through photosynthesis, to convert visible radiation into plant matter, including carbohydrates and proteins. These plant products sustain plant life and provide food for heterotrophs.

Organic matter has chemical bond energy.

Plant products such as wood also contain chemical energy which has a variety of uses. The wood can be sawed, split, and burned as firewood. Here, chemical energy is converted, with the aid of mechanical energy of the laborer, to heat and light energy in a wood-burning stove or fireplace. The path of energy conversion would be as follows:

solar energy —> chemical energy —> heat + light energy
   (light)   (wood)

Likewise, *fossil fuels*, representing preserved organic carbon compounds now in the form of coal, oil, and natural gas, that originated from ancient life forms, can be burned to produce electricity to run our appliances. The conversion path is as follows:

solar —> chemical —> heat —> mechanical energy —> electrical energy
energy   energy (coal) energy   (steam turbine)   (from generator)

Energy behaves predictably:

FIRST LAW: conservation and conversion

In spite of the apparent versatility of energy, there are laws (Section 1-E) that define the behavior of energy, called **laws of thermodynamics** [Gr. (*thermo* = heat) + (*dynamo* = power)]. The **first law of thermodynamics** states that energy can be converted from one form to another by a process; but, energy can neither be created nor destroyed. No known process in the universe can create or destroy energy. The first law thus allows for the energy conversions noted above. However, as you may have realized from observing the release of heat and water vapor from cooling towers of coal-fired power-generating stations, not all of the chemical energy of the coal is converted to electrical energy. The excess energy is not destroyed (recall the first law). Rather, it escapes into the environment as heat energy. The **second law of thermodynamics** predicts this occurrence, stating that no conversion of energy from one form to another occurs with 100 percent efficiency. The use of *high quality energy* such as radiant, chemical, or nuclear energy, capable of working on a variety of different systems, will always result in the escape of some energy as heat. Recall that heat energy is necessary to maintain life in the biosphere. However, heat represents disordered, low-quality energy which is quickly scattered into the environment and outer space, never to be recovered.

SECOND LAW: increasing heat and disorder ...

The second law adds a solemn tone to our discussion. Taken together, the first and second laws suggest that the universe as a system is becoming more disordered each day. No new sources of high-quality energy are being

... or entropy.

created, and every process acts to increase low quality energy and disorder, or **entropy**. Indeed, this process continues spontaneously without apparent human intervention, as witnessed in rusting automobiles, untidy dormitory rooms, crumbling buildings, and aging bodies. An orderly system can be created and maintained only by expending more energy and producing more entropy outside the system. [How are the first and second laws consistent with the Genesis 1 – 3 account?]

Energy laws apply to life:

Thermodynamic laws apply to the flow of energy from sunlight through the biotic world. Living organisms are highly ordered systems that must have a continual supply of high-quality energy (radiant and chemical) and heat energy from their surroundings. Radiant energy in the form of electromagnetic waves enters the biosphere from the sun. As noted earlier (Section 3-D), **autotrophs** are organisms that can utilize radiant energy from the visible range of the solar spectrum to convert inorganic compounds into energy-rich organic compounds. Terrestrial green plants and aquatic algae are the most common autotrophs. These organisms are called **photosynthetic autotrophs** because they rely on light as their energy source. **Chemosynthetic autotrophs** include certain bacteria that acquire their energy from inorganic compounds (Section 6-F). For our purposes, "autotroph" or "producer" will refer to photosynthetic autotrophs, unless otherwise specified.

...autotrophs

and heterotrophs

Organisms that must have a supply of energy-rich organic compounds (food) for their energy supply are called **heterotrophs**. Heterotrophs that acquire their energy from living autotrophs are called primary consumers, or **herbivores**; Those that feed on herbivores are secondary consumers, or **carnivores**. **Decomposers** and **detritivores** are also heterotrophic, and they obtain their energy from dead organisms. Decomposers include various species of fungi and bacteria. Vultures, earthworms, and soil insects are common detritivores. Thus, with regard to food and energy conversions, organisms have distinct ecological roles, or **niches**, which are interdependent. Collectively these niches constitute a **trophic structure** which can be pictured as a **food chain** or **food web**. Before reading further, if necessary, review Section 3-D and Figure 3-2.

Mode of energy use determines trophic niche.

5-D ENERGY FLOW AND FOOD CHAINS

Photosynthesis:

light energy to chemical energy

Autotrophs acquire their energy by light absorption and photosynthesis of glucose and other organic compounds. That is, autotrophs convert *light energy* into *chemical bond energy*. However, as shown in Figure 5-3a, absorbed light energy cannot be transferred into chemical bonds unless the *chemicals* carbon dioxide and water are also absorbed. The figure illustrates the summary reaction of photosynthesis,

$$6\,CO_2 + 6\,H_2O + light \longrightarrow C_6H_{12}O_6 + 6\,O_2$$

in which light energy is used to build complex, orderly molecules of glucose out of simple CO_2 and H_2O molecules.

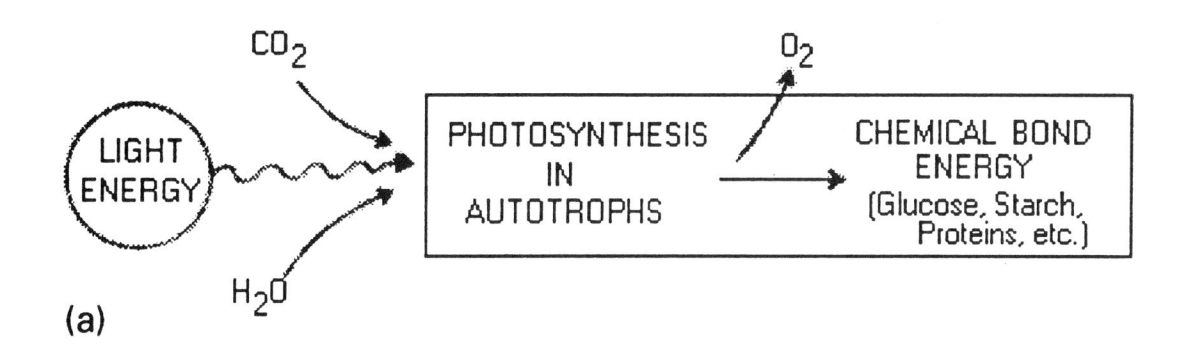

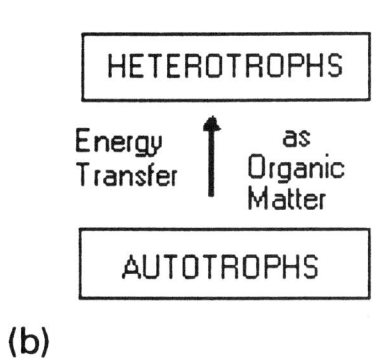

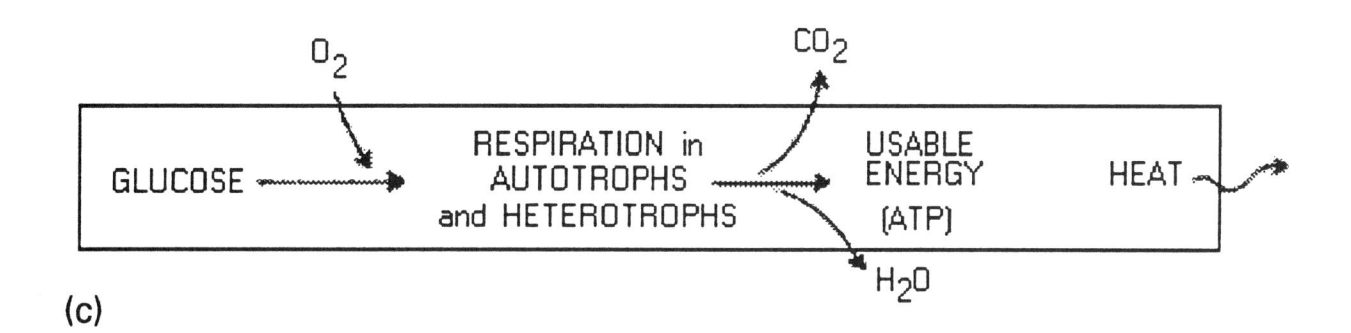

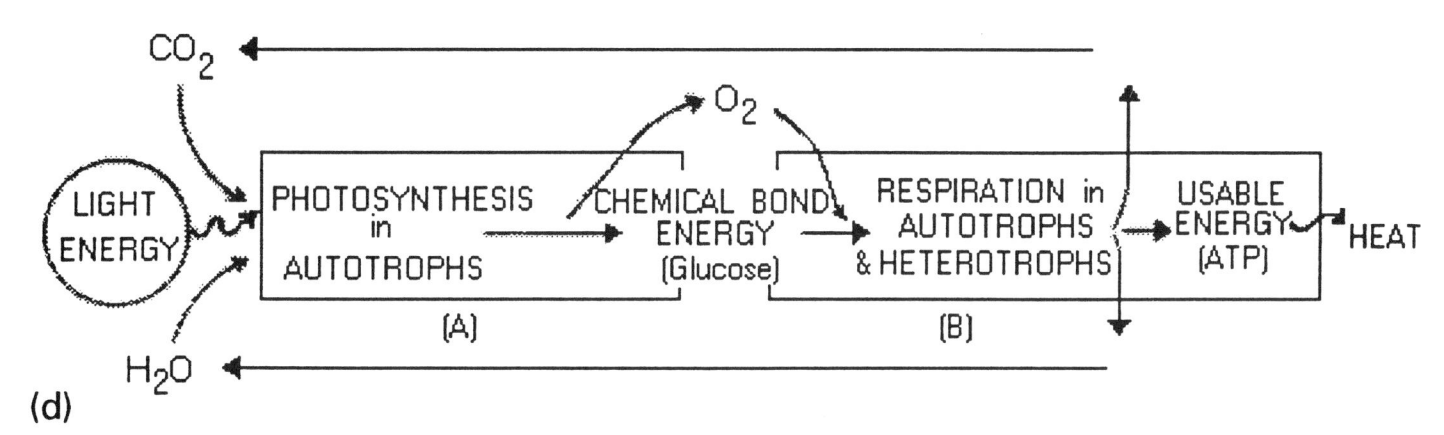

Figure 5-3. Basic processes in the flow of energy and matter in a food chain: (a) Photosynthetic conversions, (b) trophic level conversions, (c) respiratory conversions, and (d) composite model of (a) through (c).

Heterotrophs need autotrophs for "food" and oxygen.

[Return to Section 4-G and notice the complexity of carbohydrates in contrast to that of CO_2 (O=C=O) and H_2O (H-O-H).] Thus, the acquisition of energy by autotrophs is inseparably linked with acquisition of inorganic matter in the form of CO_2 and H_2O. Given light, CO_2, H_2O, and minerals, photosynthetic autotrophs form all of the organic molecules necessary for their growth. Heterotrophs, by definition, cannot create organic molecules from inorganic substances, and therefore must acquire them directly or indirectly as "food".

Consider the following principle and related experiment:

PRINCIPLE: Autotrophs and heterotrophs depend upon one another for needed energy and inorganic substances.

MATERIALS: — Sunny, south-facing window or greenhouse
— 20-gallon aquarium with tight-fitting glass lid
— small container of water
— fertile organic soil with young clover plants
— 1 small rabbit (handle with care)

EXPERIMENTAL OBJECTIVE:
To determine whether the autotrophic population (clover) can support the heterotrophic rabbit.

QUESTION: What requirements do each species have with regard to matter and energy? What can we learn about the *global* biosphere from this *local* experiment?

The green clover leaves grow as "solar collectors", self-assembled molecule by molecule, cell by cell from light, CO_2, and H_2O as shown in Figure 5-3a. Nitrates, phosphates, and other minerals are absorbed from the soil. The rabbit then eats clover leaves containing "prefabricated" carbohydrates, protein, and other organic molecules as food. Her role as herbivore is to convert plant tissue into rabbit protein, lipids, etc., using respiration to energize this bond making and remaking (Figure 5-3b).

Respiration uses O_2 to tap energy of food.

Finally, production of usable energy from glucose by respiration involves an exchange of inorganic matter with the environment. As shown in Figure 5-3c, respiration in autotrophic clover and heterotrophic rabbit involves the intake of O_2, and the release of CO_2 and H_2O. Note that heat energy is also released along with the material products of respiration. In summary, we have seen that each *energy* and *matter* transactions of metabolism occur simultaneously.

Matter cycles between photosynthesis and respiration.

Autotrophs respire, too!

A careful examination of Figures 5-3a and 5-3c reveals that the chemical products of photosynthesis are the *reactants* (i.e. requirements for a chemical reaction) of respiration. This means that there is a **cycling** of inorganic matter between photosynthesis by autotrophic clover (Figure 5-3a) and respiration by *both* clover and rabbit (Figure 5-3c). Figure 5-3d illustrates this relationship by combining Figures 5-3a and 5-3c. Note that *respiration is not a unique feature of heterotrophs! Autotrophs also perform respiration in light and darkness*. However, in daylight, the autotroph satisfies most of its energy-matter needs by photosynthesis, and respiration is suppressed. Therefore, autotrophic clover is taking in more CO_2 by photosynthesis than it releases in respiration, releases more O_2 than it uses in respiration, and forms more

Gross minus respiration is net production.

organic molecules than it consumes to "fuel" respiration. That is, the **gross primary productivity (GPP)** of clover is greater than its "expenses" for **respiration (R)**, and the difference is its **net primary productivity (NPP)**. That is,

$$NPP = GPP - R$$

Unless GPP is greater than R, the clover would not grow in mass (no NPP), nor be able to provide the oxygen and organic compounds needed by the rabbit. Thus, the cycling of CO_2, O_2, and H_2O between photosynthesis and respiration reveals yet another interrelationship between autotrophs and heterotrophs. Heterotrophs depend upon autotrophs for a net production of chemical bond energy *and* O_2, the life-supporting gas so apparently unique to the earth's biosphere.

Energy cannot be recycled; ...

matter can be.

So far we have emphasized (a) the dependence of energy flow upon the flow of matter, and (b) the interdependency between autotroph and heterotroph involving both energy flow and the cycling of inorganic substances. However, there is a distinction between the flow of energy in the ecosystem, and the flow of matter that we should also consider. As shown in Figure 6-4, energy flows through from autotroph to heterotroph and is eventually lost as heat which cannot be recycled. However, matter in the form of inorganic substances can be recycled as discussed in Section 5-H and Chapter 6.

By now you are, no doubt, asking whether the rabbit can survive as an isolated heterotroph in such a small "field" of clover. [Describe the scenario that may occur in the event of "overgrazing". How will NPP (of food and O_2) be affected?]

estimating NPP for a biotic community

To determine the area of clover necessary to sustain the rabbit, we could estimate the NPP of clover. The NPP is usually expressed as the number of grams of dry matter produced per unit area per unit time (e.g. grams/meter^2/year). One could estimate NPP of a whole clover field by collecting all of the clover plant matter from each of a number of small, randomly placed plots of known area. [Why should *random* samples be taken? What sources of error do you envision in estimating NPP for a 5-acre clover field?] One estimate of the NPP of clover is 23 g/m^2/day (5).

estimating calories

The average energy content per unit mass of plant material can be determined by combustion of a measured mass of the material in a **calorimeter** and observing the temperature rise of a known quantity of water surrounding the combustion compartment. Energy content of organic matter, including foods we eat, is measured in kilocalories (kcal). One **kilocalorie** (or kilogram-calorie) is the quantity of heat energy necessary to raise the temperature of one kilogram of water 1°C at 15°C. The kilocalorie is the same unit of energy as the Calorie (large "C") used by nutritionists, and "calorie counters". One Calorie equals 1,000 gram-calories. The Calorie is a convenient unit for trophic structure studies because it can be used to quantify radiant energy of the sun as well as chemical energy of organic molecules, and the heat released in respiration or combustion.

Energy flows from the sun.

Figure 5-4 illustrates the flow of energy from the sun as it enters the earth's atmosphere and biosphere through an imaginary "window" 1 meter^2 in area. The amount of radiant energy entering through this window on a June day at 40° North latitude is estimated to be 10,000 Cal./m^2/day. [What is your present latitudinal position on earth?]

79

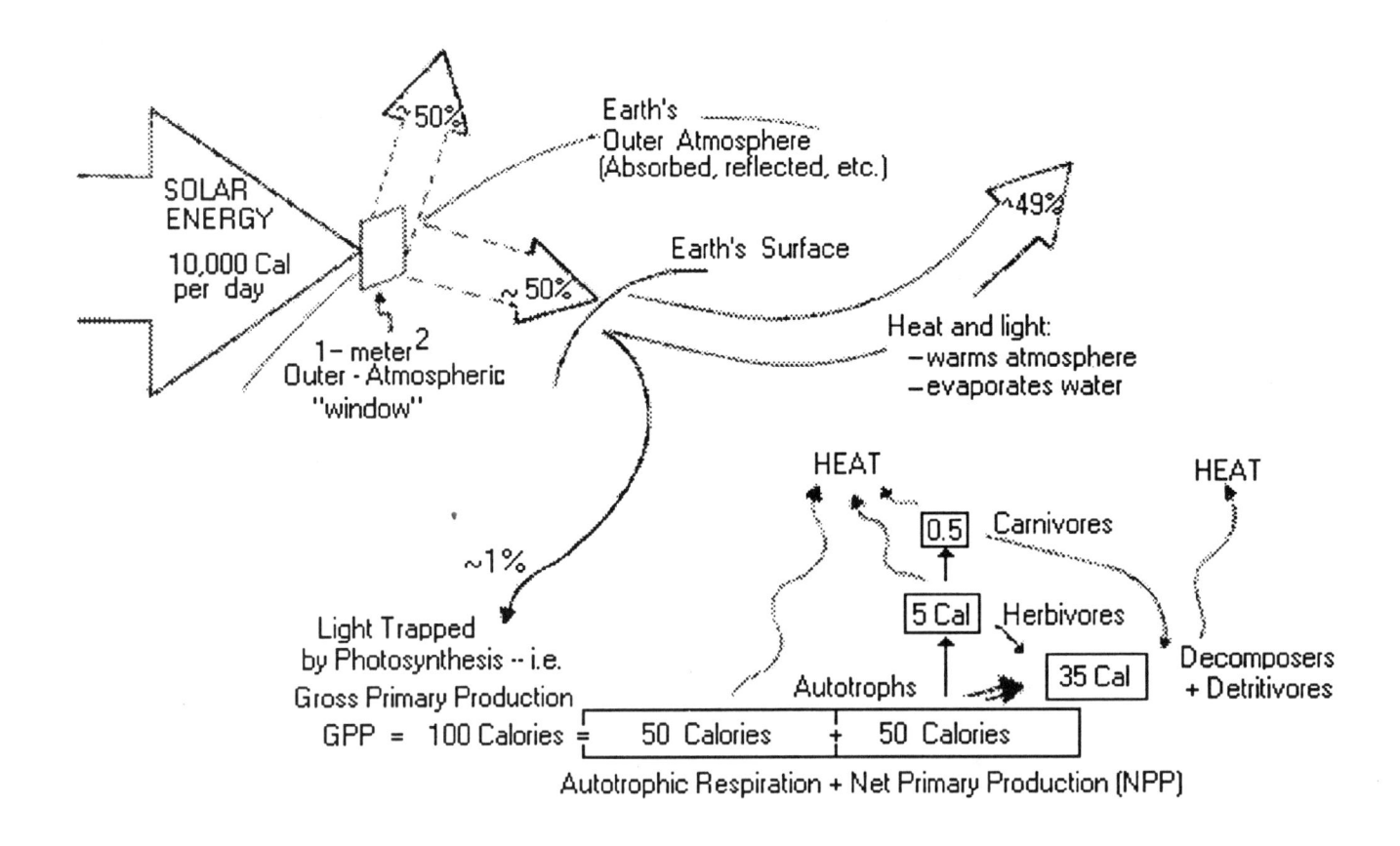

Figure 5-4. Distribution of 10,000 Calories of radiant energy, the quantity incident upon a 1-m² perpendicular plane of Earth's outer atmosphere above 40° N latitude during one day in June at solar noon. Data are from Clapham (3) and Gates (6)

Energy content is conserved.

Several "energy flow principles" are illustrated in Figure 5-4. First, notice that *energy content is conserved* as the first law of thermodynamics predicts. Eventually, the entire energy influx is either reflected, radiated as heat, or stored temporarily as chemical energy in the *biomass* of the food chain. Any energy not accounted for in these major destinations will serve to add to the thermal energy content of Earth and its atmosphere, causing *global warming*.

Solar energy is made compatible.

Second, the model in Figure 5-4 shows that the *energy composition is made compatible* with life on Earth in three ways. (a) Much of the harmful short-wave, ionizing radiation is screened out by stratospheric ozone (Section 5-B). Other wavelengths including some visible light is reflected from clouds and the ground surface. This reflected light gives the earth a special beauty, as pictures taken from the moon or outer space have shown. (b) Thermal (infrared, or heat) rays provide atmospheric and surface warmth essential for life on Earth and propel the water cycle to provide an ongoing supply of fresh water for terrestrial life. Certainly it is light and not heat energy that is converted into chemical bond energy (GPP). However, without the diurnal and seasonal exchanges of heat energy by atmosphere, soil, and oceans, photosynthesis and other life processes could not occur. Heat is low quality energy, but life needs it, too!

Autotrophs set energy income.

Third, Figure 5-4 demonstrates that autotrophs determine the food *energy income of the community*. The NPP of autotrophs is the "income" of organic or food molecules essential for survival of the heterotrophs. In other words, our rabbit's food and O_2 income is determined by the clover. The autotrophs of an ecosystem may use from 10 to 75 percent of the GPP for their own energy demands through respiration. Let's say respiration consumes 50% of GPP (Fig. 5-4) which leaves an NPP of 50% available for grazing herbivores. Herbivore populations incorporate approximately 10% of the NPP into their body structures (5 Calories in our example, Figure 5-4). The remaining 45 Cal. (90%) represent (a) ungrazed plant material which eventually dies and is consumed by decomposers and detritivores; and (b) respiratory heat loss from herbivore metabolism. Carnivores in turn incorporate about 10% (or 0.5 Cal) of the herbivore biomass into their own body constituents. The other 4.5 Cal (90%) represent (a) energy losses via respiration, and (b) herbivores that escape predation and die of natural causes after which energy passes to decomposers and detritus feeders.

Summary:

In summary, for every 10,000 Cal. of solar energy that reaches the outer atmosphere of earth, 10,000 Cal. is eventually returned, much of it as lower quality, heat energy. Thus, energy is not destroyed (first law) but it is degraded in quality (second law). Notice that only about 10% of the energy available at one trophic level is actually incorporated into the bodies of organisms at the next higher level. The second law is responsible for a significant portion of the 90% loss at each level.

5-E ECOLOGICAL PYRAMIDS

Pyramids show energy flow.

Ecologists have developed various visual models to picture relationships among the various trophic levels of a community. **Ecological pyramids** are used to illustrate relative biomass of different trophic levels (Figure 5-5).

Trophic Level	BIOMASS mg. dry mass / meter^2	PRODUCTIVITY Calories / meter^2 / year	NUMBERS Individuals / meter^2
SECOND CARNIV.	0.10	0.10	15
PRIMARY CARNIVORES	0.66	1.2	100
HERBIVORES	1.25	26.8	15,000
PRODUCERS	17.7	280	72,000,000,000

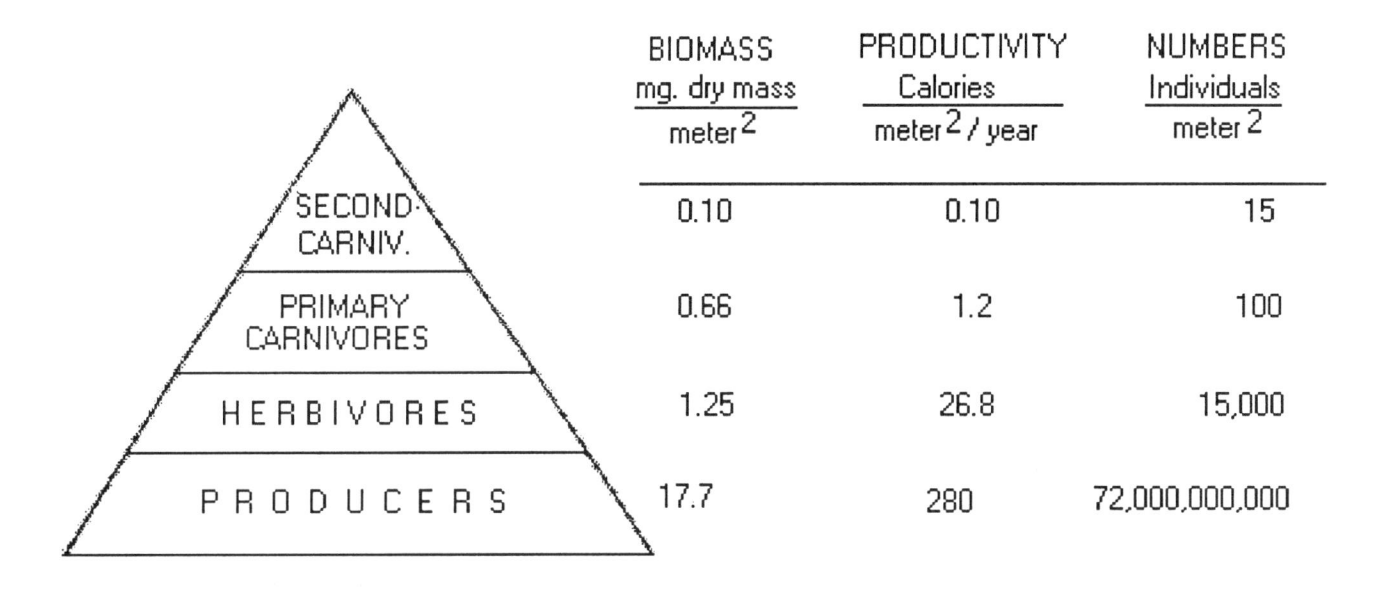

Figure 5-5. Ecological pyramids of biomass, energy productivity, and number of individuals in a shallow pond community. Data from Whittaker (13).

Biomass is the total mass of all organisms in a designated classification (e.g. herbivores) per unit area. Pyramid models are also used to illustrate energy productivity, and the relative populations at each level (Figure 5-5). Much time and patience are required to collect these kinds of data. [Briefly outline an experimental design you would use to determine the parameters in Figure 5-5 for a forest community.]

Pyramids show the effect of the Second Law.

All three types of pyramids illustrate the effect of the second law of thermodynamics upon ecosystem—i.e. the production of heat energy during metabolism of the chemical bond energy of food. Therefore, the biomass supported at a given level is usually less than the level below, and limited by the quantity stored at the lower level. For this reason, it is rare for an ecosystem to have more than four trophic levels, since there would be too little energy to support a population of strictly tertiary carnivores. In fact, many populations that feed at higher trophic levels do not feed there exclusively. Feeding at lower trophic levels permits access to additional energy and reduces energy expenditures needed to obtain food.

Some aquatic ecosystems have inverted biomass pyramids. Here, microscopic algae (phytoplankton), which can rapidly multiply, become an adequate food supply for heavily grazing, herbivorous zooplankton even though grazing keeps algal biomass low. In retail marketing terms, the lake has a high turnover of algae but low inventory. The result is an inverted pyramid with a smaller autotrophic base than the heterotrophic layers above. [Why wouldn't the *energy pyramid* (Figure 5-5) of this aquatic ecosystem also be inverted? Relate this to the retail store analogy above.]

5-F FOOD WEBS AND HOMEOSTASIS

Food webs show energy flow among populations...

Most biotic communities have more than one population occupying each trophic level. **Food web** models, such as that shown in Figure 5-6, are more convenient than ecological pyramids for illustrating energy flow patterns among individual populations of a community. Food webs are useful to illustrate two criteria for analyzing biotic communities, their *complexity* and their degree of *stability*.

and emphasize community complexity...

The complex nature of most biotic communities derives from the large number of resident plant, animal, and microbial populations; and, the multitude of possible trophic connections among resident populations. Figure 5-6 presents a simplified food web showing energy flow among a few of the more common populations of a meadow community. Detailed food webs require much tedious field work to complete, and are highly complex.

and stability.

Each community functions as a living system made up of many populations interacting with each other and with their environment in a predictable manner (Section 5-A). When the interactions among populations of a community are such that any fluctuations are steady and/or repeating, the community is said to be at **homeostasis** (Sections 2-C, 5-A). Communities are capable of self-maintenance and self-regulation and tend to resist change as long as environmental conditions (e.g. moisture, temperature, human disturbance) do not fluctuate beyond tolerable limits. The more **stable** a community is the more it is capable of resisting change and maintaining its homeostatic condition.

Figure 5-6. A simplified food web of a meadow community. Arrows indicate the direction of energy flow from producers (bottom) upward.

The many interlocking arrows of the meadow food web are a contributor to the stability of this community. Many arrows pointing to a given population indicate many options for energy (food) supply. Thus, a disease that reduces the clover population will probably not reduce the honeybee population. Instead, this would-be "shock" can be "absorbed" as the honeybee relies more heavily on dandelion. Likewise, the owl has the energy options of feeding as a primary carnivore, secondary carnivore, or tertiary carnivore, by preying upon the mouse, quail, or shrew populations, respectively, depending upon preference and size of the various herbivore populations. The food web is like a spider web. A disruption of one population (strand of the web) is detected throughout the web. However, subsequent **homeostatic adjustments** among the other populations allow preservation of the community as a whole.

homeostatic adjustments

5-G HUMAN IMPACTS UPON FOOD CHAINS

Humans occupy a unique position within the creation as image-bearers of the Creator. However, the Creator has ordained that our nutritional needs be met through dependence upon and stewardship of biotic communities that include plants, animals, and microbes as essential, functional parts.

Humans are heterotrophs...

As heterotrophs, humans depend upon autotrophic green plants directly and indirectly for food. For thousands of years, the human population remained below the 0.5 billion mark. Widely scattered hunter-gatherer and primitive agricultural societies caused minimal disturbance to natural ecosystems. However, with the advent of the industrial revolution and the use of fossil fuel (e.g. coal and oil), technology was developed with the capacity to cause major alterations in ecosystems. Through mechanization, we exchanged our dependence upon manual labor and beasts of burden, both of which had depended upon inexhaustible solar energy, for machines fueled by Earth's finite energy reserves. Machines powered by fossil fuel were used to mine coal and pump oil reserves from deeper sources at faster rates. In recent decades, this "capacity to do work" (energy), has *molded* (reworked the form of) the landscape of the Earth, and of human institutions! [List ways that energy available to you is molding your life to be different from a student in the 19th century? From a person in a Third World country?]

...with great need of energy.

Energy molds landscapes and institutions.

energy use in two types of ecosystems

How have these changes in energy usage impacted the biosphere? Let us address this question as it relates specifically to energy flow in **agricultural ecosystems**. Agricultural ecosystems are designed and managed in such a way as to facilitate conversion of sunlight into plants and animals for food, clothing, and shelter, usually for economic gain. Ecosystems in which humans have not yet made major alterations in the original food chain will be called **natural ecosystems**. [Is a meadow really a "natural ecosystem"?]

To illustrate how agricultural ecosystems are different from natural ecosystems, let us suppose that a farmer wishes to obtain high yields of clover in the meadow community of Figure 5-6. As it is, the net primary productivity (NPP) of the meadow community results from photosynthesis in leaves of dandelion, grass, and clover. However, the goal of increased clover yield leads the farmer to exclude dandelion and grass populations by plowing, cultivating, and reseeding so that clover can dominate the soil surface and capture maximum sunlight. In reality, a "pure stand" of a single population is hard to attain except in sterile laboratory cultures. However, agriculture does produce land areas *dominated* by one plant population, called **monocultures**.

channelling energy flow for profits

More *solar energy flow* to clover means a higher NPP; and, more *cash flow* to the farmer. But wait! The farmer's practice not only reduced populations of dandelion and grass, but also caused migration of quail, fox, and other carnivores away from the community. However, it is unlikely that he was able to frighten away mice and grasshoppers whose populations increase because of less predation by carnivores. We can see that a disruption of one trophic level, in this case at the autotrophic level, rapidly impacts the higher trophic levels. Creatures whose populations were controlled and in a homeostatic balance, are now seen as *pests* to be eliminated.

In order to channel the chemical energy of clover to his beef cattle, rather than to mice and grasshoppers, the farmer turns to a choice of pesticides for use in field and granary. Pesticides have their own "side effects" upon ecosystems, as we shall see in Chapter 6. With plow and pesticide, the farmer has simplified the meadow ecosystem so that the predominant energy flow in the food chain is now as follows:

plows and pesticides

SUN —> CLOVER —> CATTLE —> HUMANS

This agricultural ecosystem appears to be economically efficient but is ecologically disruptive. Thus, economic and ecological perspectives seem to be in conflict. We cannot blame modern farmers who are generally astute managers of their businesses and striving to remain solvent. But, does the *management* or *stewardship* of the house (Gr. *oikonomia*, or economy) have to be at odds with our *understanding* of the workings of the house (Gr. *oikoslogos*, or ecology)? [Please ponder this question as you read on.]

ecological *versus* economic value

Food chain principles also relate to food costs and dietary considerations. Assume that our farmer cooperates with sun, soil, and weather to enhance the growth of clover, sweetcorn, field corn, and beef cattle. As an astute businessman, the farmer will have tallied his input costs (tractor fuel, fertilizer, pesticides, grain drying costs, antibiotics for the cattle, etc.) on his computer spreadsheet and set some minimum acceptable goals of economic return after expenses. Cattle should bring a higher price because it costs the farmer more to produce a pound of beef than it costs to grow the corn to fatten the cattle. This "corn-and-cow contrast" derives from the thermodynamic laws operating in the food chain of the agricultural ecosystem. The model below shows the transfer of energy from corn to cattle or humans according to the "ten percent rule" noted in Figure 5-4.

Does a farmer really *grow* crops?

```
                     10%
    SWEETCORN ————————> HUMANS
    or CORN (100)         (10)
         \
      10% \       10%
         CATTLE ————————> HUMANS
          (10)              (1)
```

If we feed at the carnivore level, we have access to only 1 Calorie of energy for each 100 Cal. trapped by the autotroph (corn). However, if we feed as herbivores, or **vegetarians**, we have access to ten times more energy.

Good nutrition is more than calories.

Keep in mind two important points. First, proper nutrition is not judged by caloric intake alone. Men of average body weight must have at least 2,700 Cal./day, women, 2,000 Cal./day, to maintain body metabolism under moderate physical exertion (10). However, to avoid malnutrition, we must also have

Why are meat prices high?

essential vitamins, essential amino acids, minerals, and other constituents that are not often found in many vegetable sources.

A second point to remember is that the higher one feeds in the ecological pyramid, the smaller the quantity of available energy or food. The economics of supply and demand require that meat-consumers must in effect pay the farmer (and handlers) for the calories it takes to grow (and process) the meat, *including the thermodynamic losses.* The resultant higher cost of meat forces residents of poorer nations to depend upon rice, cassava, and other plant foods which cost less. Similar price differences are evident in markets of western nations. We can see that ecological principles have a major influence upon the market system.

Mechanized agriculture...

Before leaving our discussion of agricultural systems, we should briefly contrast contemporary agricultural approaches—mechanized and non-mechanized agriculture. About 5 hours of human labor are required to produce one acre of corn using our industrialized, mechanized agriculture, as compared to 500 hours required by nonmechanized agriculture. Machines, commercial fertilizers, synthetic pesticides, and genetically improved crop varieties, are among the developments that have made it possible for one American farmer to produce enough food and fiber for 120 other people at home and abroad (10).

frees us from toil in soil.

Mechanized agriculture frees us from the daily toil of growing our own food and allows us to pursue education, career, and leisure. Americans spend only 11% to 15% of their disposable income on food, compared to 60% to 80% spent on food by the estimated 1 billion who are also considered undernourished (10). Each year, an estimated 20 to 40 million persons (half of them age 5 or below) die prematurely from nutrition-related causes (10). According to Durning (6), one billion of us "live in unprecedented luxury; 1 billion live in destitution". [No "guilt trip", please, but we do need to examine this issue.]

Is mechanized agriculture the solution for the less developed countries (LDC's)? To begin, let's contrast agricultural productivity between mechanized and non-mechanized cultures as shown in the yields of cereal grains, a major world component of human and livestock food, reported in Table 5-1. [What conclusions can you draw from this data?]

negative side of mechanized agriculture:

Greater crop yields resulting from mechanized agriculture have come at great cost, namely loss of soil, water, and air quality. Between 1950 and 1984, world grain harvests increased 2.6-fold, due in large part to a 9-fold increase in fertilizer use (2, 3). However, in the five years since, only a 1% increase was recorded. This amounts to a *7% decrease in world grain production per person*! (3). Table 5-2 illustrates this downturn on a regional basis. To explain these declines, one must consider that each region is different. Periodic drought and political instabilities are involved. However, there is good evidence that, with the exception of certain developing nations such as India, substantial increases in crop yield are no longer possible by further increases in fertilizer and pesticide applications (2). Indeed, significant soil degradation due to erosion and salinization (salt buildup from irrigation) has occurred since World War II on 1.2 billion hectares of vegetated land—an area as large as India and China combined (15).

...degraded soil

Table 5-1. Average Cereal Grain Yield from Mechanized and Non-mechanized Agriculture. [1 hectare (Ha) = 10,000 m^2)

	JAPAN	U.S.A.	NIGERIA	WORLD
Total Cropland[1]	4,637	189,915	31,335	1,478,190
Cropland/Person[2]	0.04	0.76	0.29	0.28
Grain Yield[3] (avg. 1989-90)	5,662	4,300	1,118	2,638
Fertilizer Use[4]				
Average 1977-79	481	105	3	73
Average 1987-88	425	95	10	97
Pesticide Use[5]	32,000	373,333	4,000	?
Avg. # Tractors (1987-1989)	1,979,260	4,670,000	11,000	25,896,523

SOURCE: World Resources, 1992-1993. Report by World Resources Institute, 1992. Oxford University Press, New York.

[1] Area (1989) in thousands of hectares. One hectare (ha) = 10,000 m^2, or 2.47 acres
[2] Hectares of farmable land per person in specified nations, 1990.
[3] Grain (corn, wheat, rice, etc.) in kilograms/ha. 1 Kg = 2.20 lbs.
[4] Kilograms per hectare of cropland.
[5] Metric tons of active ingredient. 1 Metric Ton = 1,000 kilograms

...and limits to food production

What is the outlook for improving food production? Japan has been cited as an example of what can be accomplished agriculturally in a literate, scientific society with powerful market incentives. Japan has exceeded the USA and most other nations in grain production per hectare as shown on Table 5-1. Yet, Brown, *et.al.* (2) note that, "Japanese farmers have run out of agronomic options to achieve major additional gains in productivity". They wonder how less literate, less scientifically oriented farmers in the rest of the world will do any better.

Table 5-2. Regional and World Per Capita Grain Production. Peak Year and in 1990.

Region	Peak Year	Production (Kg/person)	1990 Production (Kg/person)	Change Since Peak Year (percent)
Africa	1967	169	121	-28
E. Europe/USSR	1978	826	763	-8
Latin America	1981	250	210	-16
North America	1981	1,509	1,324	-12
Western Europe	1984	538	496	-8
Asia	1984	227	217	-4
World	1984	343	329	-4

Source: Brown, L.R. 1991. The new world order. In L.R. Brown, *et.al.* eds. *State of the World*, pp 3-20, Worldwatch Institute Report. W.W. Norton, New York.

Summary: In summary, food is a form of energy, and energy is required to produce, process, transport, and market it. Mechanized agriculture in the western nations provides abundant food at reasonable cost, but causes major disruptions of natural ecosystems. On the other hand, millions in less developed countries are reported to be malnourished. What principles are there to govern our use of earth's resources so that suffering can be minimized and environmental disruptions reduced? Can the market system provide a just distribution of resources and opportunity? Does the science of ecology have principles to guide man's decisions? These are questions we will address in more detail in future chapters.

5-H THE DETRITUS FOOD WEB

O how can it be that the ground itself does not sicken?
How can you be alive you growths of spring?
How can you furnish health you blood of herbs, roots...?
Are they not continually putting distemper'd corpses within you?
Is not every continent work'd over and over with sour dead?
—Walt Whitman, "This Compost"

food from foul things

As Walt Whitman expressed, the soil and landscape gives us the impression of life, growth, and sweet fragrance, rather than death, scattered skeletons, and stench. The answer to this paradox lies in the **detritus food web**, a portion of each biotic community, consisting of detritivores and decomposers. Unlike the **grazing food web** which includes herbivores and carnivores which feed on living plants and animals, members of the detritus food web draw their food energy from **detritus**, organic remains of living organisms.

energy remains after death

Obviously, not all autotrophs are consumed by herbivores; nor are all herbivores captured and eaten by carnivores. Many survive to "die of old age". Each trophic level is left with an excess. Much of each year's net primary and secondary (herbivore and carnivore) production becomes dead organic matter, or detritus. The chemical make-up and energy content of live bodies do not decrease markedly when an organism dies. Therefore, animal carcasses, excreta, dead branches, and fallen leaves represent a rich energy store for the detritus food web.

According to data cited in Figure 5-4, 35% of the NPP flows into the detritus food web, while only 14% moves into the grazing food web! Studies of the deciduous (L. *decidu* = falling off) forest ecosystem in eastern United States revealed that about 75% of the annual forest NPP enters the detritus food web (1).

underworld detritivores and decomposers

On the forest floor, often hidden beneath leaf litter, logs, and rocks are seemingly countless earthworms, insects, millipedes, and other invertebrates that ingest and digest organic matter as their source of energy. Excreta from detritivores consists of smaller particles of detritus material with greater surface area which facilitates further digestion by wave after wave of decomposers. Common decomposers include many species of soil bacteria and the molds and mushrooms of Kingdom Fungi. Decomposers secrete powerful digestive enzymes from their cells, causing extracellular (L. *extra* = outside

The detritus community recycles nutrients...

while feeding themselves and being fed upon.

Explore a compost pile.

How do plants benefit from compost?

of) digestion of energy-rich organic molecules constituting the detritus. Digestion releases simpler constituents such as glucose, amino acids, and mineral nutrients. These soluble substances are absorbed by the decomposers, and used in growth and respiration, as in other heterotrophic organisms. However, not all of the mineral nutrients are absorbed by decomposers. Some become accessible to plant roots where they are taken up and reused to form new organic compounds at the autotrophic level.

In summary, there are two major influences of the detritus food web. First, detritivores and decomposers degrade organic debris and reassemble the constituents into their body structures. Unused inorganic constituents such as CO_2, H_2O, and mineral nutrients are released, and may be recycled to the autotrophs (Figure 5-7). Second, the detritivores and decomposers actually grow and reproduce themselves from "dead things" and waste. Various insects, protozoa, and worms feed upon the bacteria and fungi, and these in turn become prey for carnivorous animals such as birds, snakes, and shrews. Thus, at the carnivore level, the detritus food web blends with the grazing food web (Fig. 5-7). As organic matter moves from one trophic level to another, little by little, respiration which is thermodynamically only about 40% efficient, releases all of the chemical bond energy into the environment as heat, along with CO_2 and H_2O. Digging into an active compost pile is a great place to see the detritus food web in action and feel the heat being released. It is also the place to see the "secret" of how *dead things* are marvelously converted into what we call "soil" which is so essential to the survival of live on Earth. "Behold this compost! behold it well!" And Whitman adds: "Now I am terrified at the Earth, it is that calm and patient, it grows such sweet things out of such corruptions."

We have now reached a point of major ecological significance. That is, matter in the form of mineral nutrients, CO_2, and H_2O can be recycled via the detritus food web back to the producers. However, energy cannot be recycled because it is converted ultimately to low-quality heat energy. Autotrophs cannot grow without heat energy, but must have high-quality, radiant energy for photosynthesis. Thus, whereas energy and matter move together through the ecosystem to some extent, they part company when respiration degrades organic matter to CO_2, H_2O and minerals; and, releases heat energy. In Chapter 6, we shall examine the cycling of nutrients in the ecosystem.

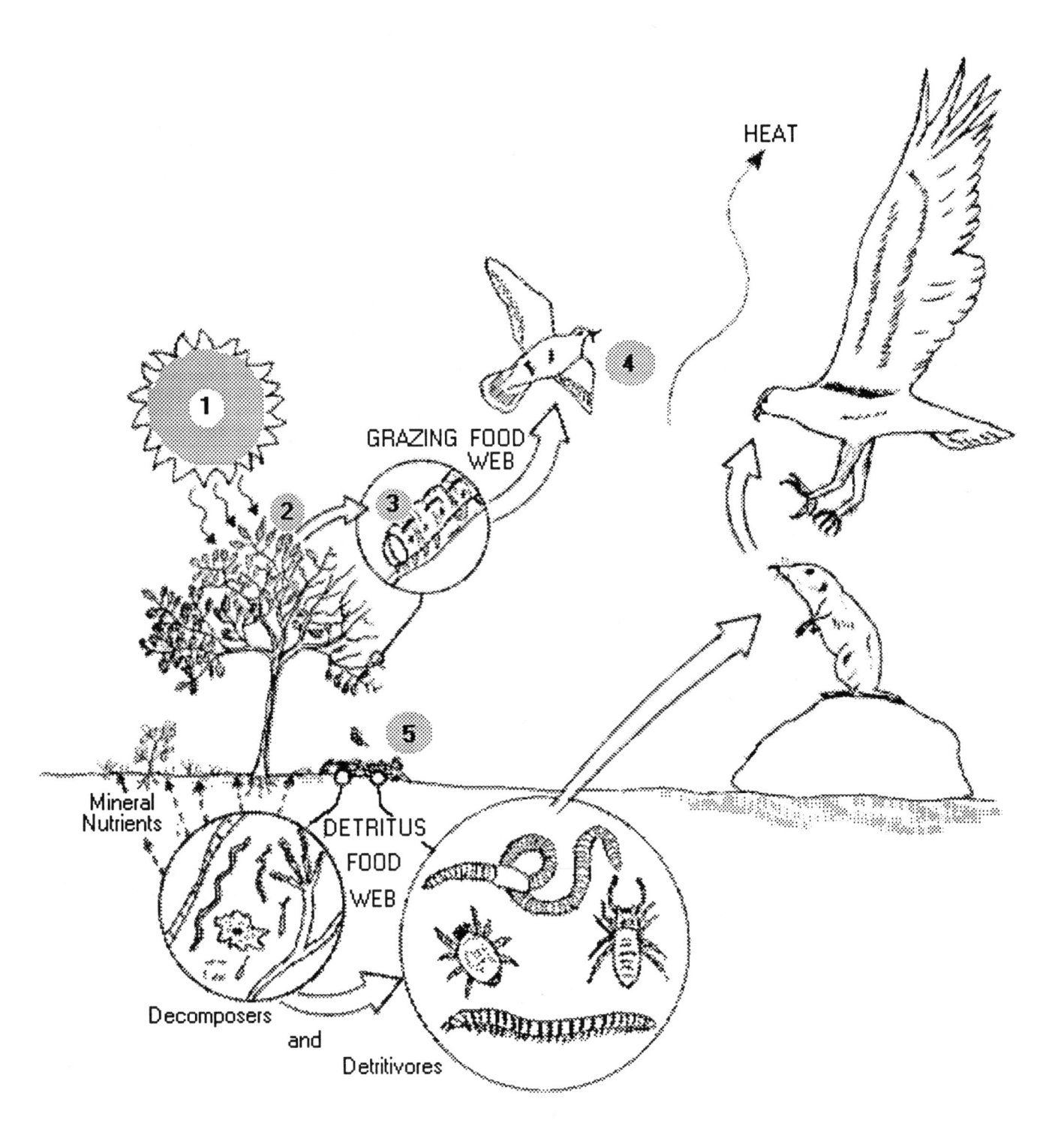

Figure 5-7. Diagrammatic view of a simple grazing food chain. Radiant energy (1) is absorbed by producers and converted to chemical bond energy during photosynthesis (2). The resulting organic matter serves as food for herbivores (3), which are in turn eaten by carnivores (4). Chemical bond energy content decreases across each trophic level due to respiratory loss of heat. Unconsumed organic material from the grazing food chain enters the detritus food web via refuse and death (5). Organic molecules are reduced to their inorganic constituents and recycled (dotted arrows); energy is released as heat and not recycled.

QUESTIONS AND DISCUSSION TOPICS

1. Consider the following locations, and generate a list of factors that influence the amount of radiant energy that will reach the surface of the earth: (a) your backyard in summer *versus* winter, (b) Anchorage *versus* Mexico City, (c) Honolulu *versus* Mexico City.

2. Theoretically, all radiant energy reaching the earth is eventually returned to outer space. Discuss various paths taken by this energy. Which requires the shortest time to return to space after reaching earth? Which path would allow long-term retention of energy on earth?

3. Why are pyramids used to describe energy flow in communities? How and where does energy enter the pyramid? How and where does it leave?

4. Make a table with the following two headings:

    NATURAL ECOSYSTEM   AGRICULTURAL ECOSYSTEM

    Compare the two with respect to each of the following:
    (a) pattern of energy flow, (b) influence of ecology vs. economics, (c) bases for judging efficiency, (d) energy sources, and (e) stability.

5. If solar energy reaching the biosphere is analogous to your monetary *income*, what source of energy is analogous to an *inheritance* you might receive? Upon which source of energy do you depend most? Other creatures? What implications do you see in this?

6. Based upon what you have learned in this chapter regarding the challenge of feeding the human population on Earth, what factors must be taken into consideration in strategy development for the future? See TOPICS #4 and #5 below.

TOPICS FOR FURTHER STUDY

1. Did you know that not all biotic communities depend upon the entry of solar energy via autotrophic green plants? Ocean floor communities are essentially detritus food webs based upon a constant 'rain' of detritus from shallower depths above. However, certain ocean floor communities near oceanic rifts (faults) can live in darkness, apparently independent of sunlight or falling detritus. These communities depend upon **chemosynthetic autotrophs**, bacteria that metabolize H_2S (hydrogen sulfide) spewed from volcano-like vents along the oceanic rifts. For further information, see the following references:

    Edmond, J.M. and K. von Damm. 1983. Hot springs on the ocean floor. *Scientific American* 248(4): 78-91.

    Grassle, J.F. 1985. Hydrothermal vent animals: Distribution and biology. *Science* 229: 713-717.

2. Is modern agriculture overcoming the curse of Genesis 3: 17-19?
    Reference:

    Brown, L.R. and J.E. Young. 1991. Feeding the world in the nineties. In L.R. Brown, *et.al. State of the World, 1991*. Worldwatch Institute. W.W. Norton, New York.

3. How much energy do different human activities require? Which city uses more energy per square meter, New York or Shanghai? What form of energy can be generated fastest on the least land

area? For answers to these and other intriguing questions, see the article by V. Smil (1984), in REFERENCES.

4. According to Worldwatch Institute (See REFERENCE #3), grain carryover stocks had dropped to 461 million tons in 1987, and further, to 290 million tons, in 1990. The 1987 carryover was enough to feed the world for 102 days, whereas the 1990 carryover would last 62 days. The 1990 bumper harvest raises the carryover enough to feed the world for 66 days, should it be needed. What would happen to world grain market prices, and hence food prices if this reserve continues to drop?

5. What is your analysis and predictions for a world in which 1 billion live in luxury, and 1 billion live in destitution. Consider the implications for your future vocation. (See REFERENCE #6)

REFERENCES

1. Bormann, F.H. and G.E. Likens. 1979. *Pattern and Process in a Forested Ecosystem*. Springer-Verlag, New York.

2. Brown, L.R., J.E. Young. 1990. Feeding the world. In L.R. Brown, *et.al. State of the World*, pp 59-78, Worldwatch Institute Report. W.W. Norton, New York.

3. Brown, L.R. 1991. The new world order. In L.R. Brown, *et.al. State of the World*, pp 3-20, Worldwatch Institute Report. W.W. Norton, New York.

4. Clapham, W.B. Jr. 1983. *Natural Ecosystems*. 2nd ed. Macmillan, New York.

5. Cooper, J.P. 1975. *Photosynthesis and Productivity in Different Ecosystems*. Cambridge Univ. Press, Cambridge.

6. Durning, A. 1991. Asking how much is enough. In Brown, L.R., *et.al. State of the World*. pp 152-169. Worldwatch Institute Report. W.W. Norton, New York, NY

7. Gates, D.M. 1971. The flow of energy in the biosphere. *Scientific American* 225(3): 88-100.

8. Graedel, T.E. and P.J. Crutzen. 1989. The changing atmosphere. *Scientific American* 261(3): 58-68.

9. Kormondy, E.J. 1984. *Concepts of Ecology*. 3rd ed. Printice-Hall, Englewood Cliffs, NJ.

10. Miller, G.T. Jr. 1992. *Living in the Environment*, 7th ed. Wadsworth, Belmont, CA.

11. Rappaport, R.A. 1971. The flow of energy in an agricultural society. *Scientific American* 225(3): 116-132.

12. Smil, V. 1984. On energy and land. *American Scientist* 72: 15-21.

13. Whittaker, R.H. 1975. *Communities and Ecosystems*. Macmillan, New York.

14. Woodwell, G.M. 1970. The energy cycle of the biosphere. *Scientific American* 223(3): 64-74.

15. *World Resources, 1992-1993*. Report by World Resources Institute, 1992. Oxford University Press, New York, NY.

Chapter 6

Nutrient Cycles

Humans alter the biosphere

As the twentieth century nears its end, humans have become the most significant force under heaven in determining the quality of life on Earth for all species. We have altered the chemical composition of the atmosphere and oceans, and in so doing, may be altering the climate of the planet. We are impacting biospheric systems that are as near to our lives and future survival as a breath of air or drink of water. Beyond self-serving reasons for caring, we have been given responsibility to exercise good stewardship of the Earth.

local actions have global impact

Therefore, it behooves us to understand how our actions *locally* combine to affect the planet *globally*. To do this, we must understand how humans and other creatures exchange energy and matter with the physical environment.

Life requires matter and energy.

In Chapter 5, we learned that every living cell must continually acquire *energy* and *matter* from the environment. Energy is necessary to drive metabolic processes. Matter (raw material) must be taken into cells to form new cytoplasm necessary for growth in volume and mass. Thus, living and growing organisms must continually acquire both energy and matter from the environment. As we shall now learn, the acquisition of energy and matter from the environment occurs simultaneously (Figure 6-1).

6-A ENERGY AND NUTRIENT ACQUISITION

Our experiment with the clover and rabbit in a closed, lighted "ecosystem" illustrated the importance of autotrophs in determining the food energy and O_2 "income", or GPP (Section 5-D). The NPP, that which remains after the clover "pays its respiratory bills", is potential "food" for the rabbit. As Figure 6-1 illustrates, the chemical products of photosynthesis are the reactants (i.e. requirements for a chemical reaction) of respiration. This means that there is a **cycling** of inorganic matter between photosynthesis by autotrophic clover and respiration by *both* wheat and rabbit.

Nutrients move in a cyclic path.

Energy cannot be recycled.

If matter can be "recycled", can energy also cycle between photosynthesis and respiration? The answer is "no"! Respiration releases usable energy for growth and maintenance, but is only about 40% efficient in doing so. The remaining 60% of the energy of each glucose molecule entering respiration is released as heat, according to the second law of thermodynamics (Section 5-C). This heat is too low in quality to drive photosynthesis which must have light energy. Therefore, *matter can be recycled, but energy flows in one "direction" from higher to lower quality* and eventually is dissipated as heat.

nutrient cycle models

It is clear that a different sort of model is needed to portray the cycling of matter in contrast to one-way energy flow. Therefore, just as we used food webs and pyramids to emphasize energy relationships, we shall now use **nutrient cycles** to explain nutrient flow. The term, **nutrient**, will be used to distinguish inorganic matter such as CO_2, H_2O, and minerals, required by

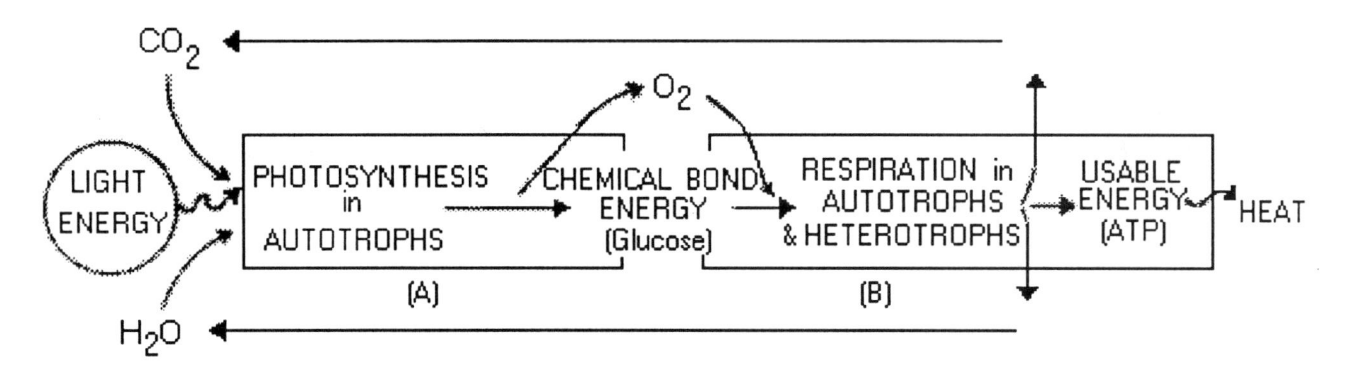

Figure 6-1. Cycling of inorganic matter between photosynthesis and respiration, and the non-cycling of energy. Respiration occurs in autotrophs in light (part A), but note that there is a *net* release of O_2, and *net* uptake of CO_2, because photosynthesis is occurring faster than respiration. Only in very low light or darkness do autotrophs become *net* users of O_2 and releasers of CO_2, like the heterotrophs (part B).

living cells. As Figure 6-2 illustrates, energy is lost as heat, and the nutrients are cycled back to a so-called **reservoir,** or point of supply in the physical environment. From there they can reenter the food web. Recall the vital role of the detritus food web in this recycling process (Section 5-H). Nutrient cycles are also called **biogeochemical cycles** to emphasize that chemicals are being cycled between life (bio-) and the earth (geo-).

6-B BIOGEOCHEMICAL CYCLE—GENERALIZED MODEL

To begin a discussion of biogeochemical cycling, we must identify the "biogeochemicals" that are cycled between reservoir and food web (Figure 6-2). Principle chemical elements that compose living matter were listed in Table 4-1. As the table indicates, approximately 99 out of 100 atoms in an average sample of living matter consists of of three elements—hydrogen, oxygen, and carbon. There are two nutrient cycles that involve these three

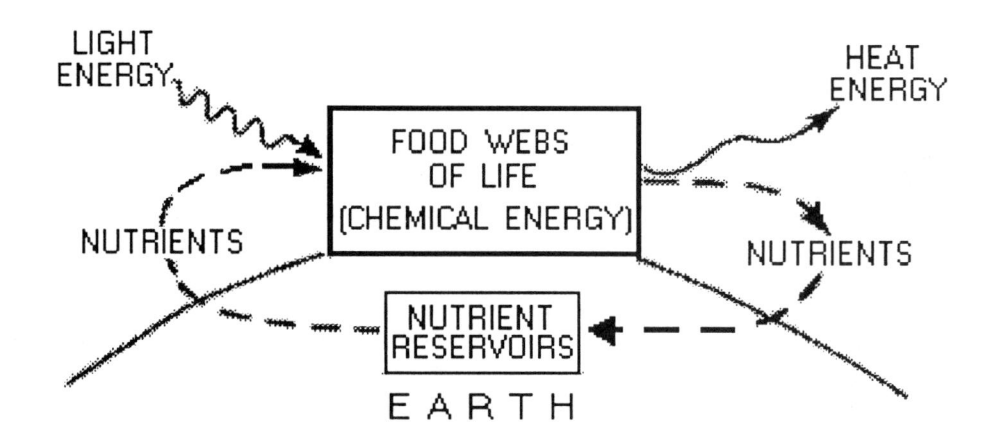

Figure 6-2. Contrast between non-recyclable energy flow through food webs, and the recycling of nutrients.

gaseous cycles

elements in a prominent way: the **water cycle** and the **carbon cycle**. The **nitrogen cycle** and the **sulfur cycle** account for the cycling of nitrogen and sulfur. These four cycles each have a gaseous form which allows the element to move within the atmosphere. [Can you name a gaseous molecular form containing each of the above elements?]

mineral cycles

Finally, we shall consider collectively the **mineral cycles**. These include elements such as Ca^{2+}, K^+, Na^+, and PO_4^{3-} that do not have a gaseous form. These are the minerals remaining in the ashes of a burned sample of plant or animal tissue. While mineral ash seldom occupies more than 5% of the body weight of an organism, nevertheless minerals are essential to life processes.

Figure 6-3 illustrates a generalized model representing the common features of all biogeochemical cycles. These common features are outlined in the figure legend for convenient reference. Refer to these general features as you are introduced to each specific cycle. Remember that these are simplified models that help us understand some of the complexity of the ecosystems of which we are a part. No one cycle occurs independently of the others.

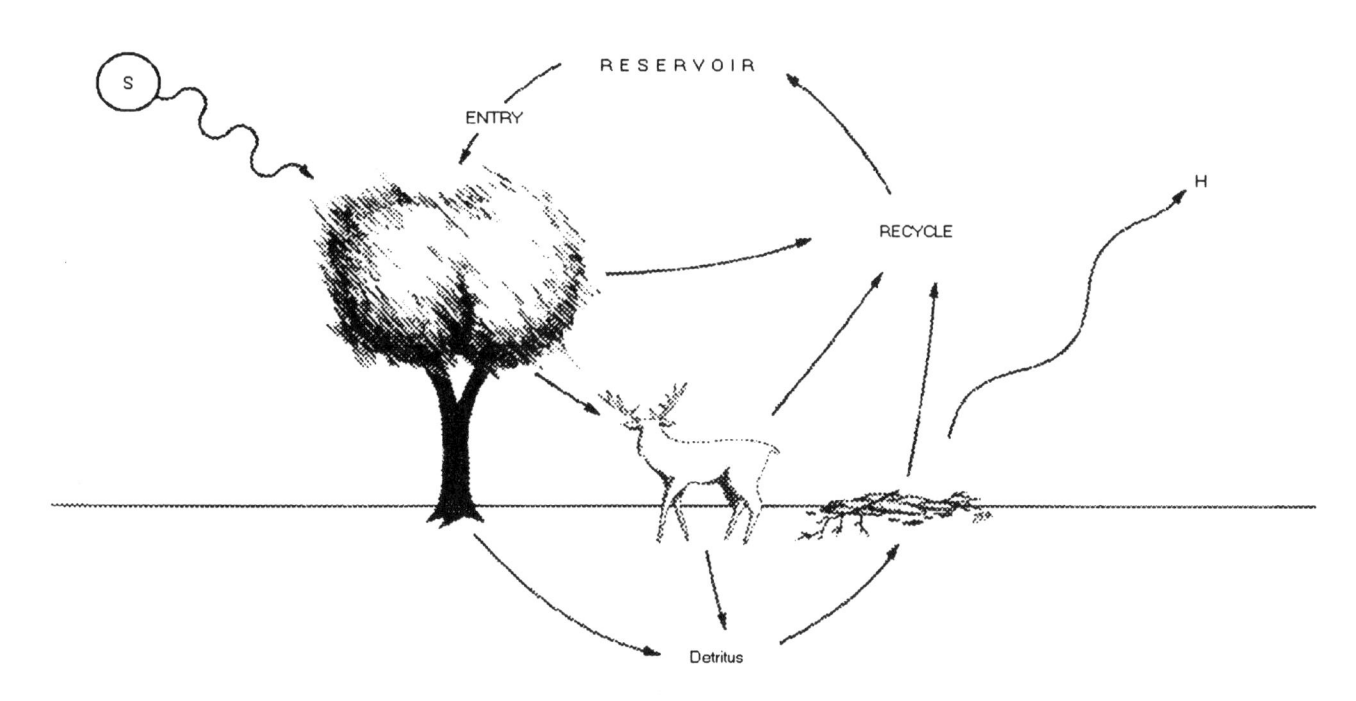

Figure 6-3. A generalized model of the biogeochemical cycles. All cycles have a RESERVOIR, a reserve of the element in the physical environment—i.e. the atmosphere, hydrosphere, or lithosphere, or any combination of these compartments (Section 3-B). There are always one or more processes (physical or metabolic) that require solar energy (S) to energize the ENTRY of the element into organic compounds. These in turn move along the grazing food chain from autotroph (A) to consumers (C). The detritus food web (not shown; see Fig. 5-7) and decomposers (D) eventually release the element from the organic form where it is free to recycle to the RESERVOIR. Heat energy (H), is a product of metabolism, and cannot be recycled because photosynthesis requires light energy.

6-C THE WATER CYCLE

The water cycle, or hydrologic cycle, represents a dynamic, solar energy-driven cycling of water among three biospheric compartments—the atmosphere, lithosphere, and hydrosphere (Figure 6-4).

solar-driven

Solar energy causes **evaporation** of water from land and sea, as well as from plants and animals, and their excretory wastes. Each year an estimated 5.2×10^{17} kilograms, or 95,000 cubic miles, of water is lifted into the atmosphere in vapor form. This giant distillation process draws water from such varied sources as salty oceans and polluted streams. The water vapor condenses on tiny molecules or dust particles, dissolves various atmospheric gases and ions within it, and falls to land and sea through precipitation. The result is a source of "fresh water" essential to land-dwellers. Only an average of 0.15×10^{17} kg. of water remains in the atmosphere as vapor and cloud droplets. [If 540 Cal. of energy are required to evaporate 1 kg. of H_2O already at 100° C, how many Calories of solar energy (as infrared or heat) are used in the water cycle each year? See Figure 5-4. Why is this an underestimation?]

Water movement on land . . .

Precipitation on land (an average of 75 cm/yr. in U.S.A.) causes two gravity-driven processes. First, if the soil is permeable, water **infiltrates**, or percolates downward, vertically into deeper layers. Infiltration serves to maintain the soil water supply for plant growth, and replenishes deep underground reservoirs, or *aquifers*, that often supply residential, commercial, and agricultural needs. As water infiltrates through soil, it often carries with it nutrients

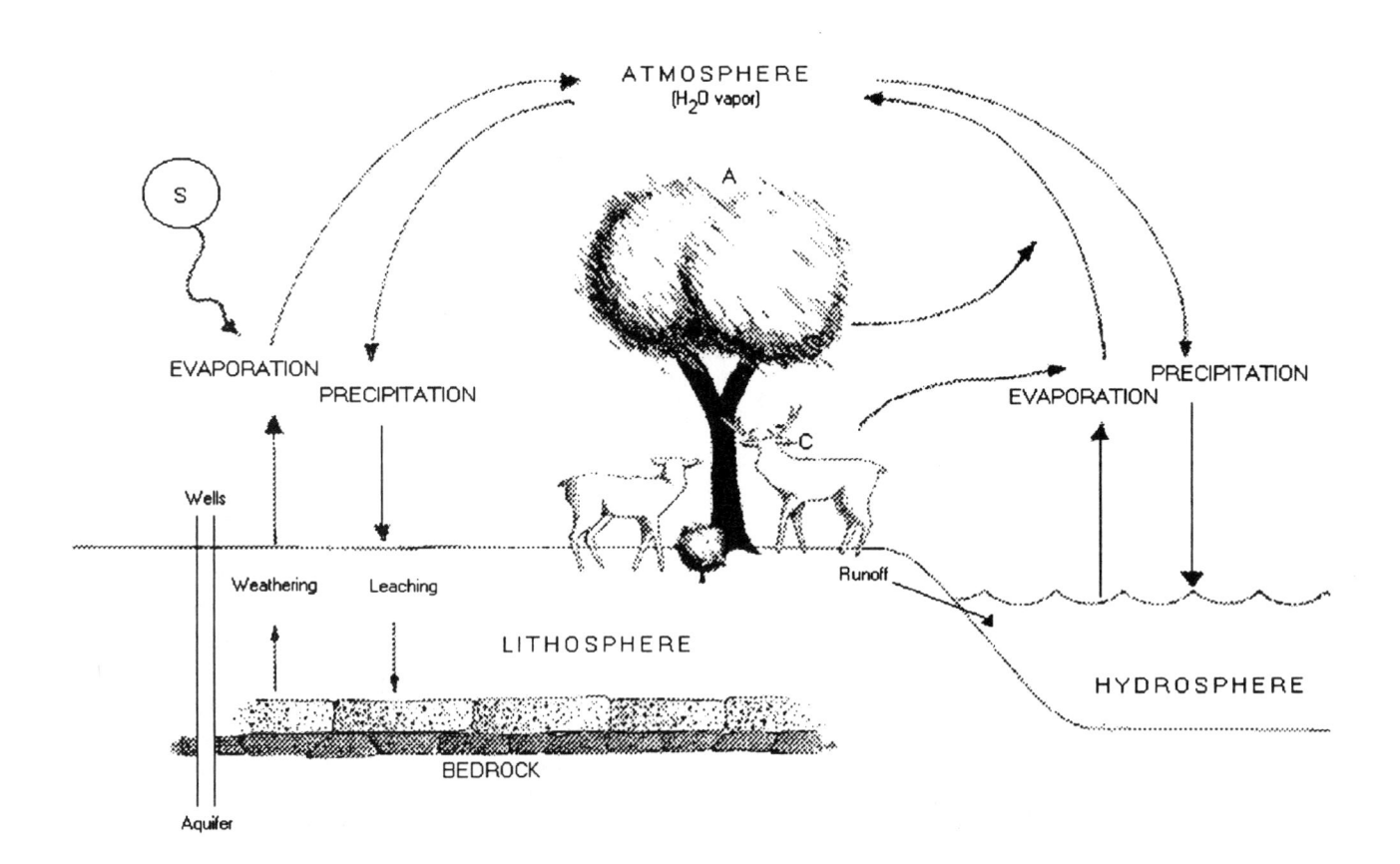

Figure 6-4. The water cycle.

performs the work of leaching ...

released from the detritus food web and decomposers, dispersing them into the plant root zone of the soil, and sometimes deeper than roots can reach. The term **leaching** refers to this downward vertical transfer of nutrients into the soil during infiltration.

The second gravity-influenced movement of water is called **surface runoff** (Figure 6-4). Water from precipitation that is unable to infiltrate will flow across the surface under the influence of gravity, and eventually enter streams and rivers. When the soil surface is not covered by vegetation or organic matter, surface runoff water will often carry soil particles along with it. This phenomenon is called **erosion**.

erosion ...

and weathering.

Water also affects the lithosphere through **weathering**, the physical (e.g. frost heaving, cracking) and chemical breakdown of bedrock and soil particles. We shall discuss this process in conjunction with the mineral cycles (Section 6-G).

In summary, the water cycle is a solar-driven cycling of water involving changes of state of water—solid, liquid, and gas. The water cycle can affect the soil, landscape, and the biotic community directly because of its role within living organisms, and indirectly because water is a medium of transport for nutrients among the soil, water, and air.

Clean water is a blessing ...

When, in "This Compost", Whitman marveled ... *That all is clean forever and forever, That the cool drink from the well tastes so good,* he reminded us of the wonder and blessing of having water, fresh, safe, and abundant. Today, there are threats to the water supply so vital to the survival of humans as well as both terrestrial and aquatic life. Freshwater amounts to only about 2.5% of global water, and is found in ice caps and glaciers, groundwater, and in streams and lakes. Only 0.014% of this freshwater is readily available to the human population (10). Every human use of water decreases its purity. [Doesn't this apply to your daily uses?] The water cycle, as Whitman realized, provides fresh water from polluted water—it is a dynamic "recycling system" which has great capacity to sustain life. However, there are several human intrusions upon the water cycle which are summarized here:

now threatened by ...

... toxic contaminants ...

a. Surface water continues to be contaminated by toxic chemicals and human waste which enter from *point sources* such as industrial effluent, or from *non-point sources,* including agricultural chemical runoff and acid rain (Sections 6-E, 6-F).

... and aquifer depletion.

b. Underground water supplies are being depleted by urban and agricultural demands. In the USA, about 20% of irrigated land is watered in excess of aquifer recharge rates (12). Worldwide, about 24% of irrigated land is damaged due to salt residues left from evaporation in arid areas. In some agricultural and urban areas, well water is spoiled when fertilizers, industrial chemicals, or landfill leachates infiltrate into ground water.

The upcoming discussion of other nutrient cycles will emphasize the importance of the water cycle in dissolving and transporting natural and human-created substances. Try to relate your *local*, daily usage of water and other resources to these *global* processes and serious concerns.

6-D THE CARBON CYCLE

global carbon balance

The element carbon not only forms the backbone of all organic molecules so essential to life, but also plays a major role in controlling the climate of the earth via the greenhouse effect (Section 5-D). Carbon exists as CO_2 gas in the atmosphere, 0.03% by volume. Like water, however, carbon has a reservoir in each of the three compartments, linked in a complex dynamic equilibrium as shown in Figure 6-5. In ways that are not clearly understood, this equilibrium lends stability to the biosphere by acting to mediate against sharp rises or drops in CO_2 levels, as we shall see.

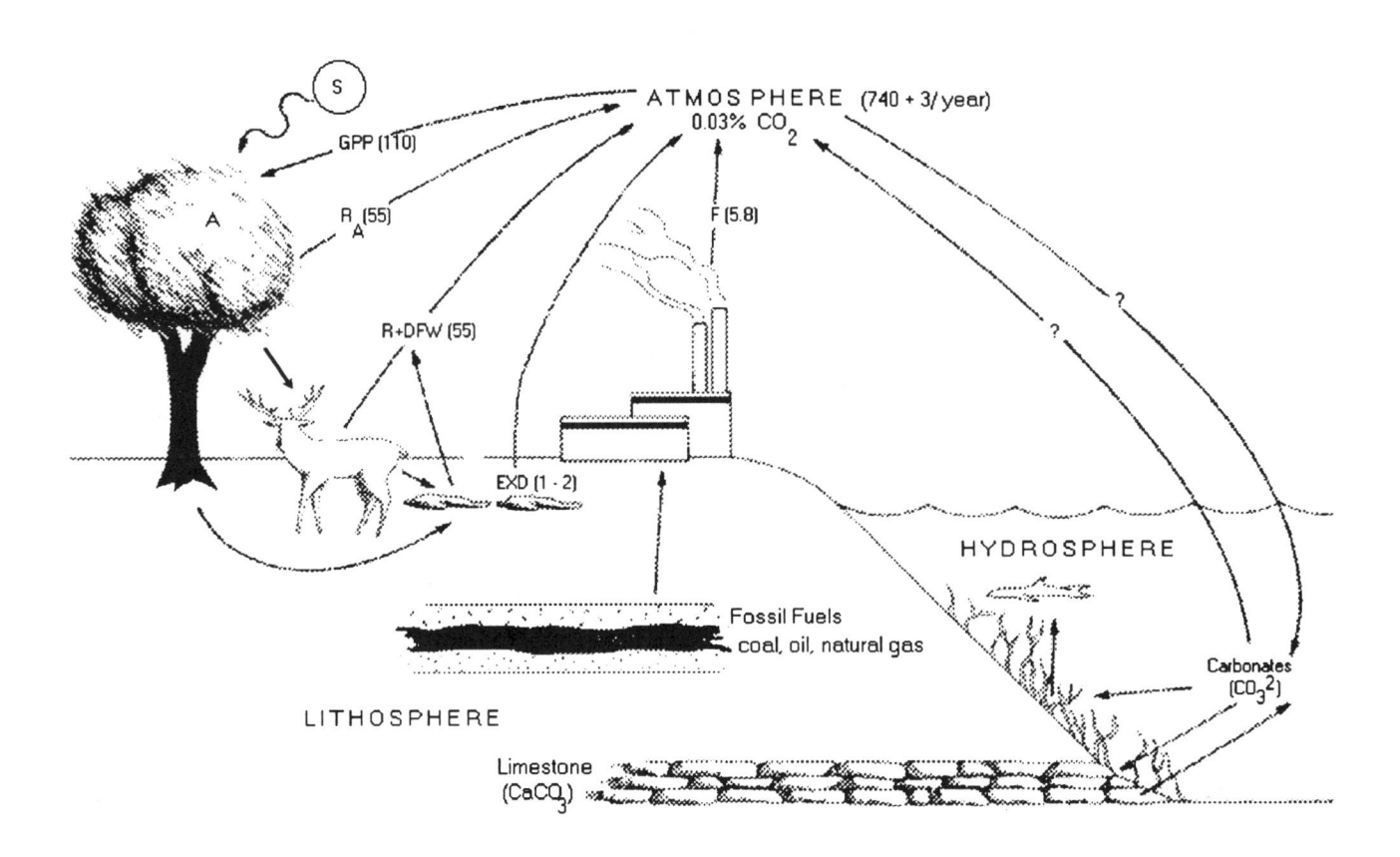

Figure 6-5. The carbon cycle. Carbon has a three-part reservoir, and enters the food web via photosynthesis (GPP). Eventually CO_2 is released from organic matter by respiration (R) of autotrophs (A), consumers (C), and the detritus food web (DFW). Numbers in parentheses are in trillions (10^{12}) of kg. of carbon/yr. Note that there may be as much as 7.8×10^{12} kg. more carbon released into the atmosphere per year from land than absorbed by photosynthesis. Two practices are believed to cause this: (a) accelerated decomposition and fires (EXD) associated with forest clearing, and (b) fossil fuel combustion (F). Atmospheric carbon increases by about 3×10^{12} kg./yr.; but, what about the remaining excess of 4.8×10^{12} kg? The oceans are believed to absorb no more than 3×10^{12} kg., leaving about 1.8×10^{12} kg. unaccounted for. Data are estimates from Schneider (16), and Smith, (18).

Both the carbon cycle and the water cycle are powered by solar energy (Figures 6-4, 6-5). Carbon enters the food web via solar-powered photosynthesis which converts CO2 into organic molecules (Section 5-D). All consumers (C) and members of the detritus food web (DFW) rely upon autotrophs for food. Carbon dioxide is released from organic carbon by respiration (R) at all trophic levels (including decomposers), by fires caused by lightning, volcanic activity, or human activities.

CO₂ levels are increasing by fuel combustion and forest clearing.

Based on data from polar ice cores (19), atmospheric CO_2 levels appear to have increased by about 25% from the preindustrial period (280 ppm) to the present (354 ppm, in 1990). Figure 6-5 and its legend present estimates of the annual flow of carbon from several of the key processes. Annual uptake by autotrophs (GPP) is similar to annual release by respiration (R). However, the human activities, namely combustion of fossil fuels, and clearing of tropical forests add additional carbon to the atmosphere each year (Figure 6-5). Forest clearing leads to increased CO_2 release from accelerated decomposition of felled trees and detritus opened up to light. Furthermore, the agricultural and urban ecosystems that replace the forests have lower NPP (i.e. net uptake of CO_2 from atmosphere per unit land area; see Section 5-D).

Human impacts on the carbon cycle cause ...

... possibility of global warming ...

Climatologists and ecologists are studying two potential effects of increased atmospheric carbon. First, carbon dioxide and related compounds, namely carbon monoxide (CO), methane (CH_4), and chlorofluorocarbons (CFC's) all contribute to the **greenhouse effect** (Section 5-D). The greenhouse effect provides the necessary heat balance to sustain life on Earth. However, there is evidence that increased concentrations of greenhouse gases may be causing a *global warming* trend. Since 1980, seven records for global average surface temperature were set—1990, 1989, 1988, 1987, 1983, 1981, and 1980! More data will be necessary to demonstrate a cause and effect relationship between anthropogenic sources of greenhouse gases and climate trends. However, computer models predict global temperature increases of 0.3° C per decade (19). Major changes in temperature and rainfall distribution are predicted if trends continue. Table 6-1 summarizes trends in anthropogenic (human-produced) gases in the atmosphere.

... ozone depletion ...

... and resultant increase in harmful UV.

Second, there is particular concern about the CFC's, which are doubly destructive. Besides being greenhouse gases as noted above, they also cause breakdown of stratospheric ozone, (O_3), into $O_2 + O·$. It is estimated that one CFC molecule can absorb 10,000 times as much infrared radiation as a CO_2 molecule (17). At the same time, the long-lived CFC molecules rise to the stratosphere where one CFC molecule may destroy over 20,000 O_3 molecules. The CFC'S are especially destructive over the polar regions where they cause "ozone holes", regions extremely low in O_3. Apparently, CFC's concentrate on surfaces of ice crystals and are more destructive to O_3 as a result. This O_3-depleted air spreads into temperate regions including the heavily populated areas of North America and Europe and lowers O_3 levels as much as 3% (15). Efforts are being made to halt use Of CFC's in aerosol products, as refrigerants, and as an expander of synthetic foams. However, due to the slow release of CFC's from foam products, and its "long life" in the atmosphere (est. 60 to 400 yrs), maximum atmospheric CFC concentrations are predicted to occur a decade after complete phaseout (i.e. 2010 under London revision of Montreal Protocol) (15). Meanwhile, the drop in O_3 and resultant rise in harmful ultraviolet (UV) radiation (Section 5-B) is expected to increase frequency of skin cancer and eye cataracts, and partial suppression of the human immune system, an effect similar to that of the AIDS virus (15). Ironically, the increased UV radiation reaching the ground accelerates the rate of ground

Table 6-1 Major Anthropogenic Gaseous Emissions into the Atmosphere.

Gas	Major Sources[1]	Emissions per Yr.[2] (Anthro./Total)	Residence[3] Time	Concentration 1889	1989	(PPB)[4] 2030
CO	Fossil Fuels, Biomass Burn	700/2,000	Months	?[5]	1-200[5]	Incr[5]
CO_2	Fossil Fuels, Deforestation	5,500/~5,500	100 Yrs	290 ppm	350 ppm	>400 ppm
CH_4	Fuel Production, Rice Fields, Landfill, Feedlot	300-400/550	10 Years	900	1700	>2200
NO_x	Fossil Fuels, Biomass Burn	20-30/30-50	Days	CLEAN	.001 to 50	.001 to 50
N_2O	Fertilizers, Deforestation, Biomass Burn	6/25	170 Years	285	310	330-350
SO_2	Fossil Fuels, Ore Smelting	100-130/150-200	Days-Wks.	.03-?	.03-50	.03-50
CFC	Aerosol Sprays, Refrigerants, Foams	~1/1	60-100 Yrs	0	~3	2.4-6

Source: Adapted from "The Changing Atmosphere" by T.E. Graedel and P.J. Crutzen. *Scient. Amer.* 261(3): 58-68. Copyright © 1989. by Scientific American, Inc. All rights reserved.
[1] Mention of "Fossil Fuels" refers to their combustion.
[2] Emissions in millions of tons (Divide by 1000 to get trillions of kg.)
[3] Average Residence Time of a given molecule in the atmosphere.
[4] Concentration is in parts per billion (PPB) unless otherwise noted.
[5] For Northern Hemisphere; INC = Increasing; CLEAN = until Industrial.

level O_3 production facilitated by the presence of NO_x gases and hydrocarbons in urban areas. Ozone is an irritant of the respiratory system, and has other negative effects (8).

6-E THE NITROGEN CYCLE

N_2 is abundant and unreactive

Nitrogen gas (N^2) occupies 79% of the gaseous composition of the atmosphere, the principal reservoir of the nitrogen cycle. This diatomic molecule contains triple covalent bonds and is very stable and unreactive. However, N_2 can be converted to nitrate (NO_3^-) or to ammonia (NH_3) if large amounts of energy are supplied under proper conditions. Autotrophs such as lawn and garden plants utilize nitrate and ammonia as sources of nitrogen.

atmospheric reactions

Nitrogen gas is converted to nitrates and ammonia by either atmospheric reactions or by biological fixation. Atmospheric lightning discharges or cosmic rays release enough energy to convert N_2 gas into nitrates and ammonia (Figure 6-6). Volcanic gases release nitrogen oxides into the atmosphere which are converted to nitrates.

biological N_2 fixation

The second means of N_2 conversion is **biological nitrogen fixation**. The reaction is summarized as follows,

$$N_2 + 3H_2 + \text{energy} \longrightarrow 2NH_3$$

and occurs in living cells of certain species of bacteria, or cyanobacteria (Section 9-B). Both groups are prokaryotic, and members of Kingdom Monera (Section 3-C). Some "N_2-fixers" are free-living species. Others such as the bacterial genus *Rhizobium* live in a **symbiotic** relationship with autotrophs. The relationship is symbiotic because neither partner can accomplish individually what the two can accomplish together. *Rhizobium* bacteria infect the roots of legume plants (e.g. clover, bean, alfalfa) and form swellings

symbiotic N_2 fixation

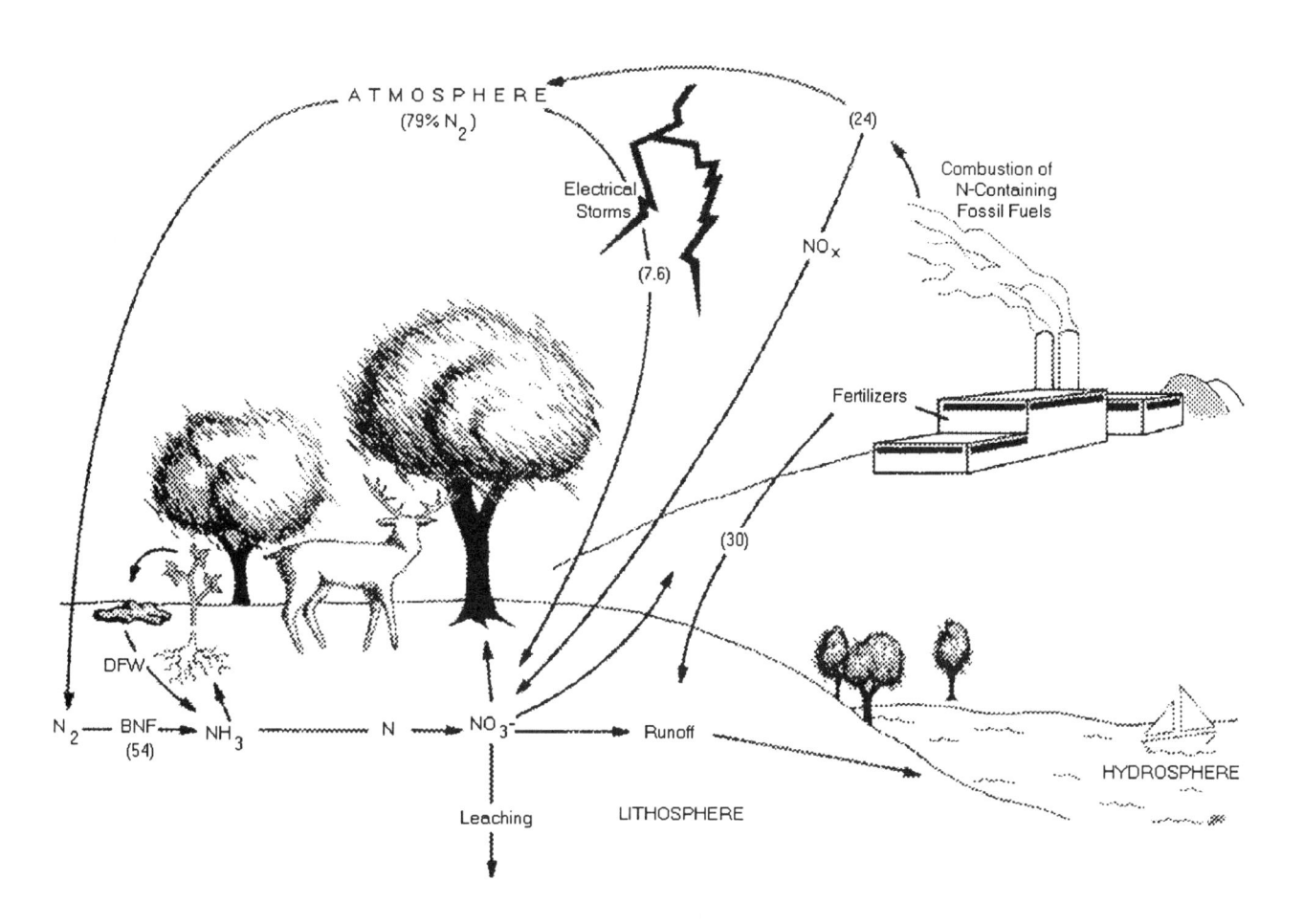

Figure 6-6. A simplified diagram of the nitrogen cycle. Nitrogen is made available to biotic communities by four processes indicated in the downward pointing arrows as described in the text. BNF = biological nitrogen fixation; N = nitrifying bacteria; D = decomposition of N-containing organic matter. Numbers in parentheses are billions (10^9) of kg. of nitrogen per year. Data from Kormondy (9).

(nodules) within which they reproduce and synthesize NH_3 from atmospheric N_2. The energy source, for their metabolism, including N_2-fixation, is the *host* plant in which the bacteria reside. Sugars from photosynthesis are transported downward to the roots and nodules. The NH_3 is produced in excess of the needs of the *Rhizobium* bacteria. Therefore, the legume host as well as other plants of the biotic community become benefactors.

Autotrophs form organic nitrogen.

Having discussed the *reservoir* and modes of *entry* of nitrogen, it should be evident that, as with carbon, so nitrogen must be converted to usable form by autotrophs. Autotrophs convert NO_3 to NH_3, and then combine NH_3 with carbon compounds to form amino acids, proteins, and nucleic acids (Sections 4-I and 4-J). Heterotrophs generally meet their nitrogen demands through ingestion of these organic compounds. Dead organisms and refuse are decomposed, and ammonia is released where it can be reused by autotrophs.

other bacterial roles in the N_2 cycle

An outstanding feature of the nitrogen cycle is the key importance of different kinds of bacteria. We have already mentioned symbiotic bacteria and decomposers. A third group, **nitrifying bacteria**, are chemosynthetic autotrophs (Section 5-D). Nitrifying bacteria utilize energy released in the conversion of NH_3 to NO_3^- (Figure 6-6). A fourth group, called **denitrifying bacteria**, convert nitrates back into N_2, thus rendering nitrogen unusable, except by reentry through atmospheric reactions or biological fixation.

Compared to NH_3, NO_3^- is highly mobile, and is readily transported in runoff water from lawns and fields, or leached to points too deep in the soil for plant roots (Figure 6-6).

Human sources of nitrogen:

An understanding of the N_2 cycle model helps us to understand how humans are modifying the biosphere with regard to nitrogen. Two practices which are literally fertilizing the biosphere with excess nitrogen are as follows:

combustion

1. Automobile exhaust, and industrial combustion of fossil fuels, wood, and other biomass release nitrogen oxides, NO_X. Unlike the CO_X's (CO_2 and CO) which are also released in combustion, NO_X's dissolve readily in water droplets and form nitric acid, a component of *acid rain* (Table 6-1). Acid rain carries nitrates into soil and waterways.

fertilizers

2. Commercial fertilizer production combines atmospheric N_2 and H_2 at great energy cost. Farmers use anhydrous ammonia and nitrate fertilizers. Lawn and garden stores also market N fertilizers.

Destinations of nitrogen:

What happens to the added nitrogen from the above human sources? Consider the following destinations:

runoff

1. **LATERALLY:** Nitrates in runoff from fields and lawns, combined with NO_3 in acid rain can disrupt aquatic ecosystems. The Chesapeake Bay is a prime example of an N-enriched system (Section 6-G, eutrophication).

leaching

2. **DOWNWARD:** High fertilizer applications in some agricultural areas have lead to nitrate leaching into underground water supplies. Pregnant women who drink such water may give birth to babies whose blood hemoglobin cannot carry adequate O_2; hence, the term "blue baby".

release into the atmosphere

3. **UPWARD:** Bacteria act upon fertilizer and livestock waste to release N_2O, nitrous oxide, a *greenhouse* gas (Table 6-1).

biological fixation yields benefits

To reduce some fertilizer costs, farmers commonly use *Rhizobium*-inoculated legume seed to encourage this essentially free source of NH_3 for crop growth. Each year, lightning storms and biological fixation account for the entry of approximately 7.6 billion and 54 billion kilograms of N respectively, into usable nitrate and ammonia (Figure 6-6). Many developing nations are unable to afford to produce or purchase fertilizers to support their food production efforts. For this reason, plants such as the winged-bean, lentils, and other tropical legumes are of great importance because they provide a high-protein diet supported by symbiotic nitrogen fixation.

6-F THE SULFUR CYCLE

reservoir in lithosphere

One might discuss sulfur cycling along with cycling of "mineral elements" (in Section 6-G). Like minerals, the principle sulfur reservoir is sedimentary rock of the lithosphere. Sulfates (SO_4^{2-}) are released into soil water from weathering rock as dissolved ions. Hydrogen sulfide (H_2S) gas is released into the atmosphere from volcanoes and from decay of organic matter. Sulfur dioxide gas enters the atmosphere from volcanoes and from fossil fuel combustion. Sulfur gases, like nitrogen gases, are quite soluble and are brought to Earth, rather than accumulating in the atmosphere as does CO_2 (See Table 6-1).

autotrophs convert S to organic form

Autotrophs require sulfates which they absorb through roots or leaves as a source of sulfur for protein synthesis. Heterotrophs acquire their sulfur as part of protein intake in food. Decomposers break the carbon-sulfur bonds of organic matter, and release sulfur as hydrogen sulfide and sulfates.

Sulfur oxides ...

Combustion of sulfur-containing fossil fuels for power and heat generation, and smelting of ores are responsible for as much as 99% of the sulfur dioxide released into the atmosphere (11). High-sulfur coal, a poor economy, and less regard for a clean environment have hampered progress in nations of the former communist bloc. On the bright side, much progress is being made by western nations and Japan in reduction of sulfur oxides.

... produce acid deposition ...

Like NO_X gases, sulfur oxide gases (SO_X — e.g. SO_2, SO_3) can dissolve in water droplets in the air and form sulfates (SO_4^{2-}) and sulfuric acid (H_2SO_4). The acidity of *acid rain* is due largely to H_2SO_4 and the nitrogen counterpart, HNO_3 (Section 6-E). Dry sulfate and nitrate salts form crystals which leave the atmosphere as particulate matter. **Acid deposition** is the combined descent of *gaseous, wet* (acid rain), and *dry* acidic or acid-forming substances to Earth. Theoretically, "pure rainfall" has a pH of 5.6, slightly more acid than milk or saliva (Section 4-E). However, rainfall in most areas of the United States, Canada, and Europe has fallen into the pH 3 to 5 range.

... possible causing forest decline ...

... and aquatic disruptions.

Acid deposition on leaves and needles of forest trees is believed to contribute to the decline and death of forests in the Appalachian Mountains, in New England, and in Europe. Among European nations, an average of 35% of forests are damaged (7). Apparently, when acid deposition lowers soil pH below 6, infiltration water leaches sulfates, calcium, magnesium, aluminum, and other ions away from plant roots and into underground drinking water, and into aquatic ecosystems. Acid deposition in aquatic ecosystems has been implicated in lowered pH and death of aquatic populations. Areas with limestone bedrock have greater stability against acid deposition because buffering provides homeostatic control of pH within the neutral range (Section 4-E).

6-G MINERAL CYCLES

Each mineral element, as defined in Section 6-B, has its own cycle—e.g. the phosphorus cycle, calcium cycle, etc. However, the similarities among these justify our considering them together in a composite model (Figure 6-7).

no significant atmospheric link

With the exception of the atmospheric transmission of wind-blown dust or mists, mineral cycles are restricted to the lithosphere and hydrosphere. The principal reservoir is the soil from which dissolved mineral ions enter the food web, either at the autotrophic level (e.g. via plant roots) or heterotrophic level. Examples of the latter case are birds foraging for minerals along roadsides, and humans taking vitamins with mineral supplements. With the exception of phosphate, the minerals are not covalently bonded to organic molecules. Rather they exist as dissolved salts in cells or ironically attached to large protein molecules. Thus, decomposition of organic matter readily releases these ionic forms back into the soil. Phosphate which is a component of bones and teeth often cycles more slowly back to the reservoir because these structures are relatively resistant to weathering and decay.

Minerals remain in ashes . . .

The mineral elements cannot escape into the atmosphere when organic matter is burned. Instead, they remain in the ash residue and dissolve in the soil water and may be absorbed by plant roots. The absence of a significant atmospheric link makes losses of minerals in runoff from land ecosystems more serious than other nutrient losses. Whereas other nutrients such as carbon and nitrogen can return to land via the atmosphere, soluble mineral elements tend to remain in ocean-bottom sediments indefinitely. The so-called 'slash-and-burn' agriculture of the tropics uses controlled burning to release mineral elements for crop growth. Minerals are otherwise very scarce in tropical soils, having been so thoroughly taken up by the lush vegetation. In mechanized agriculture, phosphate fertilizer is added along with nitrogen

and stimulate plant growth

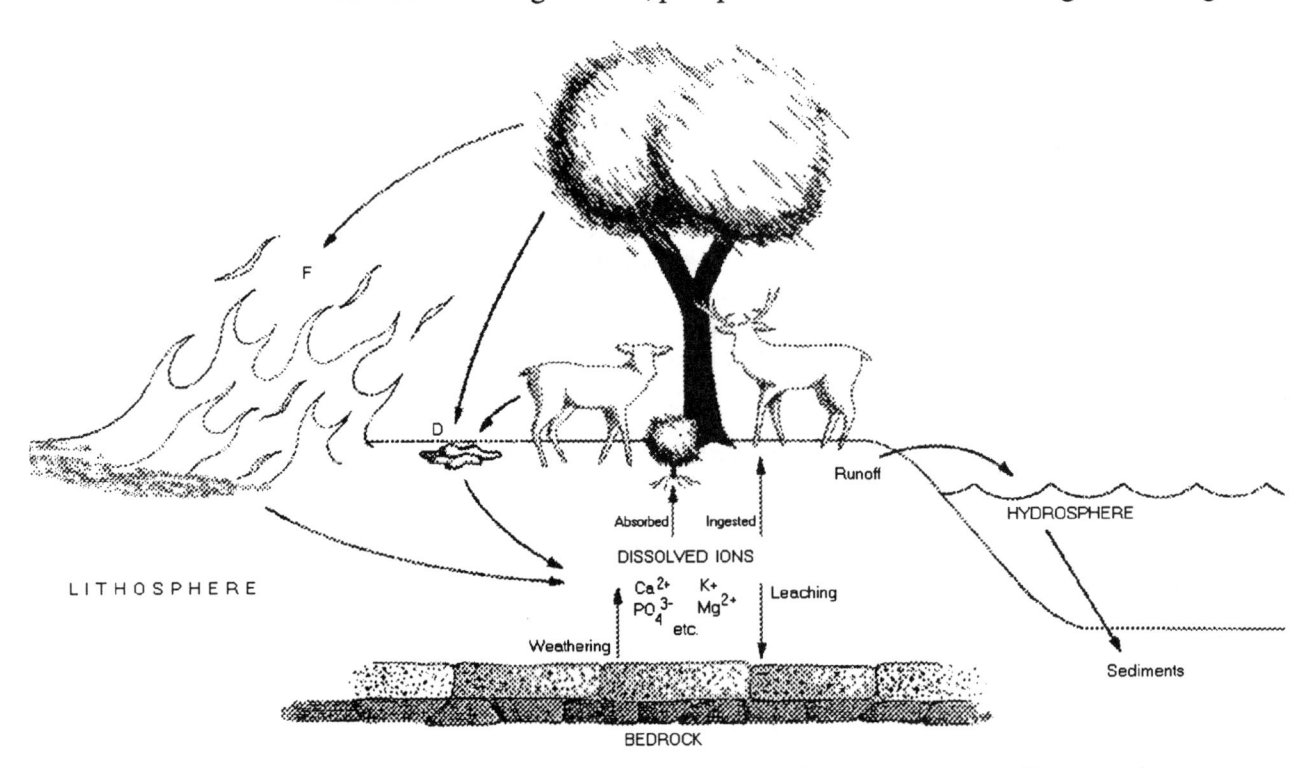

Figure 6-7. A composite diagram of mineral cycles. See text for explanation.

sources of minerals

and potassium fertilizers. Phosphate and potassium are among the fertilizers produced by mining bedrock containing these elements. Natural weathering (Figure 6-7) occurs too slowly to supply enough phosphate and other minerals to meet the needs of high-yield, mechanized agriculture in western nations.

over-fertilized aquatic systems

Surface runoff water from farm fields, feedlots where cattle manure accumulates, and from urban areas (e.g. phosphate detergents) is often rich in phosphates and nitrates. When such water reaches lakes and rivers, it causes nutrient enrichment, which in turn increases net primary productivity of algae. This process, called **eutrophication**, is often recognized by excessive growth of algae and other aquatic autotrophs. In severe cases, mats of algae on the surface may block sunlight penetration, and heterotrophs may die because of lack of oxygen. [How can excess algae which blocks light penetration lower O_2 levels?]

6-H TOXIC SUBSTANCES IN NUTRIENT CYCLES

Toxic substances include...

Isn't it amazing that the biotic community can selectively acquire about 20 out of 92 naturally occurring elements from the soil, air, and water, and incorporate them into living tissue? Even though living organisms are selective, life processes can be threatened when exposed to certain levels of chemical "look-alike's" or toxic nonessential elements. Let's consider the effects of several classes of toxic substances.

isotopes...

The release of radioactive isotopes (Appendix A) of zinc, strontium, cesium, iodine, and phosphorus into the biosphere as a by-product of atomic nuclear reactions and weapons testing is one case in point. It is now known that strontium-90 and cesium-137 can enter the atmosphere and be brought to the earth in precipitation. These isotopes are capable of releasing harmful ionizing radiation (Section 5-B) such as *beta* and *gamma* rays which may destroy covalent bonds in organic molecules. Ecologically they behave as calcium "look-alike's" in the mineral cycle. Because of isotopic similarity, the living system is unable to distinguish the harmful isotope from the non-radioactive counterpart. Once in the food web, they become concentrated within tissues to levels far above that in the surrounding environment. This, in effect, turns the tissues of living organisms into local radiation fields. The destructive effects of these isotopes when released from a nuclear reactor meltdown, or from radioactive dump sites, is a major reason for the public opposition to further development of nuclear energy.

...and heavy metals.

Mercury, lead, and other *heavy metals* are common by-products of industrial processes. These are also known to move in biogeochemical cycles and be concentrated in the food chain. The chlorinated hydrocarbon pesticides such as DDT, and the industrial ingredient known as polychlorinated biphenyl (PCB) are synthetic compounds that become concentrated in the fatty tissues of animals. Each trophic level digests and metabolizes a great quantity of the tissues of the trophic level below it (Section 5-E). Thus, if the toxic chemical is retained in body tissues rather than excreted, the concentration of that chemical will be multiplied as it moves from one trophic level to the next. This phenomenon, known as **biological magnification**, or *biological amplification* is illustrated by the data in Table 6-2. The experimental food chain features predatory insects, called lace-wings, which feed on plant sap-sucking aphids.

Table 6-2. Biological Magnification of Mercury in a Tomato-Aphid-Lacewing Food Chain, on Day Six Following Addition of 0.006 mg/liter Mercury to the Nutrient Solution.

	Mercury Concentration (mg/liter)	Magnification Factor
LACEWING	25.07	4,178
APHID	9.15	1,525
TOMATO	1.04	173
Nutrient Solution	0.006	1

*Source: A. Haney and R.L. Lipsey. 1973. *Environ. Pollution* 5:305-316.

6-1 HUMANS AND CYCLING OF MATTER

In this chapter, we have focused upon the cycling of nutrient elements between the physical environment and biotic communities. The human population has many nutritional requirements in common with other heterotrophs. Together they are dependent upon autotrophs for their food. This "food" consists of organic molecules containing *carbon, nitrogen, sulfur,* and *phosphates,* all of which must be assimilated by autotrophs prior to heterotrophic usage. Autotrophs use light energy to drive these conversions, and the resultant "food" is rich in usable chemical bond energy. As *both* autotrophs and heterotrophs use these food molecules to grow and maintain life processes, energy continually escapes as heat, never to be recycled. However, the nutrients such as CO_2, H_2O, NH_3, etc. which also escape to the soil and air, are recyclable via the autotrophs again.

Humans are different ...

Whereas nutritional needs of humans resemble those of other heterotrophs, we are different in two important ways:

in resource demands ...

1. Humans have developed agricultural and industrial systems which make *higher energy and material demands* upon the creation than other creatures. Such societies depend not only upon the solar energy-driven food web, but also depend upon fossil-fuel driven machines, petrochemical products, and fertilizers which alter the landscape, oceans, and atmosphere. These impacts, viewed through the nutrient cycle models, are summarized in Table 6-3.

and in forming nonrecyclables.

2. Humans are unique in creating *waste material that cannot be recycled* by natural biological processes of the detritus food web (e.g. polyethylene, polystyrene). As evident from Table 6-3, *acquisition and processing* as well as *use and disposal* cause ecological disruption.

In spite of these uniquenesses of human endeavor, and in spite of our apparent ability to control and dominate nature by our energy and machines, we must

Table 6-3. Summary of Human Impact Upon the Cycling of Nutrients.

CYCLE	HUMAN ACTIVITY	GLOBAL IMPACT
WATER CYCLE	Deforestation; wetland drainage; cultivation	Soil losses by erosion 24 billion tons/yr. (4)
		Forest, wetland loss
	Water usage	Surface and underground contamination; aquifer depletion; salinization
CARBON CYCLE	Coal mining, oil production	Disruption of landscapes, water tables; oil spills
	Fossil fuel combustion —e.g. industry, auto, homes	Increased CO_2 alters global climate and NPP
	Deforestation	Promotes CO_2 release by fire, increased decay; lowers NPP (CO_2 uptake)
	Chlorofluorocarbons	Destroy O_3; greenhouse gas
SULFUR CYCLE	Fossil fuel combustion and ore smelting	Acid deposition and related effects
NITROGEN CYCLE	Synthetic fertilizers, and increased biological N_2-fixation	Nutrient enrichment of Earth's ecosystems and eutrophication
	Fossil fuel combusion	Acid deposition and related effects
MINERAL CYCLES	Agriculture	Increased fertilizer usage coupled to erosion losses; eutrophication
	Toxic substances e.g. pesticides, radioactive isotoppes, heavy metals	Cycled and concentrated through food chains; global-scale transport

We cannot avoid thermodynamics.

not forget the *thermodynamic laws* (Section 5-C) which operate in the creation, and which even humans cannot sidestep. These laws may be summarized for this discussion as follows:

We cannot create or destroy, but ...

*We cannot create or destroy either **energy** or **matter**, we can only convert them from one form or another. Energy decreases in quality and cannot be recycled for use in processes that require high quality energy. As matter loses high quality energy, it becomes disorganized or dispersed. Reorganizing and reconcentrating this matter (recycling) requires great amounts of high quality energy.*

...we can and should manage.	In other words, we are not creators or destroyers of anything! We are simply managers or stewards of matter and energy, both of which tend to deteriorate. We cannot really dispose of anything without it causing problems at the deposition site or somewhere else. There is no "away"!
Distinguishing types of resources...	You may now be thinking, "if these laws are true, is there any use in trying to conserve Earth's resources?" The answer is "yes". We must distinguish natural resources into two categories—renewable and exhaustible. **Renewable resources** are those which will not be depleted in the foreseeable future *if* usage is properly managed. These include solar energy, vital gases of the atmosphere, clean water, and tillable soil. **Exhaustible resources** include raw materials and energy which are being extracted from the Earth faster than they are being formed. These include *energy resources* (e.g. fossil fuel, uranium), *mineral resources* (e.g. silica, phosphate) and *metal resources* (e.g. aluminum, iron, copper).
aids strategy to conserve.	The nature of the resource determines the conservation strategy. *Exhaustible energy resources* exist in finite supply, and are *nonrecyclable*. For example, when coal is burned, the organic molecules release heat and light energy, and the molecules are disordered and dispersed as CO and CO_2, and H_2O which enter the atmosphere. Therefore, conservation of fossil fuels must focus, not upon *recycling*, but upon *reducing* consumption and improving energy efficiency of such devices as furnaces and automobiles.

Aluminum is an example of an exhaustible resource. Table 6-4 lists annual production which has the following implications (20):

1. Mining of bauxite accounts for one-third of the energy required for aluminum production. For every ton of aluminum oxide formed from bauxite, 1 ton of caustic "red mud" is released and often pollutes the water and air.

2. In order to attract aluminum manufacturing, many countries offer lower electricity costs through subsidies paid unwittingly by taxpayers or other energy users. In 1990, 280 billion KWh of electricity were used worldwide for conversion of alumina to aluminum. Rivers are continuing to be harnessed for hydroelectric power generation in huge projects in Quebec and in Brazil's Amazon area. Large scale ecological disruption is associated with these projects.

3. It takes only 6% as much energy to produce aluminum from recycled scrap. About 25% of aluminum used in 1990 was recycled from scrap (40% of Japan's use is recycled). [What industries would lobby against aluminum recycling? Are you recycling aluminum?]

Recycling saves metals, but...

reducing usage saves metals *and* energy.

But, is aluminum *recycling* the best option to save energy and environmental degradation? As noted above, aluminum can be recycled from scrap at lower economic and ecological cost, but the *energy* required to remelt, mold, and recycle aluminum is an exhaustible resource, unless solar derived. Therefore, better conservation would call for *reducing* aluminum usage for products, such as beverage containers, with short lifetimes. For example, aluminum is increasingly used to make light-weight parts in planes, trains, and autos (140 lbs/auto), and can save more energy in fuel efficiency over a lifetime of use than that required to produce it (20). Not so when aluminum is used in manufacture of foil and beverage cans.

Table 6-4. Energy Requirements for Production and Selected Usages of Resources.

	Aluminum	Steel	Glass	Plastic
Raw Material Resource	Bauxite	Iron Ore	Silica, borates	Petroleum
Energy to Refine[1] (Million Calories/Ton)	18.8	6.5	3.5-4.5	34(11)[2]
Consumption (USA)[3] (Million Tons/yr)	4.8	—	—	25
Calories per Use of one 12-oz Container[4]:				
a) Used only once —	1750	1500	950	—
b) Recycled	670	1000	625	—
c) Reuse Ten Times	—	—	150	—

[1] Patterns of Energy Consumption in United States, 1972. Office of Sci. and Tech., Executive Office of the President, Washington, DC.
[2] Energy requirement for polyethylene and (polystyrene).
[3] Aluminum data, Reference #20; Plastic, Ref. #11.
[4] Gaines, L.L. 1981. *Energy and Materials Use in the Production and Recycling of Consumer-Goods Packaging.* Argonne National Laboratory, Argonne, IL.

Reuse *vs.* recycling

Table 6-4 reports energy requirements for processing common resources, and illustrates the energy savings possible when containers are recycled or reused. Notice that glass is most energy efficient because of lower cost to produce from silica and because glass containers can be sterilized and *reused*, a preferred strategy as opposed to *recycling* which requires more energy to crush, melt, and reshape.

Summary:

In conclusion, unlike the nonhuman members of biotic communities, we are privileged to have the technology to extract and use natural resources from the Earth. These resources may be exhaustible or renewable. Quantities of known reserves differ. The energy requirement to refine these resources also differs. Because of these differences, wise stewardship of resources should encourage us to *reduce* and *reuse* where possible. Then, instead of discarding products as solid waste when they are no longer usable, we should *recycle*.

QUESTIONS AND DISCUSSION TOPICS

1. Recalling your activities during the past week, how do you fit within each of the biogeochemical cycle models? Keep this question in mind during the next few days, and evaluate your involvements.

2. Most human practices that impact one nutrient cycle will also impact at least one other cycle. Cite three or more examples. What caution does this suggest concerning our modelling of nutrient cycling?

3. What evidence do you see of intelligent design of the Creator in the process of nutrient cycling? For example, seed and fertilizer dealers often encourage removal of lawn clippings and leaves, and suggest that fertilizer be applied in spring and autumn. How do these activities relate?

4. Budd Torcher fertilizes his garden by burning old corn stalks, autumn leaves, and grass clippings, and then spreading the ashes over the soil. Herb Mulcher mows down the dead plant material and tills it into his soil. Which will have the most recycled carbon, nitrogen, and sulfur available to their plants the following year? Explain your rationale.

5. Someone has described soil as a 'complex living community'. Attempt to justify such a description based upon your knowledge of soil as related to nutrient cycles.

6. How does the type of natural resource influence the strategy for conservation? What economic, political, and personal factors influence conservation of resources such as aluminum?

TOPICS FOR FURTHER STUDY

1. Do the benefits of using pesticides (agricultural, lawn, garden, household, etc.) outweigh the environmental and economic costs, estimated at $8,000,000,000 per year in the USA? See Pimentel, *et.al.* 1992 *Bioscience* 42(10): 750-760.

2. Examine the social justice implications of the practice of producing and marketing pesticides in less developed countries that have been banned for health reasons in the USA. See REFERENCE, Powledge (13).

3. Make a list of five different items that you purchase regularly and then dispose of (e.g. food, glass containers, metal products, etc.) For each item, what is the impact on the environment to produce it; how is it disposed of; and, what methods could be introduced to make production and disposal more ecologically compatible? See REFERENCE, Earth Works Group, (6), and Young (20).

4. Is the Earth really warming as data cited in Section 6-D may suggest? What are the global "feedback mechanisms" that some climatologists believe can oppose the upward drift in global temperature? See REFERENCE, Aufdemberge, (2)

5. What in blazes is the real effect fire on nutrient cycling and biotic communities? Inquire into the ecology of fire, and the economics of fire in a case study of the Yellowstone Fires. See REFERENCE, Romme, (14).

REFERENCES

1. Aber, J.D., R.N. Nadelhoffer, P. Steudler, and J.M. Melillo. 1989. Nitrogen saturation in northern forest ecosystems. Biosci. 39: 378-386.

2. Aufdemberge, T. 1991. Variation and fixity as seen in climatology. *Creation Res. Soc. Quart.* 28: 98-100.

3. Bormann, F.H. and G.E. Likens. 1970. The nutrient cycles of an ecosystem. *Scientific American* 223(4): 92-101.

4. Brown, L.R. 1990. The illusion of progress. In L.R. Brown, *et.al. State of the World.*, pp 4-16. Worldwatch Institute Report. W.W. Norton, Washington, DC.

5. Deevey, E.S. Jr. 1970. Mineral cycles. *Scient. Amer.* 223(3): 148-158.

6. Earth Works Group, 1990. *The Recycler's Handbook: Simple Things You Can Do*. Earth Works Press, Berkeley, CA.

7. French, H.F. 1990. Cleaning the air. In Brown, L.R. *et.al., State of the World.* pp 99-118. Worldwatch Institute Report. W.W. Norton, New York.

8. Klingsberg, C. and J. Duguid. 1982. Isolating radioactive wastes. *American Scientist* 70:182-190. This issue has other related articles.

9. Kormondy, E.J. 1984. *Concepts of Ecology*, 3rd ed. Prentice-Hall, Englewood Cliffs, NJ.

10. Maurits la Riviere, J.W. 1989. Threats to the world's water. *Scientific American* 261(3): 80-94.

11. Miller, G.T. Jr. 1992. *Living in the Environment,* 4th ed. Wadsworth, Belmont, Calif.

12. Postel, S. 1990. Saving water for agriculture. In Brown, L.R. *et.al., State of the World.* pp 39-58. Worldwatch Institute Report. W.W. Norton, New York.

13. Powledge, F. 1991. Toxic shame. *The Amicus Journal* Winter: 38-44.

14. Romme, W.H. and D.G. Despain. 1989. The yellowstone fires. *Scient. Amer.* 261(5): 37-46.

15. Rowland, F.S. 1991. Stratospheric ozone in the 21st century: The chlorofluorocarbon problem. *Environ. Sci. Technol.* 25(4): 622-628.

16. Schneider, S.H. 1989. The changing climate. *Scient. Amer.* 261(3): 70-79.

17. Shea, C.P. 1991. Disarming refrigerators. *Worldwatch* 4(3): 36.

18. Smith, R.L. 1992. *Elements of Ecology,* 3rd ed. Harper-Collins, New York.

19. *World Resources, 1992-1993* Report by World Resources Institute, 1992. Oxford University Press, New York, NY.

20. Young, J.E. 1992. Aluminum's real tab. *WorldWatch* 5(2): 26-33.

Chapter 7

Population Ecology

Each organism is linked into a *population*...

If you have ever tried to nurture an orphaned rabbit, or a bird fallen from its nest, you can appreciate that the life of every creature is a flickering candle when severed from its ties to the biotic community. Every living organism needs a continuous supply of energy and nutrients. However, these needs are met in complex ways as you may have discovered in attempting to nurture the orphaned animal. Each individual is a dependent part of a *population* (Section 2-C), and depends upon mates, family, and social groupings within the population.

...and each population is linked into a *biotic community*.

Populations of a biotic community are knit together as they interbreed, and acquire energy and nutrients. Food webs and biogeochemical cycles emphasize these community interrelationships. While no single population is totally independent of others, each population is nevertheless a unit through which energy flows and nutrients are cycled.

This chapter presents a view of life from the perspective of the *population ecologist*, whose attention is focused on the population level of organization. We shall examine the common characteristics of populations, and factors that affect population growth over time. The chapter concludes with some applications to the human population.

7-A INDIVIDUALS, POPULATIONS, AND COMMUNITIES

What is the population of Chicago? Most of us would interpret this question to mean, "What is the population of *Homo sapiens* within the geographic territory defined by the greater Chicago city limits?" However, a population ecologist would probably ask, "To which population in Chicago are you referring—humans, Norway rats, dandelions, or what?" The ecologist is accustomed to looking at the larger picture of which humans are only one part. He or she generally defines a **population** as a group of interbreeding organisms of the same kind occupying a particular space. The concept of population implies several important points which we should consider.

Populations are interbreeding organisms...

...that have an ecological role, or *niche*.

First, a population is an *ecological unit*; a structural component of the food web of a community through which energy and nutrients flow. Each population performs an important function in its community—e.g. producer, herbivore, predator, detritus feeder, etc. The function or "occupation" of a population is its ecological **niche**. Niche should be distinguished from **habitat**, which is the place of residence or "address" of a population.

...can be described numerically...

Second, a population is distinguished as a *statistical unit*. For instance, each population has a **density**, or number of individuals per unit area. The ecologist must define what an *individual* is, count the number of individuals, and divide by the area in question. The "individual" is not always easy to define as in the case of some plant species that form additional plants by sending out underground stems (e.g. certain grasses or thistles). Similarly, sponges and corals of the Animal Kingdom grow in attachment to one another. Even physically unattached individuals tend to occupy a characteristic **distribution**. They may display *uniform*, *random* (e.g. birds nesting in a meadow), or *clumped* (e.g. same bird population foraging in flocks; or grass in clumps) distribution (Figure 7-1). [Why is it critical that an ecologist consider the pattern of distribution of a population as well as its density?] As we shall see, *birth rate, death rate, immigration, and emigration* are other important statistics used to describe populations.

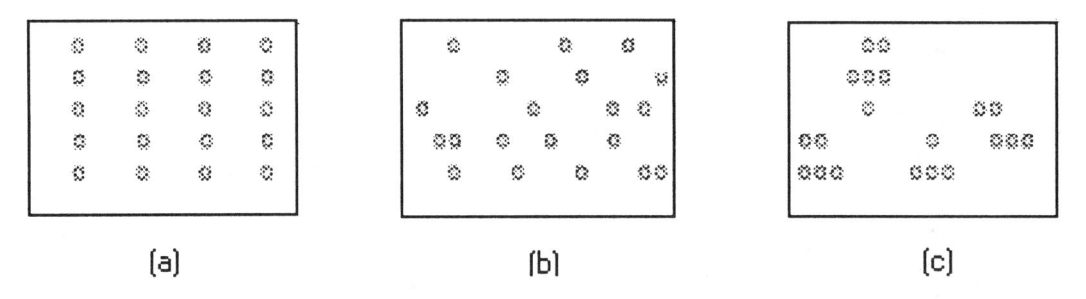

(a)   (b)   (c)

Figure 7-1. Patterns of population distribution include (a) uniform, (b) random, and (c) clumped. [If each square represents a field plot 10 meters on a side, what is the density (# individuals per square meter) of each plot sample? Can you tell which distribution makes the best use of resources?]

. . . and share one gene pool.

Finally, each population, by definition, is an interbreeding group that forms a *genetic unit*. The individuals of a population are members of one **species**, and they recognize only other members of their population as potential mates. Each member of the interbreeding group is a temporal carrier of hereditary material or genes that are shared in a so-called gene pool. The **gene pool** is the sum of all of the genetic information of the individuals of the interbreeding population. It acts as a source of genetic variation which promotes varying degrees of success in survival and reproduction among individuals of the population. Those which possess favorable **adaptations** will be most successful in survival and reproduction (Section 5-A). Thus, it is the population, not individuals, that changes genetically over successive generations under the influence of the environment.

7-B CHARACTERIZING A POPULATION

focussing in: a reductionist view

To begin this discussion, recall your experience on our "meadow community hike", in Chapter 2. We attempted to characterize the meadow as a whole system, drawn together by "threads" of flowing energy and matter. However, population biologists often take a reductionist approach (Section 1-C), and focus on the individual populations of the meadow community. One such study may attempt to characterize the meadow mouse (vole) population, interbreeding members of the species *Microtus pennsylvanicus*. Using live trapping techniques and statistical sampling, one could estimate the vole population size, density, and distribution. These data would be useful in describing the population at a given point in time. However, we would need to make repeated samplings over longer periods of time to determine patterns of population growth.

Population growth rate (*r*) is a function of birth rate (*natality*), death rate (*mortality*), and *migration* (immigration and emigration) of the population across boundaries of the community in question. Ignoring migration rates, and using a 1-year interval, the equation is as follows:

$$r = \text{birth rate} - \text{death rate}$$
$$(\text{\# births/1000}) \quad (\text{\# deaths/1000})$$

To illustrate, assume that in a population of 1,000 individuals, 650 births occurred and 150 deaths during a one-year period. Then, ignoring immigrants and emigrants,

$$r = 650/1000 - 150/1000 = (0.65 - 0.15) = 0.50$$

Because birth rate is larger than death rate, the population size at the beginning of the year (N_o) will increase by the quantity, reproductive rate (*r*) times initial population (N_o), to produce a population at the end of the interval of time of N_t, as shown below:

$$N_t = N_o + rN_o = [1000 + (.50 \times 1000)] = 1500$$

The increase of 500 individuals increases the population to 1500 at the end of one year—i.e. $N_t = N_1 = 1500$. If *r* stays at 0.50 for a second year, the population at the end of the second year (N_2) is as follows:

$$N_2 = N_1 + rN_1 = [1500 + (.50 \times 1500)] = 2250$$

Populations grow if $r > 0$.

Notice two important principles. First, as long as birth rate is greater than death rate (i.e. $r > 0$), the population will grow. Second, even if *r* stays the same over consecutive intervals, the population will increase by greater and greater increments. The increase in year #1 was 500, and for year #2 was 750. This is analogous to your earning of more interest each year as your savings grows, even though the interest rate stays the same. [Calculate the increase for year #3, assuming $r = 0.50$.]

Logarithmic growth . . .

The relationship, $N_t = N_o + rN_o$ produces increases that become greater for each successive time interval, *t*. This is called **logarithmic growth**, or exponential growth. Under favorable environmental conditions, populations tend to increase very rapidly. For example, the common housefly lays an average of 120 eggs, about half of which develop into females. When these mature, there are 60 potential pairings, each capable of producing 120 eggs, or 60 x 120 = 7,200 offspring. If you were to plot the housefly population after several more generations, it would resemble the curve in Figure 7-2. The so-called **J-shaped curve** is characteristic of logarithmic, or exponential growth.

. . . is seen as J-shaped curve,

. . . is under genetic control.

Under optimal conditions for growth, a population will theoretically express its **biotic potential**, the maximum reproductive rate. The biotic potential of a population is genetically programmed within the members of each species. If we compared the biotic potential of houseflies with elephants, we would observe that different species have quite different biotic potentials, even though both are living under optimum environmental conditions. [How would you determine the biotic potential for the meadow vole population noted above? What difficulties would you encounter?]

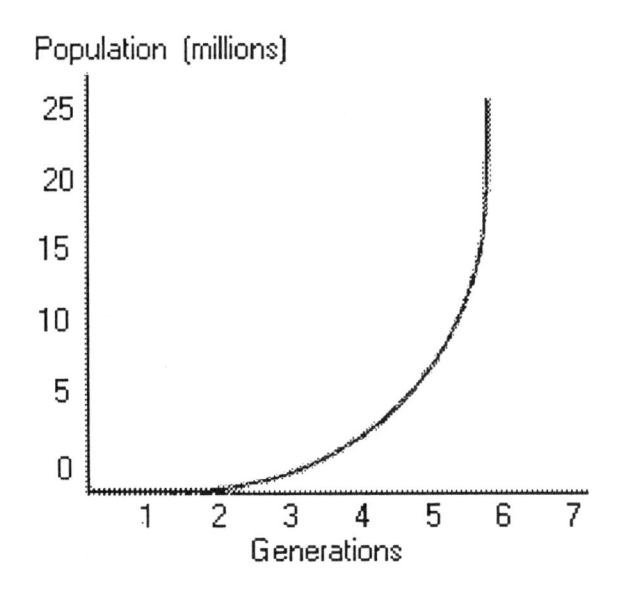

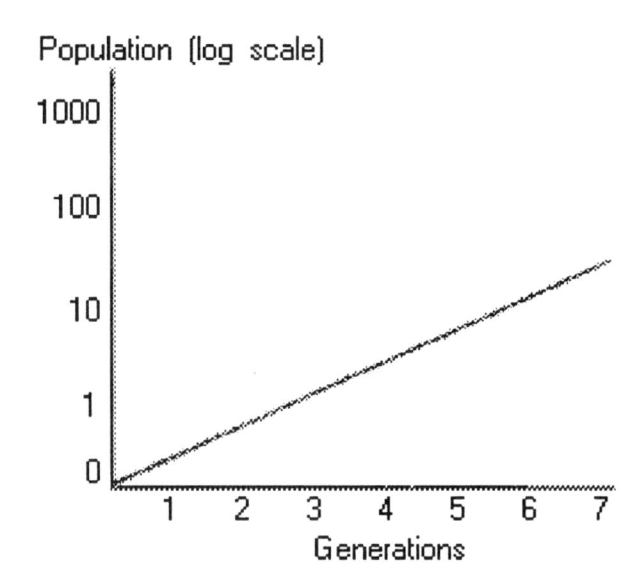

Figure 7-2. Logarithmic population growth for two hypothetical populations having different net reproductive rates (r).

What genetic factors control populations?

Table 7-1 lists several genetic factors which combine to influence reproductive rates among different species. Note that a species which reaches puberty (sexual maturity) faster, has a short gestation period, and has a large litter size will have a higher reproductive rate, other factors being equal. However, factors such as number of litters per year and length of fertility period are also significant. [How would life span influence r?]

Table 7-1. Factors That Influence Reproductive Rates in Selected Species.

SPECIES	Age at Puberty (months)	Estrus Cycle (days)	Gestation Period (months)	Avg. Life Span (months)
Homo sapiens (Human)	144	28	9	849-1380
Pan troglodytes (Chimpanzee)	120	36	8	210
Bos taurus (Cattle)	6	19	9	276
Elephus maximus (Elephant)	156	-	21	480
Oryctolagus cuniculus (Rabbit)	5.5	-	1	168
Felis catus (Cat)	15	22	2	180
Mesocricetus auratus (Golden Hamster)	2	4.5	0.5	24

[1] Source: Finch, C.E. and L. Hayflick, eds. 1977. *Handbook of the Biology of Aging.* Van Nostrand Reinhold Co. New York, NY.

Two patterns:

Recognizing that there is much diversity in reproductive mechanisms among species, ecologists have nonetheless distinguished two groups based upon their mechanisms of reproduction, competition, and survival. For our purposes, we shall call these groups "exploders" and "plodders".

... "exploders" emphasize fast growth ...

The "exploders" include such species as insects and fish among animals, and the dandelion and thistle among plants. These species produce large numbers of offspring at an early age. Short duration of pregnancy (or in plants, quick seed maturation), and a long period of fertility are other factors that favor high natality in "exploders".

... "plodders" emphasize survivorship.

The "plodders" include large mammals such as humans, and tree species such as the oaks. These become fertile at a much older age, have a longer duration of preganancy (or seed maturation), and bear fewer offspring. Therefore, comparatively lower intrinsic rates of population growth would be expected in these species. However, such species often have lower mortality rates (higher survivorship) among the offspring and a longer life span, factors that compensate somewhat for their lower natality.

We have seen that genetic factors influence biotic potential of any two populations. Two additional considerations which will determine future reproductive rates are the (a) survival rates by age group, and (b) the age distribution of members of the populations.

Survivorship curves are useful in illustrating the probability that an individual organism will reach a given age in the normal life span of a population. Greater survivorship means greater likelihood of reaching puberty, bearing offspring, and being counted in the population census in the post-reproductive period of life. Figure 7-3 illustrates survivorship curves for three general types of populations. Curves (a) and (c) represent the typical patterns of survivorship for "exploders" and "plodders", respectively. Interestingly, human survivorship data or *life tables* are widely used by life insurance companies.

Age structure shows stability and future trends.

By estimating the percentages of individuals of a meadow vole population that fall into various age categories, we could determine its **age structure** (Figure 7-4). This approach is useful in accessing the stability of a population over time. One can also predict future growth rates from the number of individuals entering, or soon to enter, the reproductive phase of life.
Figure 7-4 presents age structures of two vole populations, one expanding into a new environment, and the other, a stable population with births and deaths occurring at a similar rate. Notice the large percentage of young voles in pre-reproductive and reproductive phases of life in the unstable, expanding population.

The shape tells much.

As a general rule, wide-based, sharply tapering age structure diagrams suggest rapidly growing, unstable populations. In contrast, a slender diagram with a more uniform percentage of the total population in each age category is characteristic of stable populations. Age structure diagrams find useful application to studies of the human population in different parts of the world (Section 7-G).

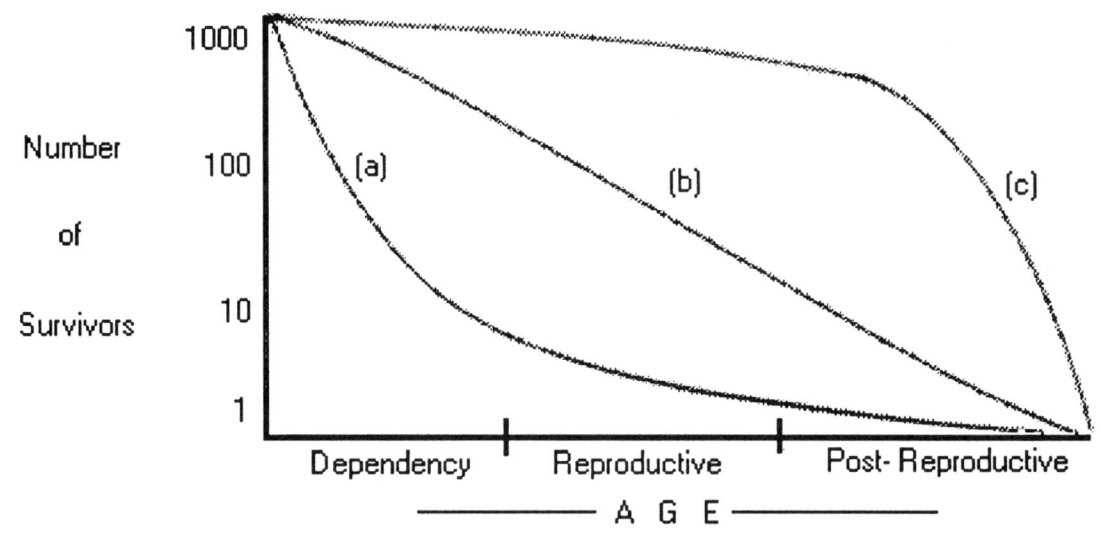

Figure 7-3. Three general types of survivorship curves based on a semi-log scale of 1000 births.

(a) High mortality rates in early ages as in many invertebrates, fish, and some annual plants that produce large numbers of offspring with low individual survivorship.
(b) Mortality rates fairly constant throughout life; typical of some insect, bird, rodent, and some perennial plant species. Human populations in less developed countries may fit curves falling between (a) and (c).
(c) Low mortality rate of young, and a tendency to live to an age near life expectancy—e.g. many large mammals including man in developed countries.

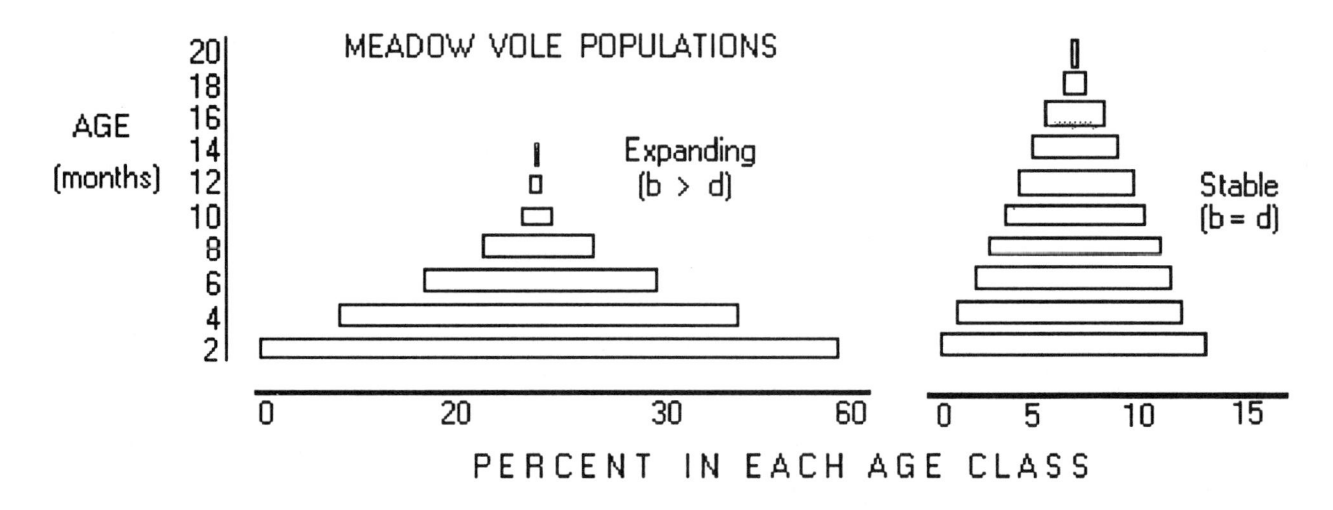

Figure 7-4. Age structure diagrams of laboratory populations of the vole, *Microtus agrestis*, when expanding at a logarithmic rate into a suitable environment (left), and when stablized at equal birth and death rates (right). Data from P.H. Leslie and R.M. Ranson (13).

7-C LIMITING FACTORS

In the natural environment, conditions that influence population growth are seldom ideal. Therefore, most populations will not be reproducing at their biotic potential, even though the rate of increase is logarithmic. As the population density increases, competition for finite resources increases. This may lead to increased mortality, decreased natality, and increased emigration from the community. Thus, what began as a logarithmic increase under more nearly optimum conditions shifts to a *plateau phase* (leveling off) as the population growth is *limited* (Figure 7-5). This explains why your house is not quickly overrun by houseflies, in spite of their high biotic potential.

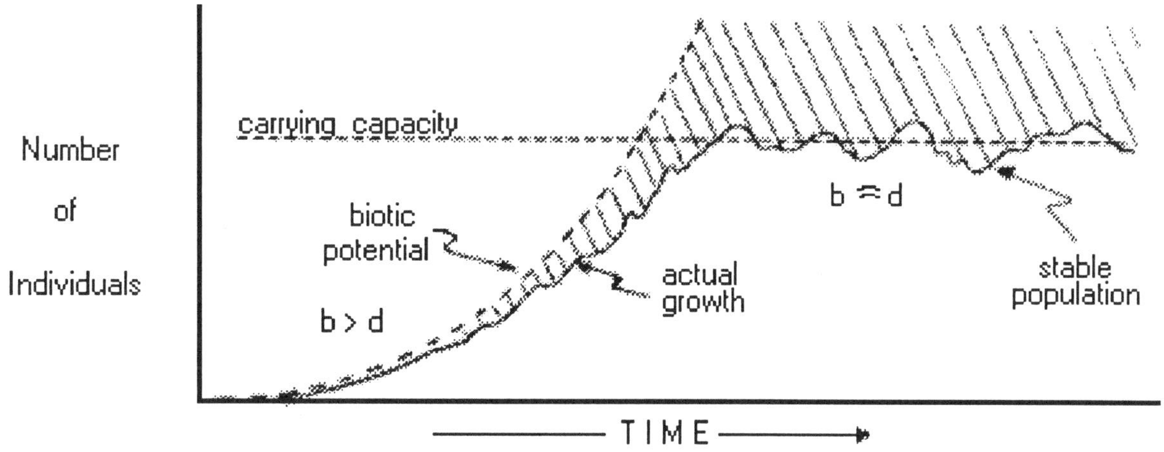

Competition leads to a plateau phase.

Figure 7-5. Actual or realized growth of a population is limited by increasing environmental resistance (shaded area) as density dependent factors become more limiting.

What does it mean for population growth to be *limited*? To answer this question, we must introduce the concept of **limiting factors**. A classic study of how environmental factors limit growth was conducted by the chemist, Justus von Liebig (1803-1873). He discovered that the growth of various crop plants in test plots could be limited by an inadequate supply of only one factor necessary for growth, such as light, water, or a single nutrient element. Thus, plants abundantly supplied with light, good soil, and nutrients, may grow poorly or die if no water is added. Water is the limiting factor in this case. Liebig manipulated the levels of various nutrients and demonstrated similar limiting effects upon plant growth. From his studies, Liebig proposed what has come to be called **Liebig's Law of the Minimum** which states: *the rate of a biological process will be limited by one factor in shortest supply*. Or, "a chain is only as strong as its weakest link". This concept is useful in both studies of individual organisms and of populations.

Liebig's Law of the Minimum

When is a given environment "full"?

Liebig's Law explains why natural populations do not continue to increase logarithmically for long (Figure 7-5). As density increases, so does consumption of environmental resources necessary for individual growth and reproduction. One limiting factor is sufficient to limit population growth. Theoretically, each environment has a **carrying capacity** (Figure 7-5), or population size of a given species that can be supported by that environment without diminishing long-term sustainability. As a population increases and approaches the carrying capacity, growth rates decrease as one or more

Carrying capacities can change

resources become limiting. This increasing resistance to population growth that is experienced as a population approaches the carrying capacity is called **environmental resistance** (Figure 7-5). The environmental resistance serves as a *negative feedback* mechanism to suppress population growth and bring it into line with the supply of available resources of the environment. The result is a *homeostatic mechanism* that operates to maintain population levels near the carrying capacity. Stable populations will fluctuate slightly above and below the carrying capacity over time as homeostatic regulation operates throughout more or less favorable seasons. Likewise, the carrying capacity will fluctuate over time depending upon environmental factors or human mismanagement. [How would you actually determine carrying capacity for a given population?]

Limiting factors that make up the environmental resistance may be of two types. **Density dependent** limiting factors influence populations in proportion to increasing density. Factors that influence populations in a manner unrelated to density are referred to as **density-independent** factors.

density-independent factors

Density-independent factors are comprised mainly of physical forces such as climatic or geologic events. Tornadoes, earthquakes, fire, extreme temperatures, or human activities such as forest clearing, pesticide application, and stream pollution are a few examples. Density-independent factors may either increase or decrease growth of a given population, depending upon its tolerance of the climatic condition involved. In each case, however, a population is affected regardless of whether density is high or low.

density-dependent factors

Living space is the most obvious density-dependent limiting factor. Animals have both physical and psychological needs for space as it pertains to nesting, shelter, and breeding. As numbers of individuals per unit area increases, population growth is often reduced through reduced fertility, reduced litter sizes, and increased mortality. Competition for light, water, and nutrients among plant populations, and competition for food among animal populations are density-dependent limits. Predation and disease also intensify as higher population density makes capture of prey more likely. Pathogenic diseases spread more readily within crowded populations. Crowding and associated stress may even cause weakened resistance to disease.

Individual limiting factors, whether density-independent or density-dependent, seldom act in isolation of one another. For instance, reduced availability of winter browse for the deer population may be caused either by increased population density or by a snowstorm that covers young tree seedlings and vegetation. Under such conditions, the deer population is more vulnerable to predation and disease. [How many interacting factors can you identify in this example?]

Populations respond to limiting factors.

Populations generally respond to environmental resistance through changes in social behavior, decreased fertility rates, reduced litter size, and increased mortality. In rabbit, deer, and fox, density-dependent stress leads to resorption of embryos in the uterus. Female armadillos exposed to a stressful environment can extend their gestation periods for up to two years by holding embryos dormant but viable prior to implantation within the womb. The result is homeostatic control of numbers near the carrying capacity. Mass starvation, general stunting, and anemia are rare in nature apart from major environmental disruptions such as the density-independent factors noted earlier. The next section surveys some of the intricate homeostatic mechanisms that regulate population levels in relationship to food supply.

7-D POPULATIONS AND NATURAL SELECTION

Two opposing forces...

Populations are theoretically maintained near the carrying capacity by the action of two opposing forces:

a. The tendency of populations to increase logarithmically in accordance with the *biotic potential*.

b. The tendency of *environmental resistance* to increase as the population approaches the carrying capacity.

...act upon populations...

In the face of these opposing forces, some organisms will die, some will survive, and others will be infertile or unable to find mates. What determines the outcome for a given individual? To answer this, we must first recall that individuals of a population are usually different from each other. This is part of the marvelous diversity that some call "the spice of life". Humankind and other species are composed of individuals that differ in morphological, physiological, and behavioral traits. The outward expressions (color, size, etc.) of these traits are called **phenotypes**. Those individuals that have phenotypes that aid in their **fitness** to obtain space, food, and nutrients; and, to acquire mates, avoid predators or grazers, and nurture offspring will be more likely to survive and reproduce. Note that two conditions are at work here, and they lead to a specific outcome, as follows:

and favor "fit" phenotypes.

GENETIC VARIATION (among members)	+	ENVIRONMENTAL RESISTANCE (phenotypes are put to the test)	→	SURVIVAL/REPRODUCTION OF ORGANISMS EXPRESSING SUITABLE PHENOTYPES (*NATURAL SELECTION*)

Populations change through natural selection.

The survival and reproduction of individuals possessing phenotypes that suit them to their environment is called **natural selection**. Because the successful competitors can survive and produce offspring, the genes for their successful phenotypes will be retained within the gene pool for the next generation. But, natural selection is continually eliminating individual units (organisms) with unfavorable traits, so that the population, as an interbreeding *genetic unit* (Section 7-A), is conserved. As a unit, the population is regulated *quantitatively* (numerically adjusted around the carrying capacity), and *qualitatively* (genetically tested to favor suitable phenotypes). Therefore, in natural selection we see the interaction between environment and population in a type of **homeostatic** feedback mechanism that conserves populations as genetic units by keeping them on track with the often changing environment. Chapter 15 continues this discussion in the context of the origin of life.

7-E INTERACTIONS WITHIN AND AMONG POPULATIONS

The previous discussion of population growth and limiting factors has been very general. Our purpose has been twofold. First, to glean some major principles and concepts that will be necessary for our future studies; and second, to provide a framework of topics which can be easily pursued in depth by those interested. In this section, we present several additional concepts with the same purpose in mind.

two types of interactions

Ecology emphasizes that no population is an "island" unto itself. Two major classes of interactions are known to influence population size and fitness—interactions *within populations* and interactions *between populations*. First, interactions obviously will occur among individuals of the same population. These so-called **intraspecific** interactions are expressed in the social behavior of animals as varying degrees of tolerance between individuals of the same species. Social behavior acts as a homeostatic mechanism to prevent population density from departing too far above or below the carrying capacity (Figure 7-5). Such behavioral conventions limit the number of animals that can occupy a given habitat, utilize its food supply, and engage in reproductive activity. We have already discussed competition as a density-dependent regulator of population growth.

Territoriality limits density.

Territoriality is a complex intraspecific interaction in which individuals defend a region, used for feeding, mating, or nesting, against other individuals. This behavior sets limits on population density and growth by a structured partitioning of environmental space and resources among members of a population. Territoriality is common in some insects and fish, most birds, and many mammals. Mammals, including deer, rabbits and gerbils, secrete chemicals called **pheromones** from special scent glands, and these are rubbed onto objects to mark territorial boundaries. Other species will bodily defend their territory. Still other species, including chickens, display a social hierarchy within their populations that involves allocation of territory, food, and access to mates. Social insects such as bees and ants display a complex social structure within colonies.

predation and parasitism

Predation and parasitism as density-dependent limiting factors, are instances of **interspecific** interactions. Here, one population is deriving its nourishment at the expense of another, the prey and host, respectively. Parasitism is distinguished from predation in that members of the two populations live in intimate contact; the one is smaller (the parasite) and consumes the other (the host) from within (e.g. bacteria, parasitic worms) or from without (e.g. lice, ticks). When the parasitic activity disrupts the homeostatic conditions of the host, **disease** becomes evident. [Is the most "successful" parasite the one that causes the most devastating disease? Explain.]

We don't fully understand interactions.

Various types of interspecific interactions are outlined in Table 7-2. As we reflect upon the myriad of types of interactions and relate them to the concepts of food web and nutrient cycles, we can appreciate the complexity of life at the population and community levels. Indeed, population ecologists have many unanswered questions relating to competition and its relationship to population control and natural selection. [What questions come to your mind as you consider these interactions?]

Table 7-2. Types of Interspecific Interactions and Common Examples of Each.

TYPE	SPECIES A	SPECIES B	COMMON EXAMPLE
Mutualism	+[1]	+	Symbolic N_2 fixation; lichens
Parasitism	+	-	Species A is parasite, B is host
Predation	+	-	Species A is predator, B is prey
Commensalism	+	0	Tree-nesting animals; epiphytic orchids
Amensalism	0	-	Many plants and soil microbes secrete chemicals that inhibit other species
Competition	-	-	Mutually suppressed growth as two species draw upon scarce resources

[1] "+" means the relationship improves fitness for this species
"-" means the relationship lowers fitness for this species
"0" indicates no known effect on the fitness of either species

7-F ECOLOGICAL SUCCESSION

"YELLOWSTONE IS BURNING!" Headlines and news coverage reported the seemingly tragic destruction of the beloved national park in the summer of 1988. How could this national gem, revered as an unchanging landmark by so many generations, now be at the edge of destruction? The answer lies in the realization that Yellowstone *had* been changing all along, though imperceptibly to many. The result was a gradual buildup of flammable detritus which, in the presence of a dry summer and high winds, fueled the fires (2).

Biotic communities change...

...in spite of human actions.

Yellowstone has seen multiple cycles of fire and recovery, operating by natural ecological principles. Widespread fires in the 1700's had produced even-age stands of lodgepole pine, a species that cannot reproduce without the heat of fire to release seeds from its cones. Then, from the 1940's to the 1970's, fire suppression policies began to protect what had become large areas of old-growth lodgepole forests. If small fires had been permitted to burn, a "patchwork" or mosaic of different-age forests would have been maintained, each with differing age and degree of flammability. In 1988, these scattered patches of younger forests, born out of previous fires, would have served as "natural fire-breaks" to confine the fire and further propogate the patchwork. But this stabilizing feature was absent.

Diversity causes stability.

Change occurs: *ecological succession.*

When the smoke of Yellowstone cleared, and the winter had passed, the spring of 1989 proclaimed that natural processes were indeed operating in predictable fashion. If ecologists could have written headlines for a spring day in 1989, they may have written — "FIRES ARE NATURAL PART OF YELLOWSTONE FORESTS". They may have added a caption stating— "Come See Rebirth of the Forest Community". This process, called **ecological succession**, is the gradual, progressive replacement of one community by another until a relatively stable, self-sustaining **climax community** is established. Succession usually operates on a scale of decades and centuries. For example, many of the Yellowstone forests have been "returned" to the conditions of the 1700's! These forests are now in a succession of changes that will not restore the 1988 forest communities until at least the 2080's (2).

Pioneers: first arrivals

Succession gives us a unique opportunity to observe populations uniting to form different communities. Germinating out of the charred earth is a succession of plant populations well adapted to barren soil conditions, and extremes in temperature and moisture. These are the **pioneer** species, able to thrive on soil enriched in mineral elements now recycled to freedom in the ashes of the burned forest (Section 6-G). The fires had set nitrogen free as well, but as NO_x gases, leaving behind soil low in nitrates (Section 6-E). Nitrogen-poor soil gives a competitive advantage to some pioneer species, especially legumes that reach for the atmosphere for their nitrogen via symbiotic N_2 fixation (Section 6-E). These "N_2-pumpers" restore soil fertility.

Pioneers are "exploders"...

Many pioneer plants are "exploders" (Section 7-B), producing large numbers of seeds, thus favoring rapid colonization of the area. Any farmer or gardener can testify to the success of such pioneers as ragweed, crabgrass, and dandelion. These small plants provide food and shelter for small rodents and birds. Detritus from these species attracts members of the detritus food chain—earthworms, millipedes, micobes, etc. Thus, the pioneer community is very effective in stabilizing barren soil against erosion, and building up the soil fertility. Ecological succession that begins with bare soil is called **secondary succession**.

and "plodders" come later.

Thanks to the pioneer community, a more favorable environment soon makes it possible for survival of species that formerly could not tolerate the harsh conditions. These newcomers include woody shrubs and tree seedlings which grow taller each year and compete successfully with the small pioneer plants for water, nutrients, and light. Populations of larger wildlife species find food and shelter as well. These larger, longer-lived, competitive species are often the "plodders" (Section 7-B).

Summary:

In summary, secondary succession leads to a climax community through a series of communities, each composed of populations that remove would-be limiting factors for succeeding populations. The succeeding populations then alter the environment so as to competitively exclude preceding populations. Only the populations of the climax community have the capacity to reproduce themselves over subsequent generations. [When a large tree falls, opening up soil space for light-loving species, is the forest still a climax forest?]

Humans cause disturbances

and succession restores.

The most common human-derived sources of disruption are associated with agricultural activities and urban landscaping. As we noted in Section 5-G, the strategy of modern agriculture is to replace a natural community, having a complex food web, many populations, and relative stability, with a simple agricultural system or monoculture. To create and sustain monocultures and manicured lawns, great amounts of energy or "human pressure" must be applied to maintain these very unstable communities. As soon as the human pressure is withdrawn, succession becomes noticeable as the abandoned field or unmown lawn takes on a more disordered appearance. Yellowstone forests appear to have overthrown the attempts of humans to order and preserve. The result is a landscape with diversity of forest communities at different stages of succession, and more stable against the onslaught of future fires. In contrast, we should note that some global biomes such as the tropical forests may respond unfavorably to fire (Chapter 8).

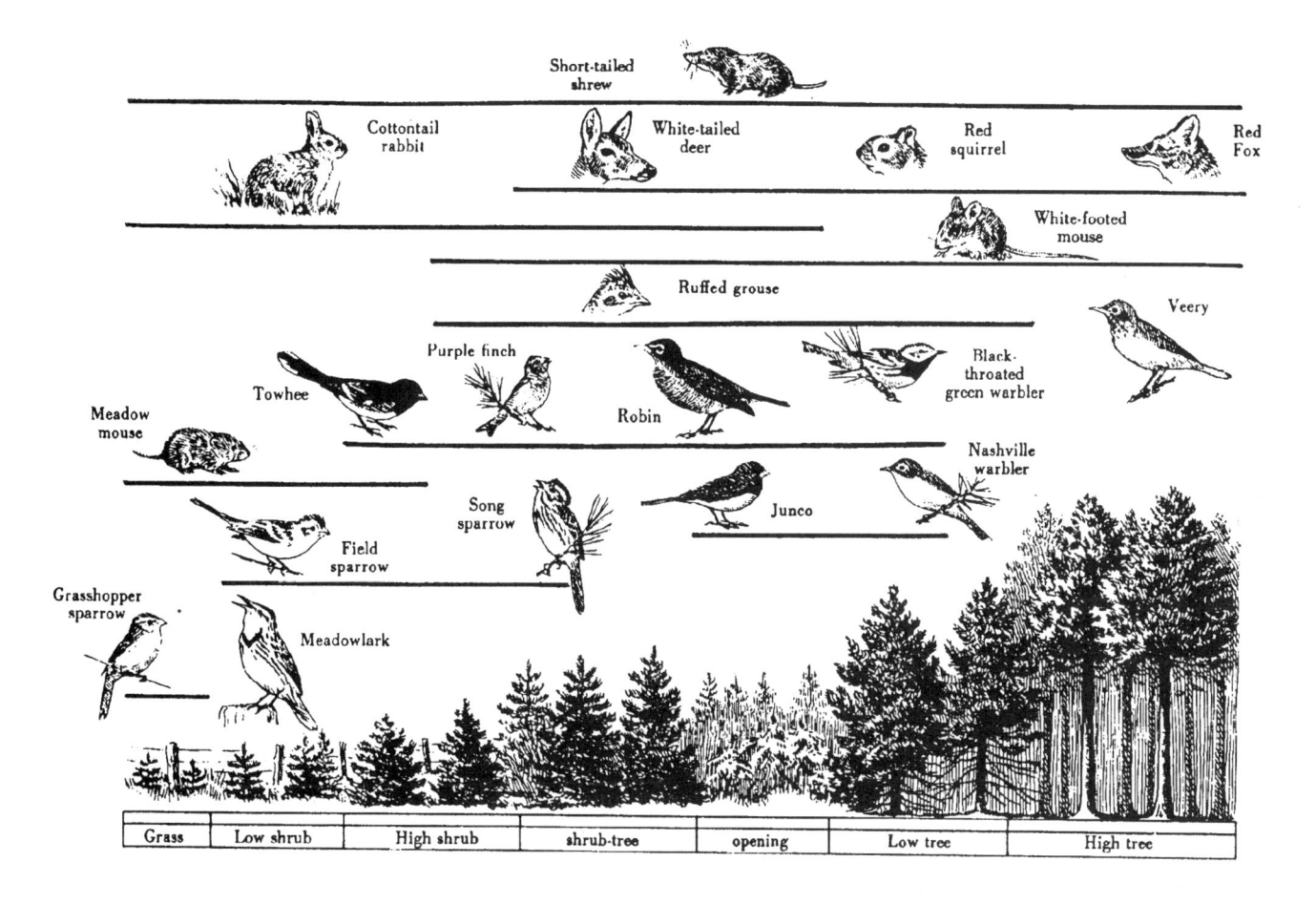

Figure 7-6. Secondary succession in Eastern North America.
Figure from ELEMENTS OF ECOLOGY, 3rd Ed. by Robert Leo Smith.
Copyright © 1992 by HarperCollins Publishers. Reprinted by permission of
Harper-Collins Publishers Inc.

Primary succession: life comes to bare rocks.

A type of ecological succession, referred to as **primary succession**, occurs when a biotic community is established on barren rock or volcanic ash where no life had previously existed. This is a very harsh environment with no soil to hold moisture, and often extreme day-night temperature differences exist. Pioneers of primary succession include the **lichens** which are symbiotic associations between various species of green algae and fungi (Table 7-2). The fungal hyphae (microscopic cellular strands) intertwine with the algal strands to provide support, and absorb minerals and water. The algal partner in turn provides chemical energy through photosynthesis. Together this symbiotic pair can withstand the severe conditions common on exposed rocks, even in mountainous regions above the treeline. Over the years the lichens may cover the rock surface and contribute organic matter which can trap rock fragments and water or wind-borne soil particles. Within this thin layer of soil, other small plants such as mosses can grow, eventually leading to a more complex vegetative cover on the rock. Centuries of time would be necessary to establish a climax forest in an area, beginning with bare rock, even when wind-blown soil and organic matter are imported.

7-G THE HUMAN POPULATION

What is man...?

Humans are a unique population.

The question of the psalmist (Psalm 8:4) must be answered by each person who attempts to apply biological principles to an understanding of the human population on Earth. On the one hand, *Homo sapiens* is a biological population, woven along with other populations into the tapestry of life, mutually interdependent for energy and nutrients. Therefore, the principles of population biology of this chapter have some direct applications to the human population. However, any answer to the question "What is man...?" must take into consideration the *objective truth* given by divine revelation (Section 1-A). In the Scriptures, we read such words as "image of God" (Genesis 1:27), and "made him to have dominion over the work of thy hands" (Psalm 8:6). Clearly, scientific data concerning both human and nonhuman populations must be interpreted in light of the Scriptures. In Chapter 8, we shall provide a comprehensive view of humanity's position in the created environment. For now, our purpose is to glean some principles from population ecology that will help us analyze human populations.

three times more births than deaths

With each beat of your heart, three more people are added to the world's population (15). For every person that dies, three are born. This translates into a net increase of over 250,000 persons per day. The human population of Earth, in 1993, was over 5.5 billion and is expected to reach 8.5 billion by 2010 (26). It took all of human history up to about 1800 to bring world population to the 1-billion mark. But, it took only 12 years to add the most recent 1 billion humans (17). One-third of todays world population are children under the age of 15, ready to enter the reproductive years. The 1990's will produce the largest generation of children ever to be born on Earth. About 90% of these children will be born in less developed countries (26). Considering that 1 billion inhabitants of the Earth live in destitution today (5), how can the planet support this upcoming generation?

One billion are in destitution.

What is Earth's carrying capacity?

Many are deeply concerned that we have exceeded the carrying capacity of our planet, as evidenced by widespread starvation and disease. Others emphasize the progress made in medicine and nutrition, and believe that the earth can support as many as twice the present population, given improved food distribution and reduced consumption of energy and mineral resources by the western world. Still others are concerned about the senseless loss of life through wars, abortion, and diseases associated with immoral lifestyles.

Don't forget moral issues.

Our first task in attempting to analyze the issue of human population is to be sure we understand some of the data provided by *demographers,* scholars who study population statistics. If you have mastered Section 7-B, the following three parameters should be understandable:

What is ZPG?

1. **POPULATION GROWTH RATE, (r)** as you may recall from Section 7-B, equals births/1000 *minus* deaths/1000, *plus* (immigrants *minus* emigrants). This parameter (ignoring migrations) is used by demographers to express the annual rate of human population growth as a percent, by nation or region. Ignoring migrations, the population of an area will increase, decrease, or reach *zero population growth (ZPG)* depending upon whether birth rate is greater than, less than, or equal to death rate, respectively. Based upon the United Nations data (26), the rate of increase for the world is as follows:

$$r = [\#births/1000 - \#deaths/1000] \times 100$$
$$r = [26.4/1000 - 9/1000] \times 100$$
$$r = 1.7\% \text{ annual increase}$$

an increase of 93 million per year

The world population in 1993 of over 5.5 billion times 1.7% annual increase (5.5 billion × 0.017) equals 93 million humans added to Earth per year, a logarithmic increase (Section 7-B).

2. **TOTAL FERTILITY RATE (TFR)** is the average number of children a woman will bear during childbearing years (considered to be age 15 to 49). TFR along with number of women of reproductive age will influence growth rate.

What influences replacement level?

3. **REPLACEMENT LEVEL FERTILITY (RLF)** is the average number of children a couple must have to replace themselves. The higher the mortality rates among infants and children of pre-reproductive age, the higher the RLF will be for that region. Note that a region that reaches RLF will not reach ZPG if the number of women entering reproductive age is increasing. [Why is it impossible for a nation to have an RLF of 2.0 ?]

Having introduced the basic population parameters, we shall now outline some of the major considerations based upon recent data. This outline is intended to assist you in integrating principles of population growth with your own field of study.

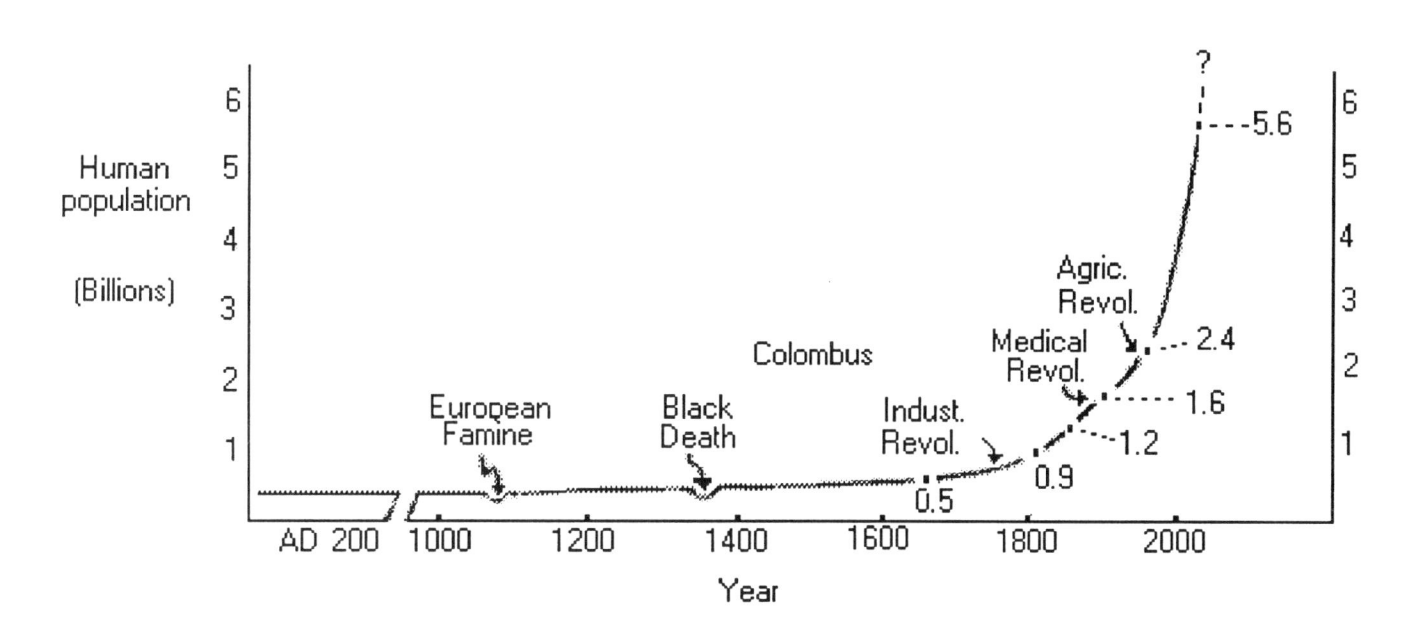

Figure 7-7. The logarithmic increase in world human population.

HUMAN POPULATION CONSIDERATIONS:

Why has human population skyrocketed?

1. After many millennia during which it was less than 1 billion, the human population began to increase logarithmically (Figure 7-7).

2. The rapid increase in world human population over the past century is not attributable to an increase in birth rates, but to a decline in death rates (Table 7-3).

3. Reduced death rates over the past century are apparently due to an alleviation of certain density-dependent limiting factors that had been restraining population growth for many centuries (Figure 7-7). Broadly speaking, this alleviation became possible following the Renaissance through increased literacy, scientific and technological "revolutions", worldwide transportation, and the harnessing of fossil fuel for mechanization.

How have we "side-stepped" limiting factors?

4. The events outlined in #3 contributed to the following specific technologies that have enabled humans to "sidestep" density-dependent limiting factors, reduce death rates, and increase population growth:

   a. Agricultural technology — mechanization, fertilizers, pesticides, and genetic improvements have increased food production and distribution; including superior nutrition in quantity (calories) and quality (protein, vitamins etc.).

   b. Medical advances — vaccines, antibiotics, surgical techniques, coupled to improvements in personal hygiene and housing, have led to reduction of contagious diseases; but, more progress is needed here.

5. It is projected (17) that in each of the next three 11-year periods, 1 billion people will be added to Earth, making a population of over 8.5 billion in 2025, *twice the population of the world when many readers of this text were born!* Over 45% of humans live in urban areas, often characterized by a poor environmental quality (26).

numerical increase *vs.* rate of increase

6. Annual rate of world population growth decreased from 1.9% in 1970 to 1.7% in 1991. However, because of the logarithmic rate of increase in total numbers of people, the actual numerical increase was 69 million in 1970, and 93 million in 1992. [How much would the 1.7% have to decrease in order to reach 1970 rates of increase?]

Table 7-3. World and Regional Human Population Statistics[1].

PARAMETER	WORLD	MDC's[2]	LDC's[2]	USA	AFRICA
PER CAPITA GNP[1] (1989 USA $)	$3,760.	$16,990.	$750.	$21,000.	$610.
BIRTH RATE[1] (Avg #/1000)	27	14	30	17	44
DEATH RATE[1] (Avg #/1000)	9	9	9	9	14
POPULATION[1] (billions)	5.38	1.22	4.16	0.25	0.68
ANNUAL GROWTH (r)[1] (1991 percent)	1.7	0.5	2.1	0.8	3.0
PCT. OF WORLD POP.[3]					
1950	100.0	33.1	66.9	6.0	8.8
1990	100.0	22.8	77.2	4.7	12.1
2025	100.0	15.9	84.1	3.5	18.8
POPULATION in 2025[3] (billions)	8.5	1.4	7.1	0.3	1.6
TOT FERTILITY RATE[1] (births/woman)	3.4	1.9	3.9	2.1	6.1
PCT. OF POPULATION UNDER AGE 15[1]	33	21	36	22	45
INFANT MORTALITY[1] (deaths/1000 births)					
1984	84	19	94	10.9	119
1991	68	14	75	9.1	102
ABORTION RATE[4] (babies/1000 women)	?	?	?	346	?

Sources: [1] *World Population Data Sheet, 1991*. Population Reference Bureau, Washington, DC.
[2] MDC's (More Developed Countries); LDC's (Less Developed Countries)
[3] *World Resources, 1992-1993*. Report by World Resources Institute, and United Nations. Oxford University, New York, NY.
[4] Number per 1,000 live births, or 24 per 1,000 women age 15-44, in 1989. Abortion Surveillance: Preliminary Analysis—United States, 1989. Division of Reproductive Health, Center for Disease Control, Atlanta, GA.

Because humans are so different...

7. Due to its heterogeneity, the world human population must be considered at least region by region, due to the differences in political systems, education, societal and religious influences, and environmental resources. These factors influence health, family planning programs, personal preferences for family size, and economic status.

... we must distinguish MDCs from LDCs.

8. The world human population is often broadly divided into two sectors: the **more developed countries (MDC's)** are those of North America, Europe, USSR, Australia, Japan, and New Zealand; and all others, **less developed countries (LDC's)** (17). The demographic statistics differ markedly between these two sectors, distinguished economically by the *per capita GNP* (Gross National Product) (Table 7-3).

9. The marked decline in death rates in LDC's was not accompanied by an equal decline in birth rates; therefore, this sector is accounting for about 90% of the annual increase in world population (Table 7-3). **Age structure** comparisons reveal that 1.7 billion children on Earth are age 15 or below; and, 82% of these are in less developed countries (Figure 7-8) (12). As these children enter reproductive age, they will account for 90% of the 1.5 billion children to be born in the present decade.

Children will have influence.

projections for after 2000

10. Even if world population growth rates declined sharply from 1.7% to 0.59%, the world population would still reach 7.6 billion in 2025. Or, according to the UN medium fertility projections, if replacement level fertility stablizes at about 2.06 in 2025, population would reach 10 billion in 2050, and stabilize at 11.6 billion just after the year 2200.

Can population increase when TFR is < RLF?

11. The population of the USA increases 0.08% or about 2 million per year, in spite of the fact that TFR (total fertility rate) is slightly below the replacement level fertility of 2.1 for the USA. The increased numbers of potential mothers from families of the post World War II "baby boom" (1950's and 1960's) has overridden the decrease in TFR. The decline in TFR in the USA is associated with the changing socioeconomic, cultural, and moral climate in which there are larger percentages of women in the work force, increased average age at marriage, legalized abortion, and increased family planning. Legalized abortion rate in the USA in 1989 was 24 babies per 1000 women (Table 7-3).

abortion rates

We must note per capita consumption.

12. It is difficult to determine whether Earth is overpopulated. The carrying capacity of Earth with respect to the human population depends upon many factors, including per capita consumption of environmental resources and global distribution of resources and technology. These, in turn, require responsible national leadership and global cooperation among nations.

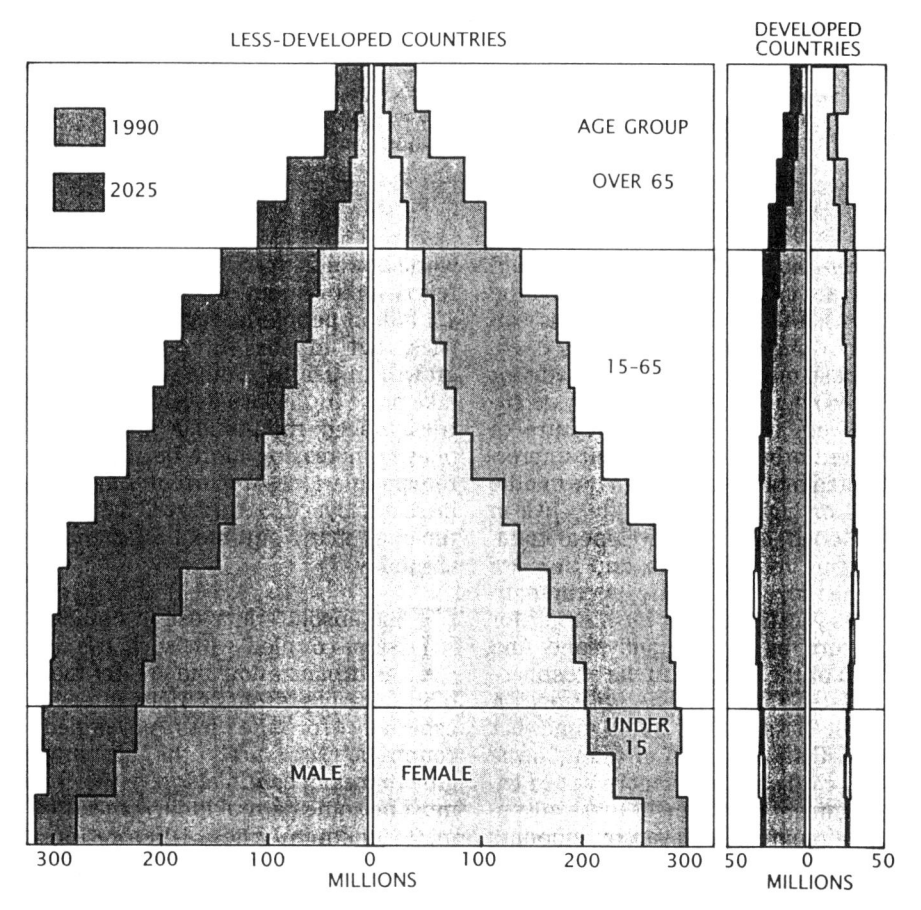

Figure 7-8. Age distribution of the populations of the less developed and the developed countries in 1990 as compared with that projected for 2025. From "The Growing Human Population", by Nathan Keyfitz. Copyright © 1989 by Scientific American, Inc. All rights reserved.

Summary:

From the above considerations, it is evident that perspectives from population biology raise serious concerns about the present and future of the human population. Many of these concerns reach beyond the limits of biological science. Humans are unique among the creatures in having a rational capacity as expressed in our science, technology, sociopolitical institutions, ethics, and morality. Therefore, while biologists may point up the threat of overpopulation, the human response will be governed by the values of men and women in positions of influence. In fact, each of us is being called to make decisions that will reflect our values and our worldviews. We must address the following questions:

We must face hard questions:

material consumption

1. What is an appropriate per capita consumption of food, energy, and resources for a meaningful life? What will the "blessed 20%" of the world population which consumes over 80% of the resources offer the other 80% who are seeking the same lifestyles?

family planning

2. Does the responsibility of dominion given in Genesis 1:27-28 include taking active responsibility to plan family size?

compassion

analyzing the issues

3. What is my individual responsibility toward those who are suffering in developing countries?

4. Can I be concerned about the deplorable loss of life due to the increasing use of abortion for "convenience" and birth control (estimated 40 million/year), and, at the same time, be concerned about the explosive increase in human population and related deaths of 12.9 million children/year under age 5 due to disease (26)?

These and other questions bring us back to the basic question that challenges our faith—"What is man...?" In Chapter 8, we will attempt to draw together several of the ecological issues with scriptural principles related to human population.

QUESTIONS AND DISCUSSION TOPICS

1. Considering the many animal, plant, and microbial populations that occupy most biotic communities, how can populations be considered as distinct units?

2. What is Liebig's Law of the Minimum? What is the limiting factor in your performance in this biology course so far?

3. What pattern of population distribution is represented by urbanization? What factors cause this distribution of the human population?

4. Distinguish the following pairs of terms:
   a. logarithmic growth and biotic potential.
   b. density dependent and density independent limiting factors
   c. survivorship curve and age structure diagram
   d. total fertility rate and natural replacement fertility

5. Would you expect the carrying capacity for each species be the same for a given environment? Why or why not?

6. Discuss the ways in which homeostatic mechanisms are involved in populations. What is meant by negative feedback?

7. How do biotic potential, environmental resistance, and genetic variability lead to natural selection and conservation of a population amid environmental changes?

8. How do social behaviors among animal populations serve as homeostatic regulators of population density? How does natural selection operate through such behaviors? Are there counterparts in the human population?

9. In what sense can predation encompass herbivory and parasitism? Are fleas parasites or predators?

10. What causes ecological succession? How are tolerance ranges and limiting factors involved? Describe succession as it would occur in an old field or abandoned city lot in your community.

11. How would pre-Noachian flood survivorship curves for the human population differ from post-flood patterns? What hypotheses can you propose to account for this? Are they testable?

12. What are the basic requirements for the life of any organism? How are these requirements met by each of the following:
    a. white-tailed deer
    b. European settlers entering the Ohio valley in 1800
    c. Cedarville residents at present

    Discuss several implications of these comparisons with regard to resource demands and the human carrying capacity?

13. Using Table 7-3, compute the population growth rate (r) for the USA, with and without the abortion rate included. Discuss the implications of having a total ban on abortion, with respect to population changes, socioeconomic factors, and responsibility of "pro-lifer's".

14. What is meant by the statement: "The long-term adaptation of a population in a given environment depends upon the differential survival and reproduction of individual members"?

TOPICS FOR FURTHER STUDY

1. Interactions between populations of two different species are fascinating. For example, the references listed below feature symbiotic relationships, chemical "warfare", and natural antibiotics. REFERENCES: #9, #19, #20, #21

2. Ecology of landscapes—How does a human population influence wildlife and vegetation by its alterations of the landscape? REFERENCES: #7 and #23

3. Has the human population reached the Earth's carrying capacity? From a biological standpoint, what are possible scenarios for the next decade or two? See REFERENCES, especially: #22 and #25,

4. To what extent should governments encourage family planning? REFERENCES: #10 and #12

5. How has abortion influenced world population demographics? Does your position on abortion influence your position on the human population "crisis"? What does it mean to be "pro-life"? REFERENCES: #16 and #18

REFERENCES

1. Begon, M., J.L. Harper, C.R. Townsend. 1990. *Ecology: Individuals, Populations, and Communities*. Blackwell Sci. Publ., Boston, MA.

2. Christensen, N.L., *et.al.* 1989. Interpreting the Yellowstone fires of 1988. *Bioscience* 39(10): 678-685. (Issue includes five other articles on ecology of Yellowstone fires.)

3. Cooper, C.F. 1961. The ecology of fire. *Scient. Amer.* 204: 150-160.

4. Daily, G.C. and P.R. Ehrlich. 1992. Population, sustainability, and Earth's carrying capacity. *Bioscience* 42(10): 761-771.

5. Durning, A. 1991. Asking how much is enough. In Brown, L.R., *et.al. State of the World.* pp 152-169. Worldwatch Institute Report. W.W. Norton, New York, NY.

6. Finch, C.E. and L. Hayflick, eds. 1977. *Handbook of the Biology of Aging*. Van Nostrand Reinhold Co. New York, NY.

7. Forman, R.T.T. and M. Godron. 1981. Patches and structural components for a landscape ecology. *Bioscience* 31:733-740.

8. Horn, H.S. 1975. Forest succession. *Scient. Amer.* 232: 90-98.

9. Huxley, C.R. and D.F. Cutler. 1991. *Ant-Plant Interactions*. Oxford Univ. Press. New York, NY.

10. Jacobson, J.L. 1991. India's misconceived family plan. *WorldWatch* 4 (Nov-Dec.): 18-25.

11. Jeeves, M.A. 1976. *Psychology and Christianity: The View Both Ways*. InterVarsity Press, London.

12. Keyfitz, N. 1989. The growing human population. *Scient. Amer.* 261(3): 118-26.

13. Leslie, P.H. and R.M. Ranson. 1940. *Journal of Ecology* 9: 27-52.

14. May, R.M. 1983. Parasitic infections as regulators of animal populations. *American Scientist* 71:36-45.

15. Miller, G.T. Jr. 1992. *Living in the Environment*, 7th ed. Wadsworth, Belmont, CA.

16. Population Crisis Committee. 1982. World abortion trends. *Population* 9: 1-6.

17. Population Reference Bureau. *World Population Data Sheet, 1991*. Washington, DC.

18. "Prolife": What does it really mean? *Christianity Today*, Jul. 14, 1989, pp 27-38, series of articles.

19. Putnam, A.R. 1983. Allelopathic chemicals. *Chem. and Engineer. News*. April, pp. 34-45.

20. Rice, E.L. 1983. *Pest Control with Nature's Chemicals*. U. of Oklahoma Press, Norman, OK.

21. Rudman, W.B. 1987. Solar-powered animals. *Natural History* Oct., pp 50-53.

22. Rushdoony, R.J. 1975. *The Myth of Overpopulation*. Thoburn Press, Princeton, NJ.

23. Silvius, J.E. 1984. The human pressure index: An integrated approach to landscape ecology. *American Biol. Teach.* 46(6): 334-337.

24. Smith, R.L. 1992. *Elements of Ecology and Field Biology*. Harper and Row, New York.

25. Wattenberg, B.J. 1987. *The Birth Dearth*. Pharos Books, New York, NY.

26. *World Resources, 1992-1993*. Report by World Resources Institute, 1992. Oxford University Press, New York, NY.

27. Wynne-Edwards, V.C. 1964. Population control in animals. *Scientific American* 221: 68-74.

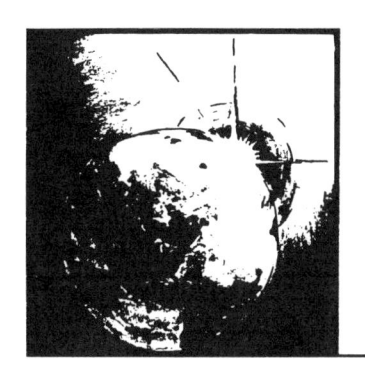

Chapter 8

Global Ecology and Stewardship

Earth is unique

As of this writing, no clear evidence exists for the presence of life outside of the biosphere of Earth! Earth is the unique habitation of millions of species. In Chapter 5 we examined the life-sustaining flow of energy from the sun through food webs, and how the atmosphere acts as both a "greenhouse" to warm the Earth, and a "filter" of harmful rays. In Chapter 6, nutrient cycles pictured how the gaseous composition of the atmosphere influences life processes, and how life processes influence atmospheric gases. In Chapter 7, we focused on **populations** as units of life. Now, in Chapter 8, we withdraw from the population level for a more wholistic view of the globe. **Global ecology** takes a broad view by focusing upon the **biosphere**, which encompasses all the ecosystems of Earth, extending into the atmosphere, hydrosphere, and lithosphere.

a wholistic view

Three worldviews:

theism

naturalism

pantheism

worldviews influence science

The intricate processes that sustain the global biosphere seem beyond the reach of our comprehension, even by modern science. Consider three worldviews that attempt to account for its existence and function. The worldview based upon scriptural revelation attributes the glorious Earth and universe to the supreme God who is its Creator and Sustainer (Colossians 1:16-17). Modern science, dominated by a naturalistic and mechanistic philosophy (Section 1-C), sees planet Earth and space as the result of time, chance, and natural laws operating in a giant mechanism of uncertain origin. However, as science continues to multiply our understanding of the Earth, many are seeing something more in this glorious "blue jewel" as photographed from space. To supporters of the **Gaia hypothesis** (19), land, oceans, air, and creatures are maintained in an inseparable "unity" of living and nonliving by the "Earth goddess" known as Gaia. In this pantheistic, "New Age" view, the ecological processes of the biosphere are not sustained by the supreme God of the Scriptures, but by a supernatural unity of the Gaia.

Biology cannot support or refute worldviews and religions, but the direction of biological science can be influenced by worldviews. We shall conclude this chapter by presenting a scriptural basis for stewardship of the Earth. But first, let's take a "global ecology view" of the biosphere and its global ecosystems from a scientific perspective. We will focus on global climate differences and how they influence geographic distribution of organisms and their adaptations.

8-A TOLERANCE RANGES AND GEOGRAPHIC DISTRIBUTION

Environmental conditions on Earth vary,...

The requirements for life on Earth are not evenly distributed across the face of the globe. For instance, temperatures may range from less than -70°C in Antarctica to more than 55°C in North Africa. Annual precipitation varies from less than 2 centimeters (cm) in certain desert regions to more than 1,500 cm in the humid tropics. One need not travel far in any direction to encounter variations in other environmental factors, including pressure changes associated with altitude or depth below sea level; and, differences in light intensity, wind speed, and salinity. These physical aspects of the environment that influence living systems are called **abiotic factors**. [A 20-minute bicycle ride would carry you a distance which, if stood up on end, would extend completely through the 5.5-mile-thick atmosphere (20).] The most significant abiotic factors are listed in Table 8-1.

... often along gradients.

Whereas it is true that the earth displays great variation in environmental conditions, it is also significant that there are *environmental gradients* involving each of these conditions. For example, if you were to go backpacking in the Cascade or Sierra Nevada Mountains, you would notice an environmental gradient, or gradually decreasing temperature as you climb in altitude from the warm base toward the peaks. A gradient of decreasing atmospheric pressure would also be evident as you became increasingly short of breath. Indeed, no matter where we go, we continually encounter gradients—indoors to outdoors, shade to sun, steamy shower to cold surroundings. Even if we stood in one location, as most plants do, we would experience gradients during the daily and seasonal cycles.

Table 8-1. Abiotic Factors and How they Affect Distribution of Life.

ABIOTIC FACTORS	GRADIENTS AS A FUNCTION OF:
light	LOCATION:
temperature	latitude
moisture	longitude
nutrients	altitude
vital gases	nearness to water
wind	water depth
fire	TIME:
atmospheric pressure	daily cycles
	seasonal cycles
	geologic history

Each population has tolerance ranges ...

Because living organisms are intimately associated with their environment, environmental gradients are of major significance. Generally, each population can withstand or tolerate only a certain range out of the entire environmental gradient for each abiotic factor. This so-called **tolerance range** will determine the geographic distribution of a given population. For example, the cactus and kangaroo rat populations are tolerant of a lower range of annual precipitation than sunflower and gopher populations. Thus, curve A on Figure 8-1 could represent cactus or pocket mouse populations, and curve B, sunflower or gopher. Note that the tolerance range for each population includes an optimum range as suggested by the greatest population in the most favorable conditions. Above and below this range, increasing stress is experienced

...which influence its distribution.

due to limiting or excess levels of the factor in question. Two populations that have similar tolerance ranges for every significant abiotic factor could presumably occupy the same environment. [Assuming that curve A (Fig. 8-1) represents pocket mouse, and curve B, the gopher, would you expect these two species to occupy the same geographical area? Explain.] If the level of only one abiotic factor in a given environment extends beyond the tolerance range of a species for a significant length of time, the species will be unable to occupy that environment. That factor would be considered a limiting factor to population growth (Section 7-C). However, many species have adaptations which enable them to tolerate and remain active during unfavorable periods. Others demonstrate adaptations that involve avoidance of unfavorable parts of the daily or seasonal cycle. [Illustrate this with reference to burrowing desert animals, or nocturnal animals.]

Summary:

In summary, the levels of the various abiotic factors of planet Earth vary as a function of geographic location and time of day or year. Each species has a tolerance range representing that portion of the environmental gradient within which it can live and reproduce. Combining these two principles, we can propose the following principle: *The regional distribution, or **geographic range**, of a species is determined by its tolerance ranges for each abiotic factor that affects its survival.* This principle, called the **law of tolerance**, is an extension of Liebig's *law of the minimum* (Section 7-C). [Can you explain how the one principle expands upon the other?]

law of tolerance

We must consider *scale*:

The law of tolerance applies to the distribution of a species at any scale we choose. **Scale** refers to a part of a hierarchy of dimensions of *space* and/or *time*. For example, on a continental scale, the range of the American Bittern would extend from Canada to Panama. However, on a *temporal scale*, the winter season of the year finds the birds restricted to a range from the Gulf states southward, due to *migration*. On a *spatial scale*, it appears from the continental map of North America that American Bitterns inhabit much of northern USA and Canada. However, when we "zoom in" to the scale of local biotic communities, the species may inhabit a local marsh community, but not a nearby meadow created by draining a marsh. In fact, wetland drainage for agricultural and urban development threatens this species by eliminating suitable habitat. Distinguishing temporal and spatial scales is important in analyzing the effects of human impact upon species. As

...whether immediate or long-term...

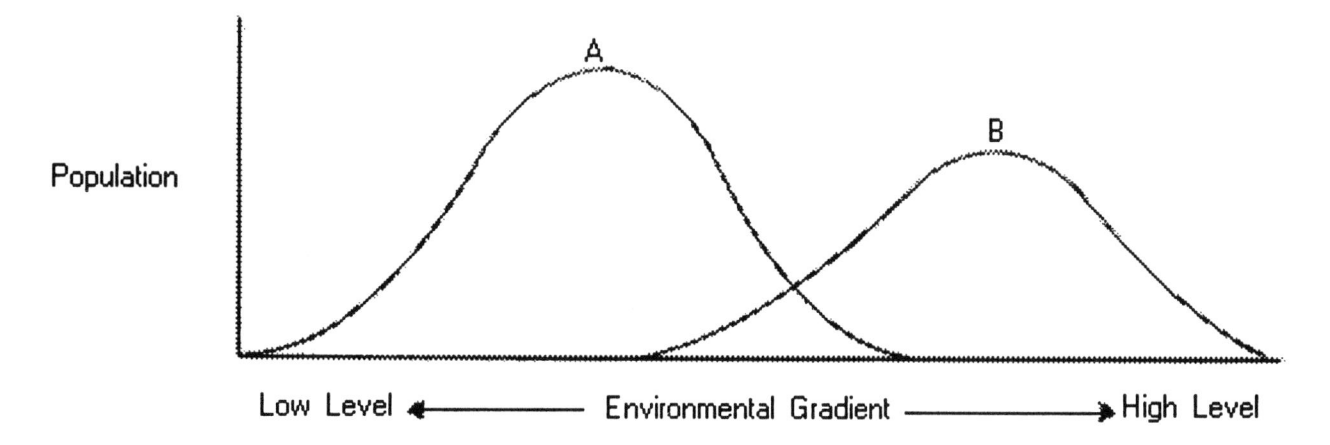

Figure 8-1. Tolerance ranges for two hypothetical species, A and B, along an environmental gradient. Species A is more tolerant of low levels of the factor (e.g. moisture) than species B.

...whether global or local.

Table 8-2 indicates, human impact on a local scale may not have immediate (short time scale) effects on the broader, continental scale. However, in time, the accumulation of local disruptions may cause global-scale disturbances.

Whereas distribution of species can be considered on several spatial scales, we shall now focus our attention on the geographic ranges of living organisms on a continental or regional scale (Table 8-2). Here groups of ecosystems which are similar in climate and resident species are collectively considered in the concept of **biomes**.

Biomes focus on a wide scale.

8-B BIOMES OF EARTH

The earth may be subdivided into two broad types of ecosystems, aquatic and terrestrial. Aquatic ecosystems may be further distinguished as freshwater (lakes and rivers), and marine (oceanic) ecosystems.

Biomes have similar climates...

Terrestrial ecosystems are classified into **biomes**, broad groupings according to major climatic patterns and the resident populations (Section 3-B). Each biome has a distinctive assemblage of plants and animals as observed in the climax stage of succession (Section 7-F). The transition from one biome to another is normally gradual because abiotic factors generally change gradually as a function of location. This in turn causes gradual replacement of one population by another according to the tolerance ranges of each population (Section 8-A). In much of the U.S., man has replaced the natural biotic communities with agricultural and urban landscapes that make biome distinctions even more difficult. After we consider the natural distinctions of biomes, you will understand the serious implications of human alterations.

Table 8-2. Analyzing Effects of Human Activities on Forests from the Scales of Levels of Organization, Space, and Time[1].

LEVEL OF ORGANIZATION	SPATIAL SCALE	TIME SCALE	PROCESSES AFFECTED	HUMAN ACTIVITIES
Biosphere	Global	Years to millennia	Energy, carbon, and water flow	Deforestation, fossil fuel use
Biome	Continents, or regions	Centuries to millennia	Natural Selection Migration, Extinction	Plant breeding, plant management, conservation
Ecosystem	10 to 10,000 hectares	Years to centuries	Nutrient cycling, NPP, competition, ecol. succession water usage	Pollution, flood control, exotic pests, fire suppression, erosion
Organism	0.01 to 1,000 m^2	Minutes to decades	Physiological processes, reproduction, death	Fertilizing, weeding, watering

[1]Modified From: Graham, R.L., *et.al.* 1990. *Bioscience* 40(8): 575-87. Used by permission.

	Temperature and precipitation are the major climatic factors that influence
...*i.e.* similar temperature and precipitation	biome distributions. As already noted in Chapter 5, every biotic community depends upon the influx of solar energy. Energy of the visible spectrum is converted to chemical energy by autotrophs, the 'breadwinners' of the whole community. The amount of infrared radiation influences the temperature.
Sun and Earth factors	Generally, as one travels from equator to pole, the solar influx to the ground decreases because the sun's rays strike the earth at an increasingly oblique angle, as shown in Figure 8-2. The four seasons of temperate regions (i.e. between the Arctic Circle and the Tropic of Cancer; or, between Antarctic Circle and the Tropic of Capricorn) are caused primarily by the combined effect of the earth's revolution around the sun and the tilt of the earth's axis which is about 23.5 degrees from the perpendicular. This relationship is illustrated in Figure 8-2.
air currents and precipitation	Precipitation is influenced to a large extent by prevailing wind patterns in relation to physiographic features such as mountain ranges and bodies of water. For example, moist air off the Pacific Ocean rises and cools as it encounters the Cascade and Sierra Nevada Mountains. Cooling air causes condensation and precipitation on the western slopes where the lush redwood forests are found. The resultant dry air, having lost much of its water vapor moves over the mountains and warms as it drops to lower elevations. It has little moisture to give, and the result is the typical desert conditions of the Great Basin in the western U.S. In contrast, eastern U.S. has sufficient precipitation throughout the year, courtesy of the moist Gulf breezes. Eastern U.S. once supported the great deciduous forests (Figure 8-3).
Superimposed continental gradients...	The grassland, or prairie biome is situated between the low-precipitation of the desert biome and the higher precipitation of the deciduous forest. Thus, an east-west precipitation gradient is superimposed upon a north-south temperature gradient. That is, the two climatic factors *interact* to produce the geographic distribution of biomes of planet Earth (Figure 8-3).

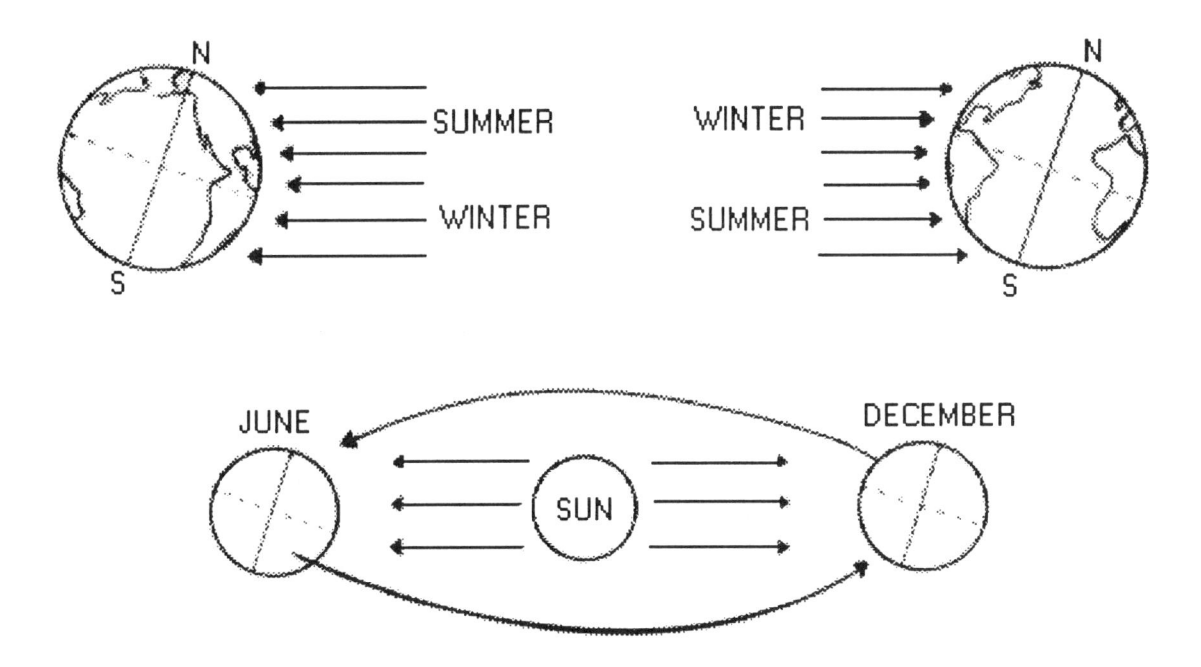

Figure 8-2. Distribution of incoming solar energy on the earth's surface as influenced by the planet's tilt and revolution around the sun.

Figure 8-3. Geographic distribution of major global biomes.
Reprinted with the permission of Macmillan Publishing Company from COMMUNITIES AND ECOSYSTEMS, 2nd Ed., by Robert H. Whittaker. Copyright © 1975 by Robert H. Whittaker.

... affect biomes.

Figure 8-4 graphically illustrates the interaction of temperature and precipitation for the North American biomes. Note that the desert biome and tundra biome both have low precipitation but differ markedly because of differences in temperature. Net primary productivity the energy "income" of each biome (Section 5-D), reflects these differences in temperature and precipitation (Figure 8-4, parentheses).

Summary:

In summary, each biome has a distinct climate as manifested in temperature and precipitation patterns throughout the annual cycle. Climate, in turn, influences the plant and animal populations that can inhabit each biome. In the following sections, we shall survey each biome, its characteristic plant and animal populations, and the adaptations that equip these populations to survive in their environment.

8-C TUNDRA

environmental extremes

The tundra (Russ. "marshy plain") is a vast, treeless expanse bordering the Arctic Ocean in North America, Europe, and Asia. The soil is completely frozen and covered with snow during most of the year. However, during the three summer months, daylight is continuous and temperatures rise far enough above freezing to melt the snow and a few inches of the frozen soil. Just

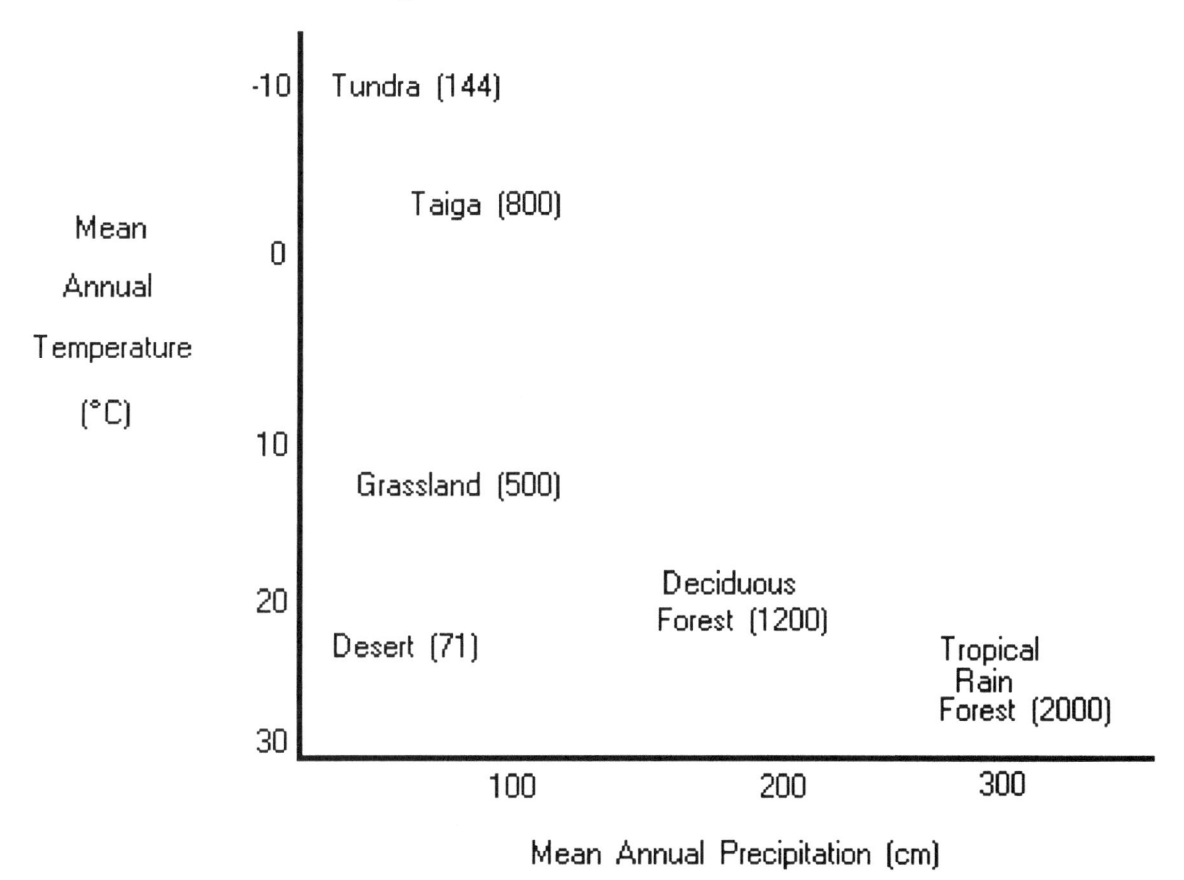

Figure 8-4. Interaction of temperature and precipitation as they influence the distribution of North American biomes. Mean net primary productivity (NPP) (grams dry matter/m^2/year) is given in parentheses.

Coordinates and data thereon, from Fig. 4.10; and, productivity data from Table 5.2. reprinted with the permission of Macmillan Publishing Company from COMMUNITIES AND ECOSYSTEMS, 2nd Ed., by Robert H. Whittaker. Copyright © 1975 by Robert H. Whittaker.

a cold desert

beneath the thawed surface layer is the permanently frozen soil, or **permafrost**. Surprising as it may seem, the tundra resembles a desert because of the scarcity of water in liquid form. Tundra plants include lichens, mosses, certain grasses, small shrubs, and dwarf trees. Dwarfism, hairy leaf surfaces, and the capacity to grow and flower quickly are common plant adaptations. [What adaptive advantage is given by each of these?]

adaptations to coldness

Principle herbivores include rodents (e.g. voles, lemmings), arctic hare, caribou (reindeer in Eurasia), and musk ox. Carnivores include the arctic fox, gray wolf, polar bear, grizzly bear, and snowy owl. Adaptations of these essentially permanent animal residents include white coloration, burrowing and feeding beneath the snow, and hibernation. [Adaptive advantages of each?] In the summer, birds and other species from biomes further south will migrate into the tundra.

simple food web

The tundra communities have a very simple trophic structure. Most of the plant biomass is contained within no more than ten species, and heterotroph populations likewise are few. The simpler food web with fewer options for energy flow (Section 5-F) causes frequent erratic changes in populations. Disruptions of the fragile tundra are of major consequence because of the low temperatures and slow rate of ecological succession.

8-D TAIGA

The taiga (Russ. "swamp forest") occupies large, trackless areas of Canada and Northern Eurasia, just south of the tundra. Compared to the tundra, the taiga has a warmer climate, increased precipitation, and a longer growing season. The absence of permafrost allows the growth of extensive needle-leafed, cone-bearing trees called *conifers*. The predominance of conifers, including pines, spruces, firs, and cedars, has given the taiga the alternate name, coniferous forest. The flexible branches and conical, 'Christmas tree' shape allow these trees to bear heavy loads of snow and ice without breaking down. Their waxy, needle-like leaves resist dehydration and freezing, while they remain on the tree year-round and photosynthesize on sunny winter days. Conifers grow in thick stands and little light penetrates to the forest floor. Dead branches and needles decay slowly. The resultant detritus accumulation, combined with high tree density, make travel on foot in such forests very difficult for humans or wildlife. The often marshy landscape is interrupted by frequent lakes and bogs.

conifers . . .

. . . in high densities.

. . . around lakes and bogs.

adaptations

Herbivores include a variety of insect populations that feed on conifer needles, buds, and bark. Birds such as the beautiful evening grossbeak and the crossbill feed on the seeds. The crossbill is so named because the tips of its bill cross each other and form a mechanism to pry open conifer cones to release seeds. The snowshoe hare, moose, beaver, and deer are herbivores that feed on low-growing vegetation and deciduous trees (e.g. aspen and birch) that grow around lakes, streams, marshes, or burned-over areas. Carnivores include a variety of bird species that feed on insects, and a variety of mammals including bear, wolverine, lynx, wolf, and shrew.

taiga in alpine regions

The coniferous forest biome extends southward along the slopes of the Rockies, Cascades, Sierra Nevadas, and Appalachians where similar climatic conditions exist due to high altitude. Indeed, some of these mountain ranges are so high that coniferous forest gives way to an alpine tundra, and snow-capped peaks. [How are cold alpine temperatures in more temperate and tropical latitudes explained by thickness of atmosphere and heat radiation?]

8-E DECIDUOUS FOREST

The deciduous forest biome (G. *decidu* = falling off) once occupied the eastern half of the United States. Before the entry of the colonists, this area was essentially unbroken forests of maples, oaks, hickories, and many other species that shed their leaves as an adaptation to harsh winters. Various *understory trees*, such as the beautiful dogwoods and redbuds, and an *herbaceous layer* of early-blooming spring wildflowers each occupy niches within the deciduous forest. The understory plants grow up and flower in early spring while sunlight penetrates the leafless canopy overhead. After the larger trees have expanded their new leaves, the forest floor darkens and only the shade-tolerant plants can grow.

More inviting forest habitats

. . . cause higher biodiversity.

Aside from a few mosquitoes, the deciduous forest is pleasant to hike through at any time of the year. It has a variety of habitats for animals such as burrowing rodents and tree-dwelling squirrels and birds. Larger mammals including deer utilize the forest for food and protective cover. Human pressure has all but eliminated many of the predator populations by hunting and extensive forest clearing. These predators included the bobcat, black bear, and mountain lion.

Detritus food web has made rich soils.

Warm summer temperatures and regular rainfall cause rapid decomposition of deciduous leaves and dead branches. An extensive detritus food web Section 5-H) is effective in degrading this organic matter and dispersing it into the soil, freeing minerals for recycling. The fertile soils of eastern United States, now used in agriculture, were formed over centuries within the cool, moist forests that covered this biome. Central Europe and northeastern China are also in the temperate deciduous forest biome.

8-F TROPICAL FORESTS

tropical forest . . .

Tropical forests occupy low altitude, equatorial regions of Central and South America, Africa, Asia, and Australia. *Tropical deciduous forests* exist in regions which have wet and dry seasons. With the exception of the southern tip of Florida, the *tropical rain forests* of the western hemisphere are restricted to Central and South America.

. . . were once circumglobal.

high species diversity . . .

and habitat diversity.

Prior to human exploitation of this biome, tropical rain forests formed a continuous ring around the globe on land north and south of the equator. They are the world's most complex biomes, and it is estimated that at least one-third of all species live within these forests. A single square mile may contain over 300 different tree species and thousands of species of insects, birds, reptiles and mammals. Uniformly high temperatures and rainfall throughout the year cause rapid and dense growth of trees. As a result, little light energy is available to support vegetation at the forest floor. Both herbivore and carnivore populations tend to inhabit specific strata within the canopy where light and food are most abundant. Habitat diversity is even greater than that of the deciduous forest. Complex communities representing all manner of insects, spiders, amphibians, and mammals may spend their entire lives within a small part of a single tree! Even some plants, including epiphytic orchids and bromeliads (members of the pineapple family), germinate and grow high in the branches of trees. Because their roots do not contact the forest floor, epiphytes obtain their water from the humid air.

limiting light and nutrients

Not only light but also mineral nutrients are often a limiting factor within the tropical rain forest. Detritivores and decomposers are extremely effective in rapidly converting organic matter to inorganic form. Plant roots may even grow up out of the ground to enclose dead leaves and other fragments of organic matter so that nutrients from the detritus are recycled effectively. Some plants have cup-shaped leaves to trap moisture and dead organisms. When the trapped detritus decays, the resultant minerals are directly absorbed by the leaf. There are many such adaptations aimed at keeping a tight nutrient cycle and preventing losses of nutrients via leaching during frequent rains.

"tight" cycling of nutrients

farming depletes soils

When tropical forests are cleared and replaced by agricultural crops, the soils are quickly depleted of nutrients. Most crop plants such as soybeans do not have the adaptive capacity to tenaciously gather nutrients like the adapted tropical plants. The crops grow poorly and the soil becomes hard-baked or severely eroded.

8-G GRASSLANDS

between forest and desert

The grassland biome, also called *prairie* (North America), *steppes* (Eurasia), *veldt* (Africa), and *pampas* (South America), is generally situated between forest and desert biomes. On the North American prairie, precipitation in summer may equal that of the deciduous forest, but decreases markedly toward the end of summer, and severe droughts are not uncommon. In pre-settlement days, periodic fires were instrumental in preventing the growth of woody plants, while releasing minerals from the burned organic matter to stimulate regrowth of grasses and legumes from seeds or underground stems. A gradient of decreasing precipitation, coupled to increasing evaporation exists as one travels from the Mississippi River west to the Rockies. Before its conversion to agricultural uses, the *tall grass prairie* occupied the eastern portion of this region. Here the big bluestem and Indian grass species formed a dense growth up to seven feet tall. Grasses are interspersed with a variety of colorful plants of the sunflower and legume family. To the west on the high plains where moisture is more limiting, grama and buffalo grasses dominate the *short grass prairie*.

adaptations to open landscapes

Herbivores of the prairie include the colonial prairie dogs, jack rabbits, gophers, and a variety of insects including grasshoppers. These species have running and burrowing habits that assist them in evading predators. Leaping behavior is also helpful in travel amid tall grasses. It is estimated that over 70 million bison once grazed the North American grassland. This number was reduced to 88 in the 1880's. Settlers replaced the bison with domesticated herbivores, such as cattle and sheep. Predators such as coyotes, weasels, badgers, and wolves remain in reduced numbers.

sodbusting

Because of reduced precipitation, leaching is minimal, and nutrients are maintained within the root zone to produce some of the most fertile soils on earth. This heritage of fertility was discovered over a century ago when the thick prairie sods were "busted" and turned over by newly available steel plows, pulled by rugged beasts of burden. Tall grasses were replaced by corn and soybeans. The short grass prairies were sown in wheat. In the drier short grass prairie, removal of the sod cover by plowing and overgrazing, combined with frequent droughts have led to extensive wind erosion of soil. The Big Drought, lasting from 1927 to 1932, set the stage for the Dust Bowl in which hundreds of millions of tons of soil were blown eastward. Prairie soil sifted into the offices of the U.S. Department of Agriculture as a reminder that

Did we learn?

something must be done. However, in spite of conservation efforts, disruption of the more delicate, arid grassland communities have caused losses of soil and vegetation, a 25 to 50% reduction in NPP, and the advance of desert conditions. Estimates project that this process, called **desertification**, has occurred on 2 billion acres (12 times the area of Texas) in arid parts of every continent since the 1950's (22). Now, an estimated 35% of Earth's land surface is classified as arid or semi-arid desert, and is home for about 1 billion people. Human and livestock populations have exceeded the carrying capacity of these fragile ecosystems, resulting in overgrazing, lowering of water tables and deforestation by fuel-wood gatherers. An estimated 230 million people in arid areas are unable to obtain adequate food, fuel wood, and shelter (22).

human pressure on the prairie

8-H DESERTS

In areas where moisture is too low or erratic (usually less than 25 cm/yr; Figure 8-3), and solar radiation influx is extremely high, grasslands give way to the desert biome. Deserts now occupy over 30 percent of the earth's land surface and are advancing into grasslands in many areas (Section 8-G). Deserts vary in aridity and in seasonal temperature fluctuations. So-called *hot deserts* include the Sahara, Death Valley, and the deserts of southwestern United States and northern Mexico. *Cold deserts* include the Great Basin of western United States and the Gobi desert of Asia.

adaptations to low moisture

Water is the major limiting factor for desert populations. When precipitation occurs it is often in torrential downpours. Desert plants, including various cacti, creosote bush, and salt bush, have a variety of adaptations to low moisture supply. These include short life cycles that are timed with moist periods, wide spacing between plants, and various anatomical features for storing water and reducing water loss.

Animal adaptations to low water levels include impervious coverings (e.g. snakes, lizards, arthropods), reduction of water loss by excretion of uric acid instead of urea, burrowing, and nocturnal habits. Mammals, which vary with desert type, include kangaroo rat, pocket mouse, jack rabbit, and gray fox.

leaching, but "in reverse"

Excessive evaporation, combined with low precipitation and leaching, cause a concentration of mineral salts at the soil surface. You may have noticed salt accumulations on the soil around potted house plants, due to evaporation, especially in the winter when the air is drier. This buildup of salts can be toxic to plants. Agriculture in many dry areas of western United States is supported by irrigation. Evaporation of irrigation water from the soil surface increases the salinity of desert soils (Section 6-C).

8-I SAVANNA AND CHAPPARAL BIOMES

We have discussed the major biomes in very brief detail. There are many variations in climate and resident populations within these categories. In addition, there are other areas that have a distinct flora and fauna that we should note. The **savannas** are grasslands with scattered trees and a variety of grazing mammal populations. These include the zebra, giraffe, and antelope, typically found throughout much of the central and southern African landscape. Areas having a Mediterranean-type climate of warm, dry summer, and cool, moist winters support a shrub forest of low-growing evergreen trees.

gallant grazers

unavoidable fires

The **chaparral** of California, and much of the land surrounding the Mediterranean (called *maquis*) are areas representing this biome. Fire is an inevitable part of the shrub forest, and resident plants are adapted to quick recovery after fire. The human population has tried to prevent fires, and consequent real estate losses, to little avail. Prolonged periods without fire simply allow a more explosive detritus fuel buildup. [Remember the Yellowstone fires? Section 7-F.]

8-J OCEANIC ECOSYSTEMS

Oceans occupy about 70 percent of the Earth's surface but net primary productivity of open ocean averages only about 100 g/m^2/year, comparable to deserts and tundra. Low fertility associated with dilute concentrations of nutrient ions is the major limiting factor in growth of autotrophic algae (phytoplankton). This is in sharp contrast to shoreline areas, estuaries, and coral reefs which are highly productive and diverse.

Interfaces ...

... draw humans.

Shoreline areas and estuaries represent an interface between the land and water, and between freshwater and saline water. Intertidal zones, alternating between exposed and submerged conditions, are habitat for a variety of marine algae, herbivores, and predators. Humans have always been drawn to the seashore, and one must visit to appreciate the immensity, beauty, and diversity of this part of the living creation.

nutrient runoff

Associated with shorelines are the salt marshes and estuaries, located at the mouths of streams and rivers. Due to nutrient-rich runoff from agricultural and urban land uses upstream, shoreline communities are highly productive, but often overfertilized to the point of food web disruption (eutrophication, Section 6-G).

coral reefs: oceanic "oases"

Coral reefs are highly productive, colorful, biodiverse communities of corals, other invertebrates, and algae that develop on the continental shelf or volcanic formations as foundations. These "oases" of high NPP (up to 5,000 g C/m^2/year), surrounded by nutrient-poor ocean, owe their existence to the ability to efficiently trap and recycle nutrients within the community. (29).

corals are symbiotic

Corals (Phylum Cnideria; Sections 9-F, 10-E) grow in sessile colonies, and are partly autotrophic and partly heterotrophic. They form a symbiotic relationship with photosynthetic dinoflagellate algae, called *zooxanthellae*. In daytime, the algae living within the corals photosynthesize organic molecules which are absorbed by coral tissues. At night, the corals feed as heterotrophs on zooplankton and thus, obtain phosphates, nitrates, and other inorganic ions needed by the coral and the zooxanthellae. Some of the CO_2 from respiratory processes is converted by the coral to $CaCO_3$ to form their calcareous skeletons. Thus, the symbiotic relationship functions in an efficient cycling of organic matter and nutrients between coral and algae. A host of other animal species, including sea urchins and fish live and prey among the colorful caverns of the reef. [Why do reef-building corals thrive only in shallow water?]

humans threaten

In recent years, it has become evident that these fragile oceanic ecosystems have become threatened, particularly under the pressures of tourism, coral rock harvesting, oil spills, and other commercial interests. The ocean is yet another resource which can yield much benefit and blessing if properly managed.

8-K HUMAN IMPACTS UPON WORLD BIOMES

Caution: Humans must recognize uniqueness of the biomes...

The introduction to this chapter emphasized that the earth is not uniform in material composition or climate. The diversity of global biomes is one manifestation of this fact. Yet as Europeans settled in Africa, Australia, and the western hemisphere, their expectations were remarkably uniform. Thus, if large trees are indicators of fertile agricultural soils in Europe and eastern North America, the same should be true in the tropical rain forest. However, tropical soils are part of a different biome, unsuited to western, mechanized agriculture. Indeed, many agricultural efforts in the tropics are creating "wasteland" with no known sustainable use thereafter. Add to this the fact that leaders of many LDC's view their tropical forests as a ready source of cash to save their struggling economies. But every tree that falls in the tropical forest carries with it untold life forms that have yet to be taxonomically classified, let alone studied for their possible agricultural, medicinal, or nutritional value. Some predict that no undisturbed tropical forests will remain by the end of this century. The global ecological impact of such a loss is of great concern to ecologists (Section 6-D, 6-I).

...and the long-term value.

Ecological distinctives:

The same scenario of disruption is occurring in other biomes as outlined in Table 8-3 (see also Table 6-3). In each instance, the disruption might be avoided if the following ecological principles would govern human activities:

soil and climate

1. *The soils and climate of Earth differ geographically, so that each biome has its own unique biotic communities.* The populations that compose these communities can exist in complex, homeostatic interdependence upon one another as energy flows through food webs and nutrients are recycled. However, biomes differ markedly in their response to human activities.

carrying capacity

2. *Each global ecosystem has an **integrity** which cannot be violated by human intervention beyond a certain limit without a collapse of the system.* Just as it is wrong to overload a beast of burden, so we must identify the limits of *reasonable demands* upon each living community and adjust our activities accordingly.

accumulated small abuses

3. *Collapse of ecosystems on a global scale are most likely to be the result of an accumulation of ecological violations occurring on a local scale.* Notice that the human practices in Table 8-3 which threaten global biomes consist of actions upon one acre at a time, or upon one forest tree at a time. But these actions on a "local scale" have accumulative effects on a "global scale".

In summary, ecology is clearly revealing the functions and limitations of global ecosystems. The question is, "What moral and ethical basis is there for being concerned about the way we treat these systems?"

8-L HUMANS AND THE ENVIRONMENT — IN SEARCH OF AN ETHIC

European roots

From his beginning, man has been in search of an understanding of his origin, his relationship to the supernatural and natural world, and his individual and corporate destiny. Wilkinson, et al (33) provide an extensive review of the influences of the Greek and Medieval philosophers, and Judeo-Christian

Table 8-3. Human Disruptions within Different Global Biomes.

BIOME	MAJOR LIMITATIONS	DISRUPTIONS	REFERENCES
TUNDRA	Easily disturbed soil; simple food webs are easily disrupted; Low temperatures cause slow recovery	Oil (pipeline), and mineral exploration and extraction	2, 29
TAIGA	Simple food webs are vulnerable to introduced species	Clearcutting timber Tree monocultures Pest outbreaks	3, 29
TEMPERATE DECIDUOUS FOREST	Steep or rolling topography	Deforestation, agriculture, and urbanization causes soil erosion, water pollution	5, 29
TROPICAL RAIN FOREST	Rapid leaching and erosion of soils; Slow recovery	Deforestation, agriculture, and urbanization	6, 16, 34
GRASSLAND	Low precipitation	Irrigation depletes underground aquifers; Overgrazing, erosion	25, 26
DESERT	Low precipitation high evaporation	Irrigation causes salt buildup; urbanization is depleting aquifers	25, 30
OCEANIC	Receives terrestrial runoff; coastal conc. of human population	Eutrophication of coastal areas; tourism, commercial fishing, spills, dumping	15, 21, 24, 32

teachings upon modern views of nature as tempered by the scientific revolution and colonization of the New World. Each of these ideas and events have influenced 20th century views of the environment.

"redeeming" the wilderness

The command given to man in Genesis 1:28, to multiply and subdue the earth, and to have dominion over the creatures was much in the minds of the pilgrim fathers as they viewed the apparently endless wilderness of North America. Like the sinful, rebellious human heart, this tangled wilderness filled with "wild beasts and wild men", according to William Bradford, must be brought under control. The clearing of forests and the construction of cities and farms "was to bring order out of chaos and to advance the kingdom of God" (33). Nature was there to be mastered; and so, with axe, plow, and gun, the pioneers went west.

an uncelebrated "triumph"

Not until the late 19th century did our nation begin to awaken to the reality that there was no more frontier. With a growing awareness of his power to eliminate both habitat and species from the face of the earth, man began to think of preservation. But at the same time a growing economy required resources from forest, mine, soil, and oil well. Indeed, the rich store of natural

pragmatism still reigns

resources and hardworking Americans combined to carry the growing nation through several periods of economic depression. What is the value of a tree, a forest, a bison, or a passenger pigeon in the face of hard times? It seemed that economics, not ecology, was the basis for putting bread on the table.

Our brief overview, discussed in detail by Wilkinson, *et.al* (33) contends that modern views of nature have been shaped by many centuries of thinking and experience. Indeed, we struggle with the same questions today as did our forefathers. In essence we too are asking how our relationship to the environment will influence our life goals, our futures, and that of our children.

Why should I care?

Should I care that the energy resources, mineral resources, forests, and wildlife of planet Earth are being consumed so rapidly that future generations may not have access to them? If I am concerned, what should be the basis for caring? What standards should govern my conduct?

Answers are plentiful; and, they come from many perspectives. We have already presented an ecological basis for understanding the creation and our responsibilities to manage *food and energy* (Chapter 5), *nutrients and material resources* (Chapter 6), *our own human population* (Chapter 7), and our *local scale activities* as they affect the Earth on a *global scale* (Chapter 8). Ecology can tell us "What's wrong?"; but, something more is required to lead us to ask, "What must I do?".

pantheistic motivations

A host of religious movements now offer theologies that emphasize human relationships to the Earth. In Section 8-A, we noted the New Age "Gaia", an outgrowth of pantheism that provides a spiritual dimension for the *Deep Ecology* movement (19). By becoming conscious of our supposed oneness with "Mother Earth", we can join in saving her and humanity. Matthew Fox's *creation spirituality* supplants Judeo-Christian theology with a combination of environmentalism, mythology, eroticism, and witchcraft. The *Green Movement* is a "complex and changing network of groups loosely joined through a common vision: to restore social, political, and ecological justice to the world" (17). "Greens" vary in the degree to which they are motivated by the Gaia and/or creation spirituality.

It is obvious that humans have a spiritual dimension which calls for a belief system to interpret the world around them. As noted in Section 1-A, a belief system forms the basis for ones *worldview*. This author is fortunate to have placed his faith in the Scriptures as being a divine revelation to mankind. I believe the Scriptures provide the only true foundation for science in general (Chapter 1), and for dealing with specific issues such as the "environmental crisis". Therefore, let us examine what the Scriptures teach concerning man's relationship to the environment.

8-M CHRISTIAN STEWARDSHIP OF THE ENVIRONMENT

Our purpose in this section is to outline several Scriptural principles and their relevance to the matter of stewardship of material resources of God's creation.

Here is a place to begin.

You are encouraged to use the outline to foster concentration and meditation on God's truth. Scriptural truth is meant to be put into action. Therefore, the outline includes parenthetical [e.g. "(8-K)"] references to topics discussed in previous sections of this text. It is hoped that you will make this an ongoing study, and that the "fruit" of your efforts will be seen in your daily lifestyle as it relates to the handling of material resources.

For additional reading, consult the REFERENCES at the end of the chapter. Stewardship references are starred(*). Christian scholars have different opinions on these issues, so you will need to evaluate each writer in light of your interpretation of the Scriptures.

I. **MEDITATION UPON SCRIPTURE**

   A. GOD

   1. CREATOR of the Universe—Genesis 1:1; John 1:1-3
   2. OWNER of the Universe—Ex. 19:5; Lev. 25:23; Psalm 24:1; 50:10-11
   3. SUSTAINER of the Universe—Colossians 1:17-18,20; Job 38-39
   4. REVEALER of Absolutes
      a. SCRIPTURE (moral laws)—Genesis 1:16-17; Psalm 19:7-14; II Timothy 3:16-17
      b. CREATION (natural laws)—Psalm 19:1-6; Colossians 1:16-17

   B. MAN

   1. RELATIONSHIP to God and Nature
      a. LIKE other biological creatures, we are...
         ...from dust (Gen. 2:7;3:19) as the plants
         (Gen. 1:11-12) and land animals (Gen. 1:24-25)
         ...given "breath of life" (Gen. 2:7) as land animals (Gen. 7:13-22)
         ...ecologically dependent on other creatures (Gen. 1:29-30)
      b. UNLIKE other creatures, we are...
         ...created in the image of God (Gen. 1:26-27)
         ...invited to worship and delight in God's truth (Psalm 19)
         ...given dominion (Gen. 1:28)

   2. RESPONSIBILITIES to God and Nature
      a. DOMINION (Gen. 1:28)—Heb. *kabash* = subdue; bring into bondage
      b. SERVANTHOOD (Gen.2:15)—Heb. *abad* = "till", "serve"; —Heb. *shamar* = "preserve", "keep"
      c. STEWARDSHIP—exercise dominion as an obedient servant (a + b) make reasonable demands that respect biotic *integrity* (Sec. 8-K)
      d. SCIENCE—pursuit of knowledge essential for stewardship (Gen. 2:19-20; Prov. 25:2)

   C. SIN

   1. REBELLION against God's laws
   2. Causes BROKEN RELATIONSHIPS between...
      a. ...man and God (Gen. 3:5-6, 10, 23)
      b. ...man and himself (Gen. 2:25 vs. 3:7-8)
      c. ...man and man (Gen. 3:12; 4:8)
      d. ...man and nature (Gen. 3:17; Rom. 8:20-21)

D. CHRIST

1. RECONCILER of all things in heaven and earth to Himself (John 3:16 ["cosmos"]; Colossians 1:20)
2. HEALER of broken relationships between...
   a. ...man and God (Rom. 5:1)
   b. ...man and himself (Rom. 6:8-12)
   c. ...man and man (John 15:12)
   d. ...man and nature (Rom. 8:19)
3. PERFECT EXAMPLE of Christian stewardship (Phil. 2:5-11)
4. PROCLAIMER of stewardship principles as a witness of the kingdom
   a. Choose whom you will serve—God or material wealth (Matt. 6:24)
   b. As you serve God, trust His "life support system"(Matt. 6:25-33)
   c. Share God's concern for creation (Psalm 104; Rom. 3:19-22)
   d. Live by God's natural laws (Prov. 6:6-9; 30:24-28; Matt. 6:26)
   e. As followers of Christ, share His compassion for the oppressed who are denied access to the essentials of life (Ex. 23:10-11; Deut. 14:28-29; Lev. 19:9-10; Isaiah 58-59:2; Matt. 25-14-46; James 2:14-17; I John 3:16-17)

II. MANAGEMENT AS STEWARDS

A. RECONSIDER your personal actions in light of the following:

1. God's Care — Be content (Psalm 23: 1; II Tim. 6: 6-9)
2. God's Kingdom — His "Economy" Matthew 6: 25-33; REF. #(1)
3. God's Commandment — Love Your Neighbor (I John 3:16-18)
4. God's Creatures — Don't deprive them of God's blessing.
   — Genesis 1:21-22; Ezekiel 34:18-19; Deut. 22:6; Rom. 8:19

B. REDUCE energy and material resource consumption, and food waste in light of the fact that...

1. We cannot create energy and matter, we are only users (5-C, 6-I)
2. We cannot "dispose of" energy or matter; they must go somewhere. (5-C, 6-C through 6-I; esp. 6-I)
3. We can only conserve exhaustable, nonrecyclable resources by reducing consumption; then, future generations can enjoy (5-C, 6-I)
4. Fulfillment in life is not conditioned upon high energy and material consumption; rather it comes by sharing and giving to those in need (7-G).
5. In light of human need in many LDC's it is Scriptural to reduce consumption, leaving more to share. (7-G).
6. Small-scale, local actions (for good, or ill) have cumulative effects on a global scale. (8-A, 8-K)

C. REUSE things as often as possible, considering the fact that...

1. We "do not need materials, but the services they provide" (36)
2. We can conserve exhaustable fossil fuel supplies by producing/purchasing articles designed for a long life, or made of materials which permit reuse. (6-I)
3. We are encouraged not to reuse things in our culture for economic reasons, not stewardship/ecological reasons.

D. RECYCLE materials which no longer provide needed service, and which cannot be REUSED (e.g. wash-sterilized container)(6-I)

E. REPRESENT a lifestyle of sound stewardship of resources as part of being Christian "salt" and "light" through your influence from the "local scale" to "global scale" (8-A):

1. IN YOUR HOME — implement the above principles daily
2. IN YOUR CHURCH — make it a body wherein individuals and families can counter materialism and "walk as Jesus did".
3. IN YOUR COMMUNITY — work with local conservation groups
4. IN YOUR PROFESSION — to make a difference by your expertise
5. IN YOUR GOVERNMENT — support candidates/legislation

"Suddenly, then, we have a real beauty. Life begins to breathe. The world begins to breathe as it never breathed before. We can love a man for his own sake, for we know who the man is—he is made in the image of God; and we can care for the animal, the tree, and even the machine portion of the universe, each thing in its own order—for we know it to be a fellow creature with ourselves, both made by the same God".
—Francis A. Schaeffer (27)

QUESTIONS AND DISCUSSION TOPICS

1. How is the concept of carrying capacity important in managing grazing herds in pastures? In planning outdoor recreational areas?

2. Based upon your study of the food webs, biogeochemical cycles, and complex interrelationships among populations, discuss the importance of preserving ecosystem *integrity* in order to preserve life as we know it.

3. Human conduct in today's world is governed by individual and corporate value systems. In each of the paired options below, which choice of values is most consistent with scriptural and ecological principles? Which choice is a part of your value system?

    a. Concern for the *present* versus concern for the *future*
    b. Concern for *man* versus concern for all of *creation*
    c. Consumption of resources based upon *wants* versus *needs*
    d. Consumption based upon *convenience* versus *conservation*.
    e. Consumption based upon *economic criteria* vs. *ecological criteria*
    f. Consumption of *exhaustable* (e.g. fossil fuels) vs. *inexhaustable* resources (e.g. solar energy, manpower)
    g. Seeking satisfaction in *material wealth/position* versus satisfaction by *submission to God* as a source of personal worth and fulfillment

4. Distinguish three current conservation practices—REDUCE, REUSE, RECYCLE; then, the relative effectiveness of each in combating the ecological problems associated with each of the following:

    a. Aluminum production (energy requirements and pollution)(Section 6-I).
    b. Food production to meet increasing world population (Section 5-G)
    c. Diminishing fossil fuel (coal, oil, natural gas) reserves (Section 5-C, Section 6-I).
    d. Increasing atmospheric CO_2 levels (Section 6-D, Section 6-I).
    e. High energy demands by MDC's lifts energy costs beyond reach of many in LDC's (Section 7-G).

5. Whereas we know that fossil fuel combustion increases atmospheric CO_2, it is less clear whether increasing CO_2 causes "global warming". Note and discuss the temporal and spatial *scales* that are involved in the "global warming" issue.

6. Many ecological issues involve phenomena such as tropical forest destruction, global warming, or ozone depletion which are hard for the individual to visualize. In addition, the predicted outcomes are often subject to debate. Why then should Christians be concerned about the environmental impact of "just little old me"? Use the concept of "scale" (Section 8-A) in your response.

TOPICS FOR FURTHER STUDY

1. Choose the biome of greatest interest to you and consult references cited in Table 8-3 to gather additional information on its populations, food webs, and unique adaptations. Then, do an in-depth study of the human impact upon that biome; and the economic, political, and ecological implications for the future. Use REFERENCES for additional information.

2. Study the history of the western environmental movement. REFERENCE: #14

3. Select one of the modern philosophical teachings that is being regarded as the answer to the human search for identity and relationship to the natural world. Analyze it from a scriptural perspective. REFERENCES: # 17, #27, #33

4. What practices in the home, on the farm, in business and industry, etc. can be modified to reflect better stewardship of the environment? Table 8-4 below presents several contrasts between ecologically unsound practices, and possible alternatives that may be more ecologically sound. Select one or more contrasts and consult REFERENCES such as #7, #12, #13, #17, #22, and #18 for discussion and additional literature citations. Based upon scientific data and scriptural principles, formulate a recommendation favoring one or the other options. What factors may oppose the practice?

Table 8-4 Management Options for Home, Agriculture, Business, and Industry.

ACTIVITY	OPTION 1	OPTION 2
Disposal of solid waste—paper, glass, metal, plastic, etc.	Dump and burn in landfill	Returnable containers: recycle material; composting
Disposal of kitchen and bathroom wastes	Pipe to wastewater plant and river	Waterless toilet, composting
Recreation	Emphasis upon electronic, motor-driven, fossil fuel-dependent activities	Emphasis upon solar-powered, bodily exercise-dependent activities
Agriculture	Monocultures Monocropping Agrochemicals	Mixed plantings Crop rotation Integrated pest management (IPM)
Lawn & Garden	Commercial pesticides and fertilizers; frequent watering and mowing	Limited fertilizing; composting, mulching

REFERENCES

1. *Berry, W. 1987. *Home Economics.* North Point Press, San Francisco, CA.

2. Bliss, L.C. *et. al.* 1973. Arctic tundra ecosystems. *Ann. Rev. Ecology and Systematics* 4: 359-399.

3. Bonan, G.B. and H.H. Shugart. 1989. Environmental factors and ecological processes in boreal forests. *Ann. Rev. of Ecol. Syst.* 20: 1-18.

4. van den Bosch, R. and M.L. Flint. 1981. *Introduction to Integrated Pest Management.* Plenum Press, New York.

5. Braun, E.L. 1950. *Deciduous Forests of Eastern North America.* McGraw-Hill-Blakiston, New York.

6. Buschbaker, R.J. 1986. Tropical deforestation and pasture development. *Bioscience* 36: 22-28.

7. *DeWitt, C.B. ed. 1991. *The Environment and the Christian: What Can We Learn from the New Testament?* Baker Book House, Grand Rapids, MI.

8. Durning, A. 1991. Asking how much is enough. In Brown, L.R., *et.al. State of the World.* pp 152-169. Worldwatch Institute Report. W.W. Norton, New York, NY.

9. Ehrlich, P.R. 1982. Human carrying capacity, extinctions, and nature reserves. *Bioscience* 32: 331-333.

10. *Elsdon, R. 1981. *Bent World*. InterVarsity Press, Downers Grove, IL.

11. Graham, R.L., *et.al.* 1990. How increasing CO_2 and climate change affect forests. *Bioscience* 40(8): 575-587.

12. *Granberg-Michaelson, W. 1984. *A Worldly Spirituality: The Call to Redeem Life on Earth*. Harper and Row, San Francisco, CA.

13. *Granberg-Michaelson, W. 1987. *Tending the Garden*. Eerdmans. Grand Rapids, MI.

14. Grove, R.H. 1992. Origins of the western environmental movement. *Sci. Amer.* 267(1): 42-47.

15. Jackson, J.B.C. 1991. Adaptation and diversity of reef corals. *Bioscience* 41:475-482.

16. Janson, D.J. 1986. The future of tropical ecology. *Ann. Rev. Ecol. Syst.* 17: 305-324.

17. *Kjos, B. 1992. *Under the Spell of Mother Earth*. Victor Books, Wheaton, IL.

18. *Land, R.D. and L.A. Moore. eds. 1992. *The Earth is the Lord's: Christians and the Environment*. Broadman Press, Nashville, TN.

19. Lovelock, J. 1987. *Gaia: A New Look at Life on Earth*. Oxford U Press. New York, NY.

20. McKibben, W. 1989. *The End of Nature*. Random House, New York, NY.

21. McLusky, D.S. 1981. *The Estuarine Ecosystem*. Halsted-Wiley, New York, NY.

22. Miller, G.T. Jr. 1992. *Living in the Environment*, 7th ed. Wadsworth, Belmont, CA.

23. *Moss, R.P. 1978. Environmental problems and the Christian Ethic. In *Horizons of Science*, C.F. Henry, ed. Harper and Row, New York.

24. Nybakken, J.W. 1988. *Marine Biology: An Ecological Approach*, 2nd ed. Harper & Row. New York, NY.

25. Reisner, M. and S. Bates. 1990. *Overtapped Oasis: Reform or Revolution for Western Water*. Island Press, Covelo, CA.

26. Risser, P.G., *et.al.* 1981. *The True Prairie Ecosystem*. Hutchinson Ross Publ. Co., Stroudsburg, PA.

27. *Schaeffer, F.A. 1970. *Pollution and the Death of Man*. Tyndale, Wheaton, IL.

28. *Sider, R.J. ed. *Cry Justice—The Bible on Hunger and Poverty*. InterVarsity Press, Downers Grove, IL.

29. Smith, R.L. 1992. *Elements of Ecology*, 3rd ed. HarperCollins Publ., New York, NY.

30. Walker, A.S. 1982. Deserts of China. *American Scientist* 70:366-376.

31. Ward, F. 1990. Coral reefs imperiled. *National Geographic,* (July)

32. Whittaker, R.H. 1975. *Communities and Ecosystems*. Macmillan Publ. Co., Inc., New York, NY.

33. *Wilkinson, L., *et. al.* 1991. *Earthkeeping in the Nineties: Stewardship and the Renewal of Creation*, rev.ed. Eerdmans, Grand Rapids, MI.

34. Wilson, E.O. ed. 1988. *Biodiversity*. National Academy Press, Washington, DC.

35. *World Resources, 1992-1993*. Report by World Resources Institute, 1992. Oxford University Press, New York, NY.

36. Young, J.E. 1991. Reducing waste, saving materials. In Brown, L.R., *et.al. State of the World.* pp 38-55. Worldwatch Institute. W.W. Norton, New York.

* Indicates Christian Stewardship Reference (see Section 8-M)

Part 3

Life Within Organisms

diversity of species and habitations

Living creatures exist in a seemingly endless array of forms and they occupy almost every conceivable location on Earth. Consider the marvelous diversity in color, size, and other anatomical and behavioral features, or adaptations essential to the survival of each species in its habitat. Any attempt to develop a science, or systematic knowledge, of this vast biodiversity of life rests upon the development of a system of classification (Chapter 3). In PART 2, we used *ecological classification* in which all living organisms were grouped according to their means of energy acquisition (Sections 3-D, 5-D). This classification system was adequate for our studies of *relationships among organisms* within ecosystems. However, our attention in PART 3 will shift to *relationships within organisms* and the **basic life processes** that sustain each living creature. For these "up close and personal" studies of such a diversity of species, we need a classification system that respects the anatomical, physiological, and behavioral features of each organism from inside out. The *taxonomic classification* system which was introduced in Section 3-C meets this need. Here, anatomically similar species are grouped into the same *genus* and similar genera occupy the same *family* and so on through *order, classes, phylum, and kingdom* (Table 3-4). Every known species is thus assigned to one of the five kingdoms outlined in Table 3-5.

shifting our emphasis

Taxonomic classification brings order.

Part 3: Preview

Chapter 9 presents a survey of the major taxonomic groups and emphasizes *distinguishing features* of each group. With this survey as background, we will study the *basic life functions* with emphasis on *representative organisms* from each taxonomic group. Thus, Chapter 10 explores the basic life function of *nutrition*, Chapters 11 and 12, *reproduction* and *genetics*, Chapters 13 and 14 discuss *responsiveness* and *homeostasis*. Finally, in Chapter 15, we end PART 3 on the topic with which we began—biodiversity. There we explore several contemporary models that attempt to account for the *origin and diversity* of life.

Chapter 9

Diversity of Life

9-A DEALING WITH DIVERSITY — TAXONOMIC CLASSIFICATION

To illustrate the importance of taxonomic classification, suppose we were to visit a natural history museum in which all "animal" species were visible within one panoramic view. Without a specific floor plan to arrange the various species according to anatomical similarities, we would be impressed primarily by the diversity among the members. We would probably observe some interesting behavioral interactions as well. However, suppose the same group of animals was classified and arranged into separate rooms, one for each animal phylum. Now we would notice certain common features among animals within each room. These features unify the species of that room but distinguish them from other rooms. Some groups would have primarily a worm-like body structure. Another phylum would have insect or spider-like features, and still others would have an internal skeletal structure. The subphylum Vertebrata contains the vertebrates, animals with a backbone. Further sorting among the vertebrates would distinguish among the vertebrate classes—reptiles, fish, birds, mammals, and so on. As sorting continues through orders, families, and genera, there would be progressively fewer species in each group.

Sorting species according to common features...

Can you imagine yourself involved in such a sorting process among all of the animal species or members of other kingdoms? This is the essence of the task of taxonomy. It should be evident that there is much subjectivity involved as to which species should or should not be included in a given genus and which genera make up a given family. Even at the phylum and kingdom levels different schemes are proposed. We must recognize that taxonomic systems are theoretical constructs that attempt to represent reality, but must be continually subjected to scientific verification.

...is the task of taxonomy.

The "species" would seem to be the only category based upon objective criteria. Recall that a **species** includes all of the natural populations that interbreed, actually or potentially, to produce fertile offspring and are anatomically distinguishable from other species. However, as we noted in Section 3-C, even species distinctions are difficult to make. Whether or not it is easy to distinguish each interbreeding population, they do exist, and the species concept is a useful approximation of reality.

difficult to define *species*

According to the creation account in Genesis 1-2, different *kinds* were created and endowed with the capacity to reproduce after their kind. This implies the same distinctions among different interbreeding groups as is suggested in the concept of species. However, we cannot be sure of the degree of correspondence between "species" and "kind." Scripture does not rule out the possibility that a "kind" is a "species," or possibly "genus," or even "family." In light

species* versus *kind

of the human subjectivity involved in defining the taxonomic groups, "kind" could be equivalent to higher taxonomic levels in some families than in others. We shall use the term "species" as defined above while recognizing the difficulties in applying the concept. [For further information, see TOPICS FOR FURTHER STUDY #3 and #5.]

registering new species

Though still imperfect, taxonomic classification, like other biological classifications, is an attempt to arrange information in a systematic manner. Thus, taxonomic classification serves biologists in both identification and study of a given species. When potentially unique specimens are discovered, say in the tropical rain forest, these specimens are compared alongside similar preserved specimens in taxonomic collections. If there is agreement among taxonomists that the new specimens are distinctly different from any in the existing collection, they are officially named and classified as a new species. However, the ultimate test of whether they are of the same species is, by definition, whether or not they can interbreed and produce fertile offspring as noted above.

inferences from representative species

Once a newly discovered species is classified, much can be inferred concerning its anatomy and physiology based upon what is already known about other species within the same taxonomic group. For our purposes, it will be possible to discuss one or a few *representative species* of each major phylum and assume general applications can be made to other species of that phylum. We shall begin to apply this principle throughout the remainder of the chapter as we briefly survey the major taxonomic kingdoms and selected phyla within each kingdom. Then, in Chapters 10 through 14, we shall discuss the manner in which representative organisms of each group perform the basic life processes common to all living things.

evolution presuppositions

Our survey of taxonomic classification will begin with the Kingdom Monera because it contains the least complex, unicellular (one-celled) organisms. Most taxonomic systems begin with the Monera because the evolution model proposes that all species on earth today are related through evolutionary descent to Monera-like ancestors. Evolutionists thus interpret taxonomic classification differently from creationists. The evolutionary interpretation is based upon the presupposition that similar form, function, and behavior among taxonomic groups is due to their close relationships through evolutionary descent. **Phylogeny** [G. (*phylum* = tribe, race) + (*geno* = descent)] is an outline of evolutionary descent of the various taxonomic groups from other groups, presumed to have occurred over aeons of time.

evolution and the "gaps"

For reasons that we shall consider later, the existence of phylogenetic relationships among the major taxonomic groups cannot be scientifically verified. The phylogenetic "trees" (Figure 9-1) have branches that are "unattached" to one another because of the existence of major "gaps" between taxonomic groups, both living and extinct (i.e. fossilized specimens only).

Creation model is also not verifiable.

Indeed, the very existence of gaps, essential for any taxonomic classification, is inconsistent with the supposed evolutionary descent of present organisms. We will hasten to mention that the creationist positions are also not scientifically testable. This is true in spite of the fact that a separate creation of distinct "kinds" of organisms (Genesis 1-2) seems to be more consistent with the distinctness now observed among kinds of contemporary organisms. Regardless of which model you presently accept, you should be conscious of your

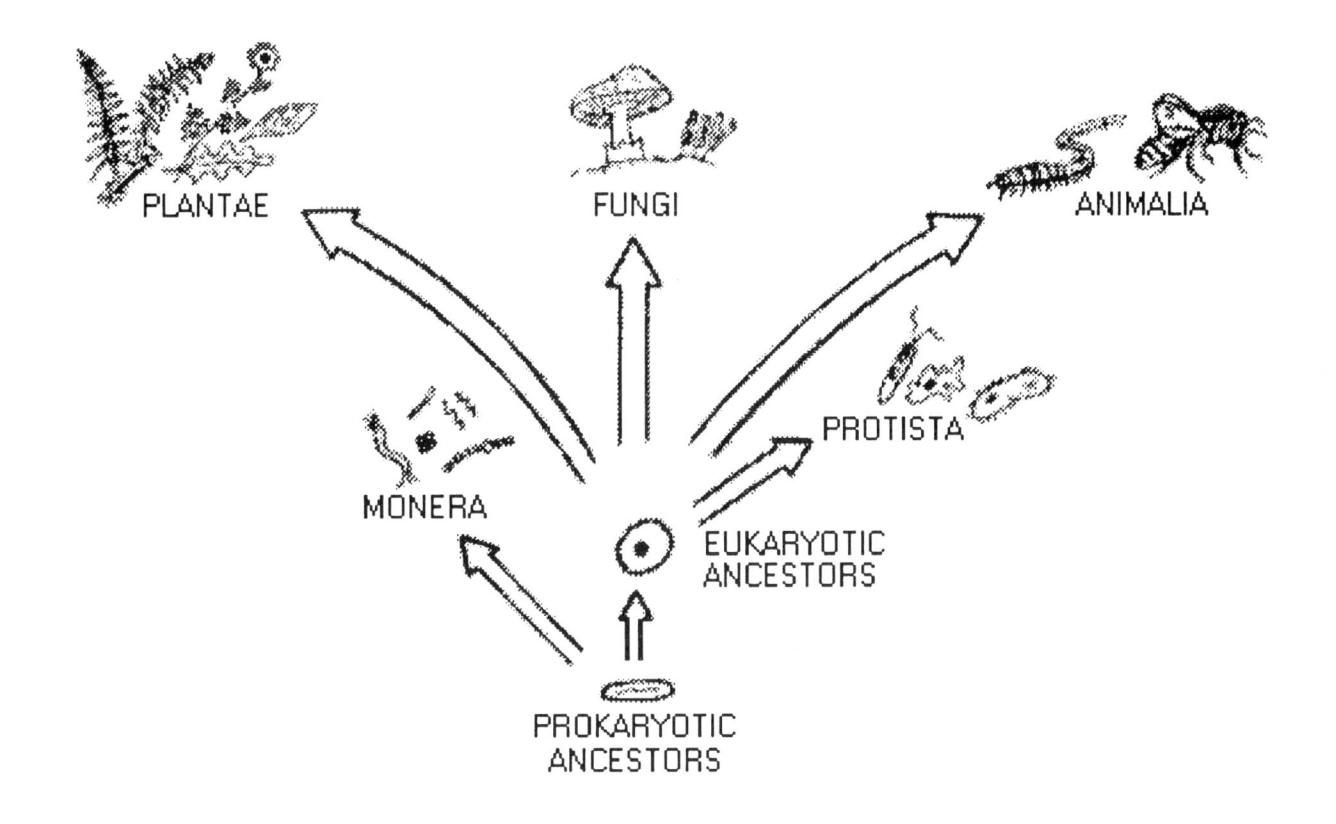

Figure 9-1. A phylogenetic "tree" showing supposed evolutionary relationships among the five kingdoms of organisms.

Watch out for presuppositions

own underlying presuppositions about the origin of life, and those held by others. Even the terminology we will use is derived from an evolution-based presupposition—e.g. prokaryote (G. *pro* = before), protozoa, protista (G. *proto* = first, original).

As we survey the taxonomic classification of species, we shall omit the use of many of the taxonomic names of the various phyla. If you wish to refer to these names, they are outlined in Appendix B, which is based primarily upon the system Whittaker (17) introduced, in 1969.

9-B KINGDOM MONERA

Kingdom Monera contains species that are distinguished by the presence of **prokaryotic** cells. As the name prokaryotic [*pro* = before + *karyon* = kernel] suggests, these organisms contain DNA that is not enclosed by a nuclear membrane—i.e. no cell nucleus, or "kernel" (Figure 9-2). Monerans also lack other organelles such as mitochondria and chloroplasts which are found in **eukaryotic** [(*eu* = true) + (*karyon* = kernel)] cells. Instead, metabolic processes occur in the cytoplasm and infoldings of the cell membrane. Moneran cells possess cell walls, often coated with a *capsule* composed of gelatin-like material. Kingdom Monera is subdivided by some taxonomists into three subkingdoms highlighted below.

Monerans lack organelles...

...but have cell membranes and cell walls.

KINGDOM MONERA:

SUBKINGDOM ARCHAEBACTERIA — "ancient" bacteria
SUBKINGDOM EUBACTERIA — "true" bacteria
SUBKINGDOM CYANOBACTERIA — "blue-green" bacteria

ARCHAEBACTERIA

methane bacteria

Archaebacteria are represented by *methane bacteria* which live in airless soils of swamps and produce methane (CH_4), or "swamp gas." Other members of this subkingdom inhabit boiling hot springs or bodies of salty water (e.g. Dead Sea). Because of their microscopic size, absence of organelles, and unusual metabolism, the Archaebacteria are regarded by adherents to the evolution model as being close descendants of the primordial cells that formed spontaneously some 3.5 billion years ago. [Did they remain essentially unchanged by evolution for 3.5 billion years while millions of other species evolved into being?] Archaebacteria have membrane lipids, cell wall components, and RNA that are not found in either prokaryotic or eukaryotic cells! Therefore, some taxonomists recommend placing Archaebacteria into a separate kingdom.

Are they really "primitive"?

EUBACTERIA

This subkingdom includes the more familiar species we call "bacteria" [*eu* = true]. True bacteria are distinguished both by *shape* and by their mode of *nutrition*. Cell shapes include **coccus** (sphere), **bacillus** (rod), and **spirillum** (corkscrew). Species having cells attached in clusters or filaments are often named with the prefix *strepto-* as in streptococcus. These shapes will appear as tiny opaque dots, dashes, and squiggles, respectively, under the high-power lens of most compound microscopes (Figure 9-2).

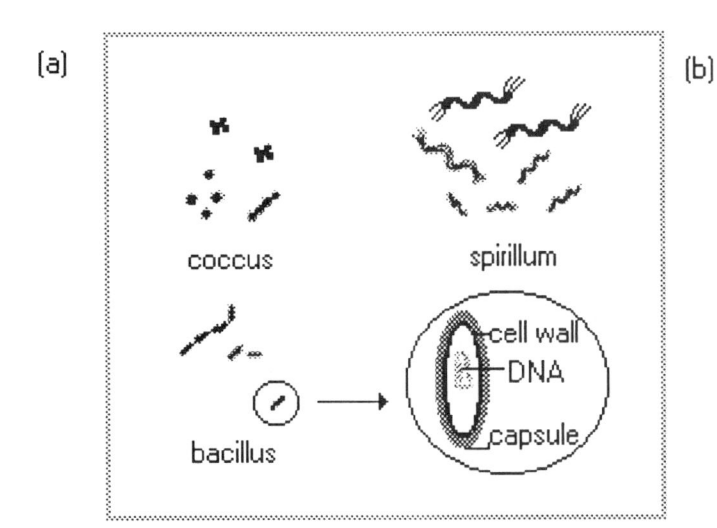

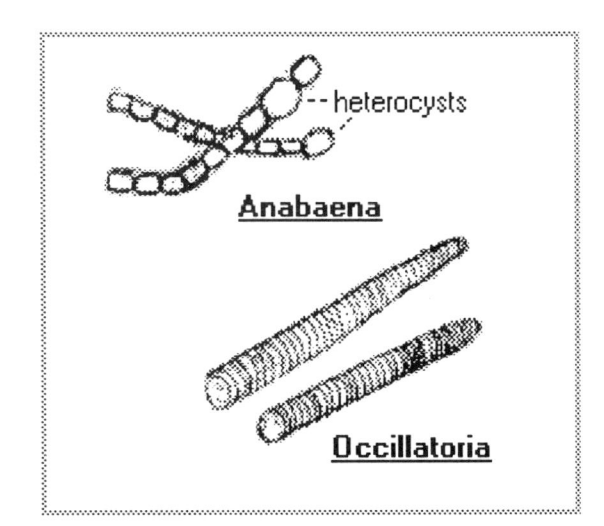

Figure 9-2. Representatives of Monera as seen under high magnification: (a) Three common shapes of bacteria; inset shows bacterial cell structure; (b) Two species of cyanobacteria. All are as seen under high magnification.

164

bacterial cell structure...

Bacterial cell shape is maintained by a **cell wall** which surrounds the cell membrane. The cell wall is a semirigid mesh of polysaccharides cross-linked by peptide chains. A gelatinous *capsule* coats the cell wall and aids in bacterial attachment to surfaces. Many thread-like **flagella** are attached to the cell membrane and extend outward through the cell wall and capsule (Figure 9-2). Rotational motion of flagella provides directional motion in response to various stimuli such as light, nutrients, or O_2.

size...

Bacterial cells range in size from 0.1 to 5 micrometers (10^{-6} meter, or about 1/250,000 inch). Though small in size, bacteria literally cover the face of the earth, its soils, water, and air. Indeed, we must sterilize liquids, objects, and surfaces if we wish to eliminate bacteria.

nutrition...

The bacteria are not only diverse in shape and in habitat, but also in mode of nutrition. Resorting to ecological classification, we find bacterial species that may be either autotrophs, decomposers, parasites, or mutualists (Section 5-A, 5-D). Autotrophic Eubacteria may be photosynthetic or chemosynthetic. [Yes, there is even diversity among bacterial autotrophs!]

bacterial decomposers...

In the process of acquiring their own nutrition, bacteria decompose organic compounds and release inorganic constituents of dead organic matter and wastes to reservoirs in the physical environment (Chapter 6). Precise strains of bacteria are "employed" deliberately to cause decay of foodstuffs. Bacterial **fermentation** (anaerobic respiration) and decay produce desirable flavor and nutritional value in the production of cheeses and other dairy products.

parasites...

All too often, we encounter parasitic (disease-causing, or pathogenic) bacteria when our bodies or our pets and plants serve as their **hosts**. Infection and disease usually result from the release of bacterial **toxins** or other substances which interfere with host cell processes. For instance, *Clostridium botulinum*, an anaerobic bacterium which becomes active inside canned foods that have been incompletely sterilized, causes *botulism*. A related species is *C. tetani* which thrives in deep wound infections and causes *tetanus*. Both species produce neurotoxins which interfere with muscle contraction and may cause death if not promptly treated.

Many bacterial species that live in soil and water can kill or suppress reproduction of other species by synthesizing and releasing organic compounds, called **antibiotics** [Gr. (*anti* = against) + (*bio* = life)]. Pharmaceutical companies use bacterial cultures as "microbial factories" to produce antibiotics such as streptomycin for treatment of infections. The bacterium *Escherichia coli* which inhabits the colon of humans and animals is among bacterial species that have been "genetically engineered" to produce a variety of useful compounds (Section 17-G).

and mutualists

Finally, bacteria are members of innumerable mutualistic relationships such as the symbiotic relationship involving N_2 fixation (Section 6-F). *E. coli*, noted above, and other bacterial species of your colon are in mutualistic relationship with you! Your digesting food supplies chemical energy and nutrients, and the bacteria synthesize certain vitamins that are absorbed by your body. When massive doses of antibiotics are administered, the complex "ecology" of the large intestine is greatly disrupted.

CYANOBACTERIA

Cyanobacteria are autotrophs...

Cyanobacteria are also called blue-green algae because they resemble the green algae of Kingdom Plantae. However, even though they do have chlorophyll and are photosynthetic autotrophs, closer examination reveals the prokaryotic cell structure, typical of Kingdom Monera (Figure 9-2). Cyanobacteria are common in eutrophic lakes and streams (Section 6-G) where they may produce toxins that kill fish and make the water undrinkable. They also form slimy coatings in birdbaths and spoutings.

...and some are symbiotic.

Certain species of Eubacteria and Cyanobacteria can perform N_2 fixation, either as free-living species, or in symbiotic relationship (Section 6-F), depending upon the species. Cyanobacteria may perform N_2 fixation in symbiotic relationship with fungi as in certain lichens (Section 7-F), or with mosses or ferns. The floating fronds of the small aquatic fern, *Azolla* contain small air chambers filled with *Anabaena*, the cyanobacterium shown in Figure 9-2,b. For centuries, this relationship has provided an important source of usable nitrogen for the rice paddies of the Orient.

Summary:

In summary, the bacteria and cyanobacteria are tiny but very successful judging from their ubiquity on land, air, and sea, and even within the bodies of other organisms. Their nutritional processes are also varied, and involve complex chemical reactions, many of which are not found in so-called "higher" organisms. Though we cannot see them without magnification, their ecological and economic importance assures that they would certainly be missed if they were to "disappear."

9-C KINGDOM PROTISTA

Most "protists" are one-celled

The members of the kingdom Protista, and all other species except the Monera, are **eukaryotic**, having cells with a true nucleus and the presence of mitochondria and other organelles. Most "protists" are unicellular.

Protista is a diverse phylum.

The existence of the Kingdom Protista testifies in part to the difficulty taxonomists have had in classifying the vast array of creatures. As already noted, Protists are prokaryotic and unicellular but, neither these nor any other features are unique characteristics that distinguish the protists from other kingdoms. For instance, almost every manner of nutrition represented in the other kingdoms is found among protists. Some individual protists such as *euglena* may be both autotrophic and heterotrophic in the presence of light, and heterotrophic in darkness.

cause for great humility

Perhaps the most significant message for us at this point is that we need to be humble and cautious in building categories and conceptual models. Indeed, we are still struggling from the old two-kingdom system in which all creatures were either "plant" or "animal." This taxonomic system was expanded when the microscope revealed the intricacies of the Kingdoms Monera, Protista, and Fungi. Shouldn't we now be asking, "What more is out there (or in there) that we have yet to see?" A classification system can discourage fresh thinking and fresh hypotheses if its limitations are not recognized. We will not belabor classification within the Protista but simply introduce a few representatives of the more notable groups.

Euglena (Phylum Euglenophyta)

Diatoms
(Phylum Chrysophyta)

Dinoflagellate
(Phylum Pyrophyta, "fire algae")

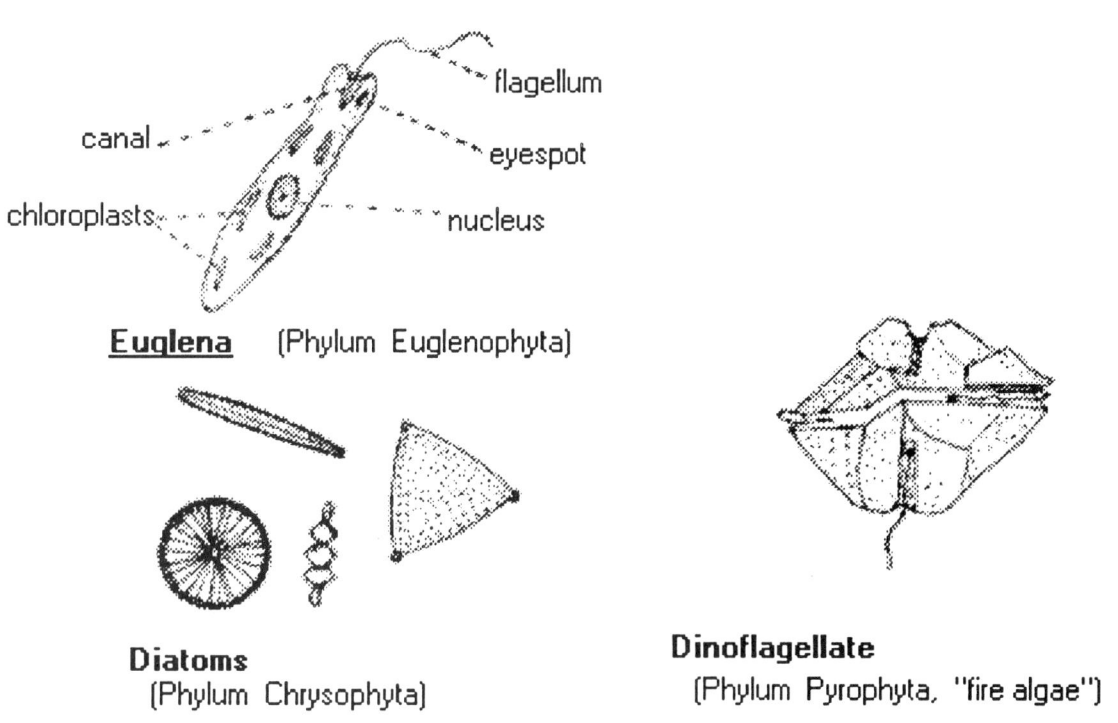

Figure 9-3. Common Plant-like Protistans as they would appear under magnification. Most are less than 0.1 mm. in size.

Kingdom Protista may be divided into seven phyla, including three *divisions* (botanical name for phyla) of plant-like protists, and four phyla of animal-like protists, or **protozoa**. [Explain the derivation of *proto-zoa*.]

KINGDOM PROTISTA

PLANT-LIKE DIVISIONS:
    PYRROPHYTA — "fire algae," dinoflagellates
    CHRYSOPHYTA — golden algae, diatoms
    EUGLENOPHYTA — euglenoids, green flagellates
PROTOZOA:
    MASTIGOPHORA — flagellated protozoa
    SARCODINA — amoeboids, radiolarians, foraminiferans
    SPOROZOA — spore-bearing, parasitic protozoa
    CILIOPHORA — cilia-bearing protozoa

PLANT-LIKE PROTISTA

Euglena: varied nutrition.

The unicellular *Euglena* is one of the most common plant-like protists, commonly found in stagnant or polluted water and in soils (Figure 9-3). It possesses one whip-like flagellum for locomotion. *Euglena* can change its body shape as it maneuvers about because it has no rigid cell wall. An orange granule called the *eyespot* located near the anterior end of the cell enables *Euglena* to detect light stimuli. [How would this enhance survival?] Some

plankton

euglenoids lack chlorophyll and are heterotrophic. Even those that contain chlorophyll become heterotrophs if maintained in darkness. Euglenoids are an important part of phytoplankton (G. *phyto* = plant + *planktos* = wandering) which consists of free-swimming and floating plant-like organisms that serve as food for *zooplankton,* fish and other larger aquatic populations.

Dinoflagella: "two flagella"

Dinoflagellates, so named for their characteristic "two flagella," are primarily marine organisms (Figure 9-3). Certain species can produce light and often send an iridescent glow from ocean waters after sundown. However, this same delightful view can spell death to humans. During summer months along the Gulf coasts, certain poisonous dinoflagellates may be consumed in large numbers by filter-feeding clams and oysters. These mollusks survive, but concentrate the deadly toxins which can be fatal to humans or fish who consume them. This is another illustration of biological magnification (Section 6-4).

diatoms have silica walls

A microscope view of a drop of pond water will reveal a variety of protists. Perhaps the least active but most beautiful are the diatoms, of Division Chrysophyta (Figure 9-3). Diatoms, like the dinoflagellates, are autotrophic and the two groups rank first and second respectively in importance as primary producers in the marine food chain. The unicellular forms of each species of diatom are encased in a uniquely shaped, silica-impregnated cell wall (Figure 9-3). The resultant jewel-like appearance makes the diatoms a geometrical wonder of creation. When diatoms die, the box-like shells sink to the bottom forming a sediment called diatomaceous earth. This material is an important ingredient in cleansers (e.g. toothpaste), filtering agents, and sound-proofing material.

PROTOZOA

Protozoa are heterotrophic protists...

but are they consumers or decomposers?

Protozoa are heterotrophic members of Protista. They obtain energy and nutrients by either *ingesting* particles of organic matter, bacterial cells, and other protozoan species or by *absorbing* dissolved compounds from their surroundings. They are often found in stagnant or polluted water, or in soils in association with bacteria and decaying organic matter, and are important members of detritus food webs (Section 5-H). Still other species are mutualistic or parasitic.

classification of protozoa

Protozoa are classified into several phyla based upon their structures for locomotion (see box above, and Figure 9-4). The phylum **Mastigophora** ("whip-bearer") is named for the whip-like flagella of its species. They are distinguished from the euglenoids of the plant-like protists (Figure 9-3) by the fact that they have no chlorophyll and are heterotrophic. Most species are free-living in aquatic habitats. An important exception are those which live in **symbiotic** relationship in the gut of termites. Without the digestive action of these flagellates, the wood-eating termites would starve to death! [How would the flagellates benefit?]

parasitic flagellates...

The flagellate *Trypanosoma* (Figure 9-4) is a parasite that causes the disease African sleeping sickness in humans. The bite of the tsetse fly injects the parasite into the bloodstream of a mammalian host. This fly is the carrier, or **vector**, of the disease which is common in Africa. The trypanosomes multiply in the bloodstream and release fever-producing toxins. The nervous system is often invaded and brain damage causes lethargy and eventually death. The disease is difficult to control because African mammals serve as a

Trypanosoma
(Phylum Mastigophora, flagellates)

Paramecium
(Phylum Ciliophora, ciliates)

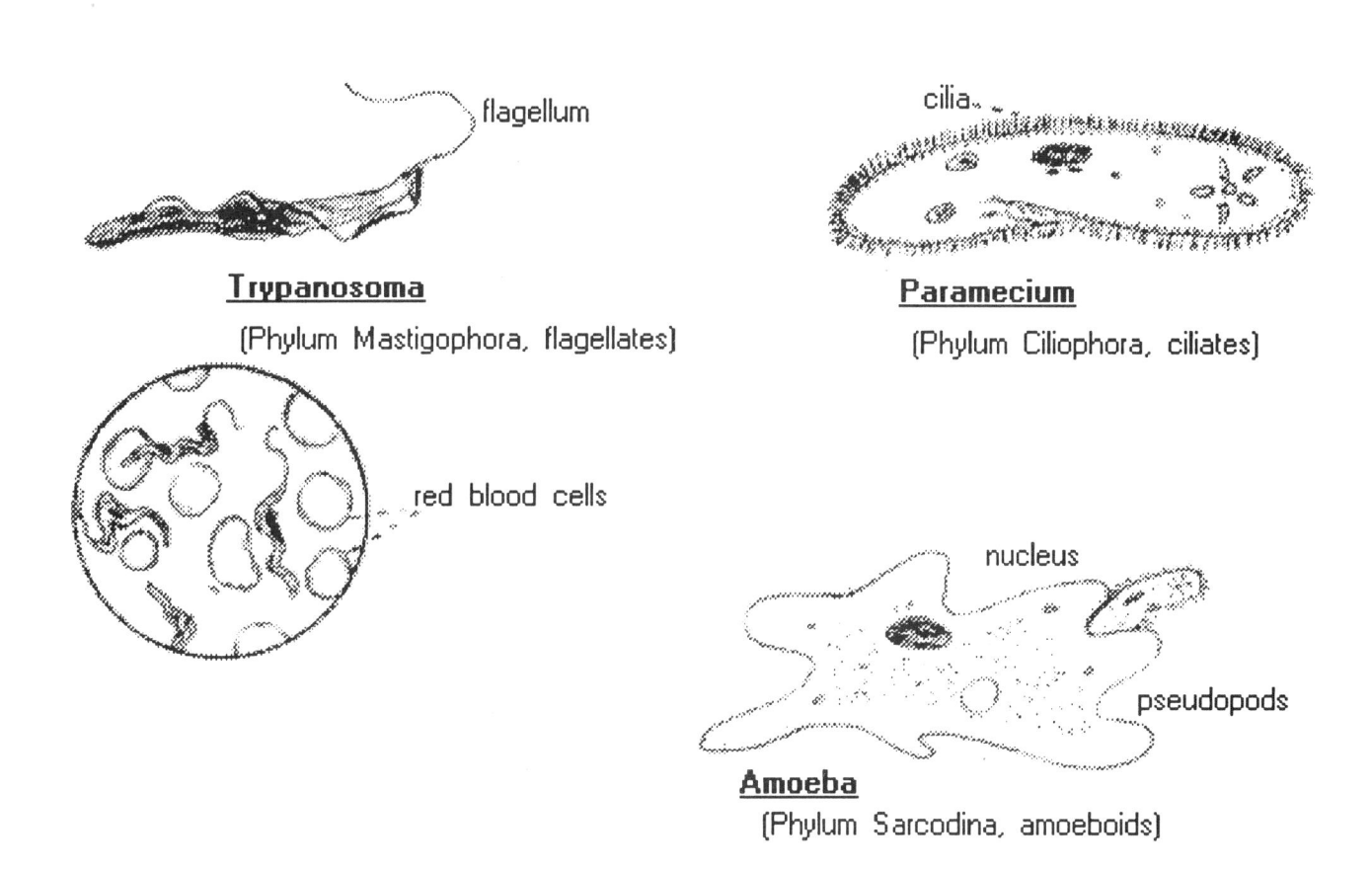

Amoeba
(Phylum Sarcodina, amoeboids)

Figure 9-4. Representative protozoans as they would appear under magnification. Most are less than 0.1 mm. in size.

reservoir of the parasite. Control is also hindered by poverty, poor communication, political corruption, and war.

and spore-formers

Phylum **Sporozoa** consists entirely of spore-forming parasites, including *Plasmodium*, which causes malaria. *Plasmodium* has no organ of locomotion. Part of its life cycle occurs within the gut and salivary glands of the *Anopheles* mosquito. The mosquitos inject the spores along with an anticoagulant into the blood of human hosts prior to drawing blood. The pathogen reproduces within human red blood cells, eventually causing the cells to burst. Nearly 2 billion people live in malaria-infested tropical and sub-tropical regions. The disease kills over 1 million people each year directly, or indirectly by lowering resistance to other diseases. Those applying for passports to malaria-prone regions must take anti-malarial drugs such as chloroquine phosphate. Resistant plasmodial strains exist in some areas prompting much research to develop a malarial vaccine.

amoeboid protozoans

Sarcodina, the phylum of amoeboid protozoans, includes the genus *Amoeba* of laboratory fame (Figure 9-4) and *Entamoeba*, a not-so favorite inhabitant of polluted water, which causes amoebic dysentery. These protozoans move by transitory lobe-like projections of the cytoplasm called **pseudopods** [Gr. (*pseudo* = false) + (*podo* = foot)]. Contrary to science fiction, there are no known amoeba large enough to engulf humans. However, amoeboids do engulf other protozoans and small crustaceans in their carnivorous habits.

ciliates

Ciliates, Phylum Ciliophora ("cilia-bearers") are propelled about with many hair-like **cilia** that protrude from the elastic membrane of these unicellular protozoa (Figure 9-4). *Paramecium* and other ciliates are considered the most complex of all protozoans. We have selected *Paramecium* as our representative of the protozoans and will discuss the basic life processes of this ciliate in detail in subsequent chapters.

Summary:

In summary, it should be obvious that the Protista is another kingdom of the "small but mighty." They are invisible to the unaided eye but ubiquitous. They are "mighty" in terms of ecological and economic significance. Protists are no longer called "simple organisms" without care to point out their incredible complexity. Proponents of evolution suggest that the kingdom Plantae evolved from ancestors of the plant-like protists; and Animalia, from ancestors of the protozoa. However, it is noteworthy that most protists have a cellular structure and metabolism that is more diversified and complex than any single cell of a "higher" plant or "higher" animal.

9-D KINGDOM FUNGI

distinguishing features of fungi

Members of Kingdom Fungi are often distinguishable by their reproductive structures, mushrooms and puffballs, which grow on the surface of soil or detritus (Figure 9-5). **Spores** are produced within these structures and serve to disperse fungal species. Mushrooms and puffballs are nutritionally supported by a complex underground network of filamentous structures called **hyphae** (Figure 9-5). Hyphae can often be seen by peeling back layers of leaf litter or compost. Other species, known as **slime molds**, move over the surface of dead organic matter with a slimy growth that advances with pseudopods like *Amoeba*.

The kingdom may be divided taxonomically as follows:

KINGDOM FUNGI
    SUBKINGDOM MYXOMYCOTINEAE — plasmodial slime molds
    SUBKINGDOM MASTIGOMYCOTINEAE — flagellated fungi
    SUBKINGDOM EUMYCOTINEAE — true fungi
        DIVISION ZYGOMYCOTA — hyphae lack crosswalls
        DIVISION EUMYCOTA — hyphae have crosswalls
            CLASS ASCOMYCETES — sac, or cup fungi
            CLASS BASIDIOMYCETES — club fungi
            CLASS DEUTEROMYCETES — imperfect fungi

Fungi are heterotrophic: specifically...

All of the nearly 47,000 species of fungi are heterotrophic, and generally derive their nutrition by *extracellular digestion* and then *absorption* of organic matter. Specifically, heterotrophic fungi include decomposers and parasites and some form symbiotic partnerships.

decomposers...

Decomposer fungi are no respecters of property or food. The mightiest tree or wooden building may be felled by the powerful digestive enzymes of fungi and bacteria (Section 5-H). Millions are spent annually to preserve wood products and foodstuffs. Before the advent of preservatives, bread mold (Figure 9-5) was a more common site with its black spore-bearing structures on aged bakery products. *Penicillium* mold, from which penicillin is derived, is commonly seen on spoiling citrus fruits. Many mushrooms are edible, and some are grown commercially. But beware! There are also very poisonous

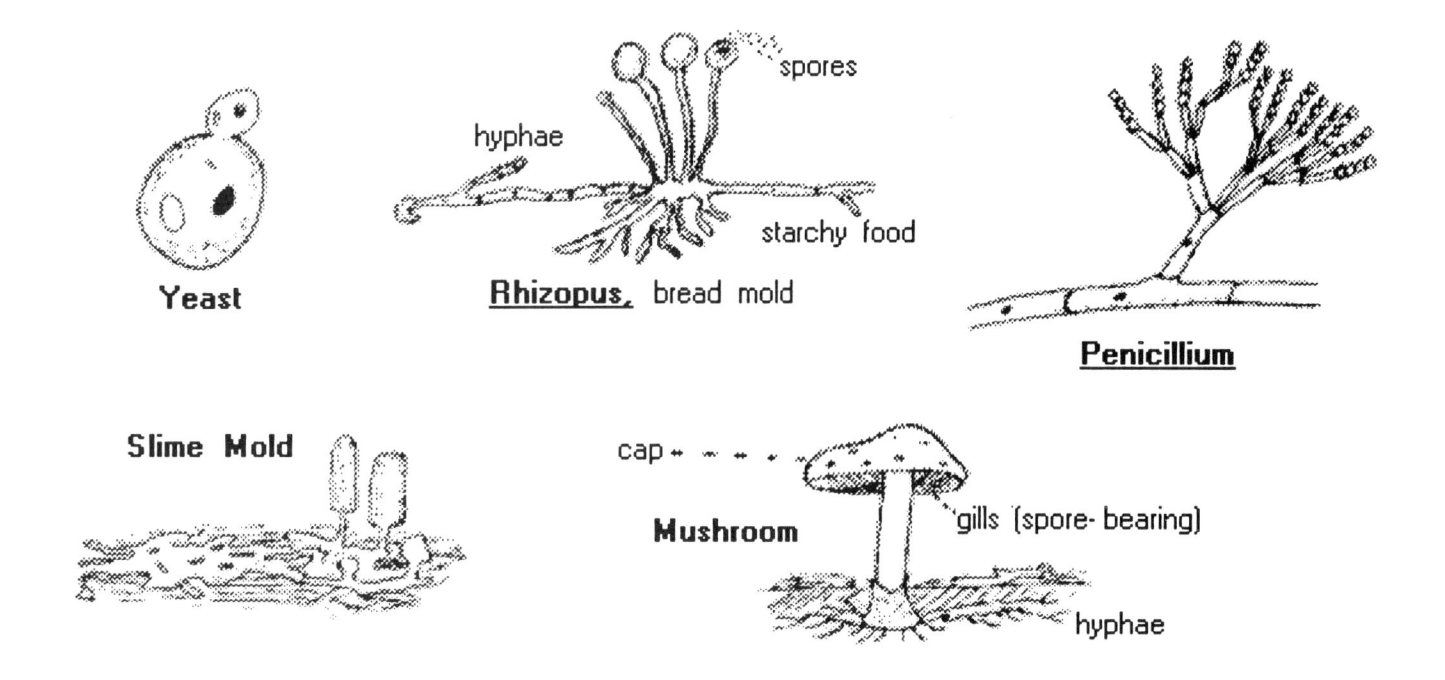

Figure 9-5. Common members of the Fungi. The yeast, Rhizopus, Penicillium, slime mold are drawn as they would appear under magnification. Note that the hyphae of Rhizopus are coenocytic, lacking crosswalls to separate individual cells.

species of mushrooms. Genuine soy sauce is made from fungi-fermented soybeans and wheat. The fungi produce a rich source of essential vitamins and amino acids, providing an important nutritional supplement for those cultures heavily dependent upon rice. Yeasts (Figure 9-5) are primarily unicellular fungi that are used in baking and in vitamin and protein synthesis.

parasitic ...

Fungal parasites may perform extracellular digestion or simply absorb materials produced by the body of their host. Fungal parasites on plants including "smuts" "rusts" and mildews, cause hundreds of millions of dollars in crop losses each year in spite of *fungicide* application and other control measures. Certain fungi are parasitic on animals. Humans suffer from parasitic yeast infections and fungal diseases of the skin such as "ringworm" and athlete's foot.

and symbiotic.

Finally, the thread-like hyphae of certain fungal species can enclose around cells of green algae or cyanobacteria to form a symbiotic relationship, called a **lichen** (Section 7-F). Lichens can withstand cold, dry conditions, and form gray-green patches common on rocks and tree trunks. Some are important food for herbivores in arctic food chains. The fungal spore-bearing structures used to identify the lichens often add bright coloration to these rugged forms. The fungal members apparently depend so heavily upon the photosynthetic green algal or cyanobacterial partner for nutrition or other factors that they are rarely observed growing independently. This has led some biologists to suggest that lichens are actually cases of fungal parasitism.

9-E KINGDOM PLANTAE

Characteristics of Plantae.

The species classified in the kingdom Plantae are eukaryotic and with few exceptions multicellular, autotrophic organisms. Differences among the major divisions (botanical name for *phyla*) relate to the degree of complexity that exists within the multicellular plant body. Generally, all of the Plantae have (1) *photosynthetic cells* often arrayed in a flattened, leaf-like structure for light absorption, (2) *anchoring and absorbing structures*, roots, rhizoids, or holdfasts, and (3) *supporting structures*, stem or stem-like structures. We shall survey the Plantae in three groups as follows: algae, bryophytes (non-vascular land plants), and vascular plants. See Appendix B for a more complete taxonomic classification of the Plantae.

KINGDOM PLANTAE:

    ALGAE:
        DIVISION RHODOPHYTA — red algae (*rhodo* = red)
        DIVISION PHAEOPHYTA — brown algae (*phaio* = brown)
        DIVISION CHLOROPHYTA — green algae (*chloro* = green)

    BRYOPHYTES:
        DIVISION BRYOPHYTA — mosses and liverworts

    VASCULAR PLANTS (major divisions only):
        DIVISION SPHENOPHYTA — horsetails
        DIVISION PTEROPHYTA — ferns
        DIVISION CYCADOPHYTA — cycads
        DIVISION GINKGOPHYTA — ginkgoes
        DIVISION CONIFEROPHYTA — conifers
        DIVISION ANTHOPHYTA — flowering plants
            CLASS DICOTYLEDONAE — dicots
            CLASS MONOCOTYLEDONAE — monocots

ALGAE

Algae are grouped by pigments.

marine algae

"Algae" is a name that is now used by most botanists to include three divisions (phyla) as named in the outline above. Each division is distinguished anatomically and by the types of photosynthetic pigments. Differences in types of chlorophyll and other pigments among the divisions cause differences in color associated with the common names *green, brown,* and *red algae*. The "algae" are included in Kingdom Protista by some botanists.

The red and brown algae are primarily marine forms and are often called **seaweeds** (Figure 9-6). The word "seaweed" to many is synonymous with "slimy mass." The slimy coating of many seaweeds retards desiccation during intermittent low tides when they are exposed to dry air. When submerged, the *blades* float and stay near the lighted surface as a result of the buoyancy of *air bladders* while *holdfasts* and stem-like *stipes* keep the algae from drifting away. Many of these species, especially the *kelps*, are commercially important in the production of gelatins, agar culture media, ice cream, and dietary supplements.

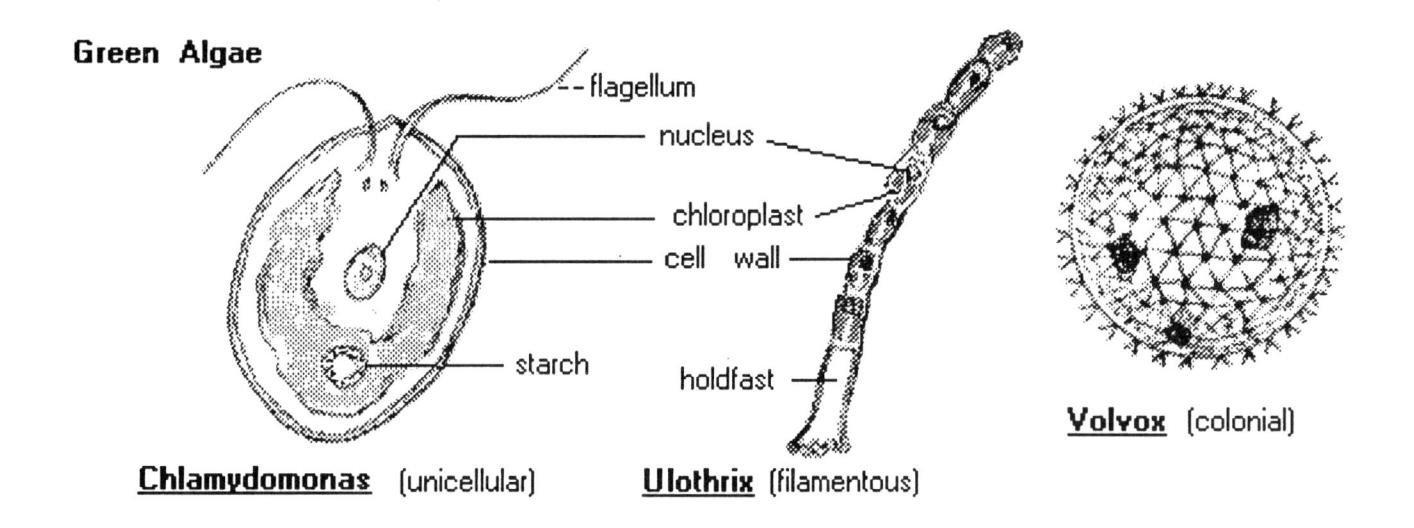

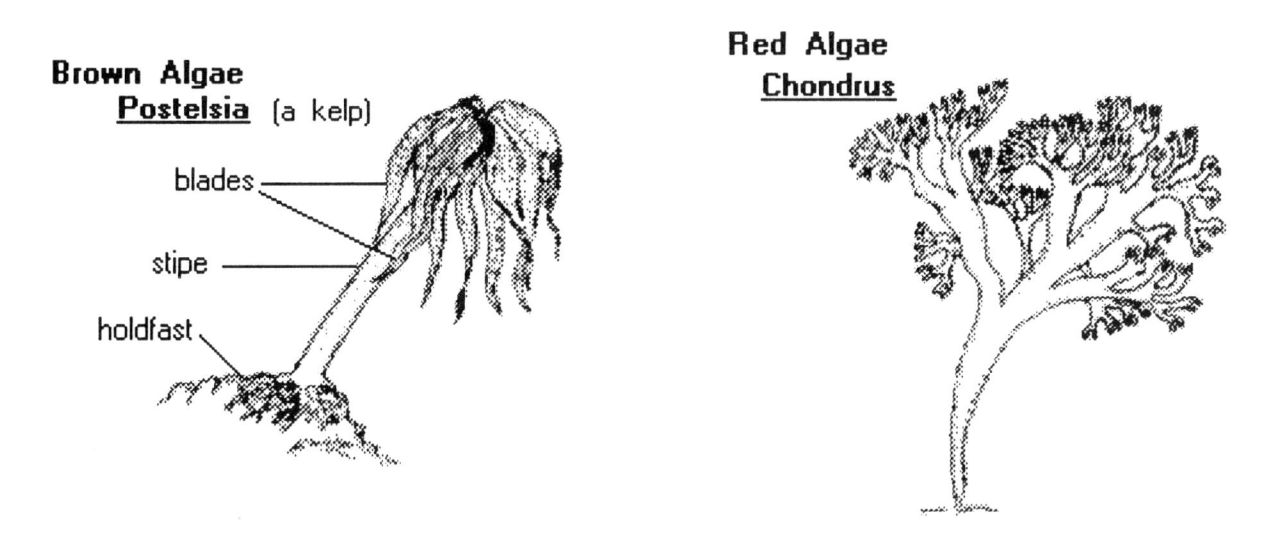

Figure 9-6. Common representatives of the algae.

green algae

The green algae are the most common aquatic non-vascular plants of freshwater lakes and streams. Many are slender and filamentous such as *Ulothix* (Figure 9-6) and *Spirogyra* and form green slimy masses especially in eutrophic lakes. Other green algal species such as *Chlamydomonas* (Figure 9-6) are unicellular with flagella and are difficult to distinguish taxonomically from the plant-like protists (Figure 9-3). [Can you see the recurrent "trials of taxonomic classification"?]

BRYOPHYTES

The bryophytes represented by the **mosses** and **liverworts** (Figure 9-7) are the simplest land plants. No doubt, you are familiar with the mosses as green mats of delicate plants covering rocks, tree trunks, and soil in moist environments. Liverworts are commonly found on rocks or shallow soil where splashing or trickling water provides abundant moisture. Bryophytes possess

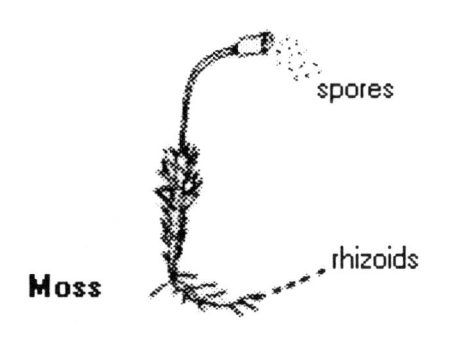

Figure 9-7. Mosses and liverworts (Bryophyta).

Bryophytes usually lack true roots, stems and leaves.

green leaf-like structures and some have short stems. They also have rootlike structures that anchor the tiny plants to the substrate. However, most bryophytes lack *true* roots, stems, or leaves as defined by the presence of vascular, or conductive tissue. The small size of the bryophytes does not necessitate an elaborate root system, or system to conduct materials for long distances throughout the plant body. However, unlike algae that live surrounded by water, the bryophytes must avoid excessive water loss by evaporation into the often dry air in which they live. Thus, the surface of these plants is usually covered with a waxy layer or **cuticle** which reduces water loss.

VASCULAR PLANTS

Vascular plants constitute the majority of the plant species of most green landscapes. These species by definition have true roots, stems, and leaves; and each of these three organs are interconnected internally by **vascular tissue** (L. *vaso* = vessel, duct) made of tubular, conductive cells called **xylem** and **phloem**. Xylem functions mainly in the transport of water and mineral nutrients from the soil via the roots to the other parts of the plant. Phloem conducts sugars and other products of photosynthesis throughout the plant. Even the stately redwoods can stand against the winds and receive water "on tap" hundreds of feet above ground because of the elaborate root anchorage and vascular system. We shall examine xylem and phloem transport mechanisms in Chapter 10.

Vascular plants have internal conduction.

spore-bearing vascular plants

Vascular plants are classified according to reproductive structures and leaf anatomy. The ferns, clubmosses, and horsetails are spore-bearing vascular plants (Figure 9-8). These have no flowers and do not produce seeds. The seed-bearing vascular plants may be divided into two classes, the gymnosperms and angiosperms.

two groups of seed plants

Gymnosperms [Gr. (*gymno* = naked) + (*sperma* = seed)] (Figure 9-9) such as cycads and coniferous trees bear seeds in cones without a protective tissue layer. **Angiosperms** (Gr. *angio* = box, case), or flowering plants, bear seeds within a protective tissue or **ovary**. The **seed** is a reproductive structure containing an immature plant or **embryo** and a food supply to initiate growth of the new generation. Angiosperms are divided into two subclasses based upon seed anatomy, the **monocotyledons** (monocots) and the **dicotyledons** (dicots). Monocots include the grasses and palms, and have one **cotyledon** or embryonic leaf. Dicots have two cotyledons in each seed as seen in beans and most broadleaf plant species. These and other distinctions between monocots and dicots are shown in Fig. 9-10.

monocots *vs.* dicots

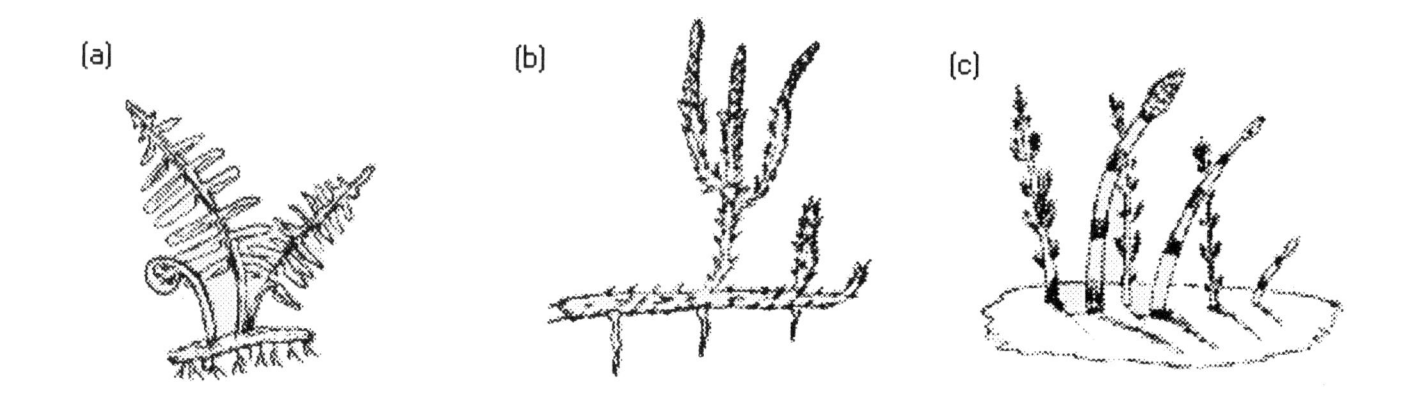

Figure 9-8. Spore-bearing vascular plants: (a) Ferns (Psilophyta), (b) Clubmosses (Lycophyta), and (c) Horsetails (Sphenophyta).

Plants have great economic importance.

Vascular plants have great ecological and economic importance in our lives. Study Table 9-1 for a moment and then imagine your present surroundings without many of these items. Table 9-1 does not even touch upon the ecosystem services provided by these plants, including food, habitat, and shelter for wildlife, and the supply of O_2 so vital to our lives as heterotrophs.

Table 9-1. Economic Significance of Non-vascular and Vascular, Land Plants.

BRYOPHYTES	Sphagnum moss—used as fuel and gardening mulch
FERNS	Floral displays, houseplants, and ornamentals
GYMNOSPERMS	**Lumber:** e.g. pine, spruce, fir, cedar, and redwood **Pulp for paper:** white spruce, pines and other conifers **Ornamentals:** pines, spruces, hemlock, junipers, yews, ginkgo, cycads **Musical instruments:** red spruce (violins; piano soundboards) **Christmas trees:** pine, spruce, fir **Chemicals:** turpentine, rosin (violin bows, baseball "grip")
ANGIOSPERMS	**Food:** seeds, berries, fruits, grain, leaves, stalks, roots, tubers, extracts, syrups, spices, chocolate **Clothing:** cotton, linen (from flax) **Lumber:** oak, maple, ash, walnut, cherry, birch, poplar **Fuel:** firewood, charcoal **Ornamentals:** many shade tree species; flowers, herbs, and flowering shrubs; lawn grasses, etc. **Beverages:** coffee, tea, cocoa **Medicines:** aspirin (originally from willow bark), opium, morphine, digitalis, quinine **Insecticides:** rotenone (derris plant), red pepper, garlic oil, lemon oil, pyrethrum (chrysanthemum flowers)

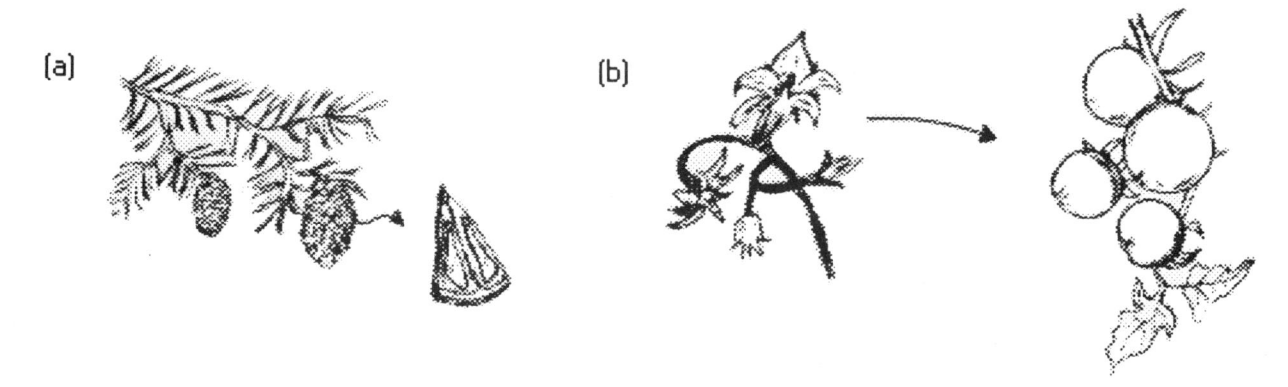

Figure 9-9. Seed-bearing vascular plants: (a) Gymnosperm (pine), and (b) Angiosperm (flowering plant).

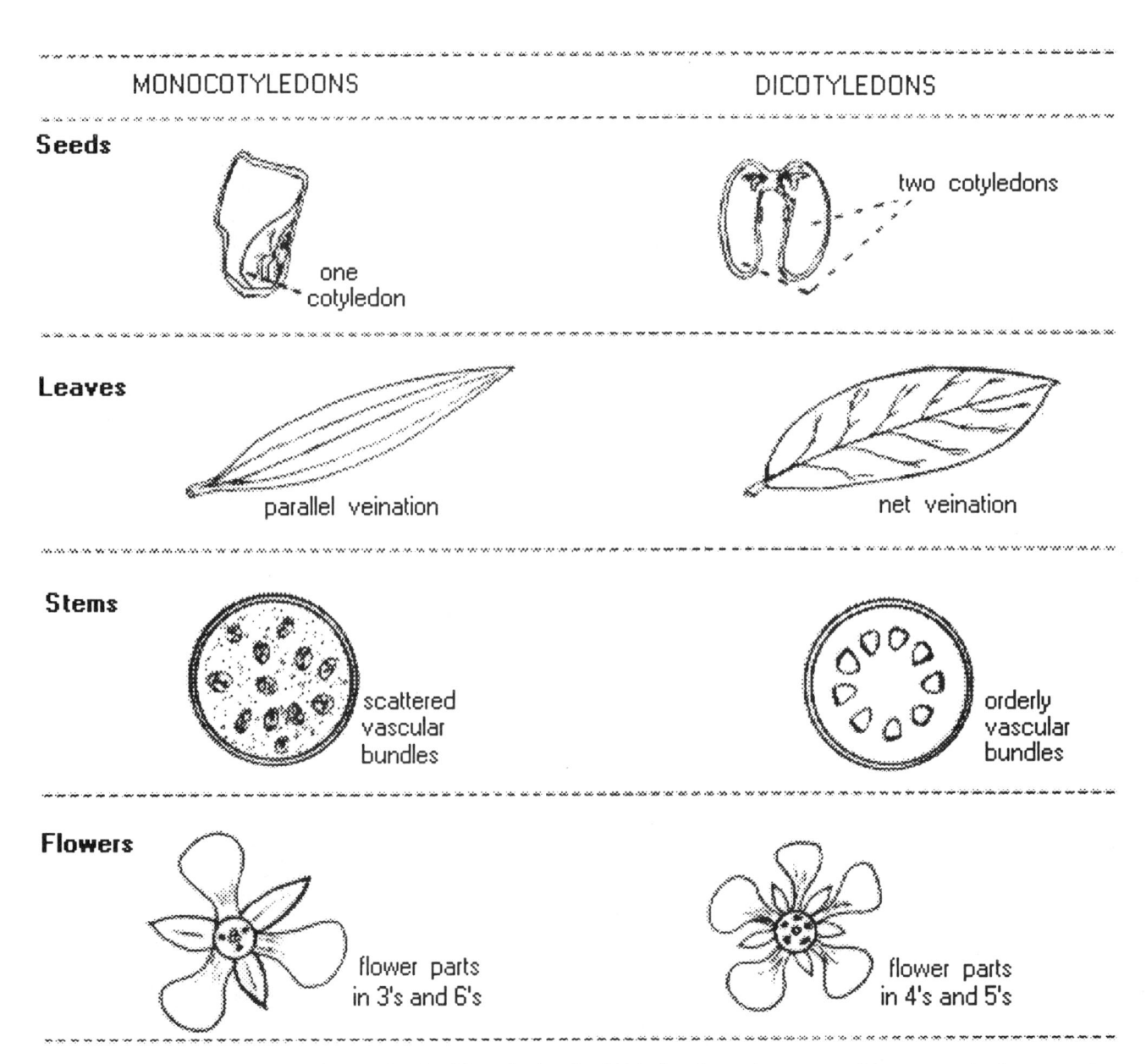

	MONOCOTYLEDONS	DICOTYLEDONS
Seeds	one cotyledon	two cotyledons
Leaves	parallel veination	net veination
Stems	scattered vascular bundles	orderly vascular bundles
Flowers	flower parts in 3's and 6's	flower parts in 4's and 5's

Figure 9-10. Morphological Distinctions between Monocotyledons and Dicotyledons.

176

9-F KINGDOM ANIMALIA

characteristics of Animalia

The kingdom Animalia contains more species than the other four kingdoms combined (Table 3-5). Indeed, hundreds of new species are being discovered every year in tropical areas. Although tremendously diverse, the Animalia are essentially *motile, multicellular,* and *heterotrophic*. Recall that the motile, *unicellular* heterotrophs are assigned to the kingdom Protista. Figure 9-11 summarizes the distinctions made among the Protista, Animalia, and the other three kingdoms. The major phyla of Kingdom Animalia are outlined below, and Appendix B presents a more detailed classification.

KINGDOM ANIMALIA:
    PHYLUM PORIFERA - sponges
    PHYLUM CNIDERIA - corals, jellyfish, *Hydra*
    PHYLUM PLATYHELMINTHES - flatworms, flukes, tapeworms
    PHYLUM ASCHELMINTHES - round worms(nematodes), rotifers
    PHYLUM ECHINODERMATA - sea stars, sea urchins, starfish
    PHYLUM MOLLUSCA - snails, clams, squid, octopus
    PHYLUM ANNELIDA - segmented worms--earthworms, leeches
    PHYLUM ARTHROPODA - jointed appendages, exoskeleton
    PHYLUM CHORDATA - cordates
        SUBPHYLUM UROCHORDATA - Tunicates
        SUBPHYLUM CEPHALOCHORDATA - Lancelets
        SUBPHYLUM VERTEBRATA - Vertebrates (See Table 9-3)

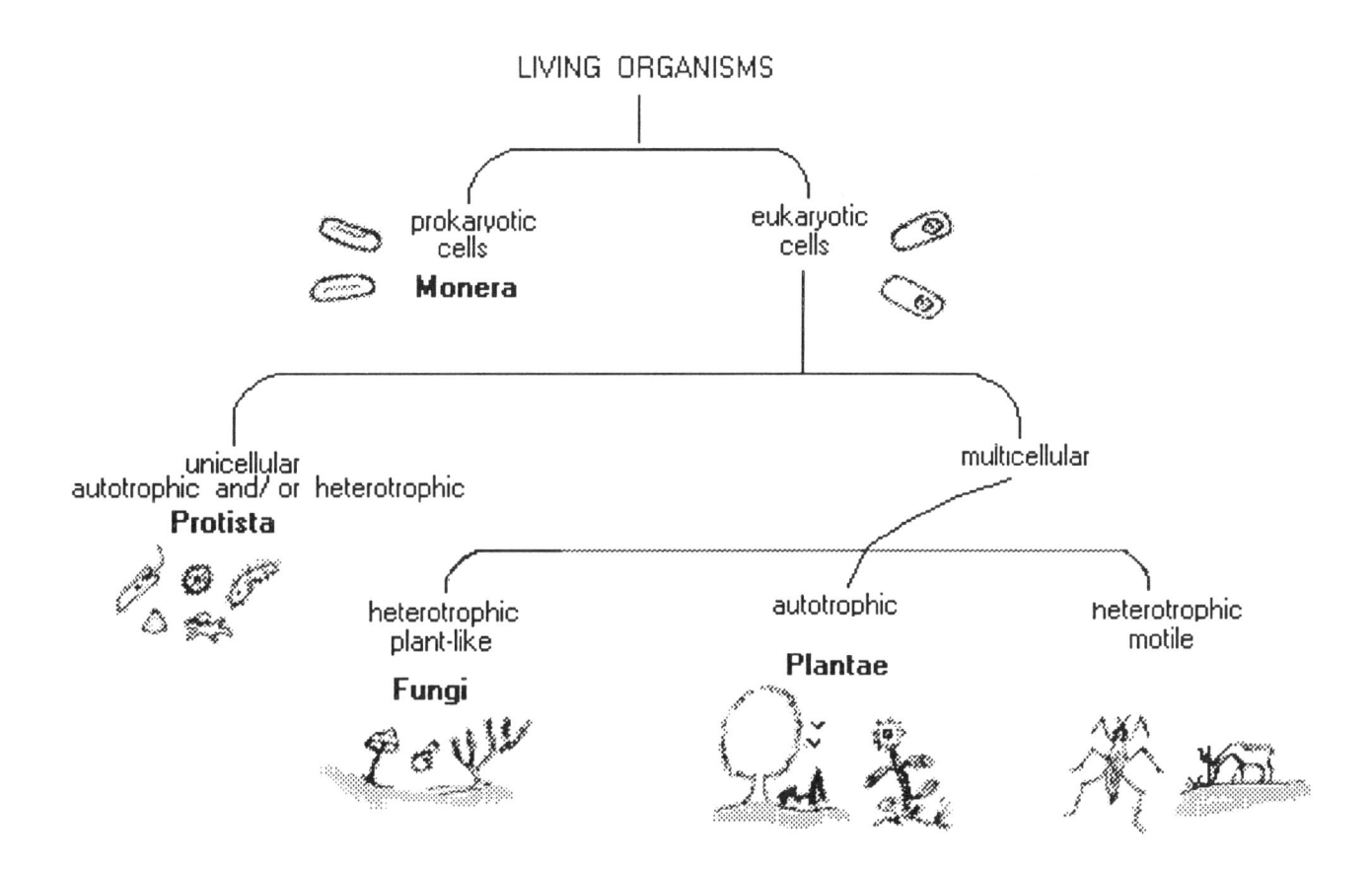

Figure 9-11. Major distinctions between the five taxonomic kingdoms.

Arthropods: largest phylum

Criteria that are useful in dividing the Animalia into the major phyla are outlined in Table 9-2. Figure 9-12 illustrates the taxonomic distinctions among eight major phyla based upon these and other criteria.

Arthropods make up the largest phylum in Kingdom Animalia representing the insects, spiders, millipedes, and crustaceans (e.g. lobster, crab, crayfish). More than one million species of arthropods are known and they occupy every type of habitat in the biosphere. Many are herbivorous and cause damage to crops and stored food while others such as lice and ticks are outer-body parasites. The other animal phyla also have ecological and economic significance. Many are important members of aquatic food chains, sources of food, and other useful products.

The Chordata is probably the most familiar phylum because many of its members closely resemble man morphologically, behaviorally, and in habitat preference. Humans can relate most readily and develop close relationships with members of this phylum. These were the last animals to be created according to Genesis 2: 18-20 and perhaps the most likely companions for Adam. Even these creatures were unable to fully meet his need for companionship (Genesis 2:21-25).

Table 9-2. Major Classification Criteria for Kingdom Animalia.

CHARACTERISTIC	AS FOUND IN PHYLA WITH LESS COMPLEX BODY STRUCTURE	AS FOUND IN PHYLA WITH MORE COMPLEX BODY STRUCTURE
Skeletal System	Absence of backbone or axial support; internal or external skeletal support	Internal skeleton organized around a centralized, axial support running length of the body (i.e. vertebral column in vertebrate animals; notochord in other chordates)
Body Symmetry	Radial—body parts arranged around a central axis through the mouth	Bilateral—distinct right and left; and, top (dorsal) and bottom (ventral) sides
Body Cavity	Have no coelom (se'-lem)	**Coelom** present—i.e. fluid-filled body cavity surrounding organs
Digestive System	Intracellular digestion with no digestive organs; digestive cavity with one opening and intracellular	Extracellular digestion; two openings to the digestive system; an "open tube" system or extracellular digestion
Nervous System	Uncentralized	Centralized nervous apparatus in anterior end (cephalization)

mammals

Table 9-3 outlines the major classes within subphylum Vertebrata. Appendix B outlines the major subclasses within the class Mammalia. Both mammals and birds are *endotherms*, meaning that they regulate body temperature by metabolically generated heat. All but a few mammals gestate their young within the protection of the mother's body. [What are some exceptions? See Appendix B.] After birth (or hatching), the young are nourished by milk from the mother's **mammary glands**. This elaborate system of nurturing offspring combined with well developed sensory and locomotion systems make land mammals well suited to the rigorous land environment.

KINGDOM ANIMALIA

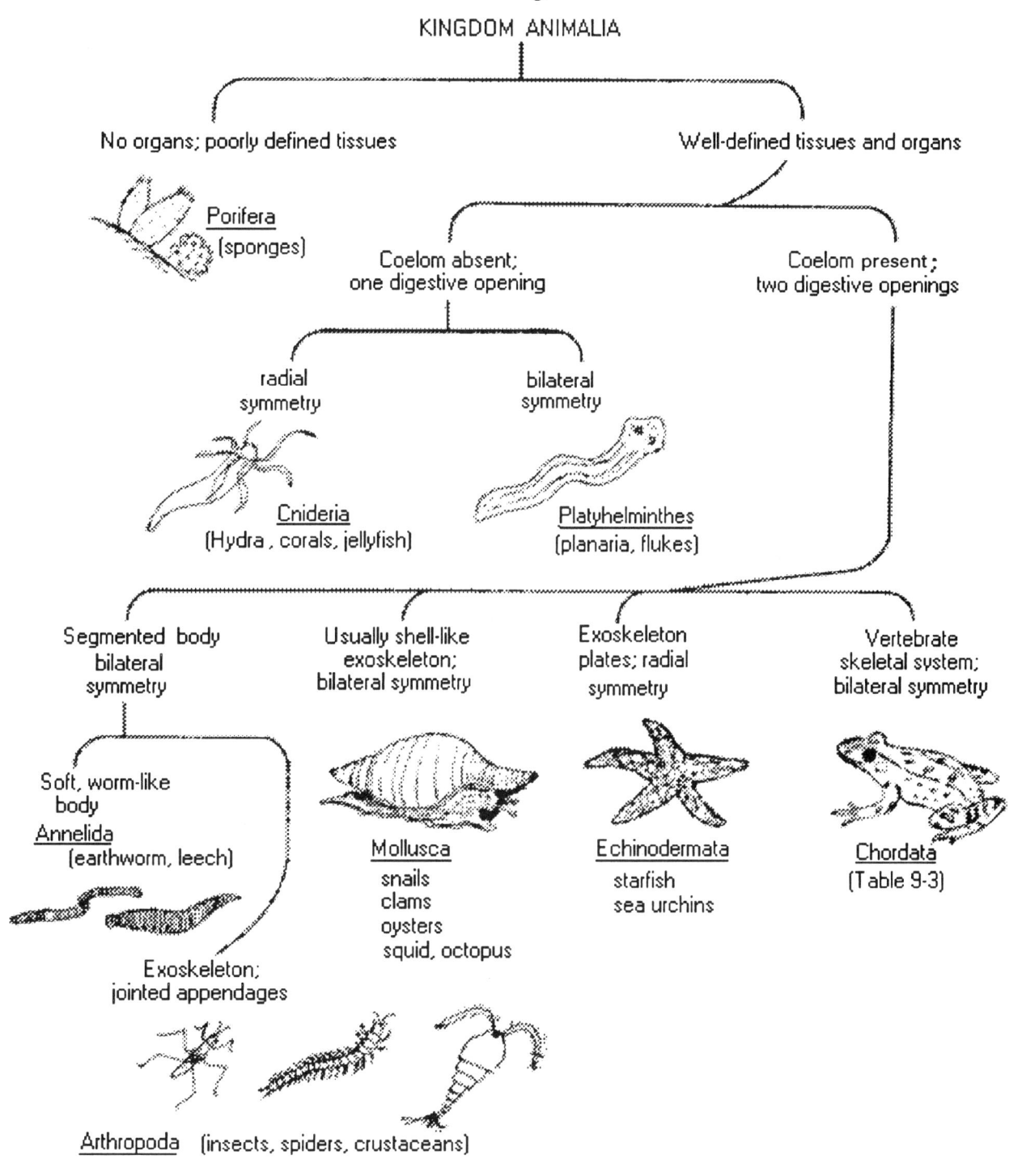

- No organs; poorly defined tissues
  - Porifera (sponges)
- Well-defined tissues and organs
  - Coelom absent; one digestive opening
    - radial symmetry
      - Cnideria (Hydra, corals, jellyfish)
    - bilateral symmetry
      - Platyhelminthes (planaria, flukes)
  - Coelom present; two digestive openings
    - Segmented body bilateral symmetry
      - Soft, worm-like body
        - Annelida (earthworm, leech)
      - Exoskeleton; jointed appendages
        - Arthropoda (insects, spiders, crustaceans)
    - Usually shell-like exoskeleton; bilateral symmetry
      - Mollusca
        snails
        clams
        oysters
        squid, octopus
    - Exoskeleton plates; radial symmetry
      - Echinodermata
        starfish
        sea urchins
    - Vertebrate skeletal system; bilateral symmetry
      - Chordata (Table 9-3)

Figure 9-12. Major phyla of Kingdom Animalia with distinguishing characteristics.

179

9-G SUMMARY

We have completed a rather brief survey of the major taxonomic kingdoms and their subgroupings. Our purpose has been first to highlight the tremendous biodiversity of planet Earth, and second to emphasize patterns of similarity within the taxonomic groupings.

Our approach: a survey of representative species.

In the subsequent chapters, we will be studying the internal anatomy and physiology of organisms. Thanks to the taxonomic classification system, we can concentrate our studies on *representative species* in each major group, and assume that, generally speaking, our knowledge of that species is applicable to other members of the group. However, this approach only applies in rather general introductory studies of life processes. Many volumes have been written by specialists who devote their lives to studying specific taxonomic groups and individual species. The more one studies individual species the more we realize that each species is a unique phenomenon, and that taxonomic classification is, after all, only a temporary aid to help us file information about a great and diverse creation. It is hoped that the general principles and concepts that we derive from our study of a few representative species will prepare and encourage you to look further into the diversity of life.

Table 9-3. Major Classes of the Vertebrates.

CLASS	DISTINGUISHING CHARACTERISTICS
AGNATHA hagfish, lamprey	Jawless fishes; gills, two-chambered heart [Gr. (*a* = without) + (*gnathos* = jaw)]
CHONDRICHTHYES shark, skate, ray	Fishes with cartilaginous skeletons; ventral mouth and nostrils; gills; two-chambered heart [Gr. (*chondros* = cartilage) + (*ichthyes* = fish)]
OSTEICHTHYES bass, trout, eel	Fishes with bony skeletons; gills covered by, operculum; two-chambered heart (Gr. *osteon* = bone)
AMPHIBIA frog, toad, newt, salamander	Four-legged; immature forms usually aquatic with gills; eggs without shells; three-chambered heart [Gr. (*amphis* = double) + (*bio* = life)]
REPTILIA snake, lizard, turtle,	Scaly skin and leathery-shelled eggs; lung-breathing; incompletely four-chambered heart alligator (L. *repere* = to creep)
AVES songbirds, penguins	Feather-covered skin; wings; hard-shelled eggs; lung-breathing; four-chambered heart (L. *avis* = bird)
MAMMALIA platypus, bats, seal kangaroo, whale, dog	Usually hair-covered skin; mammary glands; lungs; four-chambered heart; chew food in oral cavity (L. *mamma* = breast)

QUESTIONS AND DISCUSSION TOPICS

1. This chapter has provided a "survey" of the major taxonomic groups. We will be building upon this survey in weeks ahead. Prepare an outline using Sections 9-B through 9-F as headings, with subheadings for distinguishing characteristics of each major group that is discussed and include names of representative species in each group.

2. Suppose that a brilliant biologist spent her life studying what she called "twig-hopping wrens", but never referenced these birds to a species classification. Explain how this hypothetical instance emphasizes the importance of taxonomic classification in the development of scientific knowledge.

3. Both creationists and evolutionists use similar taxonomic classification systems as outlined in this chapter and Appendix B. How then does one distinguish the two views of the origin of life as we know it? Why is it impossible to scientifically test either position?

4. An evolutionist criticizes the concept of *kind* because of its biblical origin and because it is hard to define. What is your position? What might be said of the concept of *species*?

5. Review each phylum (or subkingdom) within each of the five kingdoms as outlined in this chapter and in Appendix B. Make a summary table in which you cite one common species from each group, tell its mode of nutrition, and economic importance. For example, in Kingdom Monera, there is Subkingdom Eubacteria which includes *Clostridium* species known to cause tetanus and food poisoning.

6. How do organisms such as *Euglena* challenge human attempts to build a classification? Should *Euglena* be studied in *zoology* or *botany*?

7. How do you define *zoology*? *Botany*? Why do both zoology and botany texts include studies of the Protista?

8. How do the general body forms (morphology) of each major group within Kingdoms Plantae and Animalia suit these species for survival in their respective habitats.

9. Which taxonomic group within Kingdoms Plantae and Animalia appear to be most successful on land? Explain your answer.

10. Summarize in your own words how the information from this chapter will be used in subsequent chapters of PART 2.

TOPICS FOR FURTHER STUDY

1. As is true when you learn the name of a new acquaintance, so it is when you become acquainted with a single species of plant, animal, or microbe. Only if you spend time with the person (or species) will you begin to realize how neat the person or species really is. Why not select one species and dig into some of the *natural history* of this creature? Topics can include its geographic range (why is it so distributed), food web relationships (feeding habits and what eats it), reproductive cycle, nurture of offspring, economic importance to man, impact of man, spiritual lessons from this species, etc. This could become the beginning of a life-long interest.

    REFERENCES: [NOTE: These are but a few examples of the kinds of sources.]

    Breland, O.P. 1972. *Animal Life and Lore*. Harper and Row, New York. Includes many reference citations for more detailed reading.

Gorman, M.L. and R.D. Stone. 1990. *The Natural History of Moles*. Comstock Publ. Associates, Ithica, NY. [Fascinating natural history study]

Hamerstrom, Frances. 1986. *Harrier, Hawk of the Marshes*. Smithsonian Institution, Wash., DC. [Account of a woman's study of *Circus cyaneus*, "the hawk that is ruled by a mouse".]

Institute in Basic Youth Conflicts. 1976. *Character Sketches*, Vol. I, II, III, Rand McNally and Co. Character qualities illustrated in the world of nature.

Leopold, A. 1949. *A Sand County Almanac*. Oxford Univ. Press, New York. [From "journal" entries and reflections on nature]

Periodicals: *Natural History, Audubon, National Wildlife, Scientific American, Science Digest, National Geographic* and others.

Peterson, Roger T. *Peterson Field Guide Series*. Houghton Mifflin, New York, NY. [Next time at "the mall," stop by a bookstore and select one of these; take a hike with a friend and use the guide to make some new animal, plant, or other natural history acquaintances.]

2. Make a study of a disease such as malaria, African sleeping sickness, ergotism (from fungal-infected bread), or tetanus: disease transmission, effects on the human body, and the ecological, economic, historical, and social implications of the disease.

3. Why are some orders and families of plants so much larger than others? Check on this from a creationist perspective.

   SEE "REFERENCES": Berger (1981), Howe (1979), and Stebbins (1981)

4. Since there are so many species, should Christians be concerned about extinctions (over one hundred animal species have become extinct in the last century)? What is the value of a species? What factors affect vulnerability to extinction?

References:

Day, D. 1981. *Doomsday Book of Animals*. Viking Press, New York.

May, R.M. 1992. How many species inhabit the Earth? *Sci. Am.* 267(4):42-68.

Myers, N. 1983. *A Wealth of Wild Species*. Westview Press. Boulder, CO.

Terborgh, J. 1989. *Where Have All the Birds Gone?*. Princeton U. Press, Princeton, NJ.

See also Chapter 8, CHRISTIAN STEWARDSHIP REFERENCES.

REFERENCES

1. Barnes, R.D. 1987. *Invertebrate Zoology*, 5th ed., Saunders/Holt, Rinehart & Winston. Philadelphia.

2. Berger, W.C. 1981. Why are there so many kinds of flowering plants? *Bioscience* 31(8): 572, 577-581.

3. Bold, H.C. and M.J. Wynne. 1985. *Introduction to the Algae*, 2nd ed., Prentice-Hall, Englewood Cliffs, NJ.

4. Borrer, D.J., *et. al.* 1976. *An Introduction to the Study of Insects*, 4th ed., Holt, Rinehart, Winston, Inc., New York.

5. Chapman, V.J. and D.J. Chapman. 1980. *Seaweeds and Their Uses*, 3rd ed., Chapman and Hall, London.

6. Hickman, C.P. Jr., *et.al.* 1990. *Biology of Animals.*, 5th ed., Times Mirror/Mosby College Publ., St. Louis, MO.

7. Howe, G.F. 1979. Biogeography from a creationist perspective I: Taxonomy, geography, and plate tectonics in relation to the created kinds of angiosperms. *Creation Research Society Quarterly* 16(1): 38-43.

8. Lester, L.P. and R.G. Bohlin. 1984. *The Natural Limits to Biological Change*. Zondervan, Grand Rapids, MI.

9. Moore-Landecker, E. 1982. *Fundamentals of the Fungi*, 2nd ed., Prentice-Hall, Englewood Cliffs, NJ.

10. Pennak, R.W. 1989. *Fresh-water Invertebrates of the United States*, 3rd ed., John Wiley & Sons, New York, NY.

11. Romer, A.S. and T.S. Parsons. 1986. *The Vertebrate Body*, 6th ed., Saunders, Philadelphia, PA.

12. Ross, I.K. 1979. *Biology of the Fungi*. McGraw-Hill, New York.

13. Rossmoore, H.W. 1976. *The Microbes: Our Unseen Friends*. Wayne State Univ. Press, Detroit, MI.

14. Smith, A.H. and N. Weber. 1980. *The Mushroom Field Hunter's Guide*. Univ. of Michigan Press, Ann Arbor, MI.

15. Stebbins, G.L. 1981. Why are there so many species of flowering plants? *Bioscience* 31(8): 573-577.

16. Stern, K.R. 1991. *Introductory Plant Biology*, 5th ed., W.C. Brown, Dubuque, IA.

17. Whittaker, R.H. 1969. New concepts of kingdoms of organisms. *Science* 163:150-160.

18. Woese, C. 1981. Archaebacteria. *Scientific American* 244(6):98-125.

Chapter 10

Nutrition

Nutrition: a basic life process

To this author, one of the fascinating aspects of studying living things is the challenge of tracing the "common threads" of **unity**, representing common characteristics of all species, through the "multi-colored fabric" of **diversity** among different species (Section 2-D). *Life characteristic #2* (Section 2-B) highlights one of these "threads of unity," when it states: *Living systems require a continual supply of energy and matter in usable form.* Thus, in spite of the diversity among species surveyed in Chapter 9, each must acquire atoms and molecules from its environment and *assimilate* them into precise arrangement and proportions to form organelles and cellular structure (Table 4-2). We shall use the term **nutrition** to refer to this process. Nutrition is the first of three *basic life processes* to be considered in Chapters 10 through 15. The other two basic life processes are **reproduction** and **coordination** (responsiveness and homeostasis).

10-A NUTRITION IN AUTOTROPHS AND HETEROTROPHS

Nutrition

heterotrophic nutrition . . .

versus **autotrophic.**

Nutrition may be defined as the *intake and processing of high-energy organic compounds or the intake of inorganic substances and energy necessary to synthesize organic compounds.* The definition has two parts. The first part refers to heterotrophic nutrition; and the second to autotrophic nutrition. Heterotrophs by definition are organisms that take in ready-made organic compounds from other heterotrophs or from autotrophs. This organic *food* must be broken down or *digested* (Section 4-E) into smaller molecules that can be absorbed and assimilated into constituents of the heterotroph. Autotrophic intake does not require digestion because the required nutrients, CO_2, H_2O, and minerals are small molecules that can be directly absorbed across cell membranes into cells where they are assimilated into organic molecules. Consult Sections 5-C and 5-D as necessary.

The diverse means of acquiring nutrition may be classified under three categories as follows:

1. AUTOTROPHS require only light, CO_2, H_2O, and inorganic ions as raw material and energy which they assimilate into organic molecules which form their cells and tissues. These include Cyanobacteria, plant-like Protista, and most members of Kingdom Plantae.

2. CERTAIN FUNGAL AND BACTERIAL DECOMPOSER HETEROTROPHS are more "independent" than most heterotrophs. They can synthesize complex organic molecules when exposed to only sugars, water, and inorganic ions. [Then, why aren't fungi and bacteria classed as autotrophs?]

3. **MOST HETEROTROPHIC DECOMPOSERS AND PARASITES** secrete digestive enzymes outward into their environment which is hopefully rich in organic compounds. The resultant *extracellular digestion* releases simple sugars (e.g. glucose) from complex carbohydrates, amino acids from proteins, and so on (Sections 4-G, 4-H, 4-I). The digestion products are then taken up across the cell membranes and used in metabolism and growth.

4. **CONSUMER HETEROTROPHS** (herbivores and carnivores) are the most dependent upon other organisms to prefabricate organic molecules. They must acquire in their diet, not only the major groups of organic molecules, but also essential amino acids, essential fatty acids, and vitamins. Large-bodied, multicellular consumers transport products of digestion via a **circulatory system**.

Notice that the diversity associated with nutrition involves an extremely varied means of acquiring and assimilating matter into organic compounds (carbohydrates, proteins, lipids, DNA, etc.) and into cells that are common to all life. Thus, our study of the *diverse means of nutrition*—the leaf, teeth, tentacle, and claw—should not hide the *common ends of nutrition*.

10-B NUTRITION IN THE MONERA AND FUNGI

Kingdoms Monera and Fungi comprise species that are generally unicellular or filamentous (Section 9-B, 9-D). This simple organization eliminates the need for complex ingestive, digestive, circulatory, and excretory systems. Instead, heterotrophic monerans and fungi utilize extracellular digestion and absorption as noted in Section 10-A.

absorption

Cyanobacteria and certain photosynthetic Eubacteria are autotrophic. These are unicellular and aquatic forms, so that intake of CO_2, H_2O, and minerals is easily accomplished by absorption from the surrounding water. Sugars and other organic molecules (e.g. nucleic acids, vitamins) are photosynthesized in the presence of light. Cyanobacteria may also perform N_2 fixation in symbiotic relationship with fungi as in certain lichens (Section 7-F), or with mosses, or ferns (Section 9-B).

photosynthetic bacteria

Autotrophic Eubacteria may be photosynthetic or chemosynthetic utilizing either light energy or oxidation of inorganic ions, respectively, as sources of energy to make organic constituents for growth. Among the photosynthetic bacteria are species that utilize hydrogen sulfide (H_2S) in place of H_2O in the photosynthesis of glucose. [Try to rewrite the photosynthesis equation (Section 5-D) with H_2S substituted for H_2O. The product elemental sulfur.] Chemosynthetic bacteria include the nitrifying bacteria (Section 6-E), and bacteria that derive energy from oxidation of H_2S spewed from volcano-like vents in the dark ocean floors [See TOPICS FOR FURTHER STUDY, #1, Chapter 5].

heterotrophic bacteria

Represented among the various heterotrophic species of bacteria and fungi are those that may live as decomposers, parasites, or mutualists (Section 3-D, 5-D). Nutrition in all three of these groups depends upon absorption of organic molecules from a nonliving or living source with which the microbial cell makes contact. Many heterotrophic species can live in both the presence or absence of O_2 gas and are called **facultative anaerobes**. These include many decomposers, disease-causing (pathogenic) or parasitic microbes, and microbes that conduct **fermentation** (anaerobic respiration).

Subkingdom Archaebacteria contains many species that are **strict anaerobes**, species that only grow in environments where O_2 is low or absent. Some are found within the mud of swamps where they ferment the carbohydrates of detritus and release methane and hydrogen gases ("marsh gas"). Such species are now being used in sewage treatment plants to convert human or animal waste into methane, a burnable hydrocarbon fuel (17).

parasitic bacteria

Parasitic bacteria and fungi depend upon the aqueous surroundings of the host's body as a source of nutrition. However, living host cells often have mechanisms to combat invading parasites. The parasites, in turn, will often counter this resistance by secreting toxins that kill host cells. The parasite then obtains nourishment from the dead cells by absorption and extracellular digestion if necessary. For example, the strict anaerobe, *Clostridium tetani*, thrives in deep wounds and releases a neurotoxin into the bloodstream causing fatal muscle contraction and paralysis known as lockjaw. A strict anaerobe of the same genus, *Clostridium botulinum*, exists as dormant spores on growing vegetables, but becomes active and produces a deadly toxin inside food containers that are improperly sterilized.

mutualistic bacteria

Mutualistic species, in contrast to parasites, are harmless and may improve the fitness of their partner species. Some biologists now consider some mutualistic partnerships that involve intimate nutritional co-dependence (i.e. **symbiotic relationships**) as actually being one of controlled or balanced parasitism. [How many examples of mutualism can you list? Section 6-F, 7-F, 9-B]

10-C NUTRITION IN VASCULAR PLANTS

Your success in growing plants has no doubt rested upon whether or not your plants were supplied with sufficient H_2O, CO_2, mineral nutrients, light, and warmth. Like cyanobacteria and algae, the green cells of vascular plants contain green **chlorophyll** and yellow **carotenoids**, the pigments which trap light energy necessary for photosynthesis of organic molecules. Unlike the algae and cyanobacteria, bryophytes and many vascular plants are larger, more differentiated organisms adapted to the land.

"Two worlds" of roots and shoots . . .

. . . require transport.

Land plants, in essence, occupy "two worlds". Leaves and stems are in the atmosphere where there is abundant light, but little water. The roots, on the other hand, occupy the soil where light is limiting, and water and minerals are often plentiful. Therefore, the stem and leaves collect solar energy and CO_2, combine these with water transported from the roots, and form sugars and other organic compounds. The roots, being unable to photosynthesize, receive the organic substances transported downward from the leaves for respiration and growth. In this manner, the green plant literally builds itself out of CO_2, H_2O, and minerals under the energizing power of sunlight.

van Helmont's classic study

If you were asked to estimate the percentage of the dry weight of a plant (e.g. dry leaves or seeds) that comes from the soil and the percentage from the atmosphere, what would be your estimate? The Flemish physician, Jean-Baptiste van Helmont, sought to answer this question in 1648. He planted a 5-pound willow tree in 200 pounds of dry soil. After 5 years the tree and soil were again weighed, separately. The soil weighed only about 2 ounces less, but the tree had grown to weigh 169 pounds. He concluded that, because water was the only substance he added to the plant and soil, the increase in plant biomass had arisen from water only". Nearly a century later the Englishman, Stephen Hales (1677-1761), the father of modern plant physiology, proposed that plants obtain part of their nutrition from the atmosphere. This proposal was confirmed when CO_2 was found to stimulate plant growth and photosynthesis. Later experiments in the 1940's using carbon-14 and oxygen-18 isotopes (Appendix A, Sec. A-1), demonstrated that the CO_2 contributes the six carbon atoms and six oxygen atoms of $C_6H_{12}O_6$, and H_2O contributes the twelve H atoms. Additional plant mass comes from the soil as minerals are absorbed via the roots. However, as van Helmont demonstrated, minerals account for little of the total plant mass.

LEAVES

logic of leaf structure

Leaves of plants come in all shapes and sizes and are often used to distinguish different species. Most gymnosperms (Section 9-E), including our common evergreens, have needle-like leaves with a thick, waxy **cuticle** secreted by epidermal cells that cover the leaf. These leaves can withstand the cold, dry air of the winter. Our focus will be upon the flowering plants (angiosperms), representing monocotyledons and dicotyledons (Section 9-E). These have broad, flat, thin leaves, designed for maximum light and CO_2 absorption. Recall that monocots have parallel leaf venation, and dicots have net venation.

mesophyll

Figure 10-1 illustrates the typical dicot leaf cross-section. The leaf is designed with photosynthetic tissue or **mesophyll** sandwiched between the upper and lower epidermis. The mesophyll is differentiated into two tissue types. **Palisade mesophyll** forms one or more layers of tightly packed cells located nearest the upper surface of the leaf. The **spongy mesophyll cells**, located beneath the palisade cells, are generally more loosely packed, allowing space for movement of CO_2 and O_2 within the leaf.

gas exchange

water has two roles

Carbon dioxide gas enters through pores in the epidermis, called **stomata** (or stomates). Each **stoma** (singular of stomata) is formed by two **guard cells**, specialized, chlorophyll-containing epidermal cells. Most leaves have several thousand stomata per square centimeter of leaf surface. The challenge for the plant is to absorb CO_2 from dry air without loosing too much H_2O vapor from the moist mesophyll tissues. The stomata close in response to excessive moisture loss associated with wilting. The waxy cuticle covering the epidermis reduces direct water loss from other parts of the leaf. However, if plants do not receive enough water to retain open stomata in the light, CO_2 uptake is reduced and plant growth will also be reduced. Thus, water participates *chemically* in the photosynthesis of glucose, and *physically* in the maintenance of leaves in an unwilted or **turgid** condition with open stomata. [Why would you expect a greater density of stomata on the underside of leaves of many species?]

vascular tissues:

xylem

phloem

The vascular tissue forms **veins** which extend throughout the leaf. You have no doubt observed the larger veins on the undersurface of a leaf. In our cross-sectional view (Figure 10-1), one can observe the tubular cells of **xylem** and **phloem** that constitute each vein. The xylem and phloem of the veins are connected through the stem of the leaf (**petiole**) to vascular tissue, or **vascular bundles**, of the stem (Figure 10-2). The xylem conducts water and minerals from the roots through the vascular bundles of the stem to the leaf. The phloem carries sugars from photosynthetic cells of the mesophyll downward to the roots or upward to the immature buds, or **meristematic** regions (Gr. *meristo* = divided). In this manner, the phloem and xylem perform the important role of internal transport, uniting the three organs of the plant (root, stem, and leaf).

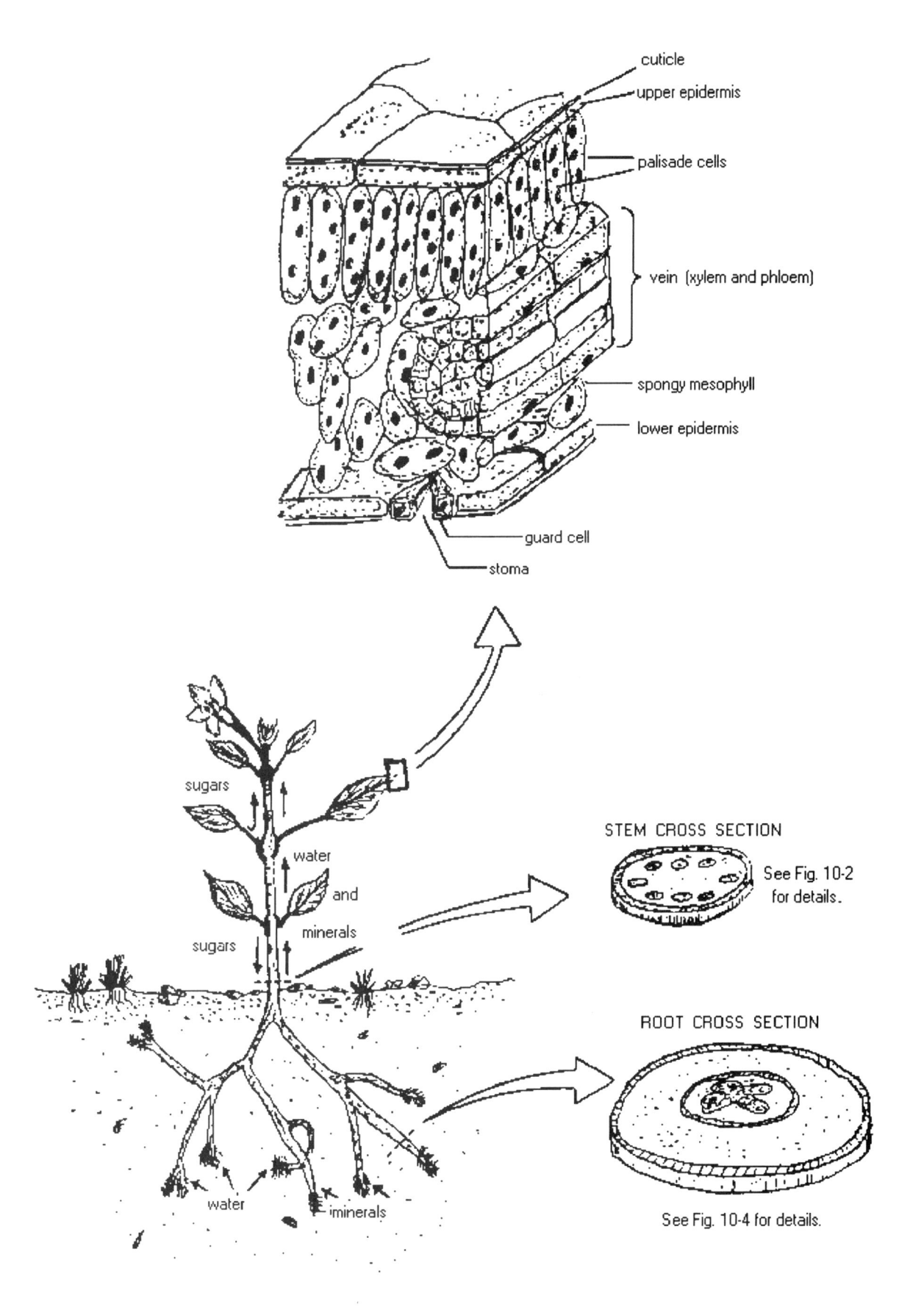

Figure 10-1. The three organs of the vascular plant, and cross-sectional views of each, with special emphasis on the leaf.

189

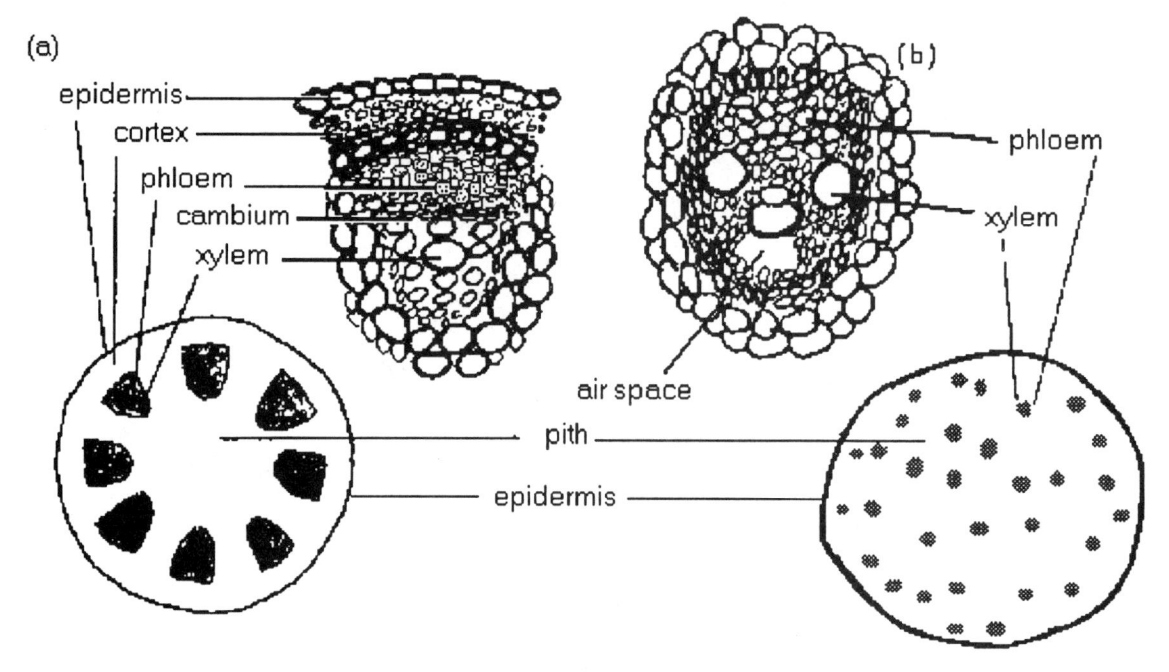

Figure 10-2. Cross-sectional views of (a) herbaceous dicot and (b) herbaceous monocot stems.

STEMS

Longevity influences stem anatomy.

Stems of **annual plants** (one-year life cycle) and **biennial plants** (two-year cycle) normally remain green and soft throughout the life of the plant. These are called **herbaceous plants** in contrast to the **perennial** shrubs and trees that have woody stems (trunks). The stems of herbaceous plants (Figure 10-2) are surrounded by protective epidermal tissue, often complete with stomata which allow gas exchange. The **cortex**, just inside the epidermis, often contains green, photosynthetic cells as well as starch-storage cells.

The most distinct tissues observed within the stem cross section are the **vascular bundles** of tubular xylem and phloem cells (Figure 10-2). Xylem cells always occupy the portion of each bundle nearest the center of the stem, phloem the portion nearest the epidermis. Each bundle is surrounded by rounded, thin-walled cells composing **ground tissue**. Monocot and dicot stems differ as summarized below:

dicot stems

1. DICOT STEMS have an *orderly* arrangement to vascular bundles separating **ground tissue** into two regions: **pith** tissue inside the ring of bundles and **cortex** between the ring and the epidermis. Pith and cortex function primarily in energy storage; the outer cortex may be photosynthetic. Xylem and phloem tissue of each vascular bundle may continue to grow because of the presence of a meristematic region, called **cambium**, which forms xylem cells toward the center, and phloem toward the cortex.

monocot stems

2. MONOCOT STEMS have *scattered* vascular bundles instead of a vascular ring. The energy-storing *ground tissue* is not distinguished into pith and cortex. Rather, it extends throughout the stem. Vascular bundles have *no cambium*, but contain an **air space** which forms when the bundle cells enlarge.

woody dicots

The woody dicot stems of shrubs and trees are similar to the herbaceous dicot stem but with two major exceptions (Figure 10-3). First, woody stems (and twigs) develop a corky protective layer, the **bark**. Second, as a result of the cambium producing additional phloem and xylem each year, the vascular bundles fuse together so as to form a complete ring of xylem toward the center and phloem toward the outside, always separated by cambium. Larger xylem cells are often produced in the spring when soil water is abundant and the woody plant is rapidly growing. The diameters of the conductive tubes formed from cambium later in summer are smaller. The result is a series of concentric rings of alternating large- and small-diameter xylem cells that can be seen as **annual rings** in a tree stump or in sawed lumber. Thus, the woody stem can grow in diameter year after year as long as the cambium remains alive. The woody plant grows in height as long as new leaf-bearing branches expand from overwintering buds each spring.

The stem transports sugars and water, and supports the leaves in an appropriate configuration for light absorption. However, unlike animals, plants have no internal or external skeletal system. Instead, each plant cell has a **cell wall** that surrounds the cell membrane. The cell wall is formed by each cell through outward secretions of organic compounds such as the polysaccharide, **cellulose** (Section 4-G), and another compound, **lignin**. Crisscrossing cellulose fibers cemented together by lignin and other compounds form a rigid wall that can withstand tremendous water pressure from within the cells. The rigid walls prevent the cell membrane from expanding under pressure and bursting. Thus, when a wilted plant is watered, water is supplied via the root xylem to every cell of the stems and leaves. As water enters each cell by diffusion across the cell membrane, a *turgor pressure* is generated from within the cell. This pressure forces the cell membrane against the inner surface of the cell wall. As each leaf cell becomes turgid, wilted leaves bend upward and unveil

stem support: cell walls . . .

and turgor pressure

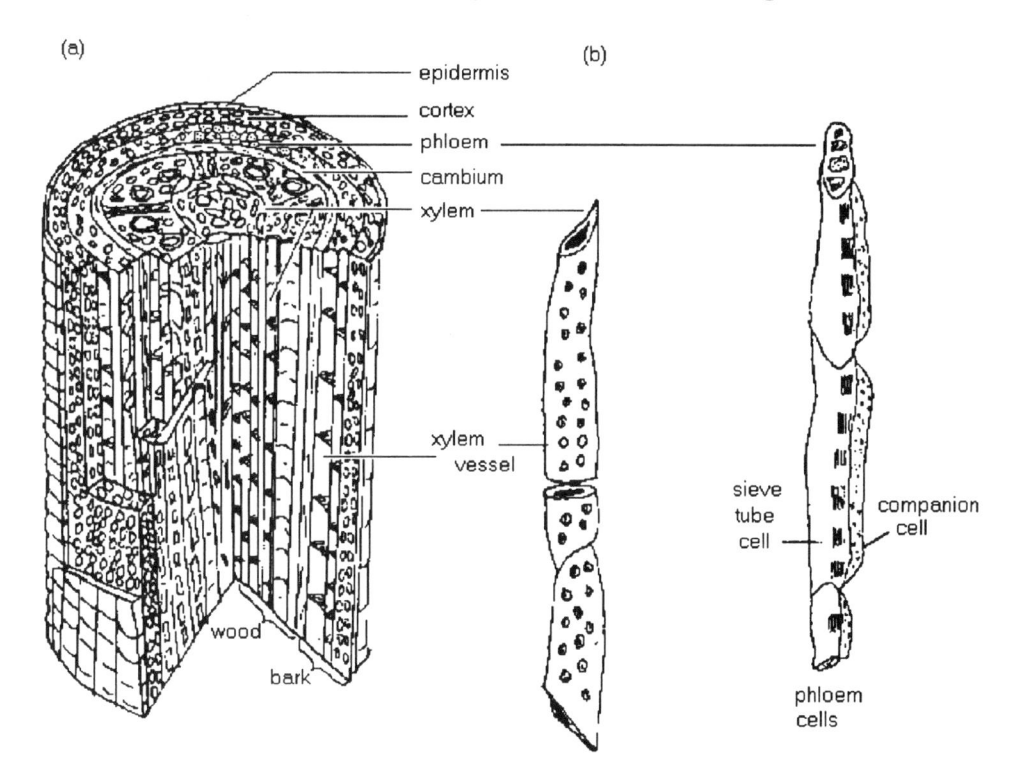

Figure 10-3. Diagrammatic view of a woody dicot stem (a) and individual xylem and phloem cells (b).

Evaporation pulls water up the xylem.

their surface area to the sunlight. Turgor within the guard cells allows stomatal opening for gas exchange. As long as soil water is available to allow a diffusion *gradient* into the root xylem, the plant will not wilt (lose turgor), and water will be transported to the leaves to replace that which is lost via evaporation. Indeed, evaporation actually serves to draw or pull water upward through the xylem from the roots. [Why would wilting slow the growth of plants?]

In woody shrubs and trees, the major structural support is provided by thick-walled xylem and phloem **fibers** within the tree trunk. The toughness that remains when wood is dry, and turgor pressure is absent, is a testimony of the great strength of woody plant stems.

ROOTS

Roots have a large absorbing surface.

Roots are often the forgotten part of plants. However, they have an important role in anchoring the plant and in absorption of water and minerals. Often the roots will occupy more volume and expose more surface area of tissue than the above ground parts. In a classic study of rye plants, a single rye plant 0.5 meter (m) tall with an above-ground surface area of 4 to 5 m² had a root system surface area of 210 m$_2$! The total length of all roots was estimated at over 370 miles! Such extensive **fibrous root** systems of rye and other grasses enable them to form a *sod* that holds soil against erosion. At the same time, the root system is tapping into every part of the soil volume for maximum contact with soil water and mineral resources.

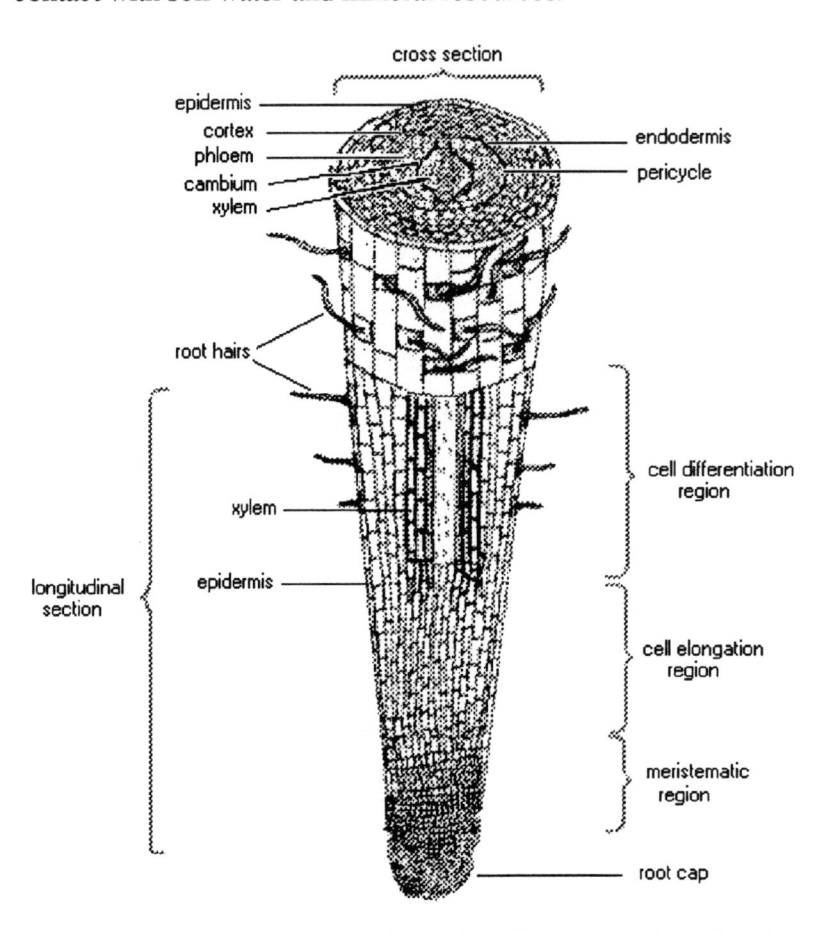

Figure 10-4. Diagrammatic view of a dicot root, showing both cross-sectional and longitudinal sectional views.

Elongation pushes...

root cap protects.

Roots push their way through the soil by elongation growth. First, new cells are produced in the meristematic region of the root tip (Figure 10-4). Each of these new cells then elongates in a direction parallel to the length of the root. This **cell elongation region** of the root is responsible for elongation growth of roots. Here, cell walls are still soft and flexible in response to turgor pressure from within each cell. The resultant cell stretching and elongation pushes the root tip downward through the soil. The **root cap** serves to protect the delicate meristematic cells from physical damage.

Root cell elongation ceases when cell walls are completely formed and no longer stretch in response to turgor pressure from within the cells. The final shape of the cell depends upon the pattern of **cell differentiation** (Figure 10-4). Epidermal cells develop fine, thread-like extensions called **root hairs**. These extensions greatly multiply the absorptive surface of the root system. Other cells develop the tubular shape of xylem and phloem. These are consolidated within the center of the dicot root, forming one large vascular bundle called the **stele**. Species that form a prominent **taproot** (e.g. carrot) have an obvious stele visible in cross-section. The surrounding cells of the cortex function in energy storage. Large taproots that function in storage are important food crops—e.g. carrot, cassava, sweet potato, sugarbeet. [Are Irish potatoes stems or roots?]

Summary:

In summary, green plant nutrition involves the gathering of inorganic substances from two different worlds. Water and minerals are absorbed via the root hairs of extensive underground root systems and translocated upward via the xylem to the stem and leaves. Here, water pressure within cell walls helps to hold the plant body erect and open the stomata. The CO_2 entering stomata is absorbed by the mesophyll cells and combined with water in the presence of light to form energy-rich organic compounds in the chloroplasts. These organic constituents are further combined with ammonium, sulfide, phosphate, and other minerals to form amino acids, fatty acids, vitamins, hormones, and other substances for plant growth. Many of these constituents, including sugars, amino acids, and hormones are transported via the phloem to tissues that are not able to perform photosynthesis—i.e. buds and growing tips of stems, and roots which grow under limited light.

10-D NUTRITION IN THE PROTOZOA

autotrophic

Nutrition in the Kingdom Protista is quite varied as you will recall from Section 9-C. The plant-like protista such as *Euglena* are autotrophic, but may absorb organic compounds from their surroundings, especially in darkness. However, protozoa are distinguished as being heterotrophic consumers, mutualists, and parasites (Section 9-C).

heterotrophic

The *Amoeba* and *Paramecium* represent those protozoa that ingest food and conduct *intracellular digestion* in organelles called *food vacuoles*. An understanding of the digestive processes of these protozoans will prepare us to discuss nutrition in the heterotrophs of Kingdom Animalia.

Paramecium as representative

The amoeboid protozoans generally feed upon microscopic pieces of organic matter and cells of bacteria or other protozoans such as *Paramecium* (Figure 9-4). At the stimulus of nearby food, the arm-like pseudopods flow outward and surround the food material. The food is then enclosed within a membrane to form a food vacuole wherein digestion occurs. This active engulfing of food by single cells is called **phagocytosis** [Gr. (*phagein* = to eat) + (*cyto* = cell)].

ingestion

intracellular digestion

excretion

Unlike the amoeboid protozoans, the ciliates, represented by *Paramecium*, have a permanent organelle designed for food intake, the **oral groove** (Figure 10-5). Water containing food particles and cells of smaller organisms is circulated into the oral groove. At the base of the oral groove is the *gullet* where food accumulates. A food vacuole forms around the accumulating food particles, moves into the cytoplasm, and begins a circulatory path throughout the cell. Tiny membranous vesicles called **lysosomes** contain digestive enzymes that are emptied into the food vacuole when the lysosome fuses with it. In this way, both the degradative process of intracellular digestion and potentially harmful substances are compartmentalized within the cell. The food vacuole also serves to distribute digested food throughout the cytoplasm as it circulates (Figure 10-5). When the food vacuole reaches a specialized site at the cell surface called the **anal pore**, the undigested contents are released to the exterior environment. Thus, the *Paramecium* and other protozoa perform food intake, intracellular digestion, and excretion, all within a single cell!

10-E INVERTEBRATE ANIMAL NUTRITION

Multicellular organisms have specialized cells . . .

with division of labor.

In Kingdom Animalia we encounter the multicellular, motile heterotrophs. The multicellular body allows *cellular specialization*. Different cells develop a specific *structure* to perform a specific *function*. All cells of one particular structure and function form a *tissue*, and in more complex animals, various tissues are assembled into the common structure and function of an *organ*. Likewise, organs that share in a common function such as digestion form an *organ system*—e.g. the digestive system. Individual cells of multicellular animals are not "Jack of all trades" like the protozoans. Instead, each specialized cell has a much restricted "freedom". It performs particular life processes essential for the organism as a whole and depends upon other cells to supply certain of its needs. This **division of labor** among cells, tissues, and organs was evident in our study of vascular plant nutrition where we observed

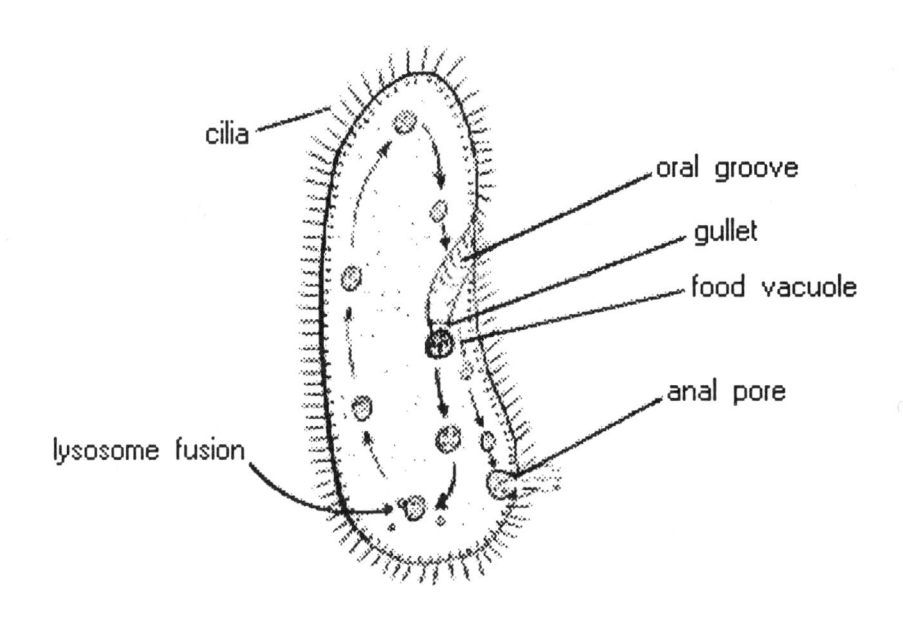

Figure 10-5. Food intake and intracellular digestion in *Paramecium*.

three major organs; leaf, stem, and root, each with specialized tissues (Section 10-C). In this section, as we survey several of the representative invertebrate animals, notice the gradual increase in nutritional complexity that accompanies increasing body size and more active life habits. Figure 9-12 will be a helpful reference during this comparative study.

SPONGES

sessile animals

Sponges compose phylum Porifera, the "pore animals". Regarded as the simplest multicellular animals, sponges are found mostly in shallow marine waters. They are sessile (nonmotile) and were once classified as plants until they were discovered to be heterotrophic.

sponges: more division of labor

Since a sponge cannot move to capture food, it must make the food come to it. The sponge is essentially a hollow sac with many pores and canals inside, each lined with many flagellated cells called *collar cells*. These cells are held in a gelatinous matrix which is supported by a simple skeletal network of *spicules*, made of either limestone, silica, or fibrous protein. Species that have fibrous protein skeletons are still sold as natural sponges, not to be confused with synthetic "sponges". Water from the surrounding environment is drawn into the body through many small pores by the beating of the flagella of the collar cells. Unsuspecting microbes and other food particles are trapped in the sticky collars and engulfed by phagocytosis (Section 10-D). Intracellular digestion occurs within collar cells and motile *amoeboid cells*. The amoeboid cells aid in transport of food to the epidermis and form the spicules by cell secretions. Water currents exit the sponge through large, centrally located pores. Oxygen intake and the removal of CO_2 and other wastes is easily accomplished via the circulating water.

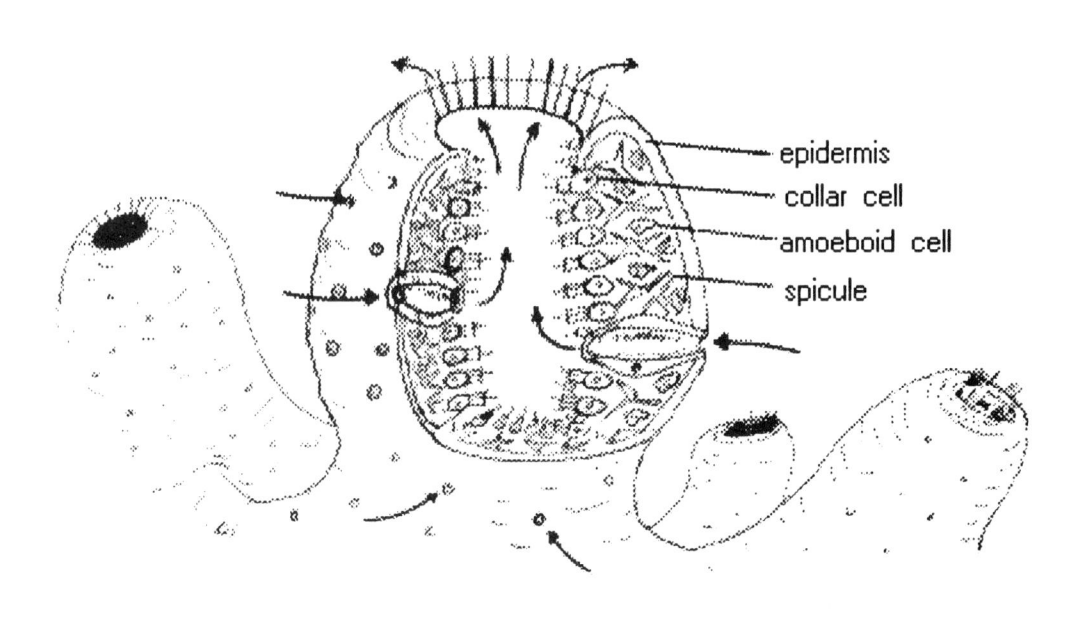

Figure 10-6. Diagrammatic view of a sponge.

HYDRA

Hydra, which is common in freshwater ecosystems, will be our representative of the phylum Cnideria [Gr. *knide* = nettle]. Cniderians (or Coelenterates) also include the corals and jellyfish species, all of which have a thin, sac-like body with one opening surrounded by tentacles. The tentacles and body surface have cells which contain organelles called **nematocysts** [Gr. *nema* = thread + *kystis* = bladder]. Nematocysts are triggered to discharge by touch or chemicals released by potential prey animals. They expel a barbed thread which wounds and injects an immobilizing poison into prey animals. Nutrition in many corals and *Chlorohydra* (green hydra) depends heavily upon a symbiotic relationship between coral and photosynthetic algae or dinoflagellates, as described in Section 8-J.

tiny predatory animals

symbiotic hydra

digestion

The stinging tentacles of *Hydra* serve to capture and immobilize prey such as small crustaceans which are then inserted through the **mouth** into the sac-like **gastrovascular cavity** (Figure 10-7). Here, **extracellular digestion** occurs when specialized **gland cells,** lining the cavity, secrete digestive juices into the cavity. As the food is reduced to smaller particles, **nutritive-muscular cells** lining the gastrovascular cavity engulf the food by phagocytosis and perform **intracellular digestion**. These cells are flagellated and serve to stir the digesting mixture within the gastrovascular cavity. Nutritive-muscular cells also have contractile fibers that function along with fluid pressure within the gastrovascular cavity to provide form and movement of *Hydra*. Undigested material is expelled outward through the mouth.

sensory cells

Hydra is more complex than the sponges. It has a distinct organization of tissues into an outer layer (ectoderm) and an inner layer (endoderm), with a jelly-like medium between. A transport system is unnecessary because of the thin body wall which allows digested food to diffuse from nutritive-muscular cells to other specialized cells. *Hydra* has a sensory system of nerve cells which, along with capturing tentacles and a digestive cavity, enable it to gain nutrition from larger prey than that of the sponges or protozoans.

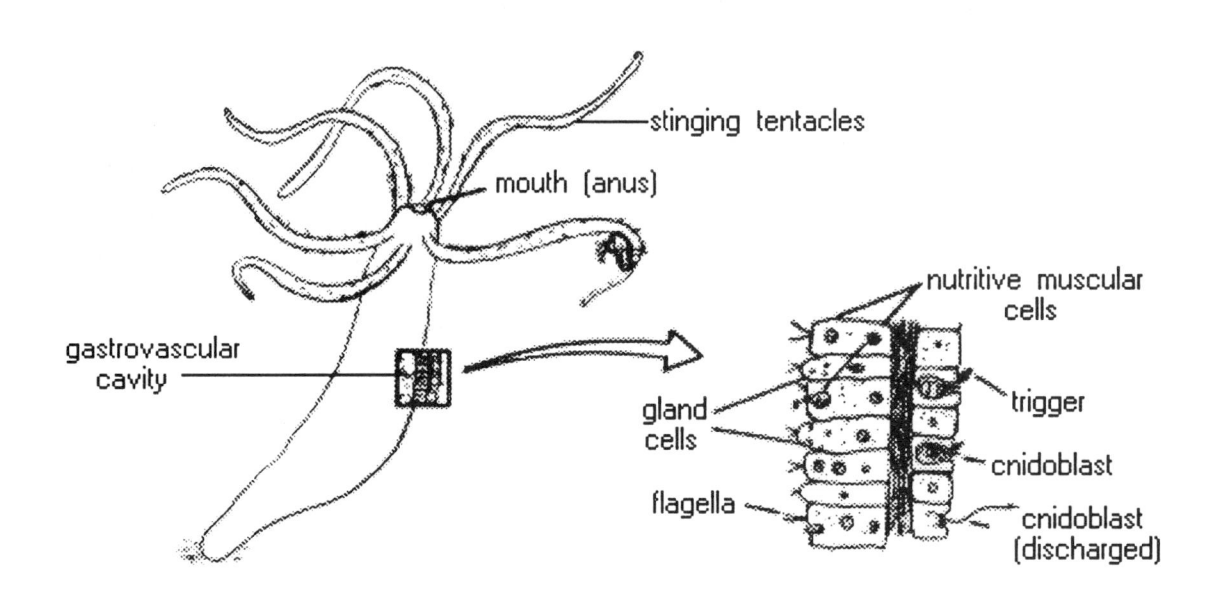

Figure 10-7. Diagrammatic view of *Hydra*.

WORM-LIKE ANIMALS

Worm-like animals are represented in several phyla, the most common of which are the Platyhelminthes (flatworms) and the Annelida (segmented worms) (Figure 9-12). The flatworms such as the **planarians** are common in aquatic habitats (Figure 10-8). They are carnivorous worms that glide over surfaces by a cilia-driven locomotion system. Unlike sponges or Cniderians, the flatworms have bilateral symmetry, and a distinct head with two **eyespots** which can detect light. The digestive system is similar to that of *Hydra*. Food is taken into a gastrovascular cavity through a muscular suction tube, the **pharynx**. The thin body design, coupled to an extremely branched gastrovascular cavity lined with phagocytic cells, allows all cells to have access to nourishment. Material that is not broken down by extracellular and intracellular digestion is expelled via the pharynx.

Eyespots detect light.

The flukes and tapeworms are parasitic flatworms. These often occupy the digestive tracts of animals and humans where they readily absorb food molecules digested by the host. Hence, they have less elaborate internal digestive structures than free living flatworms.

Annelids, including the earthworm, represent those invertebrates with a "tube-within-a-tube" body design. The digestive tube has a separate mouth and anus, and is separated from the outer body wall by a fluid-filled body cavity, the **coelom** (Section 9-F). This coelomate body design is also found in arthropods, mollusks, and echinoderms (Figure 9-12). Thus, much of our discussion of the earthworm also applies to these other phyla, even though their external anatomy is quite different.

coelomates

The digestive anatomy and behavior of the earthworm makes it an ideal detritivore (Section 5-C, 5-H). Its diet consists of soil particles and organic matter which is taken in through the mouth as the worm literally eats its way through the soil (Figure 10-9). The digestive tube is a complete **digestive system** with the following organs: **pharynx** (swallowing), **crop** (storage), **gizzard** (physical digestion by grinding), **intestine** (extracellular digestion by enzymes), and **anus** (elimination). The large, muscular body and the more efficient one-way digestive system enable the earthworm to digest large

earthworm nutrition

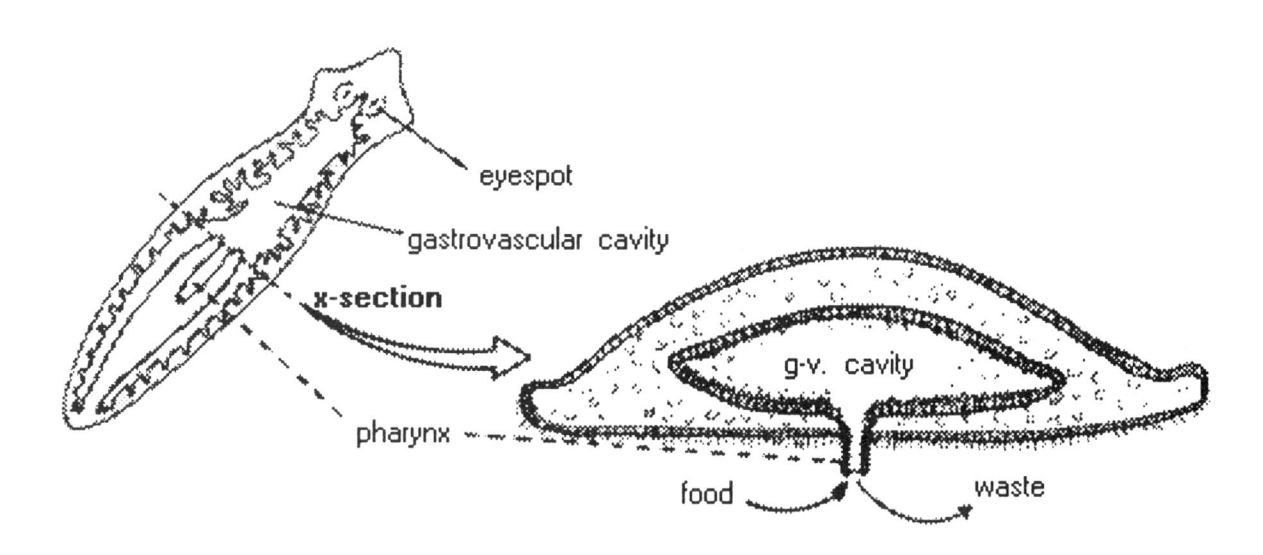

Figure 10-8. Diagrammatic view of a planarian.

circulatory system

amounts of organic matter while at the same time enriching the soil via the nutrient-rich "castings" it leaves behind. However, the more complex coelomate body design requires that digestive products released in the intestine be transported to cells which are distant from the digestive tissues. Therefore, the earthworm has a circulatory system, complete with red blood. A red protein called **hemoglobin** carries oxygen, but is not associated with red blood cells as it is in most vertebrates. The fluid circulates under the force of five pairs of "hearts" or **aortic arches**, and moves to the distant body tissues where it branches into **capillaries** (Figure 10-9). Nutrients from the centrally located digestive tract are thus transported outward to muscles and skin, while O_2 absorbed by the moist skin is transported inward. An excretory system of ducts serves to collect and transport wastes from the coelomic chamber to pores in the body surface of each segment.

10-F VERTEBRATE ANIMAL NUTRITION

Vertebrates are most complex . . .

but deficient.

The vertebrate animals have the most complex digestive systems and the most elaborate sensory and locomotor systems for locating and capturing food. The larger body mass and more active living habits of the vertebrates require the ingestion of food in bulk, and an efficient digestive system to process it. However, in spite of all the anatomical complexity of their digestive systems, vertebrates are very deficient in their capacity to synthesize complex molecules from simple carbohydrates and inorganic ions. This contrast among autotrophs, microbes, and animals was discussed in Section 10-A.

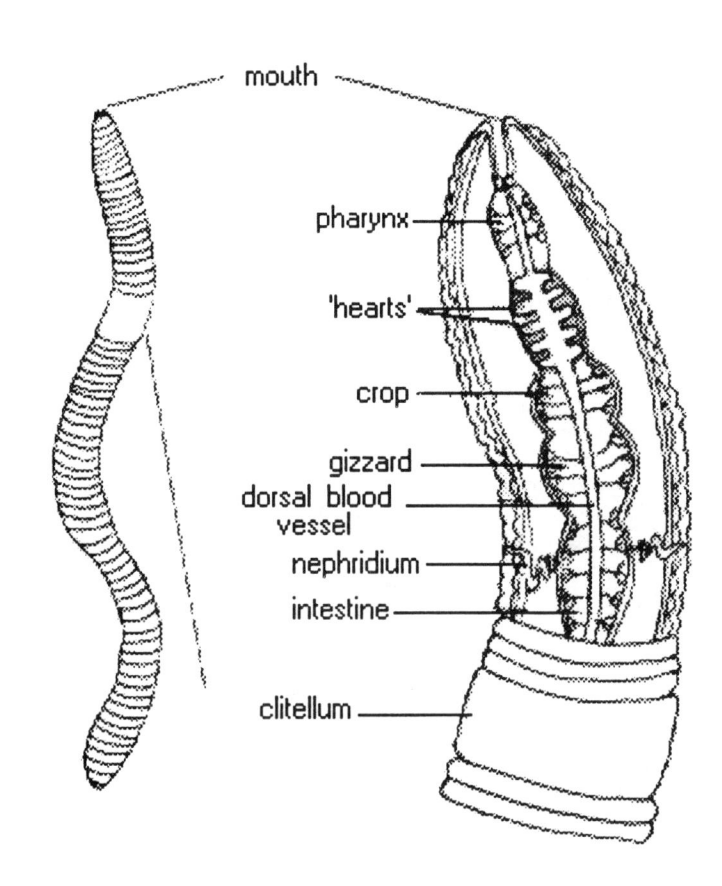

Figure 10-9. Diagrammatic view of the earthworm.

herbivores

The digestive systems of herbivores differ from carnivores, and this difference reflects their respective feeding habits. Grazing herbivores, often threatened by predators, have specialized storage compartments so that food can be consumed quickly and digested over longer periods of time. These *ruminant* animals, including cattle, sheep, and goats, have multiple digestive compartments. Cattle and sheep have teeming millions of symbiotic bacteria that digest the otherwise unusable cellulose from the cell walls of plants. The "cud-chewing" animal aids in this digestion by repeated regurgitation, chewing, and swallowing. Humans and other vertebrates that lack cellulose-digesting bacteria, gain little caloric value from vegetable fiber.

carnivores

Predators are often equipped with keen sensory perception, fast locomotion, and specialized mechanisms (e.g. talons, claws, teeth) for capturing and devouring animal prey. The lower dependence of carnivores upon plant matter is associated with a less elaborate digestive system (e.g. shorter small intestine) than that of herbivores. Because food may not be equally available in every season, many vertebrates store energy reserves in lipid form as body fat. Smaller animals such as rodents may store food in special caches for use during unfavorable seasons.

10-G   HUMAN NUTRITION

three functions:

The human digestive system in adults consists of a tube about 25 feet in length, differentiated into a series of organs. The system performs three important functions:

a. Containment and control of movement of food between meals

b. Digestion — *physically* decreasing particle size and increasing surface area; *chemically* producing smaller molecules that are soluble in digesting food, or *chyme*

c. Transfer of digestion products into the circulatory system

d. Elimination of *feces* from the body

ORAL CAVITY

Digestion begins in the mouth

The first chamber of the human digestive system is the oral cavity (Figure 10-10). Food enters the mouth where chewing action reduces particle size and increases surface area in preparation for enzyme action. The **salivary glands** secrete saliva which contains the enzyme **amylase**. Amylase hydrolyzes starch and other polysaccharides (Section 4-E). Because the salivary glands release secretions through ducts, they are called **exocrine glands**, in contrast to **endocrine glands** (e.g. the pituitary gland) that secrete directly into the bloodstream. Saliva contains three other important constituents: bicarbonate ions (HCO_3^-) which buffer the pH at around 6.5 to 7.5, a lubricant to facilitate the passage of food to the stomach, and an antimicrobial agent to prevent infections by ingested microbes.

food movement

During swallowing, the tongue forces balls of food into the *pharynx*, a muscular tube which opens to both the **esophagus**, leading to the stomach, and the *trachea* (windpipe) leading to the lungs (Figure 10-10). The latter is closed off during swallowing by muscular contractions against a flap called the *epiglottis*. Food is carried down the esophagus to the stomach by **peristalsis**, waves of alternating contractions and relaxations of muscles that encircle the esophagus. Thus, both *voluntary* (consciously controlled) and *involuntary* muscle contractions are involved in food intake. [Are decisions involving diet voluntary or involuntary?]

STOMACH

three functions of the stomach:

The stomach serves three main functions. First, like the crop in birds and annelids, it is a storage compartment. Average retention time for food in the stomach is 3 to 4 hours. Second, while you are free to pursue creative endeavors between meals, the stomach churns and mixes food by peristaltic contractions of the stomach walls. Release of chyme from the stomach into the small intestine is regulated involuntarily by the *pyloric sphincter* [Gr. (*pyloros* = gate keeper) + (*sphincter* = squeeze)], located where the stomach joins the small intestine (Figure 10-10).

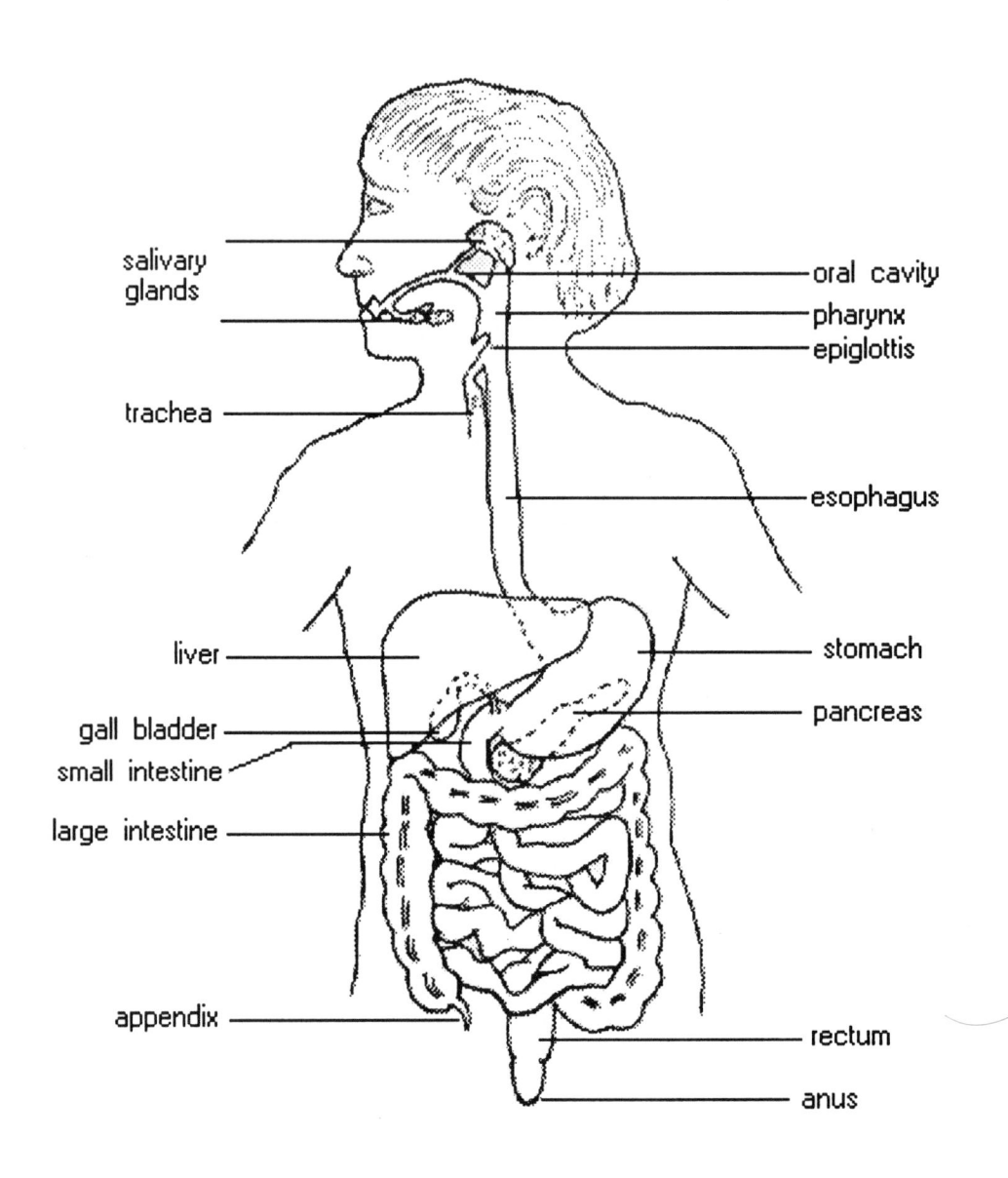

Figure 10-10. Main components of the human digestive system.

The third role of the stomach is in chemical digestion. In response to the sight or smell of food, presence of food in the stomach or anxiety, glands in the stomach wall release four gastric secretions: hydrochloric acid (HCl), *pepsinogen*, mucus, and bicarbonate. The HCl acidifies the stomach contents to about pH 2 which denatures proteins, exposing more of their peptide bonds to digestive breakdown (Section 4-I). When pepsinogen is secreted and moves away from the stomach lining into the chyme, the acidic pH converts it into an active protein-digesting enzyme, **pepsin**. By the time pepsin is formed, it has moved away from the stomach lining which is protected from self-digestion by the mucus coating and bicarbonate buffering. Peptic ulcers can result when mechanisms that control HCl secretion go awry. [Explain how stress might cause ulcers.]

peptic ulcers

SMALL INTESTINE

The small intestine is the principal organ of digestion and absorption. Partially digested carbohydrates and proteins are completely hydrolyzed to simple sugars (e.g. glucose) and amino acids, respectively.

Distinguish the roles of bile and lipase.

Lipids are insoluble in the aqueous slurry and collect in globules that are difficult for digestive enzymes to attack. However, the liver secretes an emulsifying agent called **bile** which enters the small intestine via a bile duct (Figure 10-10). The bile disperses the fat globules into small suspended droplets. Then, the **lipase** enzymes, released into the small intestine from the **pancreas**, convert the fat molecules into fatty acids and glycerol (Section 4-H). Excess bile is stored in the **gallbladder**. These digestive processes occur as peristalsis of the small intestine moves food and digestive juices along the path of the tube.

absorption of digested food

The small intestine has a major role in the absorption and transfer of the products of digestion into the bloodstream. To accomplish this function, it has a large internal surface area, due to its approximately 20-foot length, coiled within the average adult body cavity. The surface area is further increased by numerous finger-like outgrowths called *villi* (L. shaggy hair). Each villus is covered with many smaller hair-like projections called *microvilli* which further increase the absorptive surface. [Begin a list of names of tissues that are designed for large surface area—e.g. root hairs.] Each villus houses blood capillaries (Section 10-H) which form the interface between the digestive system and the circulatory system, facilitating the uptake of digestion products—sugars, amino acids, and lipids (via lymph capillaries). As peristalsis moves the chyme along this lengthy tube, and sometimes rhythmically forward and backward, there are many opportunities for digestion products to come in contact with the inner surface. [Is food really "inside" the body before it is absorbed by the villi? Explain.]

LARGE INTESTINE

water content

symbiotic bacteria

vitamin K

The large intestine (colon) absorbs water and salts from the undigested contents arriving from the small intestine, converting it into a bulk form called *feces*. Without homeostatic control of this process, the body suffers from diarrhea and dehydration on the one hand, or constipation on the other hand. Bacterial populations, living symbiotically within the large intestine, constitute over half of the dry weight of feces! These bacteria produce vitamin K and other constituents which are absorbed and used by the body.

10-H CIRCULATORY SYSTEMS

Large-bodied animals require circulatory systems.

Unlike the unicellular organisms that acquire food, perform digestion, and excrete wastes, all within a single cell, large-bodied animals require a transport system. Large quantities of food must be digested in a centralized location, and the products distributed long distances to other parts of the body. Waste products, hormones, and other compounds made in specific parts of the body must also be transported. Diffusion of substances from cell to cell is ineffective compared to transport by vascular systems.

Most circulatory systems have three major components:

three components

1. **Blood**, a fluid-like *tissue*, consisting of blood cells suspended in blood **plasma**. Plasma consists of water, dissolved ions, and plasma proteins [*e.g.* antibodies, fibrinogen (blood clotting agent)]

2. **Pumping organ** which creates pressure to move blood

3. **Vessels**, tubes which coordinate with the pumping mechanism to regulate blood pressure and distribution among regions of the body

components of blood:

Blood is a complex of tissues in a fluid medium and functions in the *transport* of nutrients, cellular waste products, gases, hormones, and antibodies. These transported components may be dissolved in the plasma or associated with blood cells. Blood cells are classified as **red blood cells** (erythrocytes; Gr. *erythros* = red), **white blood cells** (leukocytes; Gr. *leukos* = white), and **platelets**. All originate from continually embryonic tissue called *stem cells*, located in the bone marrow. Red cells lose their nuclei when mature and become disc-shaped with concave surfaces. They transport O_2 and CO_2 with the aid of the red, iron-rich protein, **hemoglobin**. White blood cells protect the body against foreign pathogens and chemical substances. *Neutrophils* and *monocytes* are white cells that move out in response to chemical signals from inflamed cells and destroy invading microbes by phagocytosis. *Lymphocytes* produce antibodies against foreign invaders. Platelets function, in conjunction with fibrinogen, to initiate clotting and plug openings in blood vessels at sites of injury.

blood vessels

Many invertebrates, including the insects, have an **open circulatory system** in which blood is pumped in vessels from the heart to body tissues. Here the blood enters the spaces around cells, exchanges materials with the cells, and then sluggishly drains back to the heart without the aid of vessels. This system is less efficient than a **closed circulatory system** wherein blood circulates faster from the heart within **arteries**, through **capillaries**, and more quickly returns to the heart via **veins**. The closed system is found in some invertebrates (e.g. earthworm) and all vertebrates, most of which are large-bodied, active animals.

the "pump"

The number of chambers in the heart of different vertebrate species also seems to be correlated with their increasing activity (Table 9-3). The four-chambered human heart has two **atria** (singular *atrium*) to receive blood that returns via veins from the body, and two **ventricles** to pump blood back out to the body via arteries (Figure 10-11).

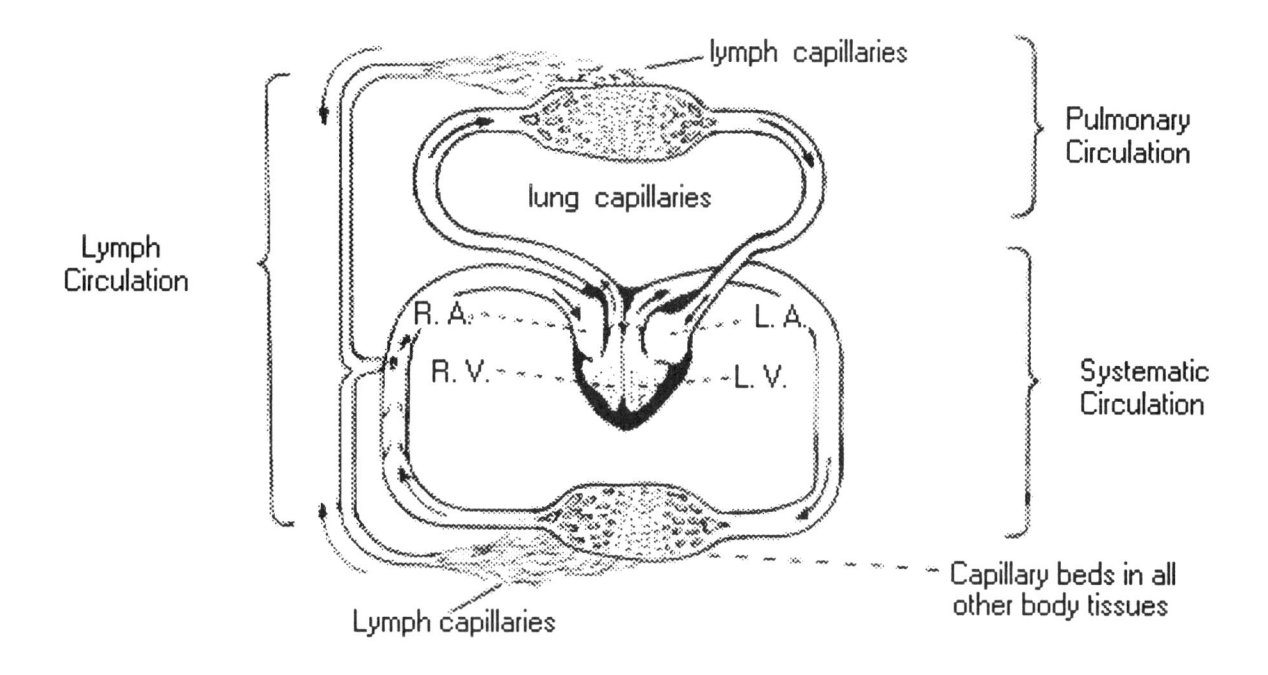

Figure 10-11. Diagrammatic representation of the relationship between pulmonary, systemic, and lymph circulation in the human body.

Arterioles regulate blood flow.

Let us now focus attention on regulation of blood flow from the heart via arteries to the capillaries. Arteries carry large volumes of blood under pressure from the heart. Arterial blood is distributed into branches called **arterioles**. Muscular arteriole walls, under involuntary control, enlarge in diameter or constrict. The resultant variable resistance to blood flow regulates the volume of blood approaching each network of capillaries, or **capillary bed** (Figure 10-11). In this way, capillaries in certain parts of the body will receive more blood at a given time than other parts. Sphincters (Section 10-G) located at the entrance to each capillary bed serve like tollgate keepers to regulate the amount of pressure-driven blood "traffic" flow off the main arterial "highway" at each "exit ramp". In this way, blood flow can be preferentially increased to metabolically active parts of the body. In inactive tissue, constriction of sphincters causes more blood to bypass capillary beds and go instead from arterioles to **venules** which converge into veins.

Exchange occurs in capillaries.

Capillary beds provide for exchange of substances between body tissue and bloodstream (Figure 10-11, bottom). Each capillary is a tube formed by a single layer of cells, and just large enough in diameter to conduct a flow of plasma which carries red blood cells in single file. Though their diameter is smaller than that of arterioles, blood pressure decreases from arteriole to the capillaries, largely due to two factors. First, the sheer number of capillaries permit many paths of flow. Second, the capillary walls are only one cell layer thick, allowing escape of blood plasma into surrounding tissues. As a result, each tissue of the body is continually "bathed" in fluids from nearby capillaries. This fluid delivers O_2, glucose, and other dissolved nutrients to the membranes of body cells.

Body tissues are "bathed".

lymphatic system

As the remaining plasma and blood cells move through the capillary bed, pressure within decreases and fluids begin to move from body cells into the capillaries. With this fluid influx comes CO_2, ammonia, and other cellular wastes. Some of this fluid enters an adjacent capillary network, called **lymph capillaries** which carry wastes and other constituents into the **lymphatic system**. This fluid, known as **lymph**, is pushed along toward larger lymph vessels under the force of one's breathing and other muscle contractions. **Lymph nodes**, located in larger lymph vessels, filter microbes and cellular debris which is destroyed by phagocytic leukocytes. Lymph then joins venous blood in its return to the heart. The lymphatic system is a major bodily defense against foreign agents and disease (Section 17-E).

Substances move in opposite directions.

We should note that substances carried by the blood may move in opposite directions depending upon location in the body. In active parts of the body, O_2 diffuses from the blood toward respiring cells whereas CO_2 moves from respiring cells into the blood. However, in the lungs O_2 diffuses into the blood from air sacs, and CO_2 diffuses to air sacs from the blood. Hemoglobin in erythrocytes has a high attraction for O_2 in the lungs and thus facilitates O_2 transport. Similarly, in the small intestinal villi, there is an net uptake of digested food molecules, while in the kidneys, there is a net loss of wastes and perhaps water from the capillaries. We will deal more specifically with homeostatic controls of nutritional and waste exchange in Chapters 13 and 14.

10-I HUMAN NUTRITION AND STEWARDSHIP

With greater complexity of body . . .

comes greater importance of wise choices.

If you were placed on a diet of sugars, minerals, and water, your health would soon deteriorate, whereas, certain microbes would thrive on this menu. Indeed, our survey from the so-called simple microbes to the multicellular animals has revealed an interesting pattern. That is, the more complex digestive systems of animals are associated with an inability to synthesize certain basic organic compounds that are essential for survival (Section 10-A). Instead of having the enzymes which are necessary to fabricate these compounds, most invertebrates and the vertebrates have the capacity to carefully select their food sources from plants and animals of their respective biotic communities. This involves sophisticated sensory and locomotor capacity and certain instincts that are often unique to each species. Humans, who alone bear the image relationship with God, have been given, in addition, the intelligence to study and understand human physiology and nutrition. We can actually learn about the physical and chemical properties of the ions and molecules we eat! Yet, in spite of superior intelligence, we seem to be uncertain about what a wholesome diet is! [Do you know?]

Human nutrition is a complex science with many unanswered questions. A detailed treatment is obviously beyond the scope of this text. Rather, we will provide some basic information and some "food for thought" in hopes of motivating you to probe further into human nutrition from the perspective of Christian stewardship of the body. Anyone seeking to have a proper diet needs answers to at least four questions as follows:

proper diet— four questions

1. Am I committed to principles of stewardship of my body?

2. What nutritional substances are required by the body?

3. How much of each substance is required?

4. What food sources provide these substances in the diet?

STEWARDSHIP COMMITMENT

Involuntary functions support our voluntary acts.

Knowledge about the human digestive system and circulatory system are of little value unless we submit the whole area of our eating and nutrition to the lordship of Christ. This chapter has hardly scratched the surface of the marvels of human digestion and assimilation. Yet, it should be quite evident that we are "fearfully and wonderfully made". If God has permitted you to have good health, you are blessed with a complex system which can physically and chemically process food, store reserves, and distribute suitable amounts through tiny vessels to each body cell. Most of this activity is controlled involuntarily and is beyond your responsibility. However, it is our responsibility to care for one important aspect of nutrition—our dietary intake.

biblical basis

It is easy to depart from a "balanced diet" — one which is governed by an understanding that food is to be thankfully received and enjoyed in a God—honoring manner (Matt. 6:11). The Bible states that in eating, working, physical exercise, and *whatever you do, do all to the glory of God* (I Corinth. 10:31), *for you have been bought with a price* (I Corinth. 6:20), *the precious...blood of Christ* (I Peter 1:19). May we seek daily the thankful, humble spirit that acknowledges God as creator, ourselves as His handiwork, and our "daily bread" as His sustenance for us, while we seek fulfillment and delight in Him, not in sensual excess of any kind.

dietary extremes:

Carelessness and disobedience to spiritual truth may lead to one of several conditions ranging from mild excesses of *intake* or *denial* of food to extreme eating behaviors. The latter include the destructive eating disorders, anorexia nervosa and bulimia. Diehl and Morris (5) present an informative discussion of these two disorders which they define as follows:

anorexia

"**Anorexia nervosa** is a physical and psychological disorder characterized by self-imposed starvation, often coupled with compulsive and excessive physical exercise." Success in causing weight loss intensifies the compulsive behavior and produces "a self-destructive downward spiral". Unless the spiral is broken, biochemical imbalances may cause heart irregularities and other potentially fatal reactions.

bulimia

Bulimia is characterized by a pattern of "binging and purging." Excessive eating is followed by "self-induced vomiting and/or use of laxatives". Repeated vomiting of acidic stomach contents damages the teeth, and irritates the esophagus and pharynx. Laxative usage may eventually cause intestinal irritation, and dehydration and salt imbalances which "can trigger life-threatening cardiac irregularities".

Seek help!

Both disorders are most common among young women who yield to an obsession to "be thin" as the perceived prerequisite to being accepted by themselves and others. Diehl and Morris' book (5), and a videotape by Meier (9) discuss signs of anorexia and bulimia. If you recognize these behaviors or tendencies toward them in your life, or in someone you know, please, read no further! Do not rest until you seek help from your institutional health service, personal physician, pastor, or professor. As a creature of God, hopefully a born-again child of God, don't grieve Him any longer by violating His principles and plan for your life. As the workings of your body testify, you are the handiwork of a loving God who has a wonderful plan for your life. What a tragedy to have read this chapter, then, out of pride, continue to destroy what God has put within you.

NUTRIENTS REQUIRED

The human body requires six classes of nutrients. In addition to *carbohydrates*, *minerals*, and *water* which are sufficient for many bacteria and fungi, humans must ingest *vitamins*, *proteins,* and *fats*.

1. **Carbohydrates** provide a source of energy in respiratory metabolism, and provide organic carbon "skeletons" for synthesis of proteins, nucleic acids, and other constituents. The majority of ones carbohydrate intake should be in the form of *complex carbohydrates* (starch, glycogen, and other polysaccharides) rather than as *simple carbohydrates* (soluble sugars, or "sweets"). We will return to this subject (Table 10-2).

2. **Minerals** are ingested in ionic form and include all of those listed on Table 4-1 except C, H, O, N, and S which must be ingested in organic form. [Which of the above classes of molecules would supply organic C, H, O, N, and S? Note that O is also acquired by inhalation of O_2 gas.] Table 4-1 describes some of the metabolic roles of mineral elements. In addition to those listed, the body also requires chlorine (for HCl of gastric juice), fluorine (bone structure), iodine (thyroid gland metabolism), cobalt (component of vitamin B_2), and zinc and copper (enzyme activators).

3. **Water** plays an important role in all living cells as discussed in Section 4-E. The health and well-being of millions of people on earth today is threatened by contaminated water. Water interacts physically and chemically with all of the other molecules of life. The water balance of blood plasma and lymph fluid is closely tied to the concentration of dissolved mineral ions, especially sodium and potassium. Levels of all of these are controlled by the kidneys.

We must obtain vitamins by ingestion.

4. **Vitamin** signifies a special class of organic molecules that are required in small quantities. Unlike most plants, humans and most animals cannot synthesize needed vitamins. Table 10-1 lists the major vitamins, their functions in the body, and likely food sources of each.

Table 10-1. Vitamins.

VITAMIN		MAJOR FUNCTION	COMMON FOOD SOURCES
WATER-SOLUBLE VITAMINS:			
B_1	Thiamine	Coenzyme in reactions involving carbon dioxide	Whole grains, pork, legumes
B_2	Riboflavin	Constituent of coenzyme FAD in respiratory metabolism	Dairy products, liver, spinach
	Niacin	Constituent of coenzymes NAD and NADP which participate in respiration and photosynthesis	Meat, liver, legumes
B_6	Pyridoxine	Coenzyme in amino acid metabolism	Most foods
	Pantothenic Acid	Constituent of Coenzyme-A which participates in respiration	Most foods
	Folic Acid	Coenzyme in amino acid and nucleic acid synthesis	Green vegetables, legumes, whole wheat
B_{12}	Cobalamin	Coenzyme in carbon transfer; synthesis of red blood cells	Meats, eggs, dairy products
	Biotin	Coenzyme in synthesis of fats, amino acids, and glycogen	Most foods; intestinal bacteria
C	Ascorbic Acid	Formation and maintenance of connective tissue; antioxidant	Citrus, vegetables, tomatoes
FAT-SOLUBLE VITAMINS:			
A	Retinol	Constituent of rhodopsin (visual pigment); maintenance of epithelial tissue	Carotene-containing vegetables, egg yolk, diary products
D	Calciferol	Calcium absorption and transport; bone formation	Cod liver oil, eggs, dairy products
E	Tocopherol	Antioxidant to protect cell membranes	Vegetable oils, cereals, peanuts
K	Naphthoquinone	Aids in blood-clotting	Green vegetables; intestinal bacteria

5. **Proteins.** Barring genetic defects, the human body can synthesize the whole host of different proteins required for cell membranes, enzymes, and other uses. However, humans and most animals cannot synthesize all of the twenty common amino acids which are the monomers, or "building blocks" of protein polymers (Section 4-I). The body cannot convert NO_3 or NH_3 into the *amino groups* of the amino acids. Nor can it even synthesize the molecular structure of 9 out of the total of 20 different amino acids needed. Therefore, the following nine *essential amino acids* must be ingested in the diet: histidine, isoleucine, lysine, methionine (Figure 4-9), phenylalanine, threonine, tryptophan, and valine.

essential amino acids

6. **Fats** are essential in our diet as a source of energy, fat-soluble vitamins (Table 10-2), and certain *essential fatty acids* (e.g. linoleic acid, Section 4-H). Saturated fats are obtained primarily from animal sources whereas vegetable oils are unsaturated fats (Section 4-H).

Cholesterol is a necessary lipid component of cell membranes, myelin sheaths, steroid hormones, and bile salts, and has been associated with *atherosclerosis*. Atheroscleros causes death when arterial walls thicken with *plaque* deposits, lose elasticity, and gradually become blocked. The body transports insoluble cholesterol in the blood in the form of *lipoproteins* that are wrapped in water-soluble proteins. Experiments have shown that diets favoring saturated fats (e.g. red meat, pork, lamb, whole milk products) cause elevated levels of cholesterol in the form of *low-density lipoproteins* (LDL's) which form plaque. However, unsaturated fats (e.g. from vegetable, poultry, fish) lead to higher levels of *high-density lipoproteins* (HDL's), which are correlated with low risk of heart attack.

a cause of high cholesterol

SOURCES OF NUTRIENTS AND QUANTITY NEEDED

The complex metabolism of the human body plus the variables such as heredity, emotional stress, environmental pollutants, and individual medical history combine to make dietary research and recommendations difficult. Vested interests of the food processing industry often influence eating habits in ways that are nutritionally unsound. Advertising influences the consumer to place a higher priority on enjoyment of eating than on nutritional value. We enjoy the best access to food but are plagued with digestive disorders that are practically nonexistent in many developing countries.

misleading information

Based upon our earlier discussion of christian stewardship, our diets should reflect good judgement, based upon generally accepted nutritional data. Table 10-1 indicates major natural sources of vitamins. The most reasonable way to acquire the essential vitamins is to eat a well-balanced assortment of foods from both vegetable and animal sources. Detailed nutritional tables are helpful in selecting quantities of each food source necessary to achieve a given caloric intake, and energy expenditure program (5, 16). Age, body size, gender, climate, and daily exertion are among factors that influence these requirements. Heredity and caloric intake during early childhood are factors that apparently influence body fat accumulation as much as ones daily caloric balance.

dietary factors

Table 10-2. Caloric Content and Recommended Distribution of Three Food Classes[1].

CLASS	CALORIES Per Gram	DISTRIBUTION OF CALORIES Recommended	American Diet
CARBOHYDRATES:	4		
SIMPLE —		10% (or less)	24%
COMPLEX —		48% (or more)	22%
PROTEINS	4	10-15%	12%
FATS	9.1	30% (or less)	42%
TOTAL INTAKE:	–	100%	100%

[1] Source: *Dietary Goals for the United States*, 2nd. ed., Select Committee on Nutrition and Human Needs, U.S. Senate (Washington, DC: US Govt. Printing Office, 1977)

energy balance

Your success in controlling or changing your body weight will depend upon your energy "balance sheet" of caloric intake *minus* consumption. However, both *total daily caloric intake* and the *distribution of calories* among carbohydrates, proteins, and fats are important, as shown in Table 10-2. Grams of each food class appropriate for your diet can be calculated from the recommended percentages and your desired Calorie intake per day.

Both shortages and excesses of nutritional components can disturb body metabolism and jeopardize health. There is no clear evidence that massive doses of vitamins exceeding the recommended allowances contribute to better health. Levels of water-soluble vitamins in excess of body needs are expelled in the urine, while an excess of fat soluble vitamins, especially A and D, can accumulate in the body and cause serious disorders.

vegetarian diet

The essential amino acids can be obtained from both plant and animal sources. Meat, however, has a larger variety of proteins and is thus more likely to provide the proper balance of all 20 amino acids. Some prefer a vegetarian diet out of concern for animal rights, others because of the greater thermodynamic efficiency achieved by feeding lower in the food chain (Section 5-G). However, vegetarians should be careful to select a wide variety of plant sources. For example, cereal grains such as rice and wheat are deficient in tryptophan and methionine, but are good sources of isoleucine and lysine. The reverse is true of beans. However, if rice and beans are combined in the same meal, one obtains all four of the essential amino acids.

Summary:

In summary, there are certain aspects of human nutrition that are still questionable, and others that are solidly accepted. Dietary approaches that emphasize *moderation*, *variety*, and *balance* are generally accepted as nutritionally favorable, and seem consistent with Christian stewardship. Stewardship of the body should also be viewed in the broader context of the impact of mechanized food production on the environment and also upon our brothers and sisters in developing countries (Sections 5-G and 7-G).

QUESTIONS AND DISCUSSION TOPICS

1. Rank the five kingdoms from greatest to least on the basis of:
   a. Nutritional metabolism with capacity to build organic molecules and cell structure from a few simple inorganic substances
   b. Average body size and anatomical complexity.
   How do you account for the pattern you observe? Which basis would you use to rank kingdoms according to supposed "evolutionary complexity"?

2. How do members of kingdoms Monera and Fungi digest food without a complex digestive system?

3. What nutritional problems are encountered by green land plants which green algae do not have? How are the problems "solved"?

4. How would removal of the phloem tissue in a foot-wide band around the trunk of a tree ("girdling") lead to death of the tree?

5. Make a table which summarizes differences in nutritional metabolism among *Paramecium*, *Hydra*, planaria, earthworm, and man. Why is intracellular digestion incapable of meeting the nutritional needs of large animals?

6. Trace the pattern and processes involved in the movement of food from chicken and mashed potatoes to your leg muscle cells.

7. Explain how you would go about determining an appropriate quantity of carbohydrate, protein, and fat intake per day for yourself.

TOPICS FOR FURTHER STUDY

1. There are many unanswered questions concerning plant nutrition. Select one of the following questions and share in the intrigue of some of the amazing aspects of plant physiology:
   a. Suction can lift water no more than 30 feet vertically from the ground. How do leaves of tall trees obtain water?
   b. Why do plants "bleed" when punctured by a needle or feeding insect?
   c. How does the sensitive plant fold its leaves in response to shock?
   REFERENCE #13; and REFERENCE #16, Chapter 9.

2. Some plants and fungi are carnivorous. How do they prey upon animals and digest food? What is the advantage of this lifestyle?
   REFERENCES #7 and #12

3. How do digestive cells digest food without self-destructing?
   REFERENCE #3; and REFERENCE #6, Chapter 9.

4. Is it biblical for Christians to learn lessons from the creatures with respect to nutritional habits? If so, what do they teach us? See, for example, Proverbs 6: 6-9; Matthew 6: 25-33. See also REFERENCE #18, Ch. 7.

5. What do many not know about proper nutrition and weight control?
   REFERENCES #15 and #16.

REFERENCES

1. Christian, S. 1986. *The Very Private Matter of Anorexia Nervosa.* Zondervan Books. Grand Rapids, MI.

2. Cohen, L. 1987. Diet and cancer. *Scientific American* 257(5): 42-68.

3. deDuve, C. 1983 Microbodies in the living cell. *Scientific American* 248(5): 74-84.

4. DeHaan, M.R. 1971. *The Chemistry of the Blood.* Zondervan, Grand Rapids, MI.

5. Diehl, P.S. and L.D. Morris. 1986. *Physical Fitness and the Christian: Exercising Stewardship.* Kendall/Hunt Publ. Co., Dubuque, Iowa.

6. Epstein, E. 1973. Roots. *Scientific American* 228(5): 48-58.

7. Folkerts, G.W. 1982. The Gulf Coast pitcher plant bogs. American Scientist 70: 260-267.

8. Martini, F. 1992. *Fundamentals of Anatomy and Physiology*, 2nd ed. Prentice Hall. Englewood Cliffs, NJ.

9. Meier, P.D. 1984. *Eating Disorders* [videocassette]. Minirth Meier Clinic, Richardson, TX.

10. Miller, D.K. and T.E. Allen. 1982. *Fitness, A Lifetime Commitment.* Burgess Publ. Co., Minneapolis, MN.

11. Nilsson, L. 1978. *Behold Man.* Little, Brown, and Co. (Color atlas of human anatomy)

12. Pramer, D. 1964. Nematode-trapping fungi. *Science* 144: 382-388.

13. Salisbury, F.B. and C.W. Ross. 1991. *Plant Physiology*, 4th ed. Wadsworth, Belmont, CA.

14. Saunders, M. 1988. *Bulimia—Help Me Lord!* Destiny Image. Shippensburg, PA.

15. Vener, A.M. and L.R. Krupta. 1985. Needed: More attention to weight control knowledge. *American Biology Teacher* 47(3): 148-153.

16. Whitney, E.N. and E.M.N. Hamilton. 1984. *Understanding Nutrition.*, 3rd ed., West Publishing Co. St. Paul, MN.

17. Woese, C. 1981. Archaebacteria. *Scientific American* 244(6):98-125

See also REFERENCES for Chapter 9: Stern, L.R. (1991); and Hickman, C.P. Jr., *et.al.* (1990) as resources on plant and animal nutrition, respectively.

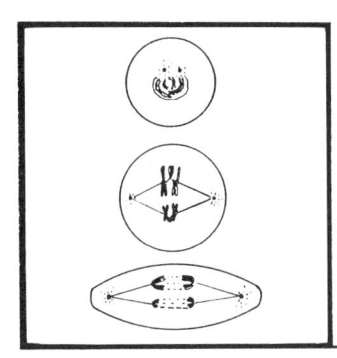

Chapter 11

Reproduction

Physical death is a fact of creation and each *individual* organism moves through a sequence involving growth and development, maturation, aging, and death. For each *species*, however, there is a solution to death of all members on a global scale (*extinction*), or on a local scale (*extirpation*). That solution is successful **reproduction**.

11-A REPRODUCTION — SIGNIFICANCE AND LIMITS

special cells to convey life

Even though individual organisms have a finite lifetime, they have been designed to pass on their life in the form of living cells to a new generation. These cells are called **gametes**, or sex cells. The **sperm** carries cytoplasm and hereditary material (DNA) from the male, and the **ovum** (egg), from the female. The fusion of sperm and egg, called **fertilization** produces a **zygote**, the first cell of a new generation. Thus, each of us began our individual lives as a microscopic living cell, composed entirely of the cytoplasmic and hereditary substance of our parents, and totally dependent upon the nurturing environment of our mother's body. This intimate beginning of a new generation is a reminder of the theory of **biogenesis**. That is, all living cells arise form pre-existing cells.

dependence upon parents

Life Characteristic #3 (Section 2-B, 2-C) is that *all living systems reproduce*. Like *nutrition* discussed in Chapter 10, *reproduction is a basic life process*. The two go hand-in-hand. Although nutrition has its place in supporting life, without reproduction, it leads to a "dead end", biologically speaking. Therefore, individuals must direct some of their nutritional resources into the production of a "living linkage" to a new generation.

the obvious orientation to reproduction

It is obvious that living organisms are oriented toward reproduction. Reproductive structures are especially obvious in multicellular plants and animals, and are associated with the fully developed adult individual. The attractive flowers and abundant fruit and seed production; and, the mating behavior that so strongly influences animals are all examples of the innate drive in every kind of organism to accomplish reproduction. From a scriptural perspective, we understand that the Creator who formed the creatures and gave them life also spoke the command that each "be fruitful and multiply upon the earth" (Genesis 8:17;9:1,7).

While we emphasize the obvious orientation toward reproduction within every species, we must hasten to point out an equally obvious limitation to reproduction. Even though most of the sexually mature organisms on earth can produce gametes, the gametes of one species cannot continue into the life of the next generation unless they fuse with gametes from the opposite sex *of the same species*. Consequently, the offspring will have the same basic body plan

the obvious barriers between *kinds*

as the parents. To summarize this principle, we could say that *there are distinct reproductive barriers between different species, and each species produces more of its own species* (i.e. reproduces).

You may recall from our discussion in Section 9-A, that the concept of *species* in biology, and *kind* in Scripture are not necessarily equivalents, and each is difficult to delineate. However, both concepts are pointing to the existence of reproductive barriers between reproductively isolated groups. Whereas creationists and evolutionists may agree that reproductive isolation exists, their interpretation of such facts differ greatly because of their differing presuppositions and world views. [Is the existence of reproductive isolation and often-associated anatomical distinctions (groups) among living and fossilized species more consistent with the evolution model or the creation model? Support your answer.]

creation and evolution views

Now that we have introduced reproduction as a basic life process of all creatures, and explained the biological limits within which it operates, let us examine the mechanisms of reproduction. Our attention will focus upon the cellular and molecular level of organization where many of these mechanisms occur.

11-B LIFE CYCLES AND REPRODUCTION — AN OVERVIEW

For most individuals, life begins as a single cell, the zygote. For unicellular organisms, including many species of the Monera and Protista, the production of more cells is not essential to growth and maturation. However, in multicellular individuals, the zygote is only the first of perhaps billions of cells that must be produced during growth and development of the individual plant or animal body. The zygote contains the hereditary material, or DNA, within its nucleus. This DNA functions like the "blueprint" in the hands of an architect. The blueprint is to the finished building what DNA of the zygote is to the mature organism. The DNA blueprint contains the "information" that determines the morphology and physiology of the adult-to-be, and influences metabolism during life of the individual.

zygote and its "DNA blueprint"

In multicellular organisms, one *copy* of the DNA blueprint originally contained in the zygote is produced for each *somatic cell* (G. *soma* = body) —i.e. all body cells except gametes. If every somatic cell has a duplicate copy of the DNA blueprint of the zygote, then it follows that the nucleus, containing the DNA, must by duplicated and divided every time cell division occurs. This process of *cell reproduction* may be represented as follows:

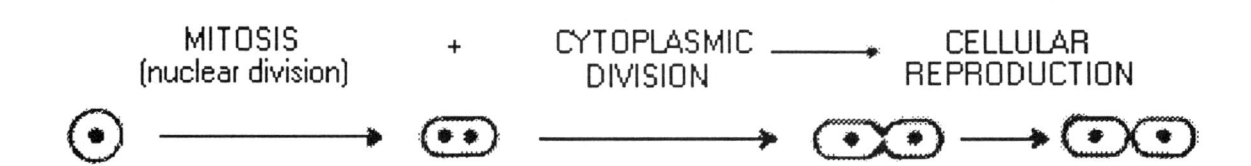

somatic cell reproduction

MITOSIS (nuclear division) + CYTOPLASMIC DIVISION ⟶ CELLULAR REPRODUCTION

mitosis: nuclear division

Cell reproduction (or cell division) involves a duplication—division of the nucleus usually accompanied by a division of the surrounding cytoplasm into two separate cells. Nuclear division in which the genetic material is duplicated and divided is called **mitosis**.

from embryo to adult

Beginning with the zygote, repeated cell divisions produce a single cluster of cells. However, very quickly, the cells begin to **differentiate**, or take on a distinct form in keeping with a specific function. You may recall that this specialization and division of labor among tissues and organs is a characteristic of the multicellular organism. A multicellular organism in its early stages of growth and development is referred to as an **embryo**. Continued mitosis, cell division, and differentiation lead to a stage in which the individual plant or animal can live independently of the parental generation. During this whole marvelous process, the DNA blueprint, in conjunction with factors in the environment of the embryo, is controlling the pattern of growth and differentiation of the individual. In Chapters 14 and 16 we will consider the genetic (DNA) control of differentiation in more detail.

adulthood — sexual maturity

Biologically speaking, maturity or adulthood is reached when an individual plant or animal is sexually mature. At this stage, the multicellular individual can enter *sexual reproduction* and produce offspring. In most cases adults are distinguished as male and female by the type of gametes they produce, whether sperm or egg (Section 3-E). Here, we encounter a new dimension in the division of labor—i.e. division of labor involving the sex organs in two separate individuals (see Section 11-G for exceptions). This is perhaps the most intricate step in reproduction. How can two individuals produce microscopic cells (gametes) and coordinate the fusion of these gametes in an often hostile environment to produce a zygote? The process by which the nuclei of specialized cells within the sex organs undergo division leading to the production of gametes is called **meiosis** (my-o'-sis).

life cycles

Fertilization, the union of sperm and egg, completes a so-called *life-cycle* when a zygote is formed. The life cycle as illustrated in Figure 11-1 summarizes the major events in the reproductive process of most organisms. Our discussion of reproduction in the following sections will focus on the various processes and stages of the generalized life cycle.

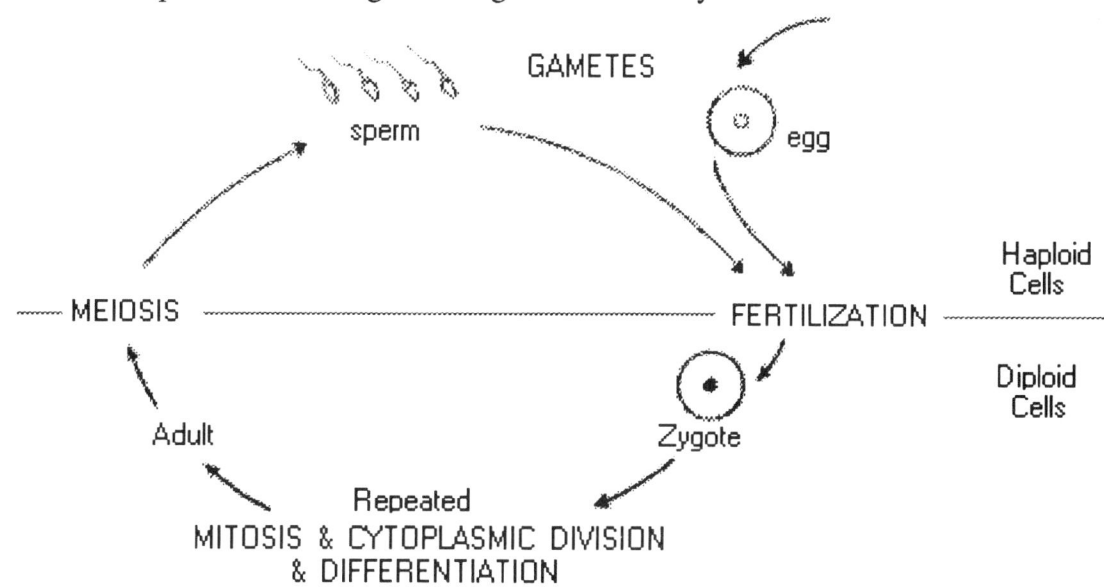

Figure 11-1. A generalized life cycle.

11-C DNA, THE INFORMATION MOLECULE

It is said that we are living in the "information age", the age in which information can be compiled, computed, and conveyed faster than men can think, or travel in body. At the heart of the revolution has been the semiconductor technology and its product, the marvelous microchip. While the microchip is amazing, it does not begin to match the "biochips" within living cells.

DNA monomers: nucleotides

Living cells each have one or more molecules of DNA, deoxyribonucleic acid. DNA is a large polymeric molecule of varying length which functions as a genetic code. That is, information is encoded by the sequence of monomers called *nucleotides* (Section 4-J). There are four different nucleotides in DNA, each distinguished by a different *nitrogen base*, a ring-like organic molecule containing nitrogen. The four nitrogen bases are *adenine, guanine, thymine*, and *cytosine*. Figure 4-17 show the molecular structure of these molecules. Another model of DNA is presented in Figure 11-2. Here, for simplicity, each nitrogen base is represented by a different shape.

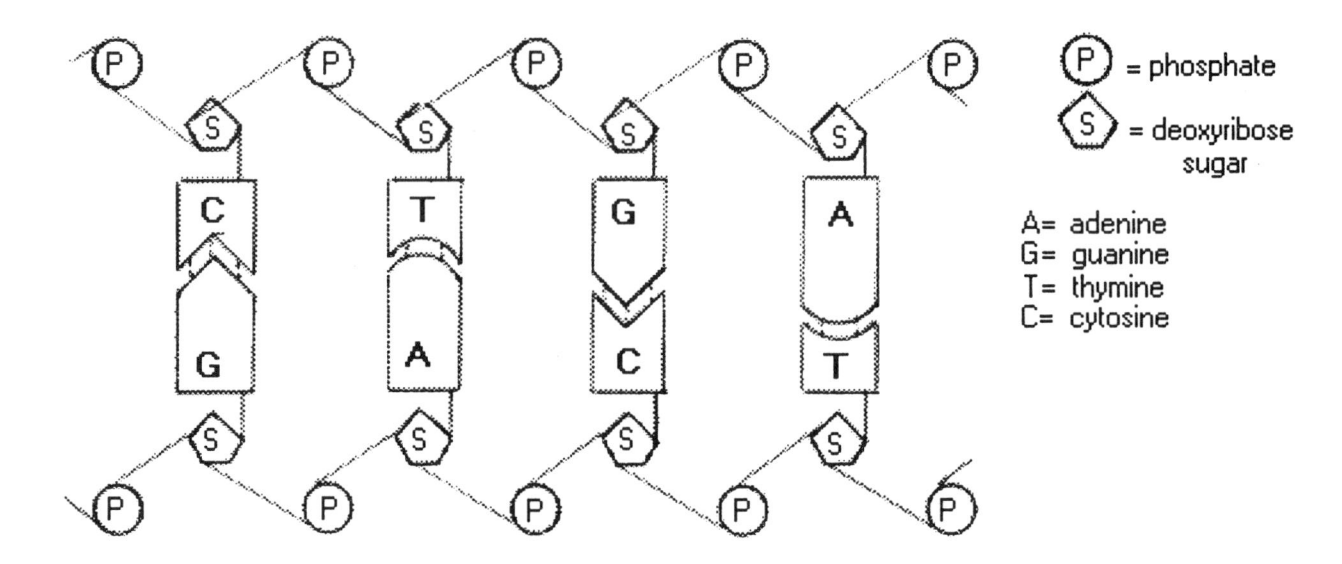

Figure 11-2. A short segment of DNA showing the complementary pairing between nitrogen bases.

pairing in the "double helix"

The DNA molecule is often described as a double-stranded spiral helix. Two strands, each containing nitrogen bases are held together by hydrogen bonds (Section 4-C) between the nitrogen bases. Note that adenine (A of one strand) always bonds with a thymine (T) from the other strand, and likewise guanine (G) with cytosine (C). The nitrogen bases are bonded to one another within each strand by a "backbone" of alternating phosphate (P) and deoxyribose sugar (S) groups (Figures 11-2, 11-3). Because of the complementary pairing between A - T and G - C, the base sequence in one chain defines the sequence in the complementary chain. The structure of DNA was first proposed in 1953 by the American James D. Watson and the Englishman Francis Crick who worked together at Cambridge University. The Watson-Crick model is based upon chemical analysis and X-ray diffraction studies. The two scientists received a Nobel Prize in 1962.

Nucleotide sequence codes information.

The nitrogen base sequence is like a four-letter "alphabet", and the sequence of "letters" produces a "code" for information storage on the DNA molecule. The alternating phosphate-sugar backbone by itself would carry little information, just as a two-letter alphabet would make few meaningful "words" unless the words were unreasonably lengthy. However, with the 4-letter alphabet of DNA, all of the information necessary to "blueprint" a whole multicellular organism can be stored in one microscopic nucleus!

In case you are not convinced, use the formula 4^n to calculate the number of different molecular codes that can be assembled in DNA. The 4 is the number of different nitrogen base pairs, and n is the length of the code sequence, or DNA molecule. For example, for a DNA molecule 3 nitrogen base pairs long, 4^3 equals 64 different code sequences. Most DNA molecules have over 100 nitrogen base pairs, and many have thousands of pairs.

Genes direct protein synthesis.

One long DNA molecule may have many segments called **genes**. Each gene is a precise unit of information to direct the synthesis of all or part of a protein molecule. Genes thus influence cell activities because proteins are influential as enzymes and membrane constituents. We will examine this process in more detail in Chapter 16. Our immediate concern is how the DNA of a single cell is duplicated and divided among two new cells.

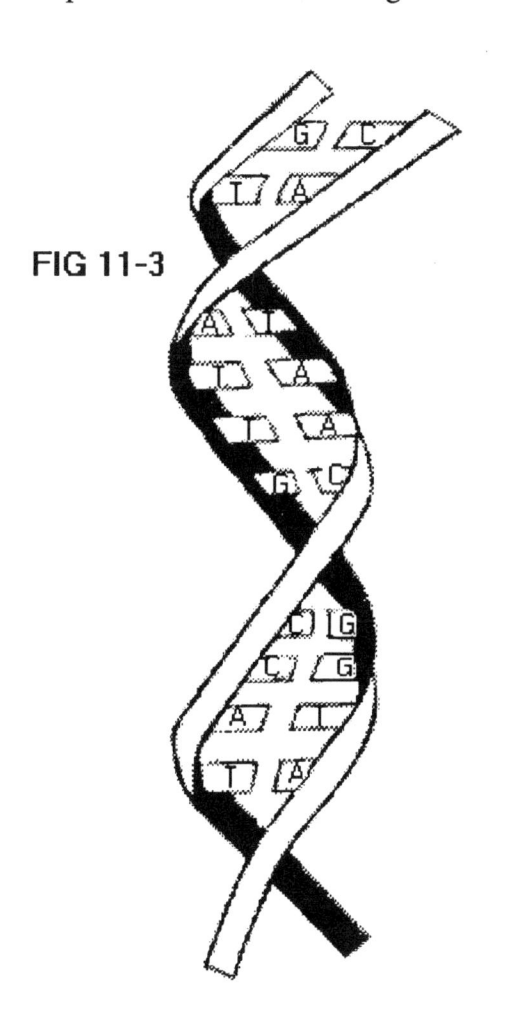

FIG 11-3

Figure 11-3. The double-stranded spiral helix of DNA. The helix resembles a twisted ladder with "rungs" of paired nitrogen bases and the uprights of alternating sugar and phosphate groups.

11-D DNA AND CHROMOSOMES

The nucleus is the control center of the cell. During most of the life of a given cell (except in Monera), the nucleus is seen under magnification as a distinct membrane-bound organelle containing a granular material. This material which darkens when chemical stains are applied is called **chromatin**. Chromatin is composed of the DNA of the cell which is normally coated with protein molecules. One or more darker-staining regions called **nucleoli** are also visible (Figure 11-7).

If you were to study cells taken from embryonic animal tissue or meristematic plant tissue under magnification you would observe the rather diffuse chromatin becoming distinct worm-like bodies called **chromosomes**. This event signals the beginning of mitosis, or nuclear division.

Chromosomes contain the DNA blueprint.

The complete genetic blueprint of each individual is contained on the chromosomes of somatic cells which vary in number depending upon the species involved. Corn plant cells from the root meristematic region (Figure 10-4) would have 20 chromosomes. Human somatic cells have 46 chromosomes as illustrated in Figure 11-4. The figure represents a **karyotype**, or pictorial representation of chromosomes arranged in a precise order. Notice that there are 23 pairs of chromosomes, called **homologous pairs**. When a zygote is formed by fertilization in humans, 23 chromosomes carried in a sperm join with 23 in an egg, making a total of 46, as shown in the karyotype. Each set of 23 chromosomes collectively houses essentially all of the genes for human inheritance, or the human **genome**. The human genome is estimated to have over 3 billion (3,000,000,000) nitrogen base pairs, representing about 100,000 genes. This number of base pairs, equal to the number of letters in 200 Manhattan telephone directories, are all contained on 23 chromosomes inside a microscopic nucleus!

human genome

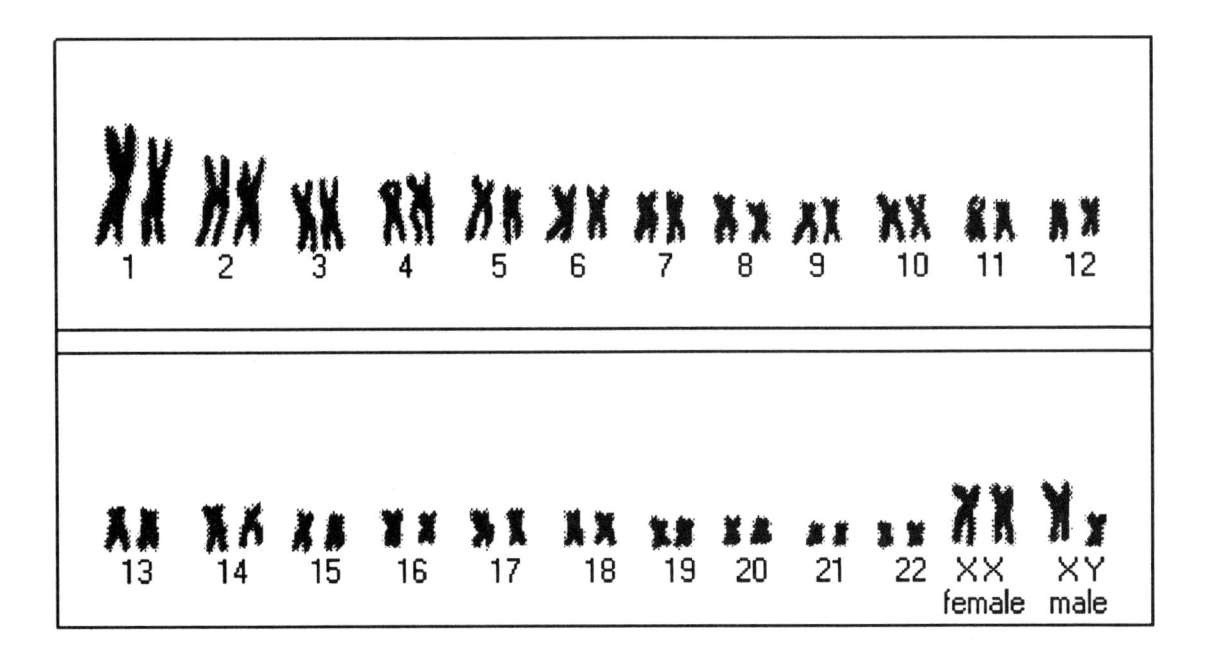

Figure 11-4. Diagrammatic view of the 46 chromosomes from a human cell.

DNA must be duplicated.

When a cell divides, each cell receives a copy of the DNA blueprint. Just as an architect may use a copier to make a complete duplicate copy of his blueprint, so the DNA must be duplicated prior to nuclear division. Each DNA molecule is duplicated according to the process illustrated and explained in Figure 11-5.

Once each DNA molecule is duplicated, another challenge arises. If fused into one piece, the 23 DNA molecules that make up the 23 chromosomes in humans would be about 1 meter long (10). To complete each DNA duplication, 46 pieces, or an average of about 2 inches per chromosomal piece, must be duplicated. Due to the microscopic confines of the cell, unless these DNA molecules are coiled up in some orderly fashion, tangling and tearing are inevitable. This problem is addressed by a mechanism that causes the freshly

DNA coils to avoid tangles.

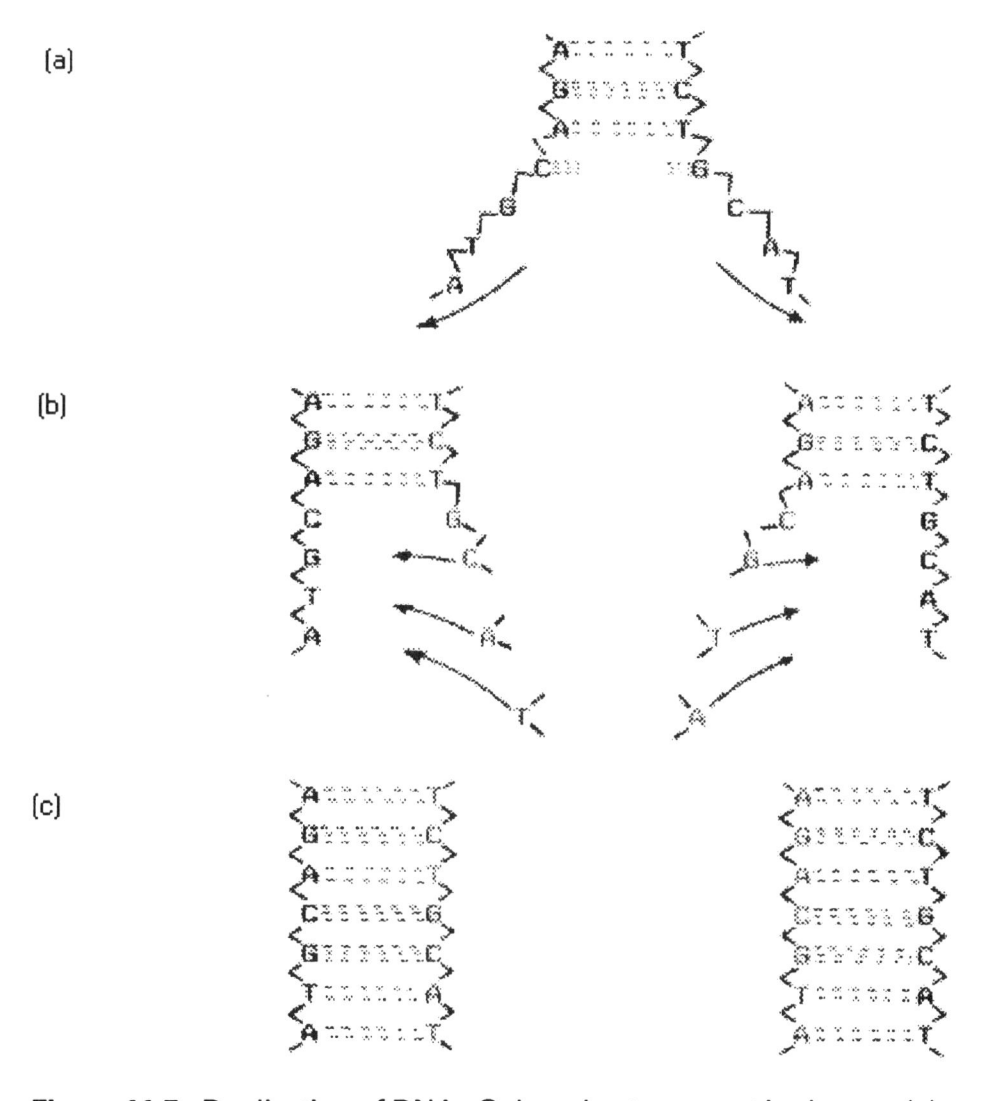

Figure 11-5. Duplication of DNA. Only a short segment is shown. (a) Hydrogen bonds between complementary base pairs are broken, allowing the two strands to separate and unwind. (b) Entry of new, free nucleotides which are assembled according to complementary pairing. They are bonded into a new strand by sugar-phosphate bonds and to the old strand by hydrogen bonding. (c) Two duplicate molecules are formed. Notice that each old strand serves as a guide to check the formation of the new, complementary strand.

nucleosomes as "spools"

**chromatids:
—copies of DNA attached by a centromere**

duplicated DNA, visible only as granular chromatin, to coil up into distinct chromosome bodies as noted earlier (Figure 11-6). All along each length of DNA, the DNA first wraps around a core of protein molecules like thread on a spool. Each protein core plus the loop of DNA is called a *nucleosome* (Figure 11-6). Each string of nucleosomes, bound by connecting DNA, coils repeatedly to form the thickened body of the chromosome. Each chromosome then appears as two attached worm-like bodies (one for each of the duplicated copies) called **chromatids**. The chromatids are attached to one another at a special region called the **centromere** (Figure 11-6).

Spindle fibers maneuver chromosomes.

The final major consideration in mitosis is a mechanism of maneuvering the chromosomes so that the chromatids can be separated from one another and gathered into two separate regions. Within each region a new nucleus, having a complete DNA blueprint can form. This is accomplished by a system of contractile proteins called **spindle fibers** that attach to the centromeres of each chromosome and draw the chromatids of each chromosome in opposite directions (Figure 11-7).

We have discussed the preliminary events of mitosis that seem to address potential problems of breakage and tangling of DNA, and loss of chromatids. Let us now view the process of mitosis beginning with one cell and leading to two cells, each with its own nucleus.

11-E   MITOSIS

Mitosis divides duplicated DNA.

The object of mitosis is to divide the nucleus of one cell so that two cells formed by cytoplasmic division each have a nucleus with a complete blueprint. Mitosis is best discussed in stages as shown in Figure 11-7. However, realize that once mitosis begins, the process occurs without significant pauses or interruptions. The entire process requires about 5 to 10 minutes in rapidly dividing cells of animal embryos, and as much as 3 hours in various plant and animal tissues.

What occurs between mitotic divisions?

Interphase refers to the period of a cell's existence between mitotic divisions. This period usually lasts for at least 8 to 12 hours. Muscle and nerve cells of adult mammals may remain in interphase for the life of the animal and never divide again. During interphase large amounts of carbohydrate, lipids, and proteins are synthesized. Then, a mechanism, not entirely understood, triggers DNA duplication which means that mitosis will follow. Much research is

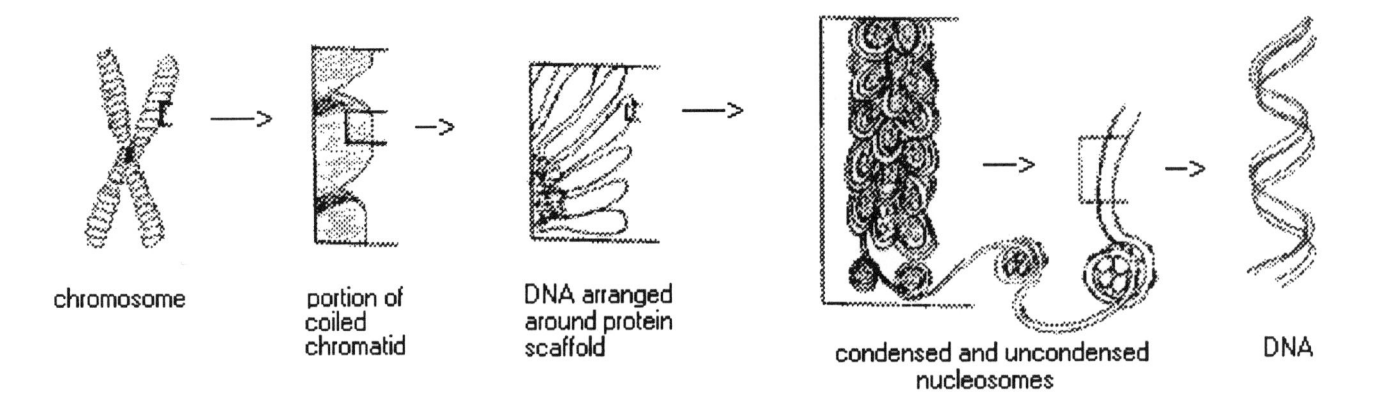

chromosome → portion of coiled chromatid → DNA arranged around protein scaffold → condensed and uncondensed nucleosomes → → DNA

Figure 11-6. Structural arrangement of DNA and proteins within a chromosome.

220

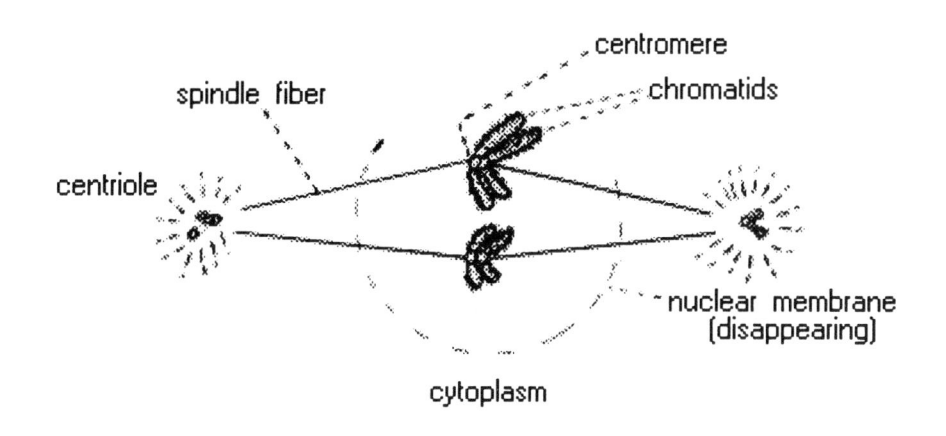

Figure 11-7. Major parts of the mitotic apparatus showing two chromosomes in late prophase.

directed at the mechanisms controlling DNA duplication and the various events of the *cell cycle*. This research is providing clues as to why cancer cells divide uncontrollably.

Chromosomes and spindle apparatus form.

Prophase (G. *pro* = in front of) involves the first visible events of mitosis, already described in Section 11-D. This phase begins with the duplicated DNA barely distinguishable as chromosomes, and ends with chromosomes fully formed and attached to spindle fibers. The nucleolus and nuclear membrane disintegrate. Centrioles to which spindle fibers are attached are positioned at opposite "poles" (Figure 11-7).

alignment on equatorial plane

Metaphase (G. *meta* = between) features all chromosomes arranged along an imaginary *equatorial plane* between the two halves of the three-dimensional spindle. Upon observing human cells in metaphase, one can count 46 chromosomes arranged along the equatorial plane. The drug colchicine interferes with spindle fiber contraction and is used in research to halt mitosis in metaphase. In metaphase, chromosomes are visible in their most distinct form. For this reason, they are extracted at this stage for kayrotyping.

Chromatids separate as "chromosomes".

Anaphase begins when spindle fibers contract, causing separation within each pair of chromatids, beginning at the centromeres. Each separated chromatid, now individually considered a chromosome, is drawn to the opposite pole of the spindle. Anaphase ends when a complete set of chromosomes (46 in humans) reaches each end of the spindle (Figure 11-8).

chromosomes uncoil

Telophase, the last phase of mitosis, somewhat resembles prophase, only in reverse. The chromosomes uncoil and gradually disappear into the granular appearance of chromatin. The spindle disintegrates, and the nucleoli reappear within each chromatin mass. These chromatin masses are then each enveloped by a nuclear membrane, forming *daughter nuclei* (Figure 11-8).

Cytoplasmic division forms two cells.

While the above nuclear events are occurring in telophase, *cytoplasmic division* also occurs. In animal cells, the cytoplasm constricts across the region of the equatorial plane until the cell is pinched into two daughter cells (Figure 11-8). However, in plant cells constriction is not possible because of the rigid cell wall. Instead, a *cell plate* forms across the region of the equatorial plane and a new cell wall and cell membrane structure is deposited. In this way, the cytoplasm is divided among two daughter cells.

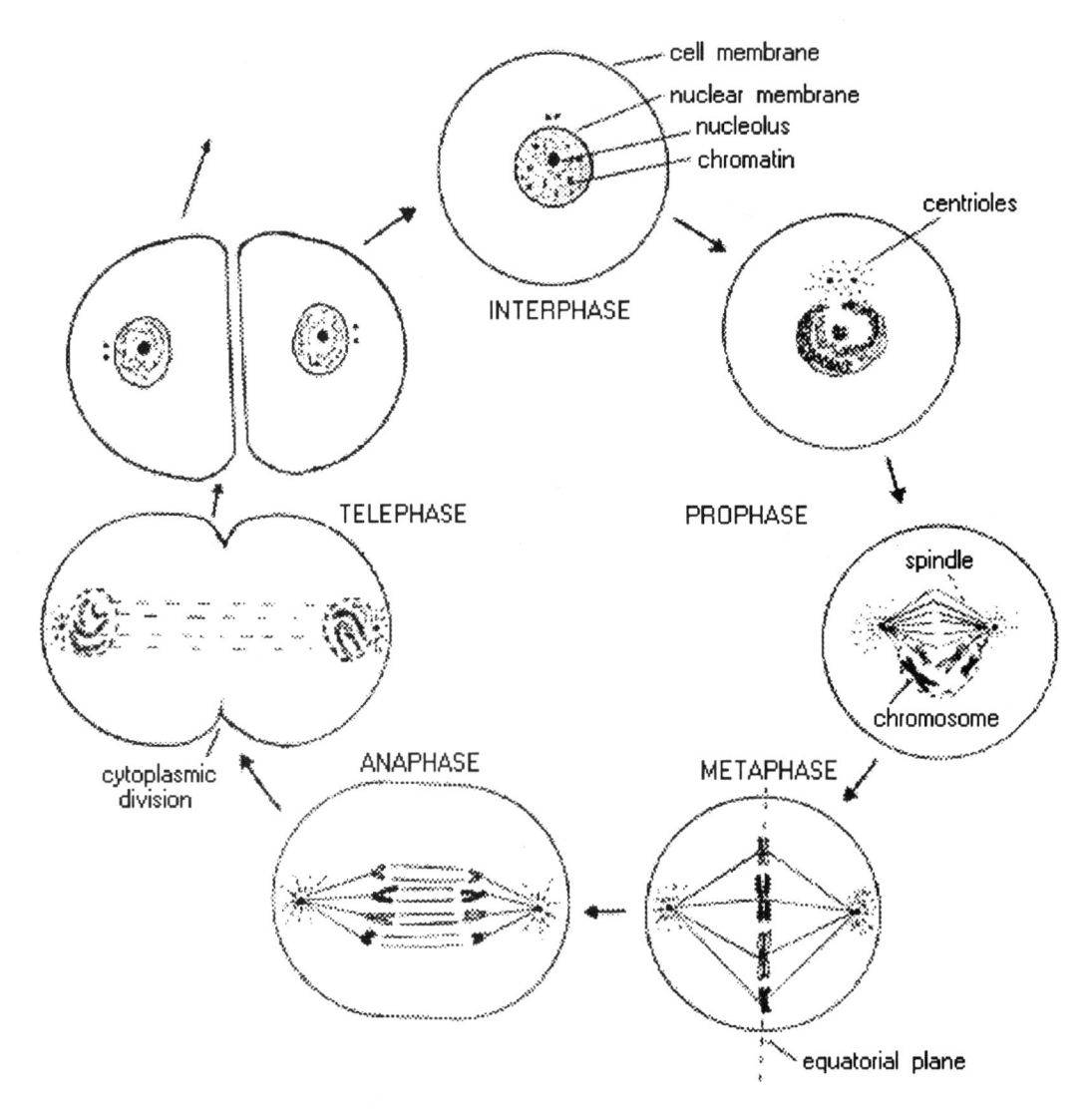

Figure 11-8. Mitosis and cytoplasmic division in a hypothetical animal cell with four chromosomes.

Cells reproduce for growth and maintenance

During the subsequent interphase, each daughter cell enlarges and prepares for another mitotic division, or differentiates into a particular structure in keeping with other neighboring cells. Some tissues, including *epithelial* tissues of skin, the linings of the digestive tract, stem cells of bone marrow, root and shoot meristems, and cambium have cells that remain embryonic and undifferentiated throughout the life of the animal or plant. In animals, these cells divide in accordance with the need to replace cells that are worn out or sluffed off. In plants, these embryonic cells allow growth in length and diameter year after year (e.g. trees).

Any questions?

As you have considered the mitotic process, it is hoped that you have accumulated a number of "how questions". We need to realize that this is a microscopic process that happens with great timing and precision millions of times from zygote to multicellular plant or animal. How is each part assembled in precise position and moved in the proper plane at the proper time? How do certain cells differentiate into a final form while others remain embryonic? [Consider these questions, add others you may have, and reflect on what this suggests about life (and death) at the cell level.]

11-F ASEXUAL REPRODUCTION

New individuals without mating.

In most multicellular plants and animals, mitosis and cytoplasmic division function mainly in the production of new cells during growth, or in the replacement of aging or injured cells. However, in most unicellular organisms, and in some invertebrates and plants, whole new individuals can be produced by cell division. Reproduction without the involvement of sex cells or fertilization is called **asexual reproduction**. Asexual reproduction appears in certain members of every taxonomic kingdom and may be classified under the following mechanisms: *binary fission, budding, fragmentation, sporulation*, and *vegetative reproduction*.

Binary fission and fragmentation occurs in prokaryotic organisms of kingdom Monera, though mitosis is not involved. The protozoa also reproduce asexually by binary fission. Figure 11-9 illustrates asexual reproduction in the Protista. Asexual mechanisms, common in multicellular plants and animals, are illustrated in Figure 11-10. Among multicellular animals, in each case, cells or groups of cells must be able to divide and differentiate so as to form the complete body structure of separate new individuals.

rapid reproduction . . .

Asexual reproduction is a rather simple form of reproduction, involving only mitosis in one individual, in contrast to the complexities of gamete production and the transfer of gametes between two individuals. Thus, one organism can enter a suitable environment, quickly populate, and even dominate an area. For example, asexual propagation by underground roots and rhizomes is common in Canadian thistle and other aggressive "weeds". In fungi, sporulation is especially effective as a means of dispersal, so important to the nonmotile plant lifestyle.

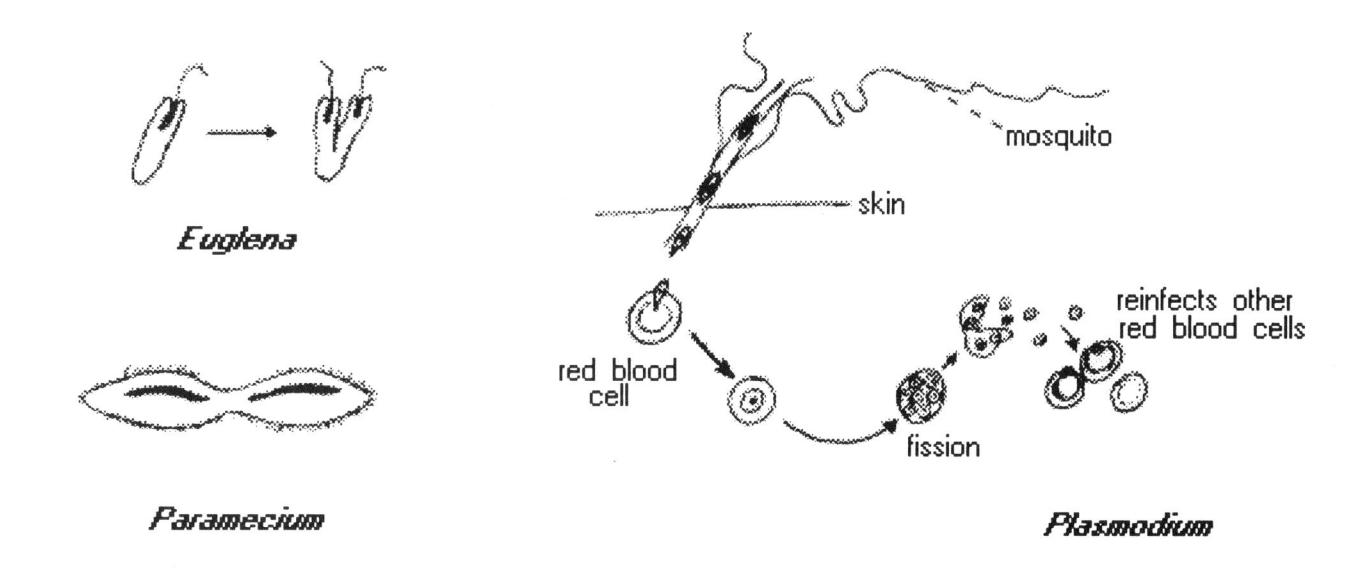

Figure 11-9. Asexual reproduction by binary fission in *Euglena* (longitudinal) and *Paramecium* (transverse). *Plasmodium*, the malarial parasite, reproduces asexually within red blood cells. Fever is associated with periodic infection and rupture of the red cells.

(a) Budding in *Hydra*

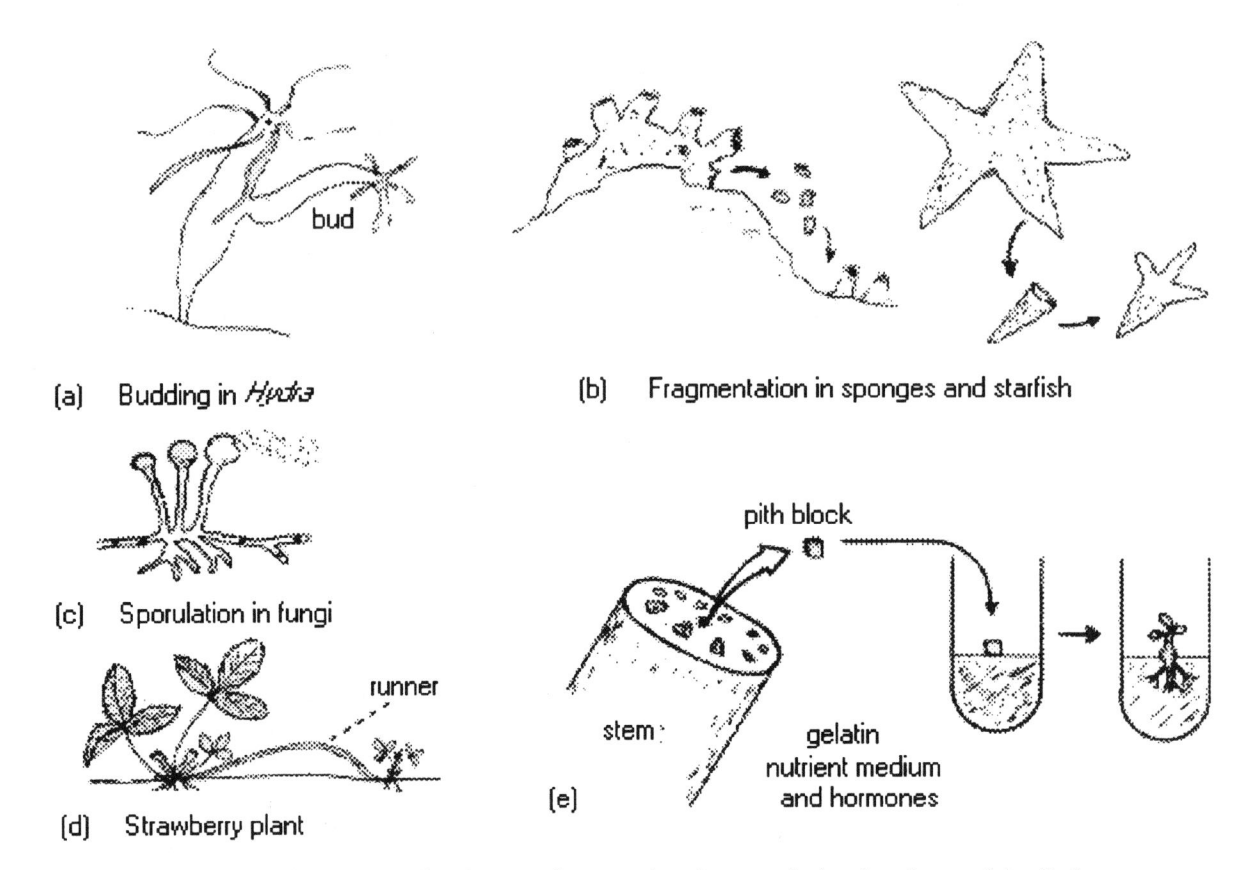

(b) Fragmentation in sponges and starfish

(c) Sporulation in fungi

(d) Strawberry plant

(e) pith block / stem / gelatin nutrient medium and hormones

Figure 11-10. Asexual reproduction and cloning in multicellular organisms.

but no genetic variation

Because asexual reproduction depends upon mitosis which can only duplicate and divide the existing DNA blueprint, the offspring produced are essentially exact duplicates, or **clones**, of the "parent". Without significant *genetic variation* from generation to generation, clonal populations will prosper only if they are adapted to the surrounding environment. If environmental conditions change so that the clone is ill-adapted, then the entire population is threatened. Most plant and animal species that reproduce asexually, also rely upon sexual reproduction in their life cycles.

11-G SEXUAL REPRODUCTION

So far we have seen that cell reproduction via mitosis and cytoplasmic division has an important role in growth and development, repair and replacement of cells, and in asexual reproduction in multicellular organisms (Figure 11-1). In every case, the nuclear blueprint is duplicated and divided so that there is no provision for genetic variation among the offspring.

Meiosis halves chromosome number.

Sexual reproduction requires . . .

Sexual reproduction is based upon a different nuclear division process called **meiosis**, which halves the chromosome number, and permits new gene combinations and genetic variability among the offspring. [Of what advantage is genetic variability among individuals of a population? See Section 11-F.] However, there is a greater complexity involved in accomplishing sexual reproduction. Let's consider three possible hurdles faced by organisms during sexual reproduction. These hurdles are gamete production, fertilization, and nurture of the offspring.

fertility ...

First, individuals must be *fertile* or able to produce viable gametes. This requires that individuals survive until sexual maturity, a time interval that varies greatly among species. Gamete production in plants and animals occurs in *sex organs*. Here, specialized cells undergo meiosis instead of mitosis.

gamete union ...

The second hurdle involves coordinating the *timing* of gamete production, and their *transfer* across distance between male and female sex organs so that fertilization can occur. Because of their immobility, plants have unique mechanisms of accomplishing this. Many plant species facilitate sexual reproduction by having both male and female sex organs on the same individual. Other species such as the ginkgo tree and American holly have separate male and female plants. Plants utilize wind, water, and nectar-seeking animals to transfer *pollen grains* containing male gametes from male to female sex organs.

Certain animals such as *Hydra*, jellyfish, and earthworm bear both sex organs on the same individual and are called **hermaphroditic** (Figure 11-11). Even though the earthworm is hermaphroditic, mating between two individuals must occur. Among animal species with separate sexes, fertilization may be external as in the case of most aquatic animals (e.g. fish, many amphibians), or internal as in many land animals (e.g. reptiles, birds, mammals). Internal fertilization usually involves the direct transfer of sperm into the reproductive tract of the female by a *penis* or penis-like organ. Many multicellular animal species display complex mating behavior which is often keyed to specific environmental events (e.g. rising and lowering of tides, seasonal changes, etc.) and internally controlled hormonal cycles. Complex hormonal control of male and female behavior and reproductive physiology is essential. We shall discuss the human female reproductive cycle in Chapter 14. Sexual behavior is a central part of intraspecific and interspecific interactions as discussed in Section 7-E.

... and nurture.

The third hurdle involves the nurture of the zygote as it develops into a self-sustaining individual. Again there are a host of mechanisms for accomplishing this among the creatures. Animals such as fish and some amphibians that practice external fertilization may have less control and protection over the early development of the offspring. Therefore, larger numbers of offspring are produced to insure that some offspring survive predation and natural destruction. The mammals, which practice internal fertilization, have the advantage of the *uterus* within the mother's body to protect and nourish the developing offspring. Nutritional investment per individual offspring is greater, and relatively fewer individuals are produced.

Mammals emphasize nurturing.

Flowering Plant — anthers, ovary

Hydra — testes, ovary

Earthworms

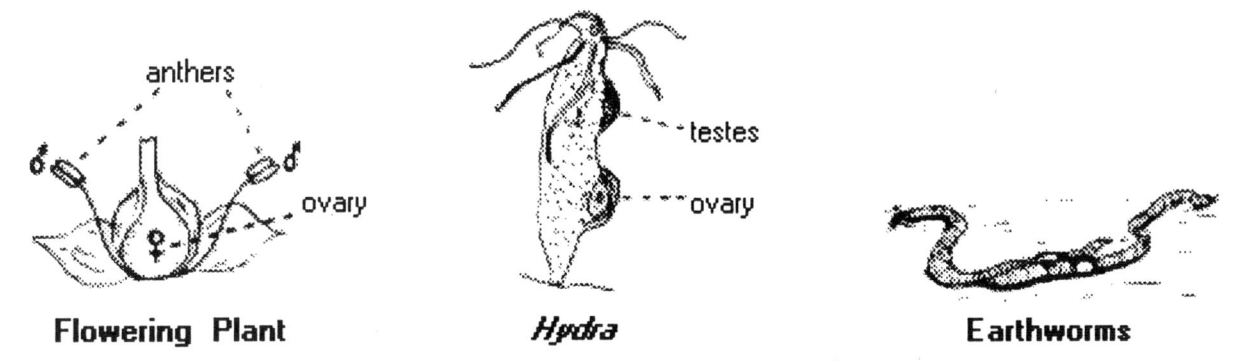

Figure 11-11. Organisms that bear both sex organs on the same individual.

In spite of the greater complexity and challenge associated with sexual reproduction, it provides a key benefit that is not possible through asexual reproduction—genetic variability among offspring. This variability as seen in new gene combinations is made possible by *meiosis* and the exchange of the resultant gametes among different individuals of a population.

11-H MEIOSIS

Meiosis differs from *mitosis* . . .

in location . . .

Meiosis differs from mitosis in three major ways. First, meiosis is a specialized form of nuclear division that occurs *only in reproductive tissues*. As shown in the generalized life cycle (Figure 11-1) mitosis and cytoplasmic division produce many cells from the zygote. Subsequent cell differentiation leads to the production of different tissues, including reproductive tissues. Certain cells within these reproductive tissues were formed by *mitosis* and cytoplasmic division, but at some point, they will undergo *meiosis* and be transformed into gametes, sperm or egg.

in effect on chromosome number . . .

Besides being confined to reproductive tissue, meiosis differs from mitosis in a second respect. Instead of *duplication division* of the chromosomes as in mitosis, meiosis involves *reduction division* in which daughter cells are formed with only half as many chromosomes. This is possible because each somatic cell has two sets of chromosomes, one set received from the male parent and one from the female, as noted in Section 11-D. Thus, in humans, the zygote and the resulting somatic cells each have two sets of 23 chromosomes each, totalling 46 (Figure 11-4). Because chromosome number varies with species, the variable, **n**, is used to indicate the number of chromosomes per set, or the **haploid number** (Gr. *haplo* = single). The chromosome number of the two sets combined, or **2n** is the **diploid number** (Gr *di* = two). Meiosis is called reduction division because it reduces the chromosome number of cells from diploid (2n) to haploid (n). As Figure 11-1 illustrates, the chromosome number must be reduced prior to fertilization or each new generation will have twice the number of chromosomes as the last. Meiosis and fertilization are like two bridges crossing from diploid cells to haploid cells and back again.

. . . in allowing variation.

The third contrast between meiosis and mitosis involves genetic variability. Unlike mitosis which duplicates the chromosomes, meiosis allows independent assortment and crossing over, leading to new gene combinations within the daughter cells.

Homologous chromosomes form pairs.

The first stage of meiosis, called **prophase I**, resembles prophase of mitosis (Figures 11-8 and 11-12). DNA duplication has occurred, and the duplicated chromosomes and spindle appear. Nucleolus and nuclear membrane disappear. However, instead of each chromosome lining up single file along the equatorial plane, each chromosome pairs off with the other member of the same **homologous pair**. Members of each pair have genes for the same hereditary traits (Figure 11-4). Homologous pairing of chromosomes in prophase I contributes to genetic variation among offspring in two ways. First, chromosomes of paternal origin (i.e. carrying the genes from the father) and maternal origin are randomly positioned along the two sides of the equatorial plane, so that it is unlikely that all paternal (or all maternal) members of each homologous pair will be on the same side. When chromosomes are later drawn apart in two directions perpendicular to the equatorial plane, there will be a "mixture" of maternal and paternal chromosomes in each daughter cell, and eventually, in each gamete. This **independent assortment** of chromosomes and genes into gametes promotes genetic variation.

Crossing over exchanges DNA between pairs.

The second contribution to genetic variation is **crossing over**. Crossing over involves the intimate contact between chromatids within homologous pairs. A chromatid from each duplicated chromosome member of the pair may break at corresponding points and exchange parts during prophase I (Figure 11-12). Crossing over thus allows exchange of paternal and maternal DNA *within* chromosomes, so that many of the chromosomes will carry both paternal and maternal DNA.

Special Diploid (2n) Cell in ovary or testis

EARLY PROPHASE I
Homologous pairing between paternal (black) and maternal chromosomes; crossing over is in progress

METAPHASE I
Crossing over is complete; homologous pairs are on equatorial plane

ANAPHASE I

PROPHASE II

METAPHASE II

ANAPHASE II

TELOPHASE II

Four daughter cells
(Each has one chromosome representing each homologous pair.)

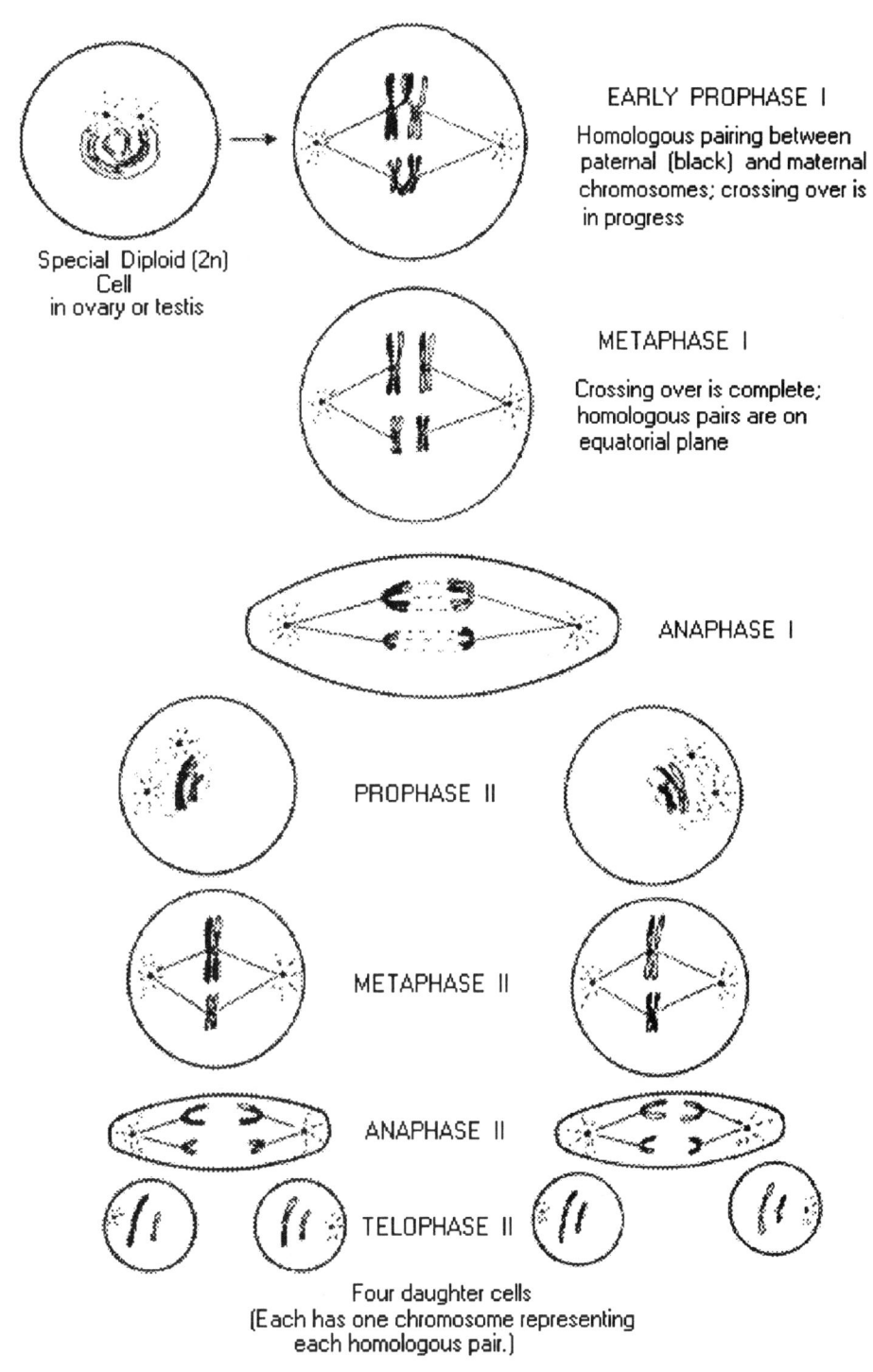

Figure 11-12. Diagrammatic representation of the major stages of meiosis in an organism with a diploid number (2n) of four. Chromosomes representing hereditary material from the paternal member of the previous generation are in black; maternal, in white.

Homologous pairs separate.

When crossing over is complete, members of each homologous pair are oriented opposite one another on the spindle. In **metaphase I**, the pairs are arranged one member on each side of the equatorial plane. **Anaphase I** then features the separation of the homologous pairs toward opposite ends of the spindle (Figure 11-12). Note that the chromatids attached at the centromere do not separate because the centromeres do not uncouple from one another.

Meiosis II divides chromatids.

Telophase I resembles telophase of mitosis (Figure 11-8) in the uncoiling of chromosomes and reappearance of nuclear membrane and nucleolus in separate daughter cells (Figure 11-12). In some species telophase I is omitted. Meiosis II is nearly a repeat of what we have just described, except that there are no homologous pairings or crossing over. In each of the two daughter cells, **prophase II** features the reappearance of the spindle, attached to each duplicated chromosome. The alignment occurs across the equatorial plane in **metaphase II**. **Anaphase II** features the uncoupling of centromeres and the movement of daughter chromatids to opposite ends of the spindle. New nuclei are formed in each of the four daughter cells in **telophase II**. Each cell is haploid, having one of each of the *kinds* of chromosomes found in the original parent—i.e. one set of chromosomes. In humans, each of these cells would have $n = 23$ chromosomes, a complete DNA blueprint. However, each haploid daughter cell has a different combination of maternal and paternal genes, which is the key to genetic variability. We shall return to this point in Chapter 12.

Meiosis yields only one ovum.

The four daughter cells produced in meiosis must undergo further differentiation to form gametes. In male animals, the daughter cells differentiate into sperm cells within the *seminiferous tubules* of the testes. Each daughter cell loses cytoplasm and forms a flagellum (Figure 11-13). Differentiation in female animals occurs within the *follicles* of the ovary. Here only one mature ovum is produced per meiotic division and it is relatively large due to unequal division of the cytoplasm (Figure 11-13). The motile, aggressive role of the sperm and the nonmotile, nourishing role of the ovum are thus provided for in this final phase of gamete production.

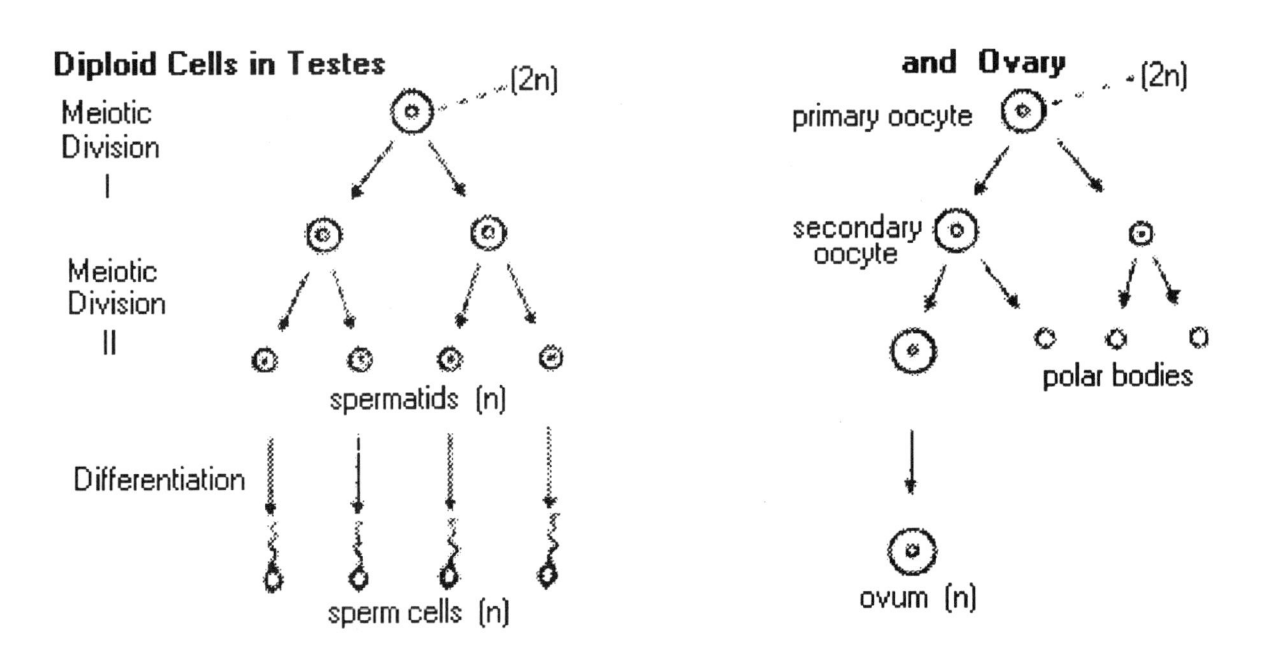

Figure 11-13. Differentiation of sperm and egg in animals.

11-I SEXUAL REPRODUCTION IN THE PLANT KINGDOM

Plants have two adaptations for dispersal:

Sexual reproduction in both Plantae and Animalia involves the processes of meiosis and fertilization. However, there are some unique variations in plants. Because plants are nonmotile, their life cycles have adaptations to promote *dispersal* to new locations. Two structures serve in reproduction and dispersal of plants. They are:

spores . . .

1. **Spores** are haploid, wind- or water-borne structures formed by meiosis. Male spores in flowering plants are called *pollen grains*.

and seeds.

2. **Seeds** consist of an embryo developed from the zygote and accompanied by nutrient and energy reserves, all packaged inside a seed coat.

In summary, spores are haploid structures formed by meiosis; whereas, seeds are diploid structures formed after fertilization. Not all plants produce seeds. Let's examine reproduction in three groups of plants.

ALGAE

reproduction in aquatic forms

In many green algae, the zygote is the only diploid structure. Figure 11-14 illustrates this pattern in which the familiar filamentous growth of the algae consists of haploid cells. Gametes are produced, not by meiosis, but by mitosis within specialized cells of the filament. The sperm can move in the aqueous environment and are attracted to the egg through a small opening in the cell wall. Fertilization produces a zygote which may enter a dormant condition during unfavorable environmental conditions. Eventually, the zygote undergoes meiosis, producing motile, haploid spores which settle down and develop into new haploid filaments by mitosis and cytoplasmic division.

Land habitats are more challenging.

The aquatic algae have a seemingly easier hurdle in getting gametes together because the motile sperm are simply released into the surrounding water through which they move to the female gamete. Because the flagellated sperm cell of most species are designed to move in an aqueous medium, the terrestrial environment offers a greater challenge. This is made doubly difficult because plants cannot move about and participate in mating behavior comparable to land animals. Let's examine the reproductive mechanisms of two different groups of land plants.

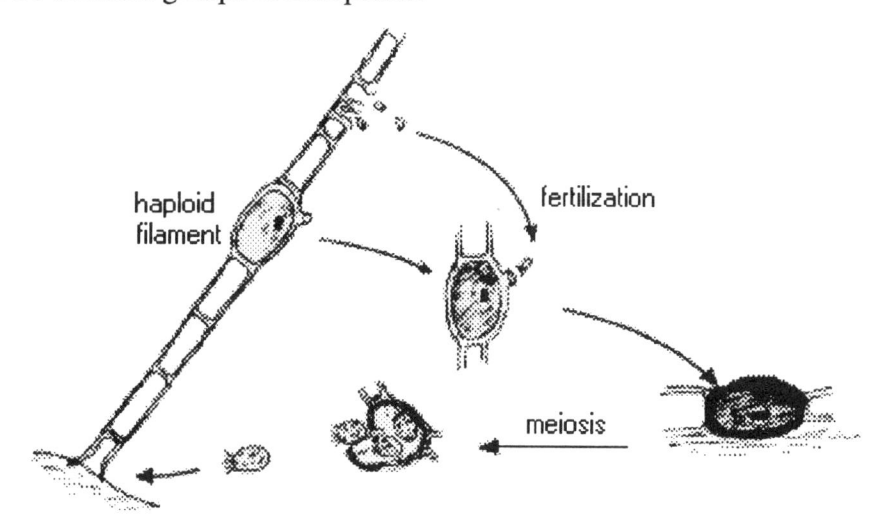

Figure 11-14. Reproduction of Oedogonium

BRYOPHYTES, FERNS, AND ALLIES

Mosses...

Bryophytes, ferns, and their "allies" are primarily land plants which produce spores but no flowers or seeds (Section 9-E). We will focus on moss reproduction as representative of this group. Mosses and other bryophytes are nonvascular land plants. Spores are produced by meiosis within tiny green capsules. Upon release from the capsule, spores are carried by wind to new locations.

...rely upon wind and water.

Upon germination each haploid spore grows by mitosis and cytoplasmic division into green moss plants. These haploid plants are called **gametophytes** because gametes are produced within the terminal areas of these plants by mitosis. Sperms may be transferred in dew or splashing rain from the male gametophyte to the eggs contained within the female gametophyte. Fertilization produces a diploid zygote which develops into the spore-producing **sporophyte** atop the original female gametophyte (Figure 11-15).

The moss life cycle illustrates how a nonmotile plant in a sometimes harsh land environment can use moving elements of the physical environment, wind and water, to complete its life cycle. Like the algae, there is a major emphasis upon the haploid (gametophyte) generation.

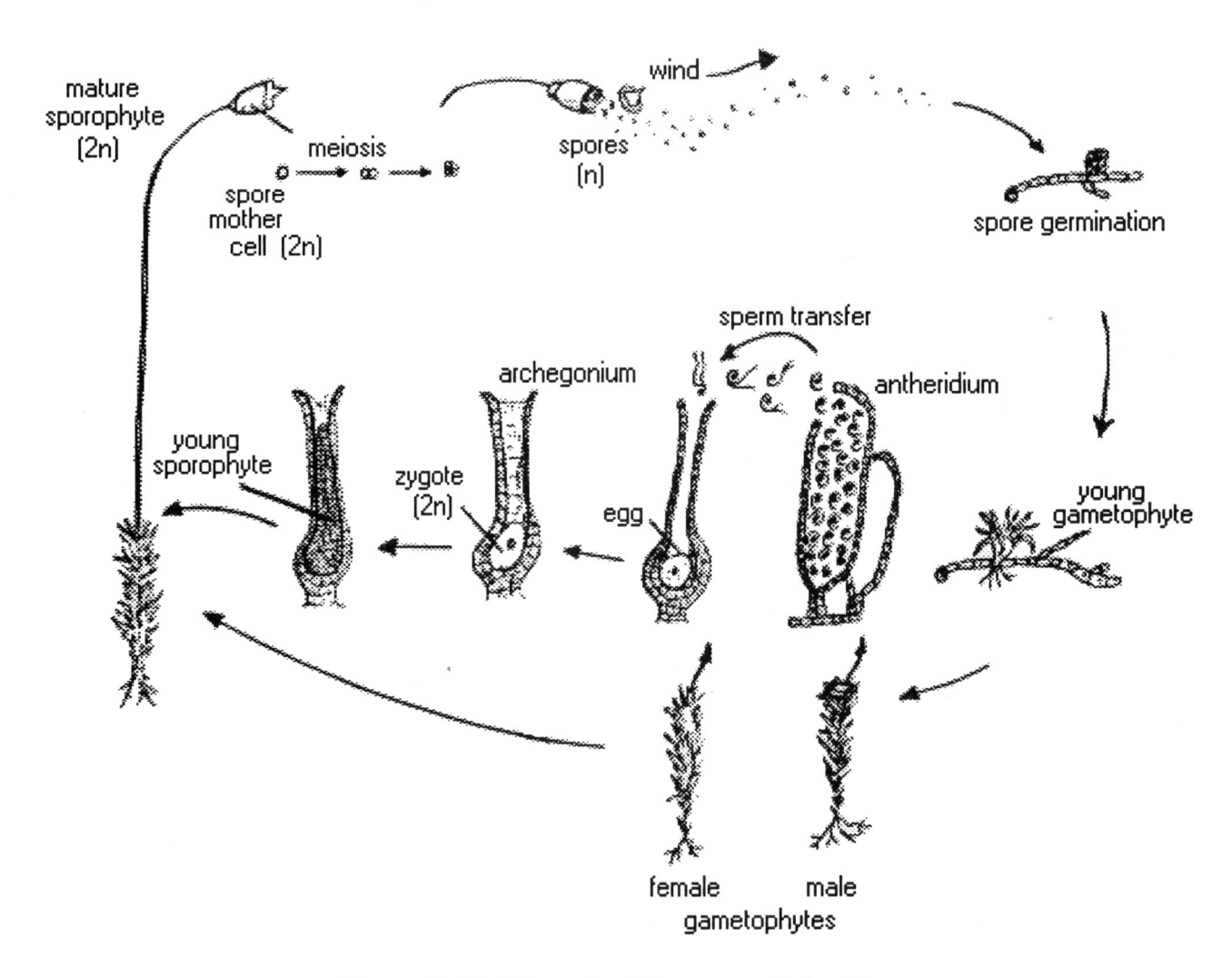

Figure 11-15. Life cycle of the moss, *Polytrichum*.

FLOWERING PLANTS

Flowering plants are primarily terrestrial vascular plants that form the most obvious component of the green landscape. Their capacity to occupy almost every terrestrial habitat is due in large part to their adaptations that facilitate sexual reproduction in spite of their nonmotile life style.

flower structure

The reproductive activity of flowering plants centers around the flowers. Here, male sex organs, the **stamens** normally surround a central female organ, the **pistil** (Figure 11-16). Some species have separate male and female flowers, either on the same plant or on separate plants (Section 11-G). Here we will focus on a species that has a *complete* flower, one with both sex organs in the same flower (Figure 11-16).

Meiosis occurs within the **anthers** (Figure 11-16), and haploid pollen grains (spores) are produced. Meanwhile, meiosis within the **ovary** at the base of the pistil forms haploid spores which, through additional mitotic divisions produce an *embryo sac* containing the egg and several other cells. Pollen is transferred by wind or animals (e.g. insects, birds) from the anther to the *stigma* at the top of the pistil. In *self pollination*, pollen transfer occurs between sex organs of the same individual plant; in *cross pollination*, the pollen transfer involves another plant. [Which type of pollination would produce the greatest genetic variation in offspring?]

pollination

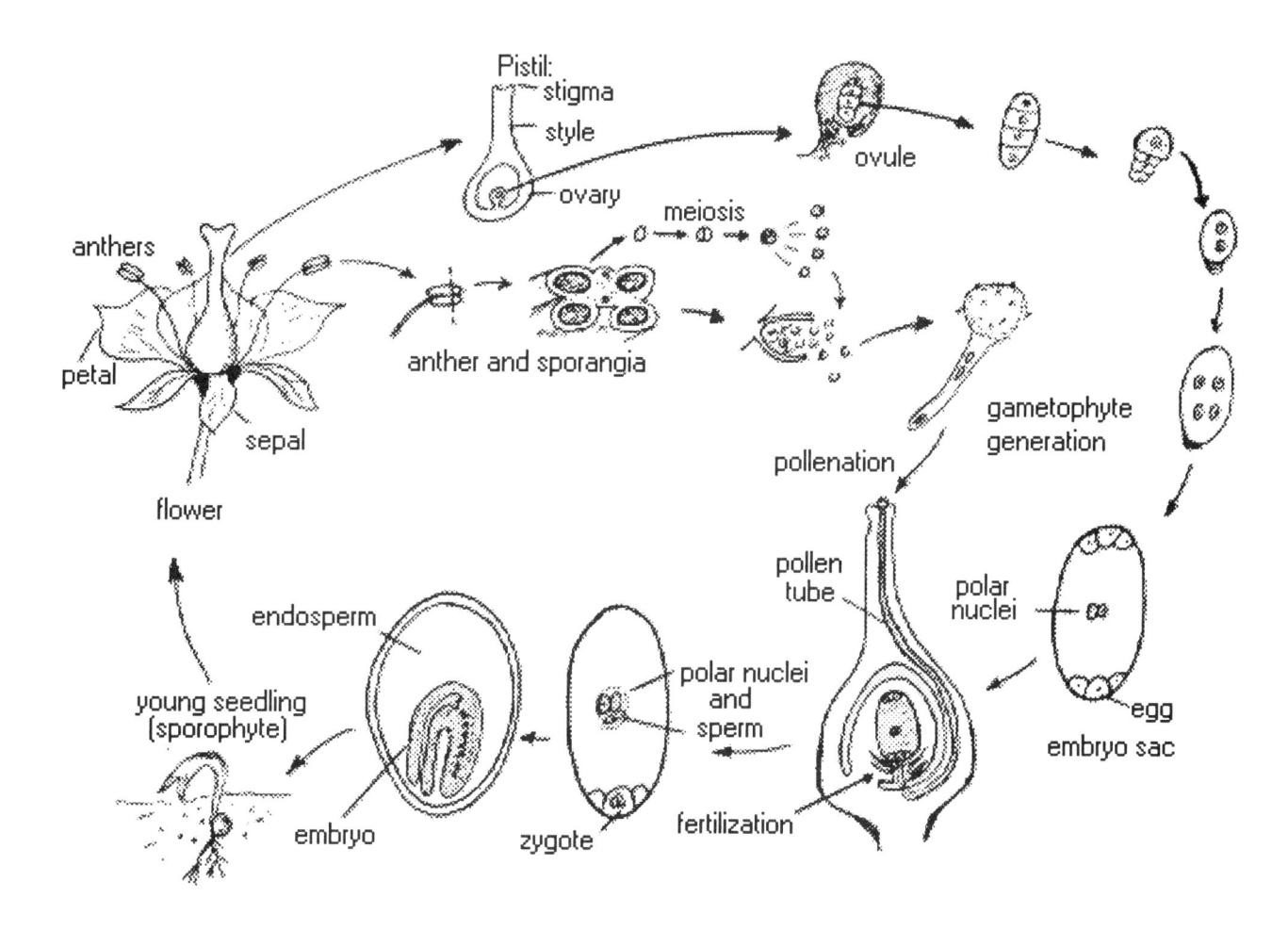

Figure 11-16. Life cycle of a flowering plant.

231

double fertilization

Interestingly, the pollen from many species may attach to the sticky stigma, but only pollen from the same species can germinate. [Reproductive barrier.] The pollen grains that germinate produce a pollen tube which grows down through the pistil to the embryo sac. Usually two sperm nuclei move within the tube to the embryo sac where one sperm nucleus fuses with the egg, forming the zygote. The zygote will divide mitotically to produce the embryo of the next generation. The other sperm nucleus fuses with one or more other nuclei of the embryo sac to initiate the formation of a nutritive tissue called **endosperm**. This tissue provides for the nutrition of the zygote and embryo. A **seed** is a plant embryo plus its food supply of starch, lipids, proteins, and minerals, all packaged within a *seed coat*.

Ovary becomes fruit which aids dispersal.

The ovary that surrounded the embryo sac before fertilization (Figure 11-16) develops into a **fruit** that surrounds the seed during and after seed development. Just as the various flower structures are designed to utilize moving wind and animals to accomplish pollination in a nonmotile plant, so the ovary is designed to produce movement for seed dispersal. Figure 11-17 illustrates a number of dispersal mechanisms. Both seeds and fruits of many species are an important part of human nutrition. [How many food items originating from seed and fruit have you eaten today?]

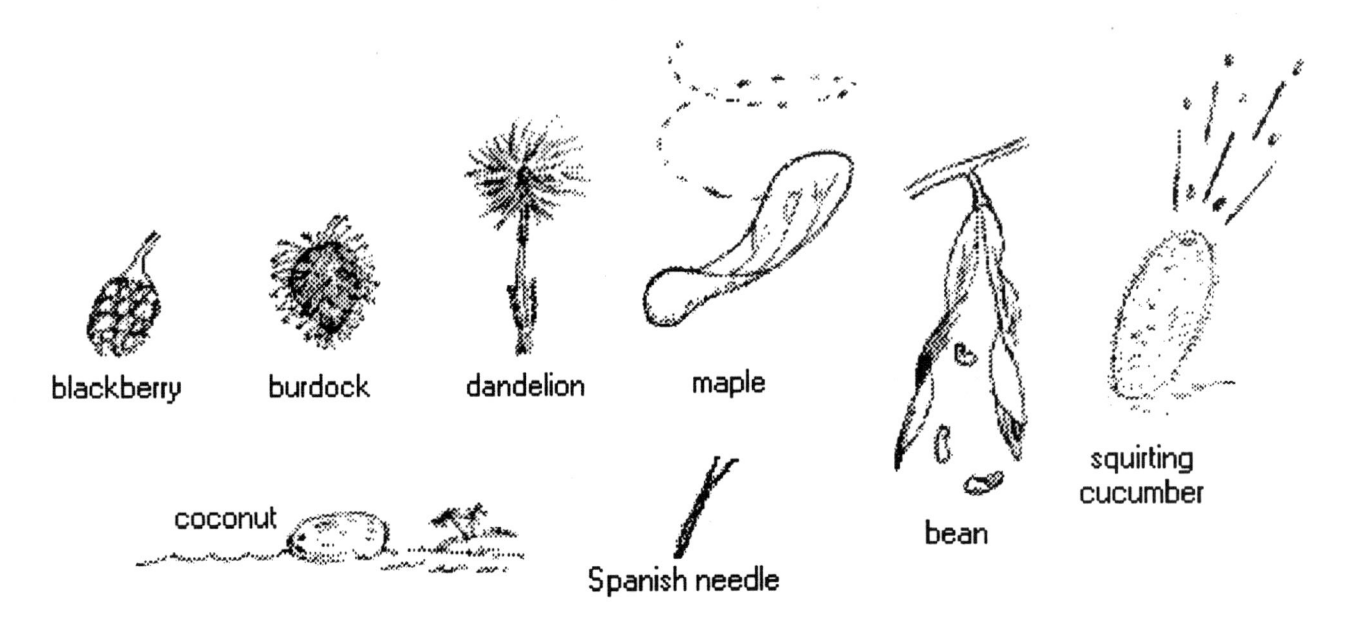

Figure 11-17. Seed dispersal mechanisms among common flowering plants. [Can you explain how each mechanism serves in seed dispersal?]

11-J HUMAN SEXUAL REPRODUCTION

Our study of reproduction has emphasized a recurring theme. Though living organisms face the prospect of death at every stage of their life cycles, reproduction and new birth point toward life and continuity for each species. Each has been endowed with anatomical, physiological, and sometimes behavioral features that are essential for fruitful reproduction. From DNA and chromosomes, to sex organs and sexual behavior, organisms have morphological, physiological, and sometimes behavioral adaptations to promote reproduction.

significance of human reproduction

Humans also have a biological nature. Like the other creatures we have received a mandate and mechanisms to multiply according to the design of the Creator (Genesis 8:17;9:1,7). However, unlike the other creatures, man has been created in the image of God and has been granted dominion over the creation (Section 8-L). With respect to reproduction this means that we are unique among the creatures. It is reserved for humans not only to participate in sexual reproduction but also to understand the spiritual and biological significance of it in relation to being an imagebearer. We conclude this chapter with a discussion of human sexual reproduction. Our discussion will have a dual purpose—first, to reveal the basic biology of mammalian reproduction using humans as an example; and second, to identify some moral and spiritual applications that are uniquely relevant to us.

dual purposes

two types of sex organs

Gametes are produced in the *primary reproductive organs*, or *gonads*, of the human male and female; namely, the **testes** and **ovaries,** respectively. These are illustrated in Figures 11-17 and 11-18. In addition to gamete production, gonads secrete sex hormones. Both male and female have *accessary reproductive organs* which function in gamete transfer, fertilization, and nurture—*e.g.* penis, gland-lined channels for transport of gametes, and the uterus. These organs do not become reproductively functional until the onset of *puberty* between ages 12 and 15, in girls; ages 13 and 15, in boys.

THE MALE

Male sex organs produce sperm...

The male sex organs serve four major functions. First, within the lining of the coiled *seminiferous tubules* of the testes, spermatids are produced by meiosis. Subsequent differentiation of spermatids (Figure 11-13) forms flagellated sperm cells with a *head* containing the genetic material, and coated with enzymes to aid in penetration of the ovum. Whereas women produce one ovum per month, males produce sperm continuously. Mature sperm are stored in a long (80-cm.), coiled tube called the *epididymis*, and can remain viable for about ten weeks. Older sperm are degraded and the products absorbed by tissues of the epididymis. The storage temperature is homeostatically maintained at about 95.6°F by adjustments in the position of the *scrotum* in relation to the main body mass (Figure 11-18), which is about three degrees warmer.

...hormones

The second function of the male sex organs involves the production of sex hormones by the testes. The most important hormone is *testosterone* which influences sperm production, maintains male sexuality, and promotes the development and maintenance of male secondary sex traits such as voice, body size, and distribution of body fat, muscles, and hair.

...and fluids.

Thirdly, the male sex organs secrete certain fluids that constitute much of the *semen*, the sperm carrying fluid. Mucus and nutrients are added from the *seminal vesicles*. The *prostate gland* secretes an alkaline fluid which raises the pH and promotes sperm motility. When sperm reach the urethra, additional mucus is added by the *Cowper's gland*.

The final function of the male sex organs involves sperm transfer to the female where fertilization occurs. We will resume this discussion after considering the female structures.

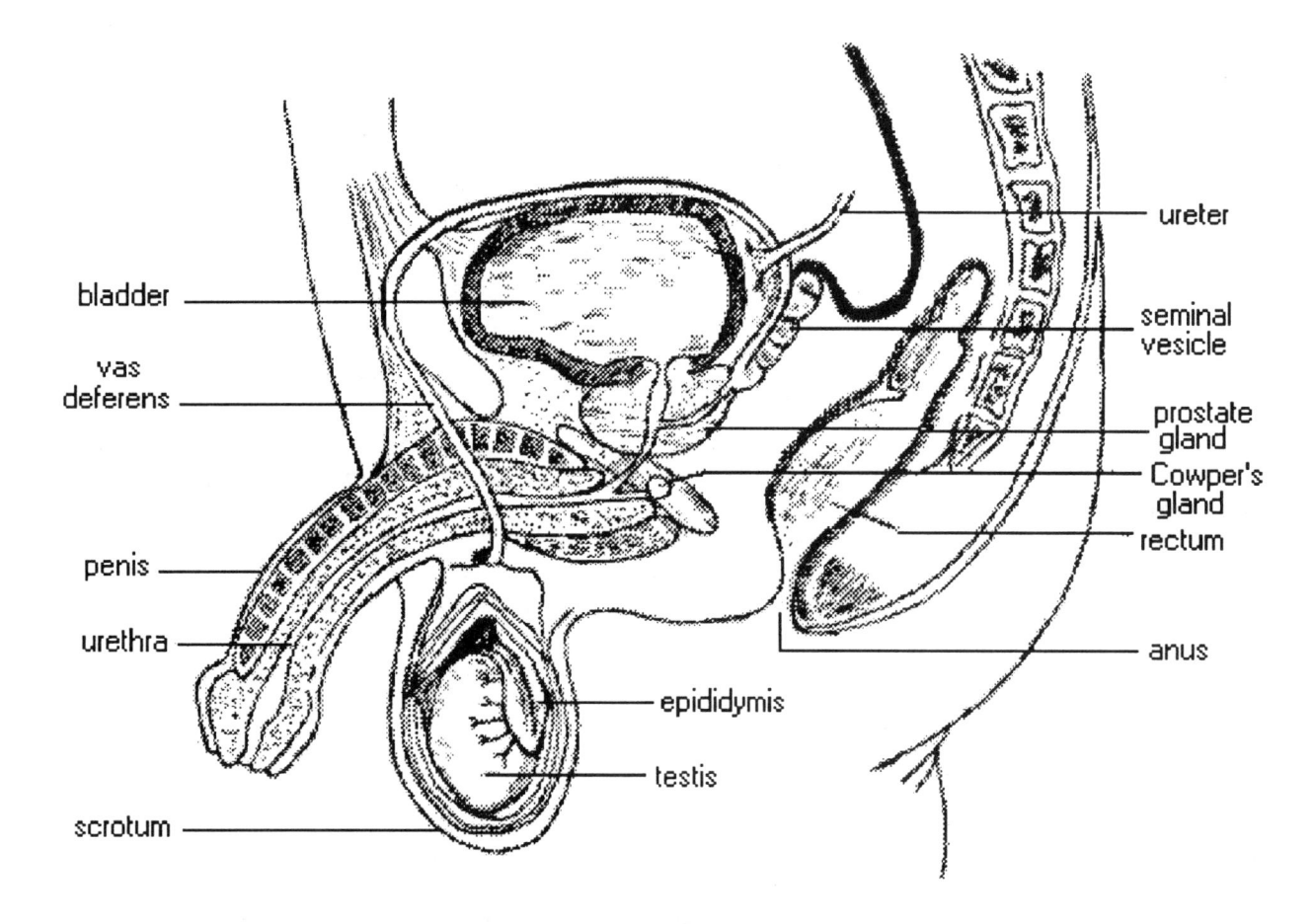

Figure 11-18. Reproductive structures of the human male.

THE FEMALE

The female sex organs perform a series of hormonal functions related to gamete production, gamete union in fertilization, and nurture of offspring within the uterus. The hormonal interactions are complex and involve gland secretions from the pituitary, ovary, and uterus. We will study the hormonal aspects of the female reproductive cycle, or *menstrual cycle*, in Chapter 14.

Meiosis I occurs before birth . . .

Prior to birth, girls have as many as 2 million diploid cells within their ovaries, called **oocytes**, which begin the process of meiosis (Section 11-H). Meiosis I is halted before completion and does not resume until puberty. Then, at the beginning of each menstrual cycle, from puberty until fertility ceases at *menopause*, one oocyte in an ovary matures within a *follicle*, completing meiosis I and beginning meiosis II (Figure 11-13). Meiosis II then ceases until the oocyte is penetrated by a sperm at fertilization.

meiosis II monthly after puberty.

oviduct and fertilization

At **ovulation**, the follicle ruptures and releases the oocyte, which enters the **oviduct** (*fallopian tube*) through an opening with the help of finger-like extensions that move over the ovary. Waving cilia move the oocyte within the oviduct toward the uterus, and it is during this time that fertilization can occur if sperm have been released through the *vagina* (Figure 11-19). When sperm contact the oocyte, Meiosis II resumes to form two daughter cells in an unequal cytoplasmic division. The larger cell with extra cytoplasm is the haploid *ovum*, ready for fertilization.

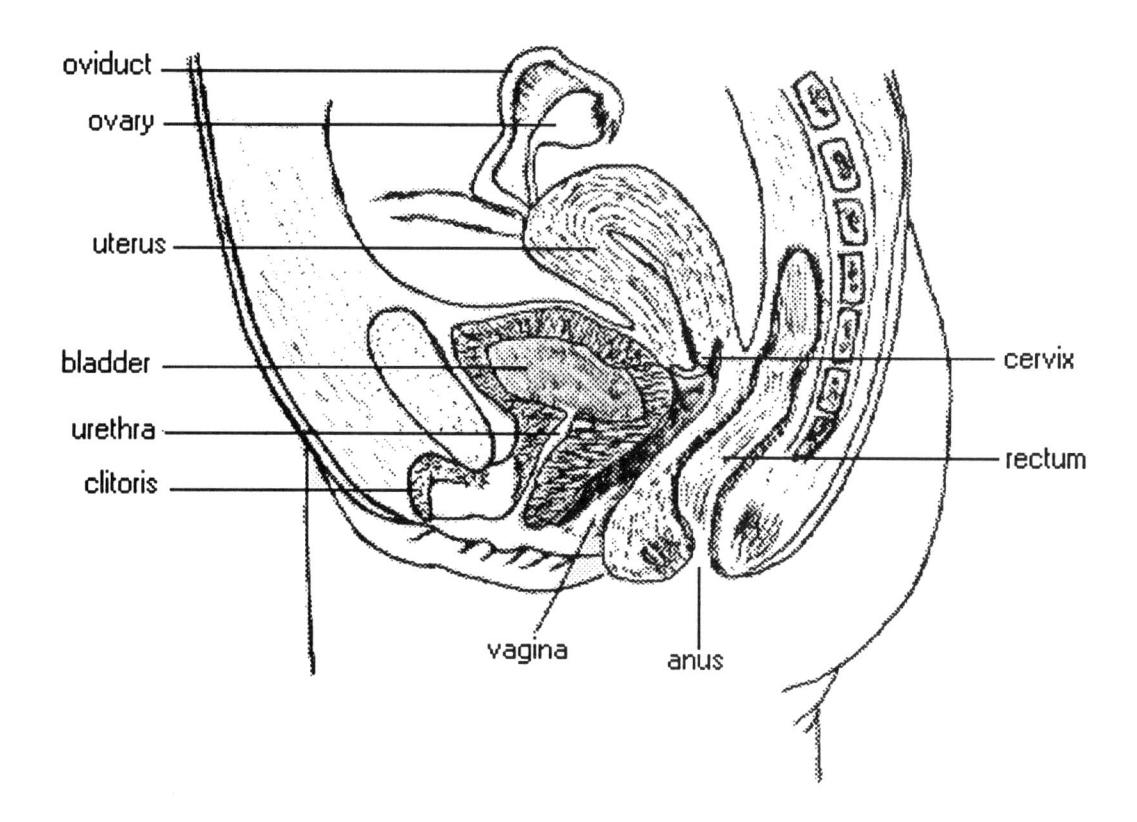

Figure 11-19. Reproductive structures of the human female.

uterine wall

During the 10 days prior to ovulation, the uterine wall has developed a thickened lining, the *endometrium* in readiness for implantation of the fertilized egg, should fertilization occur. If fertilization does not occur, the endometrium is released via the *cervix* and vagina about 12 to 18 days after ovulation. This process, called *menstruation*, lasts for 3 to 5 days and marks the completion of the 28-day (average) cycle.

SEXUAL UNION

sexual union

Male and female sex organs have complementary roles during sexual union. In response to sexual excitation, blood collects within three cylinders of spongy vascular tissue arranged around the urethra (Figure 11-18). This causes hardening and enlargement of the penis which facilitates penetration into the female vagina. During sexual union, mechanical stimulation of the penis triggers involuntary waves of peristaltic contractions which expel semen into the vagina. The breasts, vaginal walls, and clitoral region of the female body also experience stimulation. Involuntary muscular contractions within the uterus and vagina serve to conduct the semen upward toward the oviducts. These involuntary contractions and the intense sensation of release and warmth constitute the event called *orgasm*.

conception

Sperm can remain viable within the female reproductive tract for 2 to 3 days and the egg is in a fertilizable position within the oviduct for about 1 day. Therefore, *conception* can occur if sperm are released into the vagina any time 2 days prior to ovulation or one day afterward—in a "typical" cycle.

An individual life begins.

Fertilization is completed once the sperm nucleus fuses with the egg nucleus to form a human zygote. The zygote travels down the oviduct toward the uterus over a 3- or 4-day period. By the time implantation in the endometrium occurs, the zygote has undergone mitosis and cell division many times so as to produce a hollow ball of cells (blastocyst). From this point, a complex developmental sequence begins that leads to further development of the embryo and an indirect circulatory linkage between embryo and mother via the *placenta*. Although birth does not normally occur until nearly 9 months later, a new human being is beginning to develop according to his or her own genetic blueprint.

SIGNIFICANCE OF SEXUALITY

both biological and spiritual

two sources of revelation

We have briefly described the structure and function of the male and female human reproductive systems. No one can deny the major significance of these systems in procreation. Nor can anyone who has experienced the fulfillment of the sexual relationship in marriage deny that it is pleasurable. Humans, like the other mammal species, have a hormonal-influenced, sexual drive that is a dominant influence upon their behavior. However, God's purposes for human sexual relationships go much beyond its purpose in the animal world. He has revealed this to us in both the *natural revelation* (His creation), and in the *written revelation*.

humans—free of enslavement . . .

to emphasize personhood

The design of the human body conveys a message that is part of God's natural revelation. As Cosgrove (4) notes, "Our bodies were designed for sex as if people mattered." He explains that, unlike animals, we have been created free "from the enslaving cycles of heat and ovulation to follow the commands of personhood, love, fun, and spirit" (4). Only humans have the unique cognitive capacity to place the intimacy of sexual relationship within the broader context of unselfishly learning to know another person, communicating with one another, dating, courtship, enjoying romantic times together, and the thoughts of one another when separated by distance—all of this can be enjoyed during a period of abstinence before marriage. Then, speaking of the sexual union itself, Cosgrove states: "A human is the only creature who, as a rule, mates face to face...person to person...", once again emphasizing the importance of personal communion, cooperation, and sensitivity to each other. Unlike female animals who submit as "receptacles for gametes", the woman can be "a joint partner in the sex act by the very nature of her body's unique design" (4).

consequences of misuse

Unfortunately, every good gift of God can be abused when sin and selfishness reign. This is true of our sexuality in much the same way as we noted with regard to our food, in Chapter 10. Abuse of sex is a violation of God's moral principles which leads to the physical consequences of sexual unfulfillment and sexually transmitted diseases, just as abuse of food leads to loss of nutritional health or eating disorders. In spite of the clear testimony of the natural revelation, as seen in the design and function of our bodies, we need the clear testimony of His written revelation.

plan for real fulfillment:

agape **love**

The written Word provides clear testimony of the significance of human sexuality. Like other species, sexual relationship in humans was designed to allow procreation (Gen. 1:28; 8:17; 9:1,7). However, it is to be confined to the institution of marriage (Gen. 2:18-25; I Cor. 7:1-5; Eph. 5:22-23; Heb. 13:4). Within this proper context, the sexual relationship serves not only in procreation, but also as an opportunity for physical expressions of love (Gr. *agape* = unconditional love). With an attitude of loving servanthood, each mate seeks to fulfill the physical needs of the other (I Cor. 7:1-5). This involves conscious discipline on the part of each mate so that biological processes are made subject to the higher goal of bringing pleasure and fulfillment to the spouse (Prov. 5:15-19; Song of Solomon 5:10-15; 7:1-9; Eph. 5:22-23).

within marriage God's blueprint for the sexual relationship within marriage is not something that can be altered as social mores dictate. It is God's intention that humans be submissive to the moral laws as well as the physical laws of creation. Violation of either inevitably brings unfulfillment, suffering, and even death (7, 15). We are to be faithful marriage partners; and, if God entrusts us with children, we are to do our part to see that the life of Christ is reproduced in them.

**"for this is the will of God...that you abstain from sexual immorality...
I have set before you life and death, the blessing and the curse.
So, choose life in order that you may live, you and your descendants..."**
I Thess. 4:3, Deut. 30:19b

QUESTIONS AND DISCUSSION TOPICS

1. Why is the cellular level of organization so central to the life cycles and reproduction throughout the living world?

2. Distinguish the following: chromosome, chromatid, centromere, DNA, and gene.

3. With only four types of nitrogen bases, how can DNA account for such great genetic variation as seen in God's creation?

4. Distinguish the following terms: mitosis, cell division, cytoplasmic division, nuclear division, and meiosis.

5. One DNA molecule of the fruit fly is estimated to be 0.75 inch (200 mm) long, but exists within cells less than 1 mm in size. When coiled into chromatid form, the DNA shortens to about 0.001 mm in length. Explain how this event and the appearance of spindle fibers during prophase of mitosis address the logistical problems of DNA duplication and separation.

6. Distinguish the events of mitosis from meiosis. How do each contribute to the reproductive life cycle?

7. What are advantages and disadvantages of asexual reproduction? Sexual reproduction?

8. If you were to maintain a culture of a certain protozoan that continually reproduced by fission, how would you determine the age of any given cell?

9. How are the differing conditions of terrestrial and aquatic environments reflected in differences in the life cycles among the following major groups: fish, land mammals, algae, bryophytes, and flowering plants.

10. Describe the biological mechanisms and spiritual significance of human sexuality. Does the *natural revelation* by itself serve as sufficient basis for understanding the uniqueness of human sexuality? What *does* it contribute?

TOPICS FOR FURTHER STUDY

1. Some sociobiologists believe that the purpose of every creature including humans is to be an "effective vehicle for the reproduction of the master molecule, DNA". Is human behavior to be *understood* and *accepted* in terms of "selfish genes"? What is a theistic perspective?
REFERENCES: #3, #6, and #17

2. Is there a *biological* basis for distinguishing male and female, and explaining sexual behavior? This question involves a careful sorting out of evolutionary presuppositions, sociobiology (See TOPIC # 1). REFERENCE: #9

3. Delve into the human drama of one of the most influential discoveries of our century. What can you learn about the ways of science (and scientists) from these accounts?
See REFERENCES: #13 and #15

4. There is an interesting host of adaptations among plants and animals to insure that sex cells get together. Select one or more plant or animal species (or *mutualistic* relationship related to reproduction in plants and animals) and examine the reproductive mechanisms.

5. Animals in at least four phyla are known to "steal" or deliberately transplant functional cells and tissues from one body to another.
REFERENCE: #2

6. What is *in vitro* fertilization (IVF)? How is it being used, and what are the biological, ethical, and scriptural considerations involved in its use? See REFERENCES #1 and #14

7. Research some of the biological consequences of human sexual activity outside of the marriage relationship—e.g. AIDS, cervical cancer, other infections? See REFERENCE: #7.

8. Evaluate the various methods of conception control and/or birth control. What is your position on their use? Are all methods equally acceptable or unacceptable? See REFERENCE #10.

REFERENCES

1. Anderson, J.K. 1982. *Genetic Engineering*. Zondervan, Grand Rapids, MI.

2. Blackwelder, R.E. and G.S. Garoian. 1984. Animals that transplant organs. *Amer. Biol. Teach.* 46(2): 118-119.

3. Bohlin, R. 1981. Sociobiology: Cloned from the gene cult. *Christianity Today* 25(2): 16-19.

4. Cosgrove, M.P. *The Amazing Body Human*. Baker Book House. Grand Rapids, MI.

5. Epel, D. 1977. The program of fertilization. *Scient. Amer.* 237(5): 128-38.

6. Fisher, A. 1992. Sociobiology: Science or ideology. *Society* July/Aug. pp 67-79.

7. Jackson, J.K. 1992. *AIDS, STD, & Other Communicable Diseases*. Dushkin Publ. Group, Inc. Guilford, CT.

8. Jensen, K. 1982. *Reproduction: The Cycle of Life*. U.S. News Books. Washington, DC.

9. Klopfer, P.H. 1982. Mating types and human sexuality. *Bioscience* 32(10): 803-806.

10. Knight, J.W. and J.C. Callahan. 1989. *Preventing Birth: Contemporary Methods and Related Moral Controversies*. U. of Utah Press. Salt Lake City.

11. Lejeune, J. 1990. What's in the fridge? *Origins Research* 13:1
    NOTE: Source of info on genome

12. Murray, A.W. and Marc W. Kirschner. 1991. What controls the cell cycle. *Scient. Amer.* 264(3): 56-63.

13. Olby, R. 1975. *Path to the Double Helix*. Univ. of Wash. Press, Seattle.

14. Singer, P. and D. Wells. 1985. *Making Babies: The new Science and Ethics of Conception*. Scribner's Sons. New York, NY.

15. Watson, J.D. 1968. *The Double Helix*. Signet Books, N.Y.

16. White, J. 1977. *Eros Defiled: The Christian and Sexual Sin*. InterVarsity Press. Downers Grove, IL.

17. Wilson, E.O. 1975. *Sociobiology: The New Sythesis*. Harvard Univ. Press. Cambridge, MA.

    See also REFERENCES in Chapter 10: Martini, F. (1992); and Chapter 9: Stern (1991) on human reproduction, and plant reproduction, respectively.

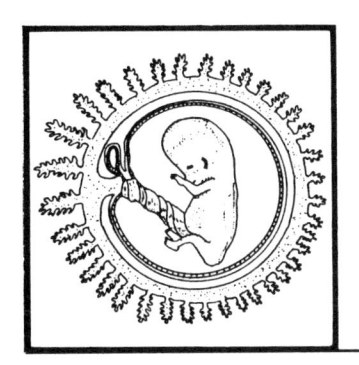

Chapter 12

Genetics

The sexual life cycle of most species features the fusion of two haploid nuclei, one from the male and one from the female parent, to form a zygote, the first cell of a new generation. The haploid nuclei each contain a complete set of chromosomes representing the genetic blueprint for the development of a new individual. As a biological rule, each species reproduces offspring of its own kind. Therefore, we would expect individuals of the new generation to resemble the parents in form and behavior. However, we would not be surprised if each of the offspring display certain obvious differences in one or more attributes, or *traits*.

Genetics has principles governing inheritance.

The branch of biology known as **genetics** is the study of genes (*i.e.* hereditary material, DNA) and the manner in which hereditary characteristics are passed from the parent generation to the offspring. In this chapter, our purpose will be to identify the principles that govern inheritance of individual differences among the offspring of two parents.

12-A   BLENDING INHERITANCE

Could characters blend like paints?

For centuries it has been assumed that physical characteristics of parents are inherited by the offspring. Yet it is only within the last century that we have begun to understand the mechanisms involved. Before anyone knew about DNA and genes, many believed that the hereditary material from one parent would somehow blend with that of the other parent during fertilization. Perhaps an analogy would have been the blending of two different colors or flowers to produce a mixture with some characteristics of each source. But this hypothesis is refuted by the fact that offspring are often not blends of the parents. Instead, offspring may often resemble one parent or grandparent and not the other. If blending of traits were the rule in nature, white and black animals would give away to gray, and red and white flowers, to pink flowers.

blending and natural selection

The blending hypothesis was a problem to Charles Darwin whose theory of natural selection was criticized as being inconsistent with blending inheritance. His theory of natural selection proposed that individuals possessing traits that were advantageous in a given environment would be better able to survive and reproduce; therefore, these traits would be passed along to the offspring and be seen more frequently in that population over time. If natural selection and resulting evolution (according to Darwin) were to work, there must be genetic variation among individuals to select from. However, if the blending hypothesis were true, new traits that arise would be blended into existing traits and evolution could not occur. Darwin never clearly answered this criticism.

Mendel makes his mark.

Meanwhile, in Austria, a man who was to be afforded no recognition for his discoveries during his lifetime, was carefully working out the principles of genetics. The man was Gregor Mendel (1822-1884), formerly a farm boy, then a substitute science teacher, and finally a monk. He spent two years studying mathematics at the University of Vienna and was able to combine this knowledge with an interest in plant breeding to identify precise mathematical relationships among inherited traits. We shall review Mendel's experiments because Mendel gives us a good example of the scientific approach, and because the principles he discovered are basic to understanding the patterns of inheritance. [Section 1-E discusses the scientific method.]

12-B MENDEL'S EXPERIMENTS

Example of good science:

... good choice of species

The success of Mendel's experiments is due in part to careful experimental design. He shared the same curiosity about inheritance patterns as others had for centuries. Yet he began by carefully defining a system to study. He narrowed his focus from all the creatures around him to one—the garden pea plant. The pea was a wise choice because the flower petals enclose the pistil and stamens making it more difficult for pollen from other sources to enter unless the flowers are opened by the experimenter and subjected to cross-pollination. This characteristic enabled Mendel to obtain *true-breeding strains* of peas that expressed the same form of a given trait generation after generation.

... good focus

Mendel narrowed his focus even further by concentrating on only seven traits. Each of these traits had two distinctly different visible expressions, or **phenotypes**. The seven traits are listed, and the phenotypes are illustrated in Table 12-1. He planted seeds from each strain and was assured that each was true-breeding. Then he narrowed his focus further by concentrating on only one trait at a time. For instance, he planted the seeds of smooth and wrinkled varieties. When flowers had developed, he removed the anthers from one variety before the pollen sacs had ripened and covered the flowers with small bags. When the pistils were mature, he removed the bags and pollinated them with pollen from the other variety. The offspring (*i.e.* seeds, destined to become plants) of such a *cross* between two true-breeding parents are called **hybrids**. Because only one trait is involved, it is called a **monohybrid cross**. For the sake of brevity, geneticists use **P** to refer to the *parent generation*. The first and second generations from the parents are the F_1 (first filial) and F_2 **generations**, respectively.

... precise genetic crosses

... good questions

Mendel found that the F_1 generation offspring were not a blend of smooth and wrinkled seed as predicted by the blending hypothesis. Instead, all seeds were smooth! The same was true in F_1 offspring involving the other traits (Table 12-1). [Remember, Mendel knew much less about genetics than you at this point, but he kept good records. Forgetting, if you can, all that you know about genes, etc, what would you conclude?] Mendel was faced with the "disappearance" of half of the phenotypes observed in the F_1 offspring. He may have asked, "Did they simply disappear, never to reappear, or were they masked over by a factor that produced the phenotypes that were observed?" Both possibilities could be formulated into a testable hypothesis. [Which hypothesis would you have selected? How would you have tested your hypothesis?]

Table 12-1. Results of Mendel's Experiments with Garden Pea Plants.

TRAIT	DOMINANT	×	RECESSIVE	=	NO. IN F_2 GENERATION Dominant	NO. IN F_2 GENERATION Recessive	RATIO
SEED SHAPE	smooth		wrinkled		5,474	1,850	2.96/1
SEED COLOR	yellow		green		6,022	2,001	3.01/1
POD SHAPE	round		wrinkled		882	299	2.95/1
POD COLOR	green		yellow		428	152	2.82/1
FLOWER COLOR	red		white		705	224	3.15/1
FLOWER POSITION	axial		terminal		651	207	3.14/1
PLANT HEIGHT	tall		short		787	277	2.84/1

... good records

Mendel chose to plant seeds representing the F$_1$ generation; and, when flowers developed, he allowed self-pollination to occur. In the resulting F$_2$ generation, he found an answer to his question. Both phenotypes for each of the seven traits were again expressed. Mendel wisely kept careful records of the number of offspring displaying each phenotype. We can summarize the results of Mendel's experiment with seed shape as follows:

P generation: *smooth x wrinkled*

F$_1$ generation: [*all smooth*], then ——> *x smooth*
(self-pollination)

F$_2$ generation: *5,474 smooth + 1,850 wrinkled*

Mendel's data from the other crosses are recorded in Table 12-1.

12-C MENDEL'S PRINCIPLES OF HEREDITY

After eight years and many additional crosses, Mendel began to see predictable ratios of phenotypes among the offspring. The following principles were proposed by Mendel, based upon his interpretation of the data (using his terminology):

two discrete factors

1. *Each hereditary trait is controlled by two discrete factors or 'particles' that are passed unchanged from one generation to another.* This was in direct contradiction to the blending hypothesis, and is the first suggestion of the existence of the **gene** as a unit of information. Mendel suggested *two factors* for each trait to account for the predictable ratios (Table 12-1). This is the first evidence that somatic cells have two sets of genes—i.e. the diploid condition (Section 11-H).

dominance

2. *When two contesting factors are present in an organism, only one factor will be expressed. The expressed factor is **dominant**, the masked, or hidden factor, is **recessive**.* Mendel did not mean to extend this principle to every inherited trait, as some mistakenly believe. He seemed to have been aware of instances of *incomplete dominance* (Section 12-E).

segregation

3. *There is a separation (segregation) of the two hereditary factors during gamete formation so that each parent contributes only **one** of the two factors for each trait during fertilization.* Thus, the pollen carries a sperm nucleus with one factor and the egg nucleus has one factor. This is the first suggestion of the haploid condition of gametes. *As a result of segregation in each mate, half of the gametes have one factor, and half of the gametes have the other factor.*

recombination at random

4. *When gametes fuse in fertilization, the two factors are combined in random fashion.* Mendel stated this because he could predict the ratio of offspring on a probability (random chance) basis.

These principles may seem rather simple to you if you have studied genetics before. However, in 1866, when Mendel published his results, these ideas were revolutionary. Perhaps for this reason, they were not given much serious attention. Or, was it because the scientific world was taken up by another publication called *The Origin of Species*, released in 1859 by Mendel's contemporary, Charles Darwin? Whatever the case, it was not until 1900, nearly 20 years after his death, that Mendel's work was rediscovered and respected for its value.

12-D MODERN APPLICATIONS OF MENDELIAN PRINCIPLES

Mendel in today's context

Gregor Mendel was the first to present evidence for the existence of *genes* as particulate or material entities. He went further to suggest that organisms have two hereditary factors (genes) for each trait. We now refer to this as the *diploid* condition, meaning that every somatic cell has two sets of chromosomes. Each trait is represented by two genes, one on each of two chromosomes representing a *homologous pair* (Section 11-H). We now know that the two genes for each trait occupy corresponding locations, or **loci** (singular = locus), on the two homologous chromosomes. The dominant and recessive pair of genes for one trait, having the same locus, are called **alleles** (short for *allelomorph* = other form)

Having drawn Mendel's principles partly into a modern context, let us consider his experiment with smooth and wrinkled seeds as illustrated in Figure 12-1 (see also the end of Section 12-B).

Learn the terminology and symbols.

Dominant alleles are represented by uppercase "S", chosen by conventional rule, as the initial of the dominant phenotype. Recessive alleles are represented in lower case using the same letter. These letters symbolize the gene combination, or *genotype*, for a given trait. Given the genotype, we can determine the phenotype as shown for smooth and wrinkled in the example above. Thus, the P generation, true-breeding plants are represented as having either the dominant allele on both homologous chromosomes (genotype SS = smooth phenotype) or the recessive on both (genotype ss = wrinkled phenotype). This condition is called **homozygous** (G. *homo* = same). Therefore, SS

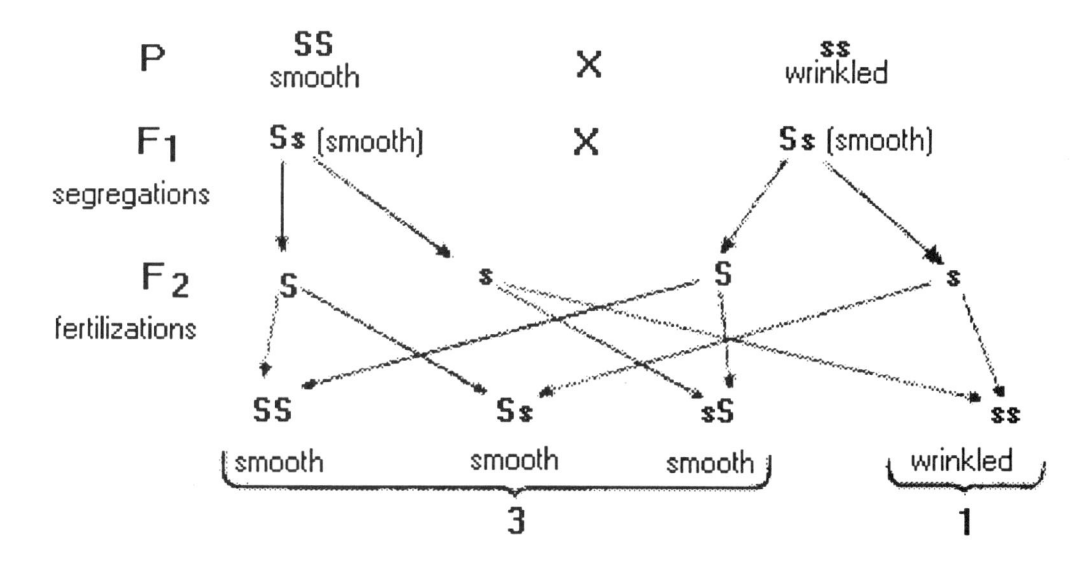

```
P        SS                  X            ss
         smooth                           wrinkled

F₁       Ss (smooth)          X            Ss (smooth)
segregations

F₂        S        s                 S         s
fertilizations

         SS         Ss               sS        ss
       { smooth   smooth           smooth }   { wrinkled }
              3                                    1
```

Figure 12-1. Symbolic representation of Mendel's monohybrid cross between smooth- and wrinkled-seeded pea plants.

Keep it straight with Punnett square.

is *homozygous dominant*. [What would you call genotype ss?] The F₁ generation offspring all have one dominant allele for smooth, and one recessive allele for wrinkled—i.e. genotype Ss. This condition is called **heterozygous** (G. *hetero* = different). Note that our symbol 'Ss' in the F₁ generation above explains the "disappearance" of the recessive allele. It didn't disappear, but was masked by the dominant allele, S. We would predict the dominant phenotype to appear in 3 out of 4 offspring of the F₁ generation, as shown above. The F₁ alleles were segregated when meiosis occurred, resulting in equal numbers of gametes with allele S and s (Mendelian principle #3, Section 12-C). Then, during fertilization there is equal probability of gametes combining to produce zygotes of SS, Ss, sS, and ss. That is, there is 1 chance in 4 that a given combination will occur each time a fertilization occurs. The so-called *Punnett square* is often used to show the gene combinations produced in fertilization. Figure 12-2 illustrates its use to clarify the events in Mendel's seed experiments. [Problem: Use Punnett squares to determine the genotype and phenotype ratios of offspring for several other traits in Table 12-1.]

dihybrid cross

If you were as curious as Mendel, you would have probably decided to perform a cross involving parent pea plants that are each true-breeding for *two traits*. This is a **dihybrid cross** and can be understood by the same principles as for the monohybrid crosses above. Figure 12-3 illustrates the results of Mendel's experiments beginning with parents (P) that are smooth yellow (SSYY) and wrinkled green (ssyy). [How do you know yellow is the dominant allele? See Table 12-1] Again, the F₁ offspring express only the dominant phenotypes, smooth-yellow. These are heterozygous for both traits. When the alleles are segregated during meiosis to produce gametes, each gamete appears to have an opportunity to receive a dominant or recessive of one trait regardless of whether it receives the dominant or recessive of the

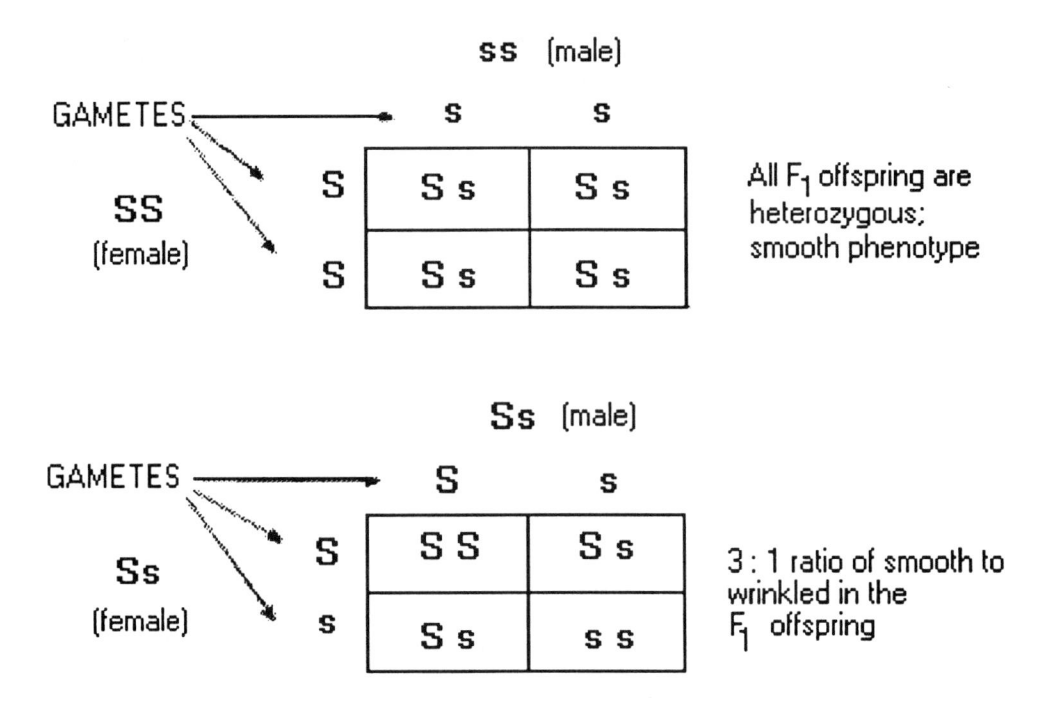

Figure 12-2. Punnett square diagrams of Mendel's seed experiments.

traits combine independently

other trait. That is, the segregated alleles for the two traits combine independently of one another when gametes are formed. This process is called **independent assortment**. The Punnett square and Mendel's data (Figure 12-3) are consistent with a 9:3:3:1 ratio of F_2 offspring. Independent assortment in gamete production, followed by equal probabilities of fusion among all types of gametes (i.e. 1 chance in 16 of a given combination of gametes forming a zygote), accounts for the observed ratio. Interestingly, Mendel did not make the *inductive jump* (Section 1-E) from his specific data to a generalization in the form of the 9:3:3:1 ratio. The ratio was proposed by those who later reviewed Mendel's data.

meiosis and independent assortment

We can explain independent assortment based upon the homologous pairing of chromosomes during metaphase I. [You may wish to review Section 11-H, MEIOSIS, before proceeding.] Let us assume that the alleles for seed shape (S and s) are on one homologous pair, and alleles for seed color (Y and y) are on a different homologous pair, as shown in Figure 12-4. We are concentrating on what meiosis does to these alleles in individuals of the F_1 generation; genotypes SsYy. [Can you see how the genotype SsYy is related to the position of alleles on the chromosomes?] Note that there are *two* pairing alignments that can occur between paternal and maternal chromosomes with respect to the equatorial plane down the middle. One alignment is as likely to occur as the other and each leads to a different assortment of alleles among the four gametes. Note that the types of gametes produced in meiosis (Figure 12-4) are each represented on the Punnett square illustrated in Figure 12-3.

P   smooth-yellow   X   wrinkled-green
         SSYY   ssyy

F_1 smooth-yellow   X   smooth-yellow
            SsYy   SsYy

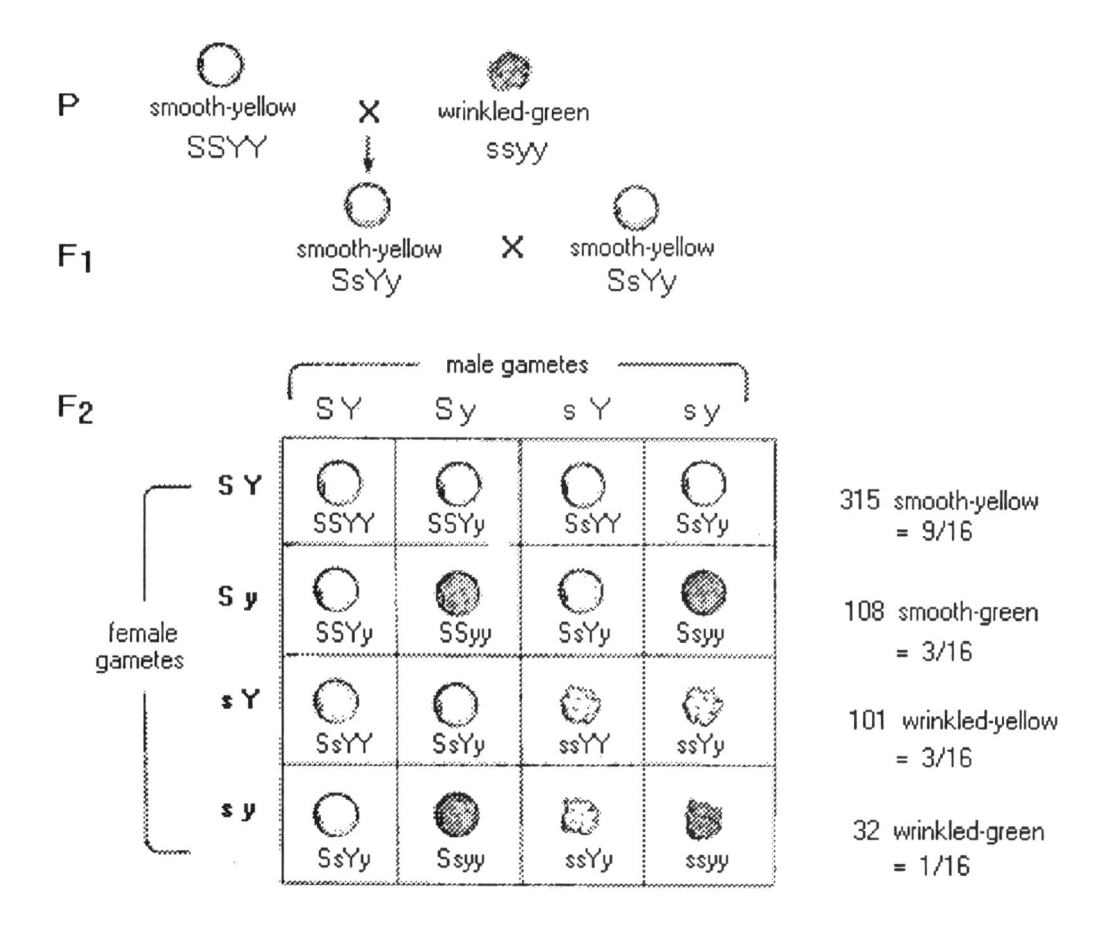

F_2

	SY	Sy	sY	sy
SY	SSYY	SSYy	SsYY	SsYy
Sy	SSYy	SSyy	SsYy	Ssyy
sY	SsYY	SsYy	ssYY	ssYy
sy	SsYy	Ssyy	ssYy	ssyy

315 smooth-yellow = 9/16

108 smooth-green = 3/16

101 wrinkled-yellow = 3/16

32 wrinkled-green = 1/16

Figure 12-3. Dihybrid cross showing Mendel's numerical evidence of independent assortment of alleles in the F_2 offspring.

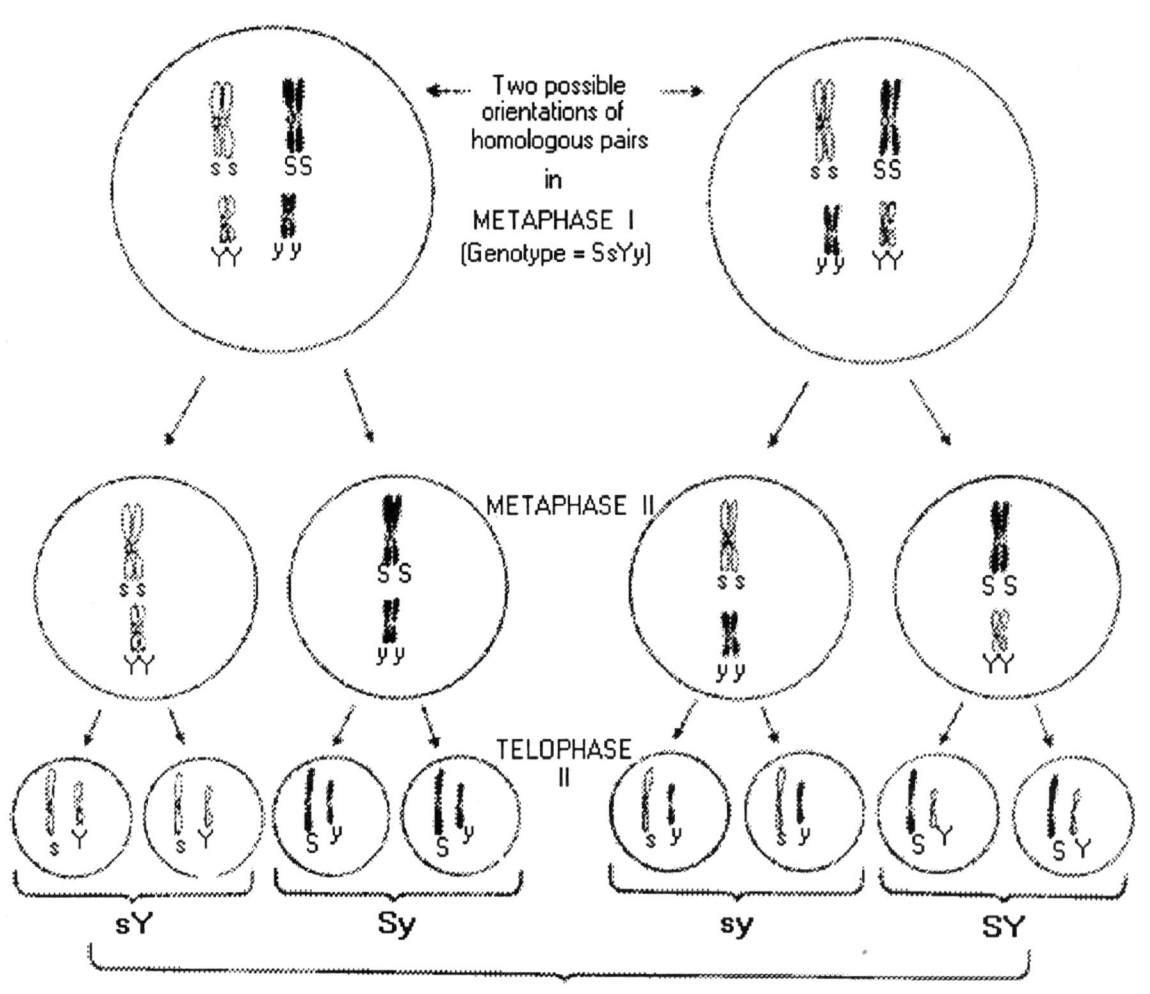

Figure 12-4. Independent assortment can be explained by the differing alignments of homologous pairs in metaphase I of meiosis and the subsequent segregation of alleles as chromosomes are drawn apart. Paternal chromosomes are *darkened* to distinguish from the maternal chromosomes.

You may recall that *crossing over* between chromatids of a homologous pair occurs during prophase I of meiosis (Section 11-H). This event is not included in Figure 12-4. However, we would conclude that crossing over had occurred if alleles on attached chromatids were different, one dominant and one recessive, instead of what is shown in metaphase I of Figure 12-4. For example, the homologous pair bearing the alleles for seed shape would be affected as follows:

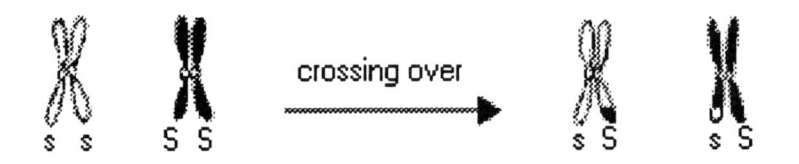

Crossing over adds more gene combinations.

Note that crossing over will allow even more genetic variation by allowing genes of a paternal chromosome to be exchanged with genes of the maternal chromosome of the same homologous pair. [Complete a new sketch similar to Figure 12-4 showing the effect of crossing over on gene combinations in the gametes.] The effect of independent assortment and crossing over in producing genetic variation would become more evident if we were to expand our focus from a dihybrid cross to consideration of all the thousands of pairs of alleles, each at a different locus on the chromosomes.

12-E OTHER HEREDITARY PATTERNS

Until now we have emphasized traits the are controlled by two alleles, where one allele has **complete dominance** over the other. That is, for a given trait, heterozygous offspring will resemble one of the parents, instead of being a "blend" of the two. As noted in Section 12-A, Mendel's discoveries dealt a major blow to the "blending hypothesis". The following discussion summarizes "incomplete dominance" and other departures from patterns we have considered so far.

INCOMPLETE DOMINANCE

"blending"?

As the term suggests, **incomplete dominance** occurs in certain traits in which one allele does not completely mask the other, in the heterozygous condition. Instead, heterozygous offspring have a phenotype intermediate between that of the two homozygous conditions. To illustrate, let's consider garden flowers such as the four-o'clocks. Here we represent white flowers as F^W and red flowers as F^R because there is no clearly dominant or recessive allele. A cross between white-($F^W F^W$) and red-flowered plants ($F^R F^R$) produces offspring that are all pink-flowered ($F^W F^R$).

Incomplete dominance also explains why a couple, one curly-haired and the other straight-haired, can have a wavy-haired child. Another trait illustrating incomplete dominance is *sickle-cell anemia*, to be discussed in more detail in Section 15-G.

Mendel discovered incomplete dominance in an eighth trait in peas, namely flowering time, not reported in Table 12-1. Because of this observation, Mendel was more reluctant than those who followed him to claim that dominance was present in every trait. Mendel teaches us the importance of being cautious not to extend a theory beyond what the data allow.

MULTIPLE ALLELES

Can a trait have more than two alleles?

Up until now we have discussed traits that are controlled by alleles at one locus on a chromosome. Because chromosomes exist in homologous pairs, a diploid organism can carry, at most, two different alleles—a dominant and a recessive allele, one on each of the homologous pairs. However, for many traits, any one of three or more alleles can be present at a particular locus. That is, where **multiple alleles** exist for a given trait, an organism could inherit any (and only) two out of three or more alleles occupying the same locus on each of two homologous chromosomes.

Human ABO blood types are distinguished by the one or two of three possible alleles a person has inherited. Table 12-2 outlines the genotypic differences and related data. Two of the three alleles, if inherited, will direct

blood antigens

the synthesis of specific proteins associated with red blood cells. These proteins are called **antigens**, specifically antigen-A and antigen-B. Persons with antigen-A (blood type A) cannot receive blood transfusions of blood type B, and *vice versa*. Transfusions that violate this relationship cause red blood cell clumping and resultant clogging of blood vessels. This condition results when other blood proteins called **antibodies** attack blood cells containing the antigen not present in their blood. Persons with type AB have both antigens and neither antibody. Therefore, they can receive transfusions from all other types as far as this category is concerned. Individuals with type AB blood have genotype $I^A I^B$, which is another example of incomplete dominance. Type O blood results from the homozygous recessive condition (*ii*). Individuals with type O blood can receive only type O blood.

Table 12-2 Multiple Alleles Involved in Human Blood Types.

BLOOD TYPE	GENOTYPE	ANTIGEN ON BLOOD CELLS	% DISTRIBUTION, USA WHITE	BLACK
A	$I^A I^A$ or $I^A i$	A	41	27
B	$I^B I^B$ or $I^B i$	B	10	21
AB	$I^A I^B$	A and B	4	4
O	i i	Neither A or B	45	48

Distribution Data From: G.E. Nelson. 1984. *Biological Principles with Human Perspectives*. Wiley and Sons, New York, NY. p. 280.

greater genetic variation

Traits that derive from multiple alleles are associated with greater genetic variation because more than two phenotypes can be present in a given population. Note from Table 12-2 that populations that have different geographic origins may differ in the *frequency* with which certain phenotypes appear. The branch of genetics called **population genetics** is concerned with the frequencies of genes within the gene pools of populations (Sections 7-A, 7-D, 12-I, 15-G).

Rh blood group

Thirty-seven blood groups, each controlled by alleles at a different locus, are listed in McKusick's catalog, *Mendelian Inheritance in Man* (12). The **Rhesus blood group**, named after **rh**esus monkeys, is one of these. So-called **Rh⁺** persons have genotype, *RR* or *Rr*, which codes for the **Rh antigen** on red blood cells. Persons with Rh⁻, genotype *rr*, lack the antigen. Unlike the ABO group, the Rh blood type is not inherited through a simple, dominant-recessive pairing.

A complication in pregnancy can occur if an Rh⁻ woman carries an Rh⁺ child. There is no direct placental exchange of blood between mother and baby. However, during delivery of the child, Rh antigen may enter her circulatory system and trigger the formation of antibodies against the Rh antigen. If there is another later pregnancy involving an Rh⁺ baby, her antibodies set out to destroy the child, causing severe and sometimes fatal anemia. This problem can now be avoided by treating Rh⁻ mothers of Rh⁺ babies with a single injection of antibodies against any Rh antigen that may have entered her system before the antigen can trigger antibody buildup within her system. [What is your blood type with respect to ABO and Rh blood groups?]

MULTIPLE GENE (POLYGENIC) INHERITANCE

two types of variation—

So far, we have been considering traits with phenotypes that display **discontinuous variation**. Plant height, seed color, Rh blood type—all of these have two distinct expressions. But, how about human height, shape, and color? Humans come in a whole range of heights, shapes, and colors. These traits exemplify **discontinuous variation**. As you might guess, these human traits cannot be a simple "either...or" system involving only one chromosomal locus, as is true of discontinuous variation. Instead, these traits are controlled by **multiple genes** (multiple *sets* of alleles), each at a different locus and each having a small effect. For example, a person with genotype AABBCC..., etc., may be extremely tall due to the many dominant alleles. Within a sample population, as the number of dominant alleles decreases down to aabbcc..., etc., there is a gradual, or continuous decrease in height.

Expression of phenotypes controlled by multiple genes within a population follows a *normal distribution*, or "bell-shaped curve". The curve for a normal distribution (Figure 12-5) records the number of individuals (*y-axis*) that are represented at each point or interval along the whole range of possible phenotypes (*x-axis*). For a given trait, the further from the mean (average) of a population, the lower the frequency of occurrence. [Why might the "extremes" be more rare in natural populations? Is there an environmental influence?]

other polygenic traits—

Other human traits that are polygenic include eye color, skin color, intellectual aptitude, and facial features. Inheritance of eye color has often been treated as a discontinuous variation involving brown and blue. However, there is a continuous variation from black through brown, green, and gray, to blue. At least two gene loci control the quantity of **melanin** pigment in the iris. Blue eyes lack melanin in the front layer. Increasing amounts of pigment in the front layers cause darker and darker eyes.

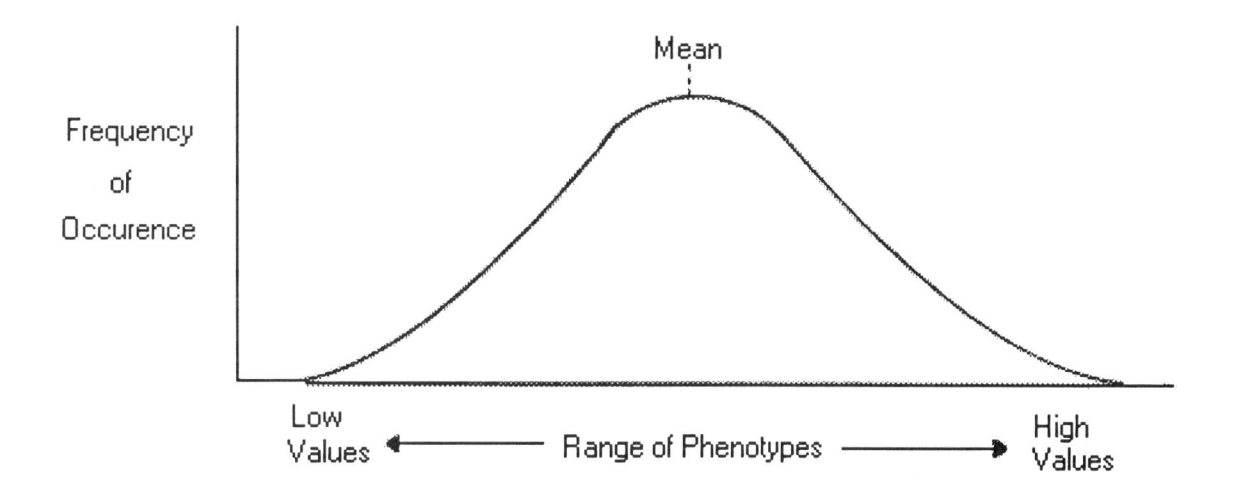

Figure 12-5. Normal distribution of the number of individuals at each point or interval along a range of phenotypic expressions. Frequency distributions of such phenotypic expressions as birth weight, height, skin color, and standardized test scores may resemble this 'bell-shaped curve'.

Human skin color also varies in proportion to the amount of melanin, which in turn, is polygenic (Section 12-I). In contrast, cases of **albinism** in humans result from a recessive gene which blocks all melanin synthesis by coding for a defective enzyme which blocks the pathway. Albinism illustrates a case in which a single pair of alleles mask the expression of multiple genes at other loci on the chromosomes.

LINKED GENES

Mendel's theory of independent assortment was also tested by later studies. For instance, certain groups of traits did not follow the pattern of inheritance one would predict if independent assortment were occurring (Section 12-D). Then it was discovered that the genes involved were located on the same chromosome, and tended to be inherited together unless separated by crossing over. Such genes are called **linked genes**.

Linked genes have loci on the same chromosome.

The only data published and discussed by Mendel are consistent with the idea that all traits are inherited independently. However, studies of the pea plant since the time of Mendel indicate that three out of the seven traits studied by Mendel (Table 12-1) have loci on the same homologous pair, and two other traits are linked on another pair. Genes for the remaining two traits are inherited on separate chromosomes.

Gene linkage significantly influences human heredity. The important **X** and **Y** chromosomes will be discussed in the next section as they relate to *sex-linked traits*.

12-F CHROMOSOMES AND HEREDITY

post-mendelian discoveries

When Gregor Mendel died in 1884, his published work lay dormant in libraries. Meanwhile, other researchers discovered slender, microscopic bodies (chromosomes) in cell nuclei. Moreover, the number of chromosomes in gametes was found to be only half the number in somatic cells. In 1887, August Weismann and others confirmed the existence of meiosis as the process that reduces the chromosome number and separates them among different gametes. By then, others had rediscovered what Mendel had already published, even as his reports were discovered.

Within a few years, Mendel's principles were explained on the basis of observed chromosomal activity, forming a **chromosomal theory** of heredity. As we have seen, chromosomes constitute the visible, material substance of the hereditary material. Each chromosome is a highly coiled mass of DNA and associated proteins. Genes represent segments of DNA like words linked into a sentence.

We have come a long way

Today, a century after the birth of the chromosomal theory, biologists are progressing in one of the most ambitious research projects in history—the **Human Genome Project**. This 15-year project, involving collaborating teams of researchers from many countries, aims to map the locations of each of the approximately 100,000 genes of the human genome and to determine the sequence of the some 3 billion nitrogen base pairs (Section 11-D). At the time of this writing, over two thousand genes have been mapped, including chromosome #21 and the Y chromosome in their entirety (5). Figure 12-6 is a human *chromosome map* showing the locations of several genes. Note that each chromosome can be identified by a combination of size and pattern of banding. Bands are developed by stains and reflect variations in the relative amounts of A-T *versus* C-G base pairings (Section 11-C).

Human Chromosomes

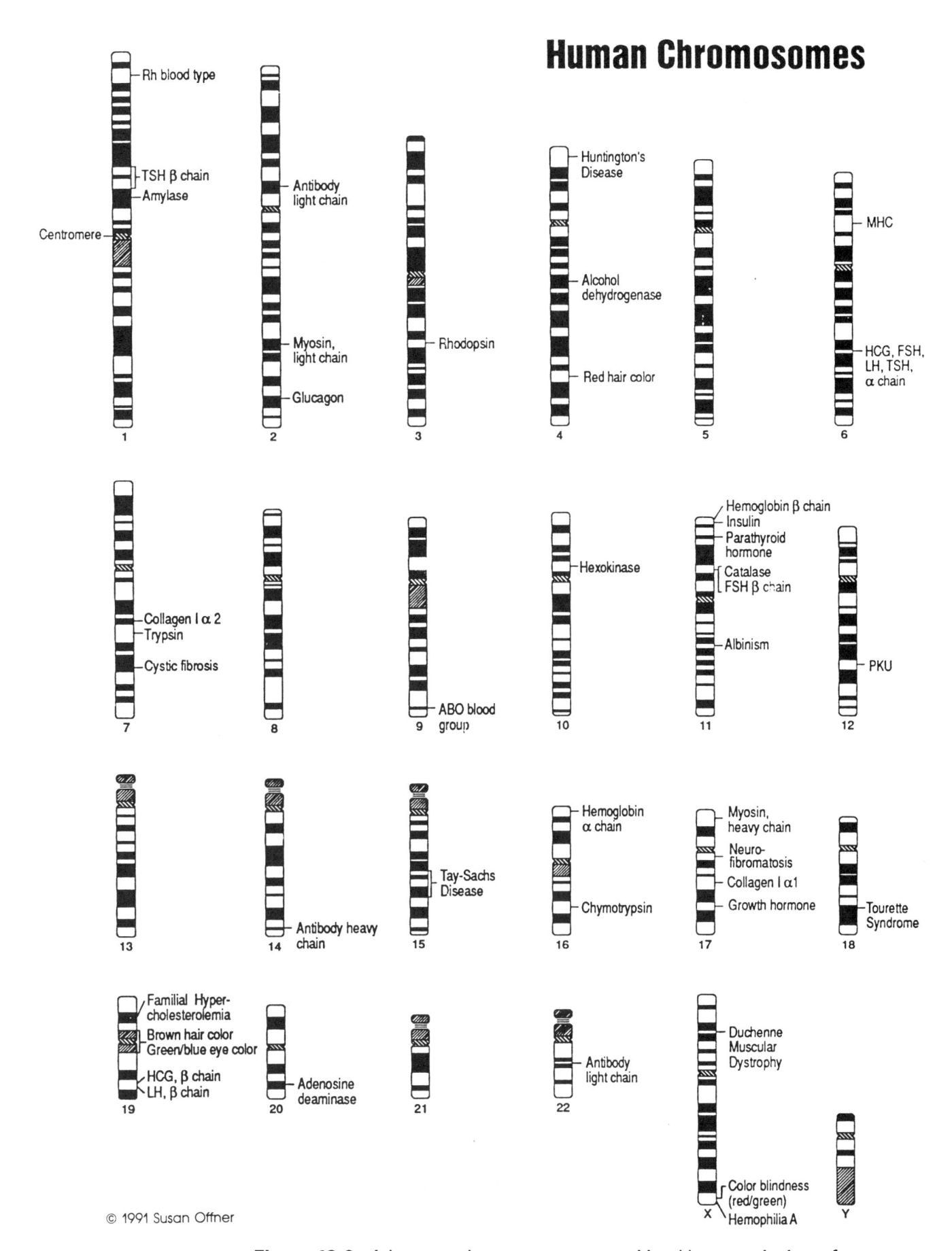

Figure 12-6. A human chromosome map. Used by permission of Susan Offner, Ph.D., Milton High School, Milton, MA. See REF. (14).

253

X-LINKED INHERITANCE

autosomes and sex chromosomes

As shown in Figures 11-4 and 12-6, all chromosomes except the **X** and **Y** chromosomes are matched pairs equal in size within each pair. These chromosomes which are the same in males and females are called **autosomes**. However, the **X** and **Y** chromosomes differ in size from each other, and males have XY and females, XX. This pair constitutes the **sex chromosomes**. The Y chromosome is much shorter and has only a few gene loci, among which is the "male-determining" gene. During the second month of human development, this gene apparently triggers the development of testes, which in turn produce male hormones that coordinate subsequent development. Absence of the Y in females allows ovary development instead of testes. In males, which have XY, if a Y-containing sperm unites with an egg, the offspring will be a boy; X-carrying sperm will produce a girl. [Show by Punnett Square the probability of a child being boy or girl.]

Is it a boy or a girl?

Most mammals have the same sex-determining mechanism as humans. In birds, the situation is somewhat reversed and the females have the differing-size chromosomes. In bees, the females are diploid, and males are haploid.

You will recall that the heterozygous condition means that an organism is carrying a recessive allele that is masked by a dominant allele (except in incomplete dominance). This condition is only possible because there are two sets of chromosomes (and alleles). However, in the human male, because of the short Y chromosome, the X chromosome carries the only genes for important traits. Therefore, recessive genes passed from the mother to a son via the X chromosome (father donates the Y) will not be masked.

X-linked traits

The so-called **X-linked traits** of major significance include red-green color blindness, hemophilia, and certain types of muscular dystrophy. Hemophilia is a condition in which blood is very slow to clot, and many hemophiliacs die before maturity. If a hemophilic man (*h*) marries a "normal" woman (*HH*), none of the sons (*H*) will be hemophilic because the father donates only the Y chromosome (no *H* or *h* allele on the Y chromosome). However, all of the daughters will be carriers (*Hh*). When these daughters marry, they can bear hemophilic sons, regardless of their husbands' genotypes. [What is the probability of this occurring?] Queen Victoria was a carrier of hemophilia and the appearance of the trait among her descendants has been widely studied.

Do females use both X chromosomes?

In some cases, female carriers of recessive X-linked traits such as hemophilia may be as much affected as males, or have symptoms intermediate in severity. One hypothesis, now called the *Lyon hypothesis*, was developed in the 1960's by geneticist Mary Lyon (1). She proposed that only one of the two X chromosomes in human females is permitted to express its genes. She was aware that at two to three weeks after conception, one of the two X chromosomes in each of the embryonic cells (paternal X in some, maternal in others) coils up as a visible structure and remains coiled so that genes cannot be expressed. This dark structure, barely visible under high power light microscopy, is called a *Barr body*, after its discoverer, Murray Barr. The embryo now contains a mix of cells, some able to express alleles from paternal X chromosomes, and some from the maternal. This pattern follows consistently through the millions of daughter cells descending through repeated mitosis, so that some regions of the female body may be influenced by the father and some by the mother. Where the female is heterozygous for traits on the XX pair, phenotype may differ by bodily region. The Lyon hypothesis would thus explain the variable hemophilia and muscular dystrophy symptoms among female carriers. A more obvious phenotypic evidence for the Lyon hypothesis is in female calico cats that have the heterozygous condition ("yellow allele" on one X, and "black allele" on the other). A "mosaic" of yellow and black result in regions of the body representing hair-producing cells descending from cells where one or the other X chromosome was inactivated.

AUTOSOMAL INHERITANCE

Most of the genetic traits governing bodily form, function, and behavior are inherited through the 22 pairs of autosomal chromosomes. Return to Figure 12-6 and notice a number of familiar traits and the chromosomal locus of each. Note that some govern outwardly visible phenotypes such as color or debilitating disease symptoms. Others are responsible for protein (*e.g* enzyme hexokinase, or antibody) or hormone (*e.g.* insulin) synthesis. Many of the loci are of interest because of their role in major diseases as mutated genes, as discussed in the next section.

12-G   GENETICS AND HUMAN HEALTH

challenging accepted theory

In 1947, while performing genetic experiments on maize, geneticist Barbara McClintock observed that pigment streaks and blotches appeared in Indian corn seeds in patterns which were not explainable by mendelian and chromosomal theory. She seized upon this discrepancy between *new observation* and *accepted theory* as an invitation to scientific discovery. Her critical analysis and careful experimentation enabled her to play a role in further *unifying these theories*, and *expanding their predictive power*, which is a key to scientific progress. Her hypothesis was that genes could actually move within or among chromosomes! Specifically, small segments of DNA can "jump into" other DNA sequences and garble the genetic code. In maize seeds, the "jumping gene" interrupted pigment synthesis in some cells and not others, producing splotches and streaks in corn seeds.

Was Barbara too radical?

Barbara McClintock's theory was a bit too radical for the mid-1900's. Therefore, her paper's "collected dust" until the 1970's when discoveries in molecular biology vindicated her. She was awarded a Nobel Prize in 1983. [Does this incident sound similar to an earlier one in this chapter? Is skepticism in science good or bad?]

GENE MUTATIONS

Mutations:

...source

"Jumping genes" are now called *transposons*, segments of DNA which may be inserted into other genes, perhaps causing an altered phenotype. Transposons are one of several types of **gene mutations**. Gene mutations may occur during DNA replication and consist of a change in nucleotide sequence, either by addition, deletion, or change in a single nucleotide pair (Sections 11-C, 11-D). Some gene mutations are *spontaneous* whereas others are *induced* by factors in the environment.

...consequences

What are the consequences of an altered *sequence* of nucleotides in terms of cellular function? As explained in Chapter 17, DNA influences protein synthesis. For present discussion purposes, the "chain of command" is as follows:

DNA $\longrightarrow$ RNA $\longrightarrow$ PROTEIN $\longrightarrow$ METABOLIC ROLE
nucleotide   nucleotide   amino acid   e.g. enzyme, membrane
sequence   sequence   sequence   protein, or hormone

How do mutations disrupt?

In this "chain of command" any alteration in the DNA code sequence will be transmitted as an "error." However, the processes that involve RNA and protein synthesis may not perceive the error. Instead, the altered code directs the insertion of wrong amino acids into one or more positions in the protein molecule (Section 4-I). The altered protein may be unable to properly function as an enzyme, membrane protein, or hormone. For example, the "jumping genes" inhibited pigment production in maize seeds when they were transposed into a gene that codes for an enzyme needed for pigment systhesis.

We can understand more clearly the "masking" of *recessive* alleles by *dominant* alleles by understanding the "chain of command" as it relates to mutations. Often recessive alleles are mutations of the dominant allele, and are expressed only when the functional allele is absent. The resultant disorder occurs simply because the mutated alleles coded for a nonfunctional enzyme or membrane protein. The blocking of melanin synthesis and production of sickle-cell hemoglobin are respective examples (Section 12-E). [Note that dominant alleles may also be deleterious (see Table 12-3).]

What determines severity of a mutation?

The severity of the effect of a mutation depends upon when and where it happens. The most serious consequences are observed when a mutation occurs in reproductive cells and a gamete containing the mutation is viable enough to fertilize an egg. The parent would not be affected but the offspring would have the mutation duplicated by mitosis throughout every cell of the developing embryo. Hemophilia, noted earlier, is a result of a recessive mutant allele located on the X chromosome. On the other hand, a mutation occurring in a somatic cell would affect only the individual. Somatic cell mutations that occur early in embryo development may be more serious than those that occur in an adult. [Why?]

mutation rates and repair

Fortunately, mutations are the exception and not the rule. Estimates of the frequency of gene mutations are in the order of 1 per one million replications! The average rate of mutation for an individual gene or locus on a chromosome is probably one for every 100,000 cells. The existence of special **DNA repair enzymes** is one reason why mutations are not more frequent. These enzymes apparently "inspect" the DNA strands and repair irregularities such as broken sugar-phosphate bonds or improperly paired nitrogen bases. Repair may consist of removing the "bad strand" and splicing in a "good" one.

mutagenic agents

The probability of mutation increases when individuals are exposed to certain **mutagenic agents** in the environment. *Ionizing radiation* in the form of X—rays, gamma rays, and ultraviolet (UV) rays, can penetrate living cells, strike the atoms of nucleic acids or other molecules, and cause breakage of covalent bonds. The sun is a source of these high-energy waves, but not all are screened out by atmospheric gases (Sections 5-B and 6-D). Radioactive isotopes, which are a by-product of nuclear reactors and weapons testing, are also emitters of ionizing radiation. Viruses cause mutations by inserting their own DNA into genes. In fact, some transposons are suspected of being remnants of viral DNA.

In recent years, the concern about toxic chemicals in the environment is prompted in part by evidence that many industrial chemicals are **carcinogens**, or cancer causers. Certain carcinogens may cause cancer by activating certain cancer-causing genes, called **oncogenes**, that exist in normal human cells. Increased exposure to UV radiation may activate oncogenes and cause skin cancer.

reverse mutations

Most gene mutations are not beneficial and may be so deleterious that they are *lethal* (cause death) and therefore are not passed to the next generation. However, advances in molecular biology are such that the concept of mutations is being revised. For example, recent studies identified so-called "colossal genes" which result from insertions of "repeat sequences" of nucleotides into a gene (16). Thus, myotonic dystrophy which causes muscle atrophy is caused by a mutation involving an increased number of cytosine-thymine-guanine (CTG) "repeats". Indeed, Dutch researchers have just reported the first evidence of *reverse mutations*, "spontaneous correction of a deleterious mutation upon transmission to unaffected offspring" (2). Their study involves among others, a son who inherited an abnormal chromosome #19 with 150 to 500 CTG repeats from his father. However, absence of abnormal repeats in the son's chromosomes (and absence of myotonic dystrophy symptoms) suggests the reverse mutation had occurred. [Suggest a possible cause of the repeats and the reverse mutation.]

Gene mutations will be important in our future discussion of genetic variation, natural selection, and evolution. Table 12-3 summarizes several common genetic disorders attributed to gene mutations.

CHROMOSOMAL ABNORMALITIES

abnormalities involving whole chromosomes

Chromosomal abnormalities (or chromosomal mutations) represent losses, additions, or reorientations of relatively large sections of the DNA, usually involving many genes. During mitosis or meiosis, pieces of chromosomes or whole chromosomes may become disattached and wander free or even attach to another chromosome. When cytoplasmic division occurs, one daughter cell may end up with an extra chromosome or part of a chromosome, while the other daughter cell is missing this genetic information.

When gametes containing chromosomal abnormalities enter fertilization, the offspring will have visible genetic defects. For example, **Down syndrome**, a condition of mental retardation, altered physical appearance, and shortened life span, occurs in children conceived when a gamete has an extra chromosome #21. **Turner Syndrome** results when a child is conceived with one X chromosome (X_); Barr bodies are absent. The child is a sterile, abnormally developed female. **Klinefelter Syndrome** occurs when males are conceived with more than one X chromosome (e.g. XXY), and results in mild mental retardation and low fertility. Women with an extra X chromosome (XXX) usually have no detectable effects.

Table 12-3. Common Genetic Disorders Associated with Gene Mutations.

DISORDER	BIOLOGICAL EFFECTS
AUTOSOMAL DOMINANT (2,000)*	
Huntington Disease	Nervous system degeneration; beginning at about age 40; discovered after children are born.
Chronic simple glaucoma	Major cause of blindness if not treated
Hypercholesterolemia	High blood cholesterol
Polydactyly	Extra fingers or toes
AUTOSOMAL RECESSIVE (1,000)*	
Cystic Fibrosis	Alteration in sweat glands and bodily mucus; mucus accumulates in respiratory system
Galactosemia	Inability to digest "milk sugar", lactose (defective enzyme for converting galactose-1-P to glucose-1-P); treat by milk-free diet
Phenylketonuria (PKU)	Defective enzyme to converting phenylalanine to tyrosine; newborns now "genetically screened"
Sickle-cell Anemia	Defective hemoglobin (O_2-carrying protein); red blood cells burst into sickle shape; incomplete dominance; more common among blacks
Tay-Sachs Disease	Fatal brain damage affects infants of East European Jewish ancestry
Albinism	Inability to synthesize melanin pigments
X-LINKED (250)*	
Color Blindness	Inability to distinguish certain colors
Fragile X Syndrome	Mental retardation; tandem "repeats" (16)
Hemophilia	Defect in blood-clotting mechanism
Muscular Dystrophy	Muscular degeneration; one of several, inherited
MULTIFACTORIAL	Many genes interacting; environmental influences in some; uncertain of transmission in many
Spina Bifida	Open spine
Hydrocephalis	Water on the brain (with Spina Bifida)
Diabetes mellitus	Abnormal sugar metabolism

* Indicates number of confirmed or suspected disorders catalogued.
Data From: *Genetic Counseling*. 1987. March of Dimes Birth Defects Foundation.

DIAGNOSIS AND COUNSELING

Consider history of occurrence.

Increased attention has been given to the development of techniques for early detection of genetic defects. Such techniques provide a basis for **genetic counseling**. In cases involving simple mendelian inheritance, recommendations can be made to couples based upon the history of occurrence of the disease or defect in previous generations. Childbearing after age thirty increases the chances of bearing children with chromosomal abnormalities due to the increasing "age" of oocytes (Section 11-J).

Learning early about defects . . .

If a couple wishes to proceed to have children knowing there is a high probability of having a diseased child, they may elect to have *prenatal diagnosis*. Two commonly used techniques are **amniocentesis** and **chorionic villus sampling (CVS)**. These methods involve insertion of a syringe to remove

calls for moral decision-making

fetal cells from the amniotic fluid or from the chorionic membrane formed by the fetus, respectively (Section 14-G). The fetal cells are then examined for detection of over a hundred known genetic diseases. Karyotypes would reveal chromosomal abnormalities. DNA analysis can detect gene mutations for such disorders as Tay-Sachs, spina bifida, and sickle-cell anemia. Based upon the results, the couple may elect to continue the pregnancy, or request an abortion.

The moral and ethical implications accompanying the increased usage of prenatal diagnosis should be of major concern. We can now learn what the "phenotype" of a baby will be before birth, and decide whether the person should live. This technology extends the concept of *phenotype* (observable characteristic) to the "scale" of the chromosome, nucleotide sequence, or metabolic product.

correcting defects

Coupled to prenatal diagnosis, the development of *gene therapy* provides the option in some cases to "correct" the defect. Once again, our ability to answer the moral and ethical question, "Should we?" is far behind our ability to answer the question, "Can we?." [See TOPICS FOR FURTHER STUDY, # 1.]

12-H HEREDITY AND ENVIRONMENT

What makes us like we are:

genes?

"Why am I the way I am?" Have you ever asked yourself this question? The *Peanuts* character, Lucy once answered this question by claiming she was born with "crabby genes". No doubt Lucy is partly right. The genes inherited by an individual plant, animal, or person have tremendous influence on survival and the capacity to live a "normal" life.

... or environment?

On the other hand, we have seen indications that the environment may influence the expression of genes. One obvious example is the effect of environmental mutagens on genes. In Section 12-E, we observed that traits such as human height and intelligence are influenced by polygenic inheritance, but the genetic blueprint prescribes only the *potential* of a person to grow in height or develop intellectually. In actuality, a person's value system, self discipline, and access to opportunities for spiritual, physical, and intellectual development will influence the degree to which his or her genetic potential is realized. These values and opportunities depend upon the family and community dimensions of our environment. All organisms are products of their inheritance *and* their environment. Several other examples of this principle may be helpful.

Phenotype depends upon genes *and* environment.

The synthesis of chlorophyll in green plants is under genetic control. Homozygous recessive plant seedlings die because they are unable to complete the synthesis of this green pigment necessary for photosynthesis. [How can they live as long as they do?] This condition is called **albinism**, because the leaves and stems are white in the absence of chlorophyll. Even though normal plants have the genetic potential to synthesize chlorophyll, they too will lack the green coloration if allowed to grow in darkness. The presence of light is necessary for the full expression of genes that control chlorophyll synthesis.

Environment influences gene expression.

Himalayan rabbits and siamese cats normally have light-colored fur around the main trunk of their bodies, while the face, ears, tail and legs have dark-colored fur. This contrast occurs because the gene for black fur color can express itself only below a certain body temperature, such as that of the body extremities. Like the genes for chlorophyll synthesis, the gene for black fur color depends upon environmental conditions for its phenotypic expression. [How could you demonstrate that temperature influences fur color in siamese cats?]

"nature *vs.* nurture"

Distinguishing the influence of heredity and environment has not been easy to establish at the morphological or physiological levels. Distinctions are even more difficult in the behavioral sciences where parameters such as parental love, opportunity, or emotional stress are difficult to define and study in controlled experiments. The so-called "nature *versus* nurture" debate continues.

There is another dimension in which environment influences "why organisms are the way they are." This dimension involves the environment acting over a longer scale of time, upon generations of organisms as members of populations, through natural selection. The next section examines this topic.

12-I POPULATION GENETICS

Through the *Peanuts* character, Lucy, we have contemplated the question "Why am I the way I am?" This question can also be addressed from a perspective of population genetics which is concerned with gene frequencies in gene pools. In Section 8-A and Table 8-2, we noted that biological processes operate in different scales of *space* and *time*. Thus, Lucy's individual use of CFC-containing aerosol spray in her home and school would not cause an ozone hole over her neighborhood. However, on a larger *scale* of space and time, many users in many neighborhoods over decades appear to be affecting the ozone layer. Likewise, within the small scale of Lucy's lifetime, she will have to be content with her phenotype—her height, skin color, hair texture, etc. Granted, she can dye her hair, exercise, and visit the tanning booth. In this way, as noted in the previous section, the environment will influence the extent of development of genetic potential. But, as individuals, we live within a relatively short scale of time, a lifetime, and barring mutations or gene therapy, the genotype of an individual remains that which was inherited at conception from the parents.

product of our parents

If we expand our scale from *individuals*, namely Lucy, to the human *population*, including Lucy's family, we can expand our answer to "why we are the way we are." If Lucy prefers darker skin, the probability of Lucy having more melanin pigment will be determined by the frequency of genes favoring melanin production in the gene pool of which Lucy's family is a part (Section 12-E). Again, Lucy can darken her skin by tanning. Melanin production and the resultant "tan" will increase as a protective screen against UV radiation exposure from sunlight (or tanning booths). But, genetic processes operating beyond the scale of Lucy's and your lifetimes will also determine inherited levels of skin melanin content.

product of gene pools

gene frequencies

Within the human species, higher skin melanin levels are common in *populations* (or races) in tropical areas that were geographically isolated from other populations. Genetically, this means that there is a higher frequency of alleles favoring melanin production in the gene pools of dark-skinned populations. Melanin production in humans is controlled by 2 to 5 sets of alleles (polygenic). [How foolish of humans to base racial prejudices upon so few alleles out of 100,000 genes in the human genome!] Had Lucy been born within one of these geographic populations, the probability is high that her skin would be dark. However, because of the development of world travel and commerce, unfortunately including slavery, intermarriage has allowed formerly separate

Why are you the way you are?

gene pools to combine. In some regions, frequencies of alleles for dark skin have changed as well. Lucy's birth into a population affected by intermarriage would favor her inheritance of fewer alleles for melanin production. Thus, the phenotype of each individual organism is the result of environmental factors operating both within the local scale if lifetime and neighborhood; and, factors that have operated over generations of individuals of a gene pool through natural selection.

Here, we will end the discussion of natural selection, and resume in Chapter 15, for two reasons. First, natural selection is often confusing before some careful discussion of the "origin of life". Second, natural selection may be viewed as a homeostatic mechanism, operating at the scale of populations over time. Homeostasis will be introduced in the next two chapters before we return to natural selection.

12-J SCRIPTURAL PRINCIPLES AND HUMAN GENETICS

God's purposes:

The question, "Why am I the way I am?", may be one that you have often asked with a degree of disappointment or sorrow. At a time when the media and culture encourage cosmetic, surgical, and coming genetic means of "being all that you want to be", we must reflect on several principles from the Scriptures. The following are given (written in first person; substitute your name for personal application) with references and minimal comment for you further study:

God knows me

1. God knew me as a person and chose me before I was conceived (Jeremiah 1:5; Isaiah 49:1); and, before my genetic blueprint was translated into bodily characteristics (Psalm 139: 16), He knew His purpose for my life, both in the scale of time, and of eternity (Isaiah 49:5; Ephesians 1: 3-6; 2: 4-9.

creation groans

2. I live in a creation that groans under sin (Romans 8: 19-23), the effects of which are also inherited spiritually (Romans 5: 12) and biologically (Exodus 20: 5).

no pride

3. Because it is God that formed me through the genetic and environmental heritage (i.e. family, community) which He willed for me (Psalm 100:3), there is no justification for me to harbor individual or racial pride or inferiority in comparison to others (II Corinth. 10: 12; Galatians 3: 28-29).

no judging

4. Because God is acquainted with each person, and made them for His purpose, it is not for me to judge the worth of another person based upon their appearance (phenotype), abilities, or disabilities (Exodus 4:11; Isaiah 45:9).

self-acceptance

5. Biblical self-acceptance, based upon the above principles, should lead me to accept the physical, mental, and emotional characteristics which are unchangeable, and to adjust my nutrition, physical appearance, and daily schedule in such a way as to glorify God, my Creator (I Corinth. 6: 19-20; II Corinth. 3: 5-6). [For example, negative effects of disorders such as galactosemia, PKU, diabetes, etc. can be minimized by proper diet and medication.]

be informed

6. If God permits me to marry, my spouse and I should exercise proper stewardship of our bodies, become informed of our family histories of genetic disorders, and avail ourselves of genetic counseling when appropriate so that we are prepared to make wise decisions regarding childbearing. [May we avoid the tragedy of bearing a child harmed within the environment of the womb by the careless life of a mother or father. (See Principle # 2)]

QUESTIONS AND DISCUSSION TOPICS

1. In what ways was Gregor Mendel's approach to scientific investigation exemplary? What factors may have contributed to the poor reception of his results?

2. How does meiosis and fertilization account for the various gene combinations of gametes and genotype ratios among F_1 and F_2 offspring?

3. What is meant by genetic variation? How do independent assortment and crossing over promote genetic variation?

4. How did Mendel's principles prepare biologists (and you) to understand other hereditary patterns that seemed to be exceptions to his principles (e.g. incomplete dominance, linkage, multiple alleles)?

5. From your understanding of how genes influence phenotype through such human disorders as sickle-cell anemia, albinism, PKU, and galactosemia, how would you explain the difference between *dominant* and *recessive* alleles? How are the meaning of these and of *phenotype* changing with the advent of genetic diagnosis?

6. Why is there greater genetic variation in the category of ABO blood types than in the category of Rh +/- blood types? Distinguish multiple allele inheritance from polygenic (multiple gene) inheritance.

7. Why does inheritance of X-linked traits affect males differently from females? Explain with examples.

8. Distinguish gene mutations from chromosomal abnormalities. Give examples of each. What factors promote mutations?

9. Explain how environment can affect heredity in two different *scales*. Which influences your life more, your genes or your environment?

10. Distinguish the causes of color patterns in calico and in siamese cats.

GENETICS PROBLEMS

Genetics problems are useful study tools to help you develop a better understanding of the principles we have been discussing. The following skills are required in approaching genetics problems. Suggestions and problems are included to illustrate the use of each skill:

SKILL A: *Given information describing phenotypes, you should be able to write out as nearly as possible the genotypes of the individuals in question.*

1. What are possible genotypes corresponding to the following pea plant phenotypes (See Table 12-1): smooth, green-seeded plants; red-flowered, short-stemmed plants; tall plants with red, axial flowers.

SKILL B: *Given phenotypes and/or genotypes, you should be able to determine the possible kinds of gametes each individual will produce and the probability of producing any one of these kinds (i.e. 1 divided by no. of possibilities).*

2. How many different kinds of gametes can be produced during meiosis in each of the following individuals?

   a. *Ss*   b. *TTaa*   c. *TtSs*   d. *RrSsYy*

3. What is the probability that a gamete with the alleles *Sy* will be produced in individuals with each of the following genotypes:

   a. *SsYy*   b. *ssyy*   c. *SSYy*   d. *SSyy*

SKILL C: *Given phenotype or genotype information, you should be able to determine gamete types, assemble the various kinds of male and female gametes on a Punnett Square, and determine the probability that a given genotype and/or phenotype will appear in the offspring.*

4. Given the following crosses between parents where red flower color is dominant to white, and axial position is dominant to terminal, what is the probability that the offspring will have the genotype or phenotype specified in column #2 below?

Parents	Genotype of Certain Offspring
a. RR x Rr	Rr
b. Aa x Aa	aa
c. Aa x Aa	Aa
d. RrAa x RRAa	RRaa
e. RrAa x RrAa	red, terminal flowers
f. RrAa x RrAa	white, terminal flowers

SKILL D: *Given the numbers of offspring that have each of two or more phenotypes, you should be able to determine the phenotype ratio and "work backward" using a Punnett square to determine the probable genotypes of the parents.*

5. In rabbits, spotted coat color is dominant to solid color, and long hair is dominant over short. Determine the probable genotypes of the parents of the following litter totals:
   a. 56 spotted and 17 solid-color
   b. 26 long, spotted; 10 short, spotted; 9 long, solid; and, 3 short, solid

6. A man and wife have children, each having one of the four blood types in Table 12-2. What are the genotypes of the parents?

TOPICS FOR FURTHER STUDY

1. Prenatal diagnosis now makes it possible for parents to be made aware of genetic disorders such as Down syndrome and sickle-cell anemia before the child is born. Research these techniques (Section 12-G), and the medical, moral, and ethical implications of their use?
   REFERENCES: #7, #8

2. What is your genotype for the genetic traits listed in Table 12-4, p. 265? Consult REFERENCES #12 and #13 from which the data was obtained.

3. What is the Human Genome Project, and its political, social, and ethical ramifications?
   REFERENCE #11

4. How have modern molecular biology and genetics changed the meaning of the concepts of *phenotype, dominant,* and *recessive*? See Section 12-G. REFERENCE: #10

5. Sociobiology claims that human social behavior (e.g. reciprocal altruism, parent-offspring conflicts, alternate reproductive tactics, religious belief) can be traced to genes which were at some time favored by natural selection to improve human survival and reproduction. REFERENCES: #6 and #19

Table 12-4. Less Familiar Human Hereditary Traits.

TRAIT	MODE OF INHERITANCE
Achoo Syndrome (light-induced sneezing)	Autosomal Dominant
Cleft Chin (chin dimple)	Uncertain
"Wet" Ear Wax (Cerumen) Type	Autosomal Dominant
Hairy Elbows	Dominant (?)
Hairy Nose Tip	Uncertain
Chinese Restaurant Syndrome (monosodium glutamate sensitivity)	Uncertain
Cannot Smell Musk (pentadecalactone)	Autosomal Recessive
Cannot Smell Skunk (N-butylmercaptan)	

REFERENCES

1. *Basic Genetics: A Human Approach.* 1983. Biological Sciences Curriculum Study. Kendall/Hunt Publ. Co. Dubuque, IA.

2. Brunner, H.G., et.al. 1993. Brief report: Reverse mutation in myotonic dystrophy. *N. Engl. J. Med.* 328(7): 476-480.

3. Cunningham, J.C. and L.M. Taussig. 1989. *A guide to Cystic Fibrosis for Parents and Children.* Cystic Fibrosis Foundation, Tuscon, AZ.

4. Edlin, G. 1990. Reducing racial and ethnic prejudice by presenting a few facts about genetics. *Amer. Biol. Teach.* 52(8): 504-506.

5. Ezzel, C. 1992. Two human chromosomes entirely mapped. *Sci. News* 142(14): 212.

6. Fisher, A. 1992. Sociobiology: science or ideology? *Society* 29(5): 67-79.

7. Fuchs, F. 1980. Genetic amniocentesis. *Scientific American* 242(6): 47

8. *Genetic Counseling.* 1987. March of Dimes Birth Defects Foundation. Available upon request to your local chapter.

9. Hartl, D.L. 1980. *Principles of Population Genetics.* Sinauer Assoc., Inc. Sunderland, Mass.

10. Heim, W.G. 1991. What is a recessive allele? *Amer. Biol. Teach.* 53(2): 94-97.

11. Lee, T.F. 1991. *The Human Genome Project: Cracking the Genetic Code of Life.* Plenum Press. New York, NY.

12. McKusick, V.A. 1988. *Mendelian Inheritance in Man*, 8th ed. Johns Hopkins University Press. Baltimore, MD.

13. Mertins, T.R. 1988. Human genetics and McKusick's *Mendelian Inheritance in Man*. *Amer. Biol. Teach.* 50(5): 262-265.

14. Offner, S. 1992. A plain English map of the human chromosomes. *Amer. Biol. Teach.* 54(2): 87-91.

15. Peters, J.A. ed. 1959. *Classic Papers in Genetics*. Prentice-Hall, Englewood Cliffs, NJ. Contains English translation of G. Mendel's 1866 paper.

16. Rennie, J. 1993. DNA's new twists. *Sci. Amer.* 266(3): 122-132.

17. Selkiowitz, M. 1990. *Down Syndrome: The Facts*. Oxford University Press. Oxford, England.

18. Verma, R.S. and A. Babu. 1989. *Human Chromosomes: Manual of Basic Techniques*. Pergamon Press. New York, NY.

19. Wilson, E.O. 1975. *Sociobiology: The New Synthesis*. Harvard Univ. Press, Cambridge.

Chapter 13

Nervous Systems and Hormonal Systems

Buz-zzz-z-z-itt!! O-o-ohh, alarm! Oh, Monday morning! Oh, dread! That was the snooze alarm and I'm two snoozes late! Quick, out of bed, onto the cold floor, and I dash to the shower. Always too hot or too cold; but it wakes me up. A quick donut, a cup of coffee, and I'm just two minutes late for my eight o'clock speech class.

"What? Did I hear my name?"
"Yes, that's you, isn't it?" he says.
"Oh, my!! A speech? Today?"
"Yes, yours is the first of our impromptu speeches."
Impromptu speech?! Oh, boy! Impromptu is right!

environmental stimuli

Perhaps you can identify with this individual; maybe so clearly that your own body responded vicariously to the scenario. Although not always this stressful, our environment is continually placing demands upon our lives in the form of **stimuli**, environmental factors to which organisms are receptive. Our awakening student, though a bit late, was receptive to *sound, sight, touch, smell*, and *taste*. Living organisms are constantly called into **responsiveness** as a basic part of being alive. *Life characteristic #4* (Section 2-C) states: *living organisms respond to stimuli from the environment, and maintain internal homeostasis.*

Responsiveness involves bodily coordination . . .

and homeostasis.

Responsiveness occurs when certain body cells detect changes in the environment, and others produce appropriate internal and external adjustments to the changes. These adjustments often involve the **coordination** of different organ systems so that internal metabolic conditions remain within tolerable limits in the midst of environmental fluctuations. When such conditions are maintained, the processes involved are said to be at **homeostasis**. For instance, the sounding alarm was an "environmental fluctuation" that brought arousal of the sleeping student. Immediately, there were increased demands upon the body including greater demand for energy. This means that blood oxygen and sugar levels must increase, thus requiring homeostatic adjustments among many organs and glands. [Can you name several of these organs and glands?]

So far we have separated the basic life processes into separate chapters—nutrition (Chapter 10), reproduction (Chapter 11), and genetic continuity (Chapter 12). The study of responsiveness and homeostasis will emphasize how these functions are coordinated so that, in reality, each organism is *one living system*.

There are two coordination systems.

In this chapter, we examine two types of coordination systems—nervous systems and hormonal systems. In Chapter 14, we will see how nervous and hormonal systems function to maintain homeostasis.

There are at least two scales of responsiveness.

The scenario of our speech student demonstrates the need for responsiveness and homeostasis on a *scale* of minutes after "reality hit"? When we extend the scale from organism to population, and response times of minutes to generations, we will see that populations are also "responsive" to environmental changes. Survival and reproduction of individuals and the population will depend upon whether individuals have the morphological, physiological, and behavioral traits, or **adaptations**, that promote fitness—capacity to respond to environmental factors successfully. In Chapter 15, we shall examine the manner in which populations, as genetic units, respond to environmental fluctuations and maintain homeostasis over generations of time. Life characteristic #5 (Section 2-C) emphasizes adaptation as another basic feature of all living things.

13-A COMMON CHARACTERISTICS OF COORDINATION SYSTEMS

Though nervous systems and hormonal systems differ in many ways, both function in the coordination of life processes. To accomplish this, both systems form a network that penetrates throughout the other tissues and organs of the body.

network of neurons . . .

Nervous systems are unique to the kingdom Animalia and consist of networks of specialized cells called *neurons*. Hormonal systems of both kingdoms Animalia and Plantae utilize conductive tissues that also extend throughout the body (e.g. vascular tissue of plants; bloodstream of animals). These extensive networks of neurons and vessels function to conduct messages which serve to coordinate the various body structures and their function. The neurons of the nervous system carry electrical *impulses*, whereas circulatory vessels carry chemical substances, called *hormones*.

to carry impulses

As we examine the nervous and hormonal systems, we will often refer to the tissue level of organization, as we have in the previous paragraph. Each *tissue* is a group of specialized cells that performs the same function. For example, in vascular plants we encountered only three tissue types— *dermal tissue* (outer cellular envelope), *vascular tissue* (xylem and phloem), and *ground tissue* which includes all cells between dermal and vascular tissues. Table 13-1 provides an outline of the major tissue types of the vertebrate animal body. Whereas tissue classifications are helpful learning aids, each organism lives because of the complex interaction of many tissue and organ systems to form a marvelous whole. As you study this chapter and the next, this concept should become increasingly evident.

13-B NERVOUS SYSTEMS — INTRODUCTION

Nervous systems differ among animals.

Every animal phylum except Porifera, the sponges, has some type of nervous system, composed of specialized nerve cells called **neurons**. However, when one compares such representatives as *Hydra* (Cnideria), planaria (Platyhelminthes), earthworm (Annelida), and more complex animals including the vertebrates, it is evident that there are major differences in nervous systems (Figure 13-1). The *Hydra* is capable of only rather simple responses associated with tentacle movement, and extension and contraction of the gastrovascular cavity. The **nerve net** of hydra consists of a network of nerve cells, each capable of detecting stimuli and causing contraction of nearby muscle cells. There is no centralized arrangement of neurons, or *brain*, to integrate

Table 13-1. Four Major Tissue Types in Vertebrate Animals.

TISSUE TYPE	STRUCTURE AND FUNCTION	SPECIFIC TYPES
Epithelial Tissue	Forms exterior and interior cellular linings of body and organs: digestive tract, coelom; respiratory, reproductive and circulatory systems; fastened from beneath by noncellular *basement membrane* of collagen	*squamous*—outer epidermis, and lining of mouth, esophagus *glandular epithelium*—forms glandular secretions *sensory epithelium*—detection of light, smell, taste, and sound *reproductive epithelium*—production of sex cells
Connective Tissue	Serves to bind body parts together; various tissue types, many associated with a non-cellular rigid or fluid matrix	*collagen*—fibrous protein of ligaments, tendons, cartilage, basement membranes; and with minerals, forms bone *adipose tissue*—fatty tissue
Muscle Tissue	Contractile tissues responsible for movement of the skeletal system and organs	*skeletal muscles*—cylindrical striated muscle cells; voluntary control *smooth muscles*—spindle-shaped non-striated; forms walls of internal organs, blood vessels, hair erector muscles, iris, uterus *cardiac muscle*—wall of heart; involuntary, striated
Nervous Tissue	Excitable cells that compose the nervous system	*neurons*—nerve cells *glial cells*—'nerve glue' cells associated with neurons of brain and spinal cord; source of brain tumors

among incoming signals and to decide upon appropriate responses. The planaria (Figure 13-1) does have a simple brain in the anterior (head) region and can generate specific commands to muscles. The earthworm and arthropods have an even more complex brain with clusters of neurons forming a **ganglion** that runs the length of the body. A greater array of commands can be integrated to service the more complex muscular and skeletal systems of these segmented invertebrates.

human brain

The vertebrates are the most complex animals and have both centralized and peripheral nervous systems. Of all the creatures, humans have the most complex brain; and, it is the least understood of all the organs. That the brain is even partially understood is a testimony of its own marvelous function. Right now, your brain is learning about itself!

The central importance of the human brain becomes evident as we examine the following functions of the nervous system:

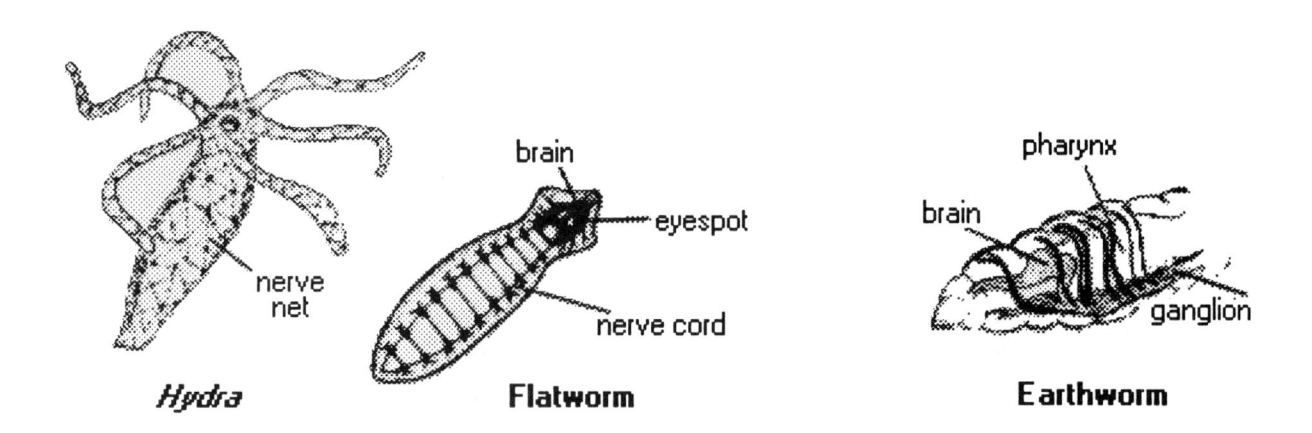

Figure 13-1. Nervous systems of invertebrate animals of varying complexity.

1. *Sensory receptors* in sense organs, including the skin, monitor environmental changes and generate electrical impulses.

2. *Integration* of electrical impulses permits judgements to be made as to which bodily responses (if any) are appropriate.

3. *Response* to stimuli occurs when impulses are sent to voluntary and involuntary muscles and to the glandular system of the body.

transcending the physical

Though many creatures have "brains", the human brain is unique in that it enables us to transcend our own physical nature through *rational thought* (e.g. abstract reasoning, problem solving, creativity). The collective activity of the various regions of the brain permit us to have *aesthetic appreciation*, to *exercise faith* and *offer worship* to our Creator, and to distinguish between *moral* and *immoral* courses of action.

centralized design

The **brain**, the largest organ of the nervous system, is a dense mass composed of about 1 trillion cells, including 100 billion neurons. The **spinal cord** represents a narrow downward continuation of the brain. This cable-like mass of neurons is protected by a fluid-filled cavity that is surrounded by the bony structure of the vertebral column. The brain and spinal cord together make up the **central nervous system** (Figure 13-2).

structural organization

functional organization

The spinal cord serves as the primary linkage between the brain and the other parts of the nervous system. Smaller bundles of neurons, called **nerves**, connect the various parts of the body with the spinal cord. These are called **spinal nerves**. In addition, **cranial nerves** such as the optic nerve, connect certain organs directly to the brain. The spinal nerves and cranial nerves make up the **peripheral nervous system** (Figure 13-2). The peripheral nervous system can be further divided into the **sensorimotor** (or somatic) **nervous system** and the **autonomic nervous system**. The sensorimotor nervous system provides for voluntary (conscious) control of the skeletal muscles, and supplies information from the sense organs to the brain. The autonomic nervous system exerts involuntary control over the internal organs and glands of the body. It has two divisions, the *sympathetic* and *parasympathetic* divisions (Figure 13-2).

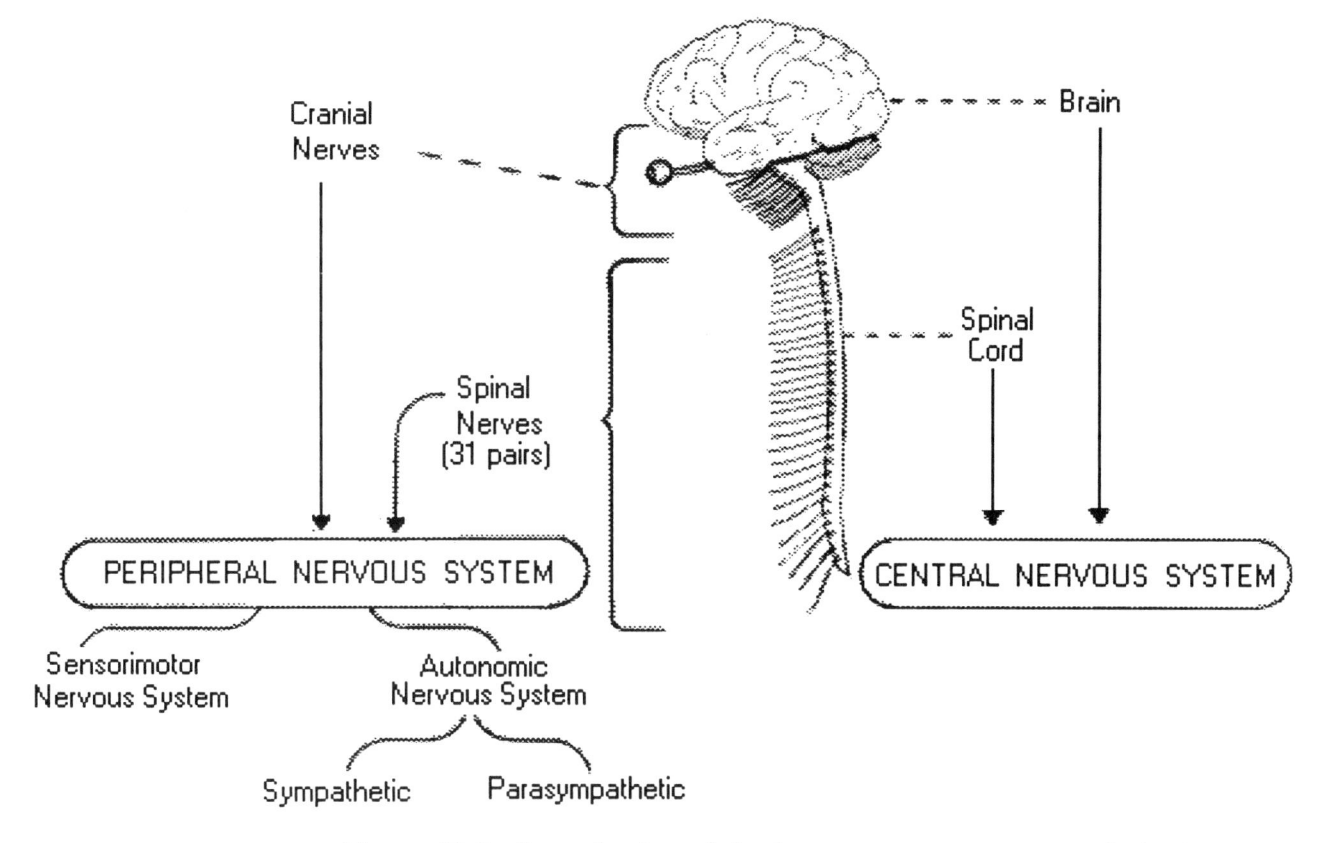

Figure 13-2. Organization of the human nervous system. Spinal nerves are shown for the left side of the body only.

We shall now discuss the various portions of the human nervous system. However, remember that the divisions such as central and peripheral or voluntary and involuntary are only arbitrary and for our convenience.

13-C NEURONS

Santiago Ramon y Cajal, the father of modern brain science, described them as "the mysterious butterflies of the soul, the beating of whose wings may some day—who knows?—clarify the secret of mental life." (7) The human brain contains 100 billion of these "butterflies", or **neurons**, nerve cells that receive and transmit signals (Table 13-1). Most neurons have a *cell body* from which extends a network of branches, called **dendrites** (G. *dendro* = tree). The main extension from the cell body is the **axon** which has a similar branched ending. Associated with neurons are the nearly 1 trillion **glial cells** (glee'-al). Some glial cells wrap their cholesterol-rich membranes around axons, forming the *myelin sheath* (Figure 13-3). Other glial cells provide support, nutrients, and excretory services for neurons.

neurons—

nerve cells

three types of neurons

There are three principal types of neurons (Figure 13-3). **Sensory neurons**, as the name implies, detect stimuli and carry electrical signals to the brain and/or spinal cord. The dendrites of each sensory neuron, carrying **receptors**, are located within the surface tissues of the skin and other sense organs of the body. Each sense organ has its own specialized sensory neurons which are activated by a particular environmental stimulus — light, touch, odor, taste, temperature, and others. Sensory neurons also give "updates" on the condi-

| SENSORY NEURON | INTERNEURON | MOTOR NEURON |

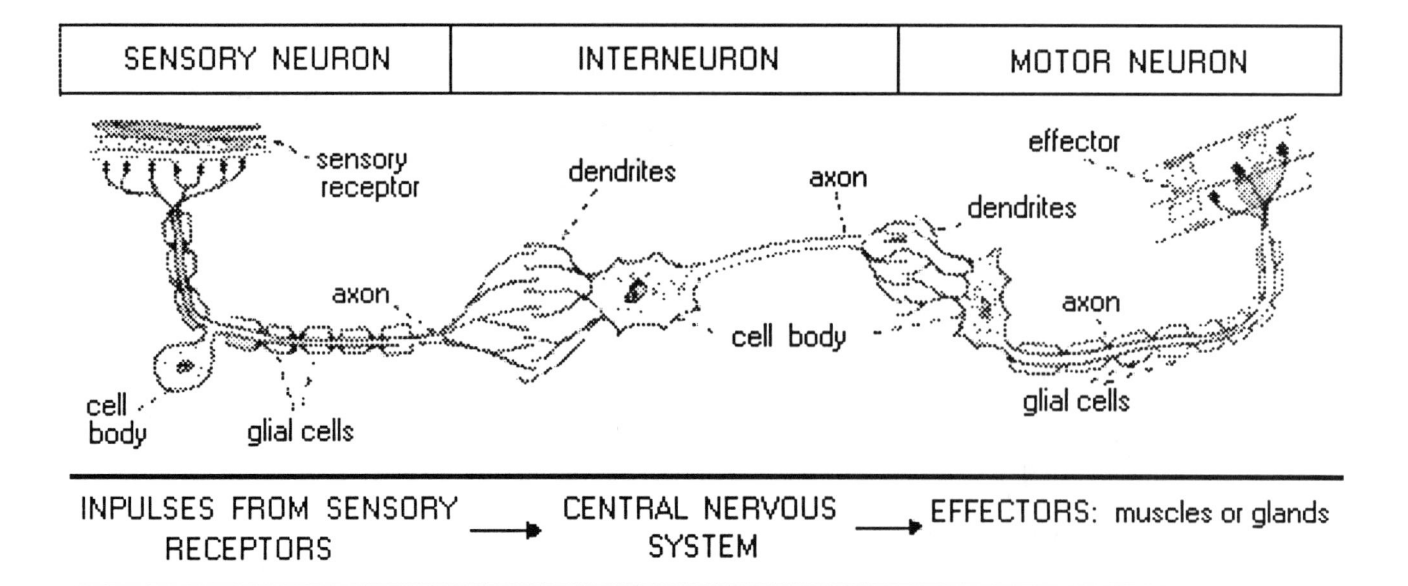

INPULSES FROM SENSORY RECEPTORS → CENTRAL NERVOUS SYSTEM → EFFECTORS: muscles or glands

Figure 13-3. Three types of neurons and their role in transmission of impulses.

The brain and spinal cord are composed of billions of **interneurons** which serve as a complex "switchboard" to relay impulses from sensory neurons to other parts of the central nervous system; and if appropriate, to relay impulses outward via motor neurons (Figure 13-3). As their name implies, interneurons communicate only with other neurons. Interneurons receive impulses from the **axon** portion of sensory neurons via their own branched dendrites. In turn, interneurons relay impulses via their axons to the dendritic ends of motor neurons. In most neurons, the dendrite and cell body receive information and pass it along via the axon.

Motor neurons relay impulses from the brain or spinal cord outward to **effectors**. Effectors are muscles and glands which *contract* and *secrete*, respectively; hence, "effecting" a response to a stimulus detected by the sensory neurons. Motor neurons and associated effectors are not activated at times when you are simply gathering information from your surroundings. However, if you, like the sleeping student, hear the alarm and decide to get up out of bed, then motor neurons and effectors become involved. In either case, one can decide how to use the sensory information and exercise *voluntary* control over skeletal movements.

interneurons are different

Sensory and motor neurons differ from interneurons in structure as well as function. First, they are generally much longer than the interneurons because many extend from as far as your toes and fingertips to your spinal cord. Though motor neurons are the longest cells of the human body, their extreme thinness reduces their requirement for space, mass, and energy. Second, sensory and motor neurons, unlike interneurons, are almost entirely covered with a myelin sheath.

Traditionally, it was believed that humans receive their full complement of brain cells during embryonic growth, and therefore, there were no embryonic cells (*stem cells*) remaining to replace damaged cells. It is true that damage to

Can some brain cells still divide?

nerve tissue is often not repairable. However, Andrusko (1) reports that researchers have recently discovered stem cells in mice brains capable of dividing and producing neurons and glial cells. The report offers encouragement that Parkinson's, Huntington's, Alzheimer's, and other neurodegenerative diseases may some day be treatable. The discovery also suggests that there are other options besides "harvesting" tissue from aborted babies.

13-D IMPULSE TRANSMISSION

electro-potential difference

Tiny electrodes attached to an oscilloscope will reveal differences in electrical charge when one electrode is placed inside an axon membrane, and the other, placed outside the membrane. This difference, called the **membrane potential**, is about 70 millivolts (mV) when the axon is not stimulated. Since the outside is more positively charged than the inside, the membrane potential is given a negative sign, -70 mV. This is due largely to a higher concentration of sodium ions (Na^+) outside the cell than inside (Figure 13-4). Potassium ions (K^+) tend to be more concentrated on the inside. However, total positive charge is still less on the inside because of the presence of other negatively charged ions (e.g. Cl^-).

difference in Na^+ and K^+ across membrane

Cells work to maintain the resting potential.

The membrane potential of an unstimulated neuron, -70 mV, is referred to as the **resting potential**. However, the neuron is not resting. At all times it maintains this resting potential by actively importing potassium (K^+) and ejecting sodium (Na^+) ions. This process is called **active transport** and is accomplished by special membrane proteins called *sodium/ potassium pumps*. These pumps seem to act as "revolving doors" driven by chemical energy (i.e. ATP) (Figure 13-4).

As long as a neuron maintains its resting potential it is ready to transmit a nerve impulse. Impulses are generally initiated with a stimulus (in sensory neurons) or by an impulse relayed from another neuron (*i.e.* interneurons

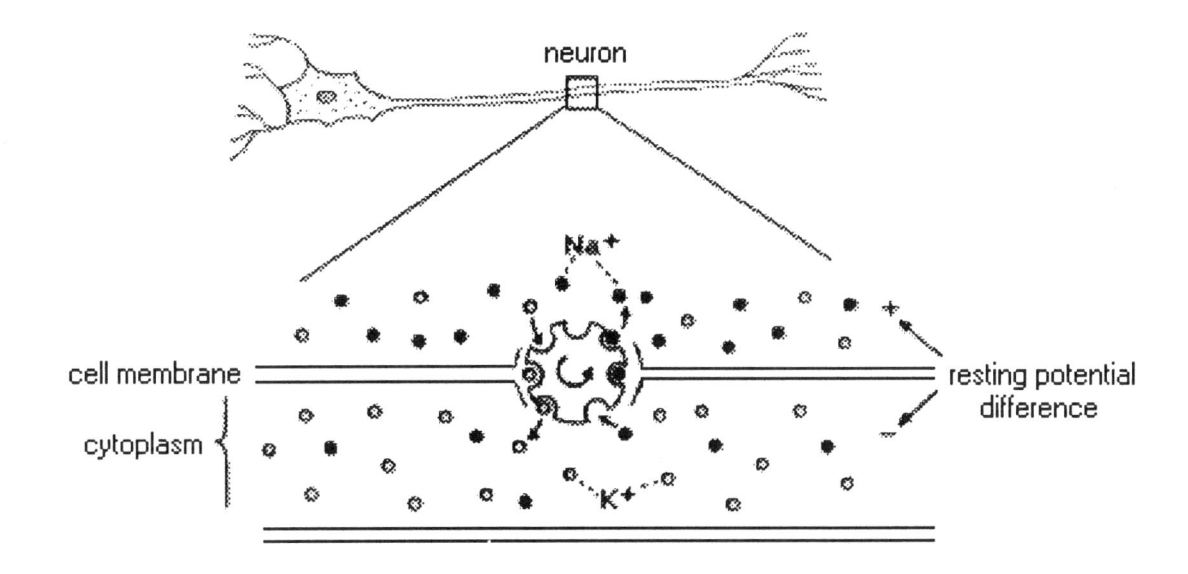

Figure 13-4. Diagrammatic view of a neuron showing the activity of the sodium/potassium pump in expelling sodium ions (Na^+) and importing potassium ions (K^+) to maintain a resting potential.

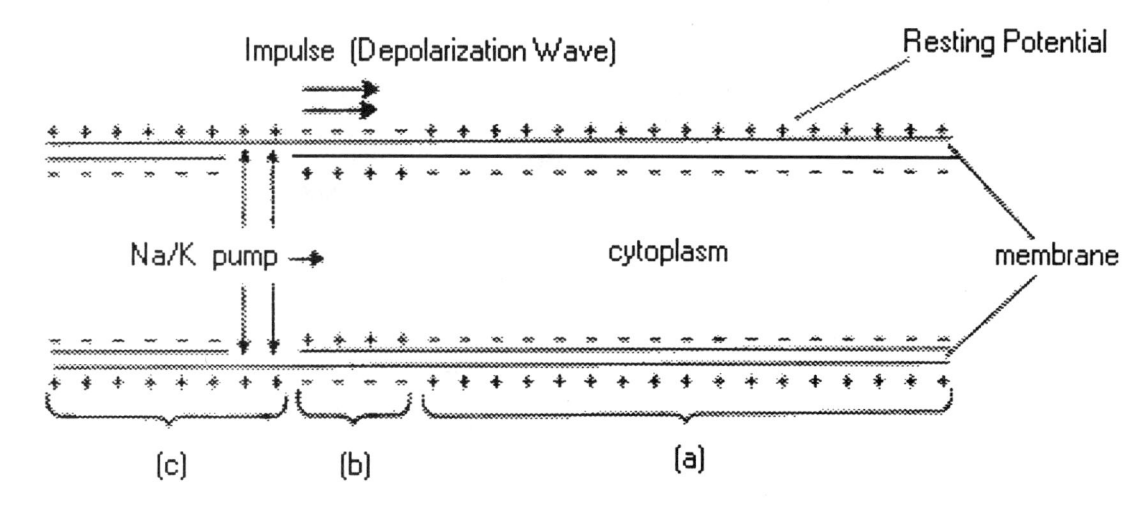

Figure 13-5. Diagrammatic view of impulse transmission. Upon stimulation, the resting potential (a) is discharged, producing an action potential in which sodium leaks into the cell and potassium leaks outward (b). The action potential passes along the axon and the Na/K pump restores the resting potential "behind" the discharge (c).

impulses are *action potentials*

and motor neurons). An **impulse**, or *action potential*, is a change in membrane potential caused by a sudden leakage of Na$^+$ across the membrane into the cytoplasm. The inrush of positive charges causes the discharge or depolarization of the membrane from -70 up to 0 mV, and on to +30 mV, inside. An action potential is initiated at the junction of cell body and axon on an *all-or-nothing* basis. That is, for a given stimulation of a neuron to be registered as "information", the stimulation must cause a rise in membrane potential to some *threshold*, or minimum voltage necessary to cause a more widespread inrush of Na$^+$ and associated rise to +30 mV. The rise or "spike" of an action potential would appear as an approximately 1-millisecond "blip" on an oscilloscope.

Why is it all-or-nothing?

action potentials can "jump" node-to-node

Once an action potential is generated, it moves along the axon (Figure 13-5). In this way, an electrical message is transmitted from one point to another. The Na$^+$/K$^+$ pump works in the "wake" of the depolarization wave to restore the resting potential in readiness for another impulse (Figure 13-5). The resultant short duration allows a second one to follow within 5 milliseconds! In axons with myelin sheaths, action potentials can "jump" from node to node, thus increasing speed to over 100 meters/sec. Impulses are the "beating wings" of Cajal's butterflies.

... but needs chemical transmitter to cross synapse

The junctions between neurons, or between a neuron and muscle or gland cells are called **synapses** [G. *syn* = together + *apsi* = juncture] (Figure 13-6). In other words, the synapse separates the axon endings of a *presynaptic neuron* from the membrane of a *postsynaptic neuron*. Action potentials (impulses) cannot "jump" across a synapse from one cell to another. Instead, when an impulse reaches the bulb-like endings of a presynaptic cell, it induces cytoplasmic *vesicles* to release **transmitter substance** (or neurotransmitters) across the membrane into the *synaptic cleft*. Transmitter molecules diffuse across the narrow (20 X 10^{-9} m) cleft and bind to receptor sites on the *postsynaptic membrane*. This triggers a change in postsynaptic membrane permeability to ions in such a way as to cause either (a) an *excitatory effect*

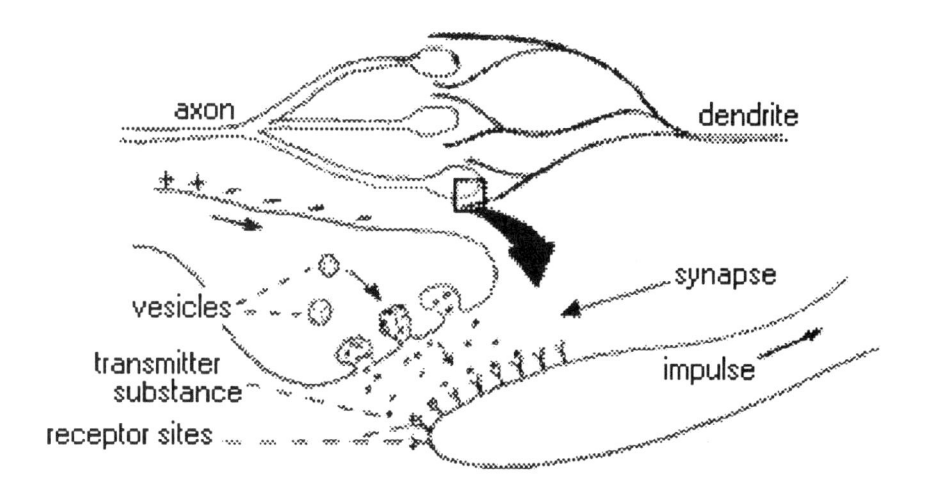

Figure 13-6. Transmission of an impulse across a synapse when chemical transmitters are released by the axon and contact the dendrite of another neuron.

upon the postsynaptic cell by eliciting an action potential, or (b) an *inhibitory effect* by stabilizing the neuron at its resting potential. Whether or not an action potential is generated in the postsynaptic cell is an all-or-nothing situation which depends upon whether or not the integrated signals of chemical transmitters bring the membrane potential to threshold voltage. When synaptic transmission is finished, chemical transmitters are either inactivated or escorted rapidly back into the presynaptic cell by *transporters*, membrane proteins which function as "re-uptake pumps".

one-way transmission

and integration

The synapse is a crucial control point in impulse transmission for a number of reasons. First, the synapse provides for one-way impulse transmission allowing each neuron to be specialized to "receive signals" at the dendrite and cell body region, and "send information" at the axon region. Second, the synapse allows each neuron to function as a unit to "integrate up to 1,000 synaptic inputs" (7) and to determine whether the output to the postsynaptic membrane will be excitatory or inhibitory. "Each neuron is a sophisticated computer." (7)

understanding neurotransmitters, ...

Fifty or more chemical transmitters are now known, including **acetylcholine**, **norepinephrine**, and **dopamine**. Specific transmitters have been identified in relationship to such experiences as thirst, hunger, anxiety, pleasure, and psychotic depression. Acetylcholine functions at synapses of motor neurons with muscles (or glands) (Figures 13-3 and 13-8). Impulses reaching the axon endings of a motor neuron in the muscles trigger release of acetylcholine into the neuromuscular cleft. There the chemical transmitter binds to receptors on the muscle cell membranes where it produces an excitatory effect. The resultant action potential in the muscle cell triggers events leading to muscle contraction (Section 13-F).

...neurotoxins,

A **neurotoxin** from *Clostridium botulinum* (Section 9-B), if ingested from improperly canned foods, inhibits acetylcholine release necessary for muscle contraction. Victims die when breathing is inhibited. A neurotoxin from the related species, *C. tetani*, interferes with acetylcholine function and causes muscles to lock in contracted state (tetanus). Certain insecticides, patterned after the chemistry of World War II nerve gases, cause death via violent muscle spasms by inhibiting cholinesterase, the enzyme that normally removes acetylcholine from the cleft.

... and drugs.

Drugs are chemicals administered to produce some physiological and behavioral change, including pleasure or emotional relief. They bring about these effects through mimicking the function of natural chemical transmitters, or by altering transmitter metabolism and release from the presynaptic membrane. Others act by blocking either postsynaptic cell receptors or transporters and causing transmitters to accumulate in the synaptic cleft. The latter causes incessant firing of impulses, often leading to a temporary "high". *Stimulants*, including nicotine, caffeine, cocaine, and amphetamines, increase bodily alertness and activity, but lead to depression. Cocaine and amphetamines both stimulate the *pleasure centers* of the hypothalamus. Cocaine blocks the action of transporters, while amphetamines resemble norepinephrine and dopamine and stimulate pleasure centers. Here, drug dependence occurs when the brain decreases its own production of transmitters, relying instead on administered drugs. *Depressants* such as barbiturates have a variety of effects depending upon dosage and degree of addiction. They are taken to produce emotional relief, sedation, or sleep. Experimentation with drugs is a serious violation of scriptural principles, as discussed in Section 10-I. [See TOPICS FOR FURTHER STUDY #1 and #2.]

13-E   THE CENTRAL NERVOUS SYSTEM

The central nervous system comprises the brain and spinal cord, the latter serving as an extension of the brain. The spinal nerves branch out all along the length of the spinal cord, providing a wider access of communication between central and peripheral nervous systems (Figure 13-2).

human brain

The human brain is so complex and amazing that whole shelves of books and thousands of research reports are being published on this organ. The brain weighs about three pounds, has a gelatin-like consistency, and is composed of an estimated 100 billion neurons and 1 trillion glial cells! This number is in "the same order of magnitude as the number of stars in the Milky Way [but numbers of cells alone] ...cannot account for the complexity of the brain." (7) The diversity of cells and their arrangements are also important.

parts of brain

The three major parts of the brain are the cerebrum, cerebellum, and the medulla (Figure 13-7). The **cerebellum** coordinates movement. The *limbic system*, which influences emotions and long-term memory, lies within the cerebellum. The **medulla** is really an enlargement of the spinal cord and all communication from the spinal cord to the brain pass through it. The medulla controls such involuntary functions as breathing, digestion, and heartbeat through the autonomic nervous system (Section 13-G).

probing the brain

The most obvious part of the brain is the **cerebrum** which is surrounded by the *cerebral cortex*, a gray, convoluted, outer layer. The cerebrum has a left and right hemisphere, separated by a deep fissure. Each hemisphere is further divided by fissures into various lobes. Detailed maps showing localized functional areas of the cerebral cortex have been made using implanted electrodes, or by studying the bodily functions impaired in stroke victims. These functions include each of the five senses as well as motor control, reading, and communication skills.

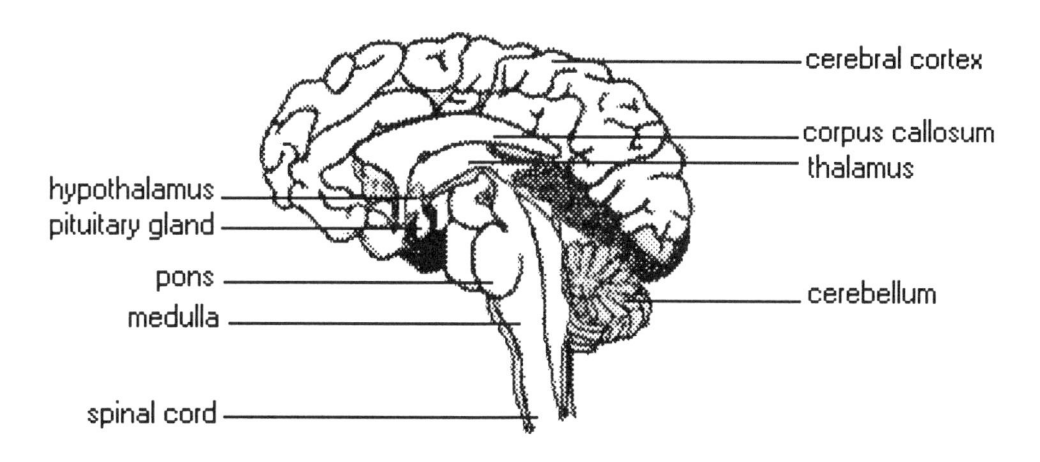

Figure 13-7. Cross-section of the human brain showing major structures.

left and right hemispheres

Communication between cerebral hemispheres is via a complex mass of neural fibers called the *corpus callosum* (Figure 13-7). When this connection was surgically severed in patients suffering from epilepsy, it was found that each hemisphere functions independently. The left hemisphere controls the right side of the body, and vice versa. Thus, in right-handed persons (about 89% of humans), the left hemisphere is dominant and slightly larger. The left hemisphere is specialized for language, speech, and analytical skills, whereas the right hemisphere is associated with form identification, artistic design, and emotional expression. In spite of these specializations, there is still an active functional integration between the two hemispheres via the corpus callosum. The **thalamus** (Figure 13-7) also functions in information relay within the brain.

The brain has an awesome capacity to collect and process sensory information, and issue commands. Perhaps it could be compared, not to one, but to an array of computers operating in parallel fashion, displaying complex intercommunication. There is also evidence for a hierarchial structure in which some systems monitor and direct others.

hypothalamus: neuro-glandular interface

The **hypothalamus** (Figure 13-7) receives sensory information from the other parts of the central nervous system. In turn, it helps to regulate the internal environment of the body through control of body temperature, heart rate, and blood pressure. The hypothalamus controls these and other internal body processes through its influence upon the autonomic nervous system (Section 13-G). The hypothalamus is also physically connected via neurons and capillaries to the **pituitary gland** (Figure 13-7), an important endocrine gland. As we shall see, the hypothalamus is a key interface between the nervous system and the endocrine system. It is also the seat of certain basic physical drives as hunger, thirst, and sexual desire.

What is the "mind"?

The tremendous complexity of the human brain is evident even from our brief survey. Indeed, you may be asking, "Can my **mind**, the seat of my thoughts, memories, intellectual skills, and emotions really be explained by concepts such as neurons, impulses, transmitters, and hormones?" Because discussions of the brain and human mind have philosophical implications, your answer to that question also depends upon your worldview. [What "thoughts" do you have about this?]

13-F THE SENSORIMOTOR NERVOUS SYSTEM

The *cranial nerves* and *spinal nerves* extending outward from the brain and spinal cord, respectively, constitute the **peripheral nervous system**. As noted in Section 13-B and Figure 13-2, the peripheral nervous system can be divided into the **sensorimotor nervous system** and the **autonomic nervous system**, for ease of discussion.

sensing and responding to stimuli . . .

The sensorimotor (or somatic) nervous system consists of all of the sensory neurons associated with sensory organs, and motor neurons associated with the skeletal muscles and certain glands. Each sense organ and the *dermis* just beneath the *epidermis* of your skin is lined with the receptor endings of sensory neurons for detection of environmental stimuli. Table 13-2 lists the major types of stimuli which humans and vertebrate animals can detect. Stimuli may be classified as either a form of energy or a chemical substance. Stimuli may originate either from the internal or the external environment of an organism. In order to detect a given stimulus, an organism must have specialized sensory neurons with receptors for that particular stimulus.

from inside and outside body.

Sense organs and their specialized sensory cells represent a truly marvelous characteristic of the living condition. These built-in precision instruments are also essential as they enable each animal to locate sources of food, to accomplish sexual reproduction, and to avoid unfavorable environmental conditions and danger.

Table 13-2. Stimuli and Sensory Receptors.

CLASS OF STIMULUS	EXAMPLES	SENSE ORGAN	SENSORY STRUCTURES
Light Energy	visible light	Eye	light-sensitive cells (*rods, cones*) and sensory neurons to process visual information
Mechanical Energy	sound waves body motion	Ear Ear	Sensory hair cells moved by membrane vibration (sound) or shifting fluid (body motion)
	touch	Skin and other parts of body	Pressure-sensitive sensory receptors
	pain	Skin and visceral organs	Naked nerve endings or modified nerve endings
Heat Energy	temperature	Skin	Naked nerve endings or modified nerve endings
Chemical Substances	odor	Nasal chamber	Chemicals dissolve in olfactory sensory cells
	taste	Taste buds of the tongue	Chemicals dissolve in taste taste receptor cells
	substances in body fluids (e.g. CO_2)	Chemoreceptors in blood vessel walls	

distinguishing type of stimulus

When the appropriate stimulus reaches a sensory structure, impulses are generated and fired off to the central nervous system for processing. But how can the brain distinguish an impulse generated by light as opposed to one from sound, odor, or others in Table 13-2? The answer is that sensory nerve pathways are devoted to carrying impulses generated by specific stimuli to specialized regions of the cerebral cortex.

distinguishing *intensity* of stimulus

The all-or-nothing basis for transmission of impulses (Section 13-D) means that the brain cannot determine the intensity of a stimulus (loudness, brightness, etc.) by the amplitude (height) of the action potential spike. Instead, intensity is distinguished by the *frequency* of action potentials being transmitted by a given neuron (number per unit time), the *duration* of the succession of spikes, and the number of neurons carrying impulses.

"screening out" stimuli

There are probably a number of environmental stimuli around you right now that you have been "screening out". Stop a moment and check your peripheral vision, background noises, and other information that has been screened. This feature is made possible, at least in part, by certain chemical transmitters that cause threshold levels of neurons to increase. Once in this mode, a sensory neuron would require a significant change in magnitude of a stimulus before it fires and causes conscious awareness on our part. The benefits of this mechanism are obvious during periods of mental concentration and sleep.

muscular and glandular responses

The sensorimotor nervous system not only monitors the environment via sensory structures but can also execute commands to muscles and glands via motor neurons. The sensory component monitors environmental stimuli, and if locomotion is required as a response to these stimuli, the motor component is activated through motor neurons to the skeletal muscles. This is generally considered as being *voluntary* or conscious control, in contrast to the *involuntary* control by the autonomic nervous system. However, these two peripheral systems work in close harmony. Some involuntary activities can be controlled after a period of training. [Can you name several?]

paths of sensory and motor neurons

Under the direction of the brain, each of 31 pairs of **spinal nerves** controls a separate region of the left and right sides of the body (Figure 13-2). Each pair contains both sensory and motor neurons that provide sensorimotor coordination for a particular skeletal region. Each spinal nerve branches into a *dorsal root* and *ventral root* at the point of attachment to the spinal cord (Figure 13-8a). Sensory neurons make up the dorsal root, and motor neurons, the ventral root. Neurons composing the dorsal and ventral roots, and neurons running up and down the spinal cord are myelinated, and appear as *white matter* (Figure 13-8). The non-myelinated interneurons are located within the spinal cord; hence the name, *gray matter*.

The sense organs and movements of the head, eyes, tongue, jaws, *et cetera* are directly controlled by the brain through eleven (11) pairs of **cranial nerves**, connected to the lower surface of the brain (Figure 13-2). The **vagus nerve** is the only cranial nerve that does not serve the head. Instead, it allows the brain to exercise direct control over the internal organs through the autonomic nervous system (Figure 13-10).

reflex arc

Certain sensorimotor systems are "wired" with fewer synapses between sensory and motor neurons, thus allowing a rapid response or **reflex**. The *knee-jerk reflex* illustrates the simplest **reflex arc**, involving only one synapse between sensory and motor neurons (Figure 13-9). A light tap below the kneecap stretches a tendon attached to a muscle in the thigh. This muscle

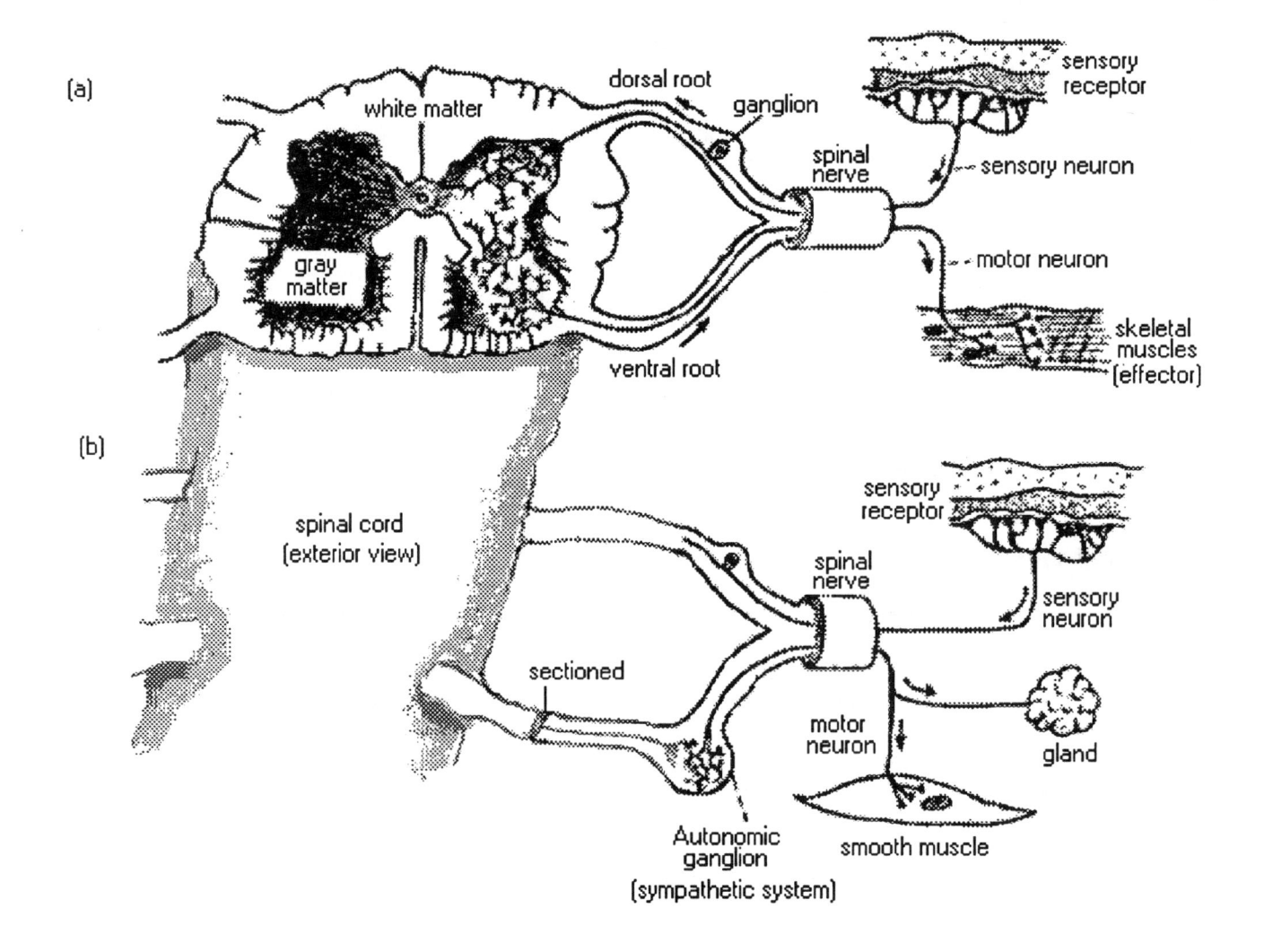

Figure 13-8. Connections between the spinal cord and spinal nerves. (a) Sensorimotor nervous system. (b) Autonomic nervous system (sympathetic division).

monitoring muscular balance

stretch (stimulus) is detected by *stretch receptors* that coil around the muscle fibers. Action potentials are fired off to the spinal cord where one synaptic jump generates action potentials that return on the motor neuron, and the "knee jerk" response occurs. Your stretch receptors also act in maintaining body posture and balance (positional homeostasis). If you lean back on the rear legs of your chair while balancing with your feet, you can "feel" the degree of muscle stretching as you maintain balance. Gravitational receptors in your inner ear are also involved.

Figure 13-9 also illustrates one of many connections between the skeletal system and the muscle effectors. Bones are connected to one another at joints by tough, flexible connective tissue called **ligaments** (Table 13-1). The muscles are attached to bones by similar connective tissue called **tendons**.

muscular contraction

Muscle cells are composed of contractile proteins, *actin* and *myosin* in parallel arrangements that form contractile units called *sarcomeres*. The sequential arrangement of sarcomeres with their alternating dark bands of myosin filaments give muscle a striped appearance as illustrated in Figure 18-1. Movement results when parallel myosin and actin filaments slide past each other in

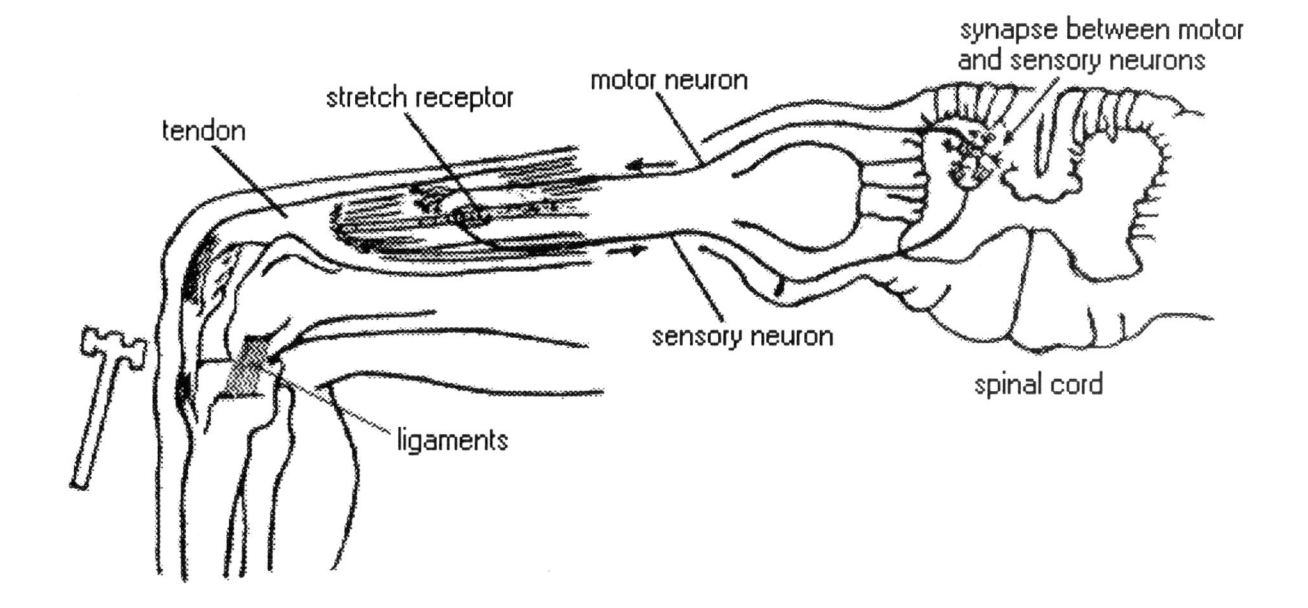

Figure 13-9. A reflex arc illustrated in the knee-jerk reflex.

opposite directions. Muscle contraction is triggered by action potentials from motor neurons.

Movement of the skeletal systems of vertebrates involves an amazing array of mechanical devices. [See TOPICS FOR FURTHER STUDY # 4.] For example, the mechanism for bending and straightening the leg at the knee joint involves the principle of the pulley (Figure 13-9).

13-G THE AUTONOMIC NERVOUS SYSTEM

Organization designed . . .

The **autonomic nervous system** specializes in the control of *involuntary activities* of the internal body organs, and responses associated with varying degrees of physical and emotional stress such as increased heart rate and sweating. The system consists of (a) sensory neurons from the internal organs, (b) the motor neurons that control the smooth muscles of organs, and (c) cardiac muscles of the heart (Table 13-1). Effectors under autonomic control include not only the muscles we have just listed, but also many *endocrine* (ductless) and *exocrine* (duct-containing) glands.

. . . to maintain internal homeostasis.

The principle of internal homeostasis is clearly illustrated by the autonomic nervous system. The system maintains internal stability at levels of activity appropriate for external demands upon the body. The autonomic nervous system is divided into two parts which are coordinated with one another but have opposite effects upon the same body organs. The **sympathetic** division serves to activate body organs and glands for emergencies or times of stress and strong emotional involvement. This *fight-or-flight response* results from a rapid stimulation of all organs involved in body alertness, energy supply and oxygen supply. Sympathetic neurons cause this stimulation by releasing *norepinephrine* at their endings (Section 13-D). Chances are, the sympathetic system was instrumental in the response of our "sleeping student" to the late alarm and the surprise in speech class. The **parasympathetic** division serves to restore body processes following periods of stress. The chemical transmitter, *acetylcholine*, is released at parasympathetic nerve endings.

two divisions: opposite effects

281

autonomic is "wired" unlike sensorimotor

Figure 13-10 illustrates the neural connections between the central nervous system and the two divisions of the autonomic nervous system. The contrasting effects of the two divisions on each organ and gland are also listed. The basis for involuntary control can be seen in the fact that two instead of one motor neurons are involved in the connection between brain or spinal cord and each organ or gland. This contrast is illustrated in Figure 13-8 (a. versus b.). In the sympathetic division, the cell bodies of the additional motor neurons form a massive elongated cluster known as the *sympathetic ganglia*, laying nearby and parallel to the spinal cord. Figure 13-8 shows a cross section of spinal cord and ganglion, while Figure 13-10 illustrates a longitudinal view.

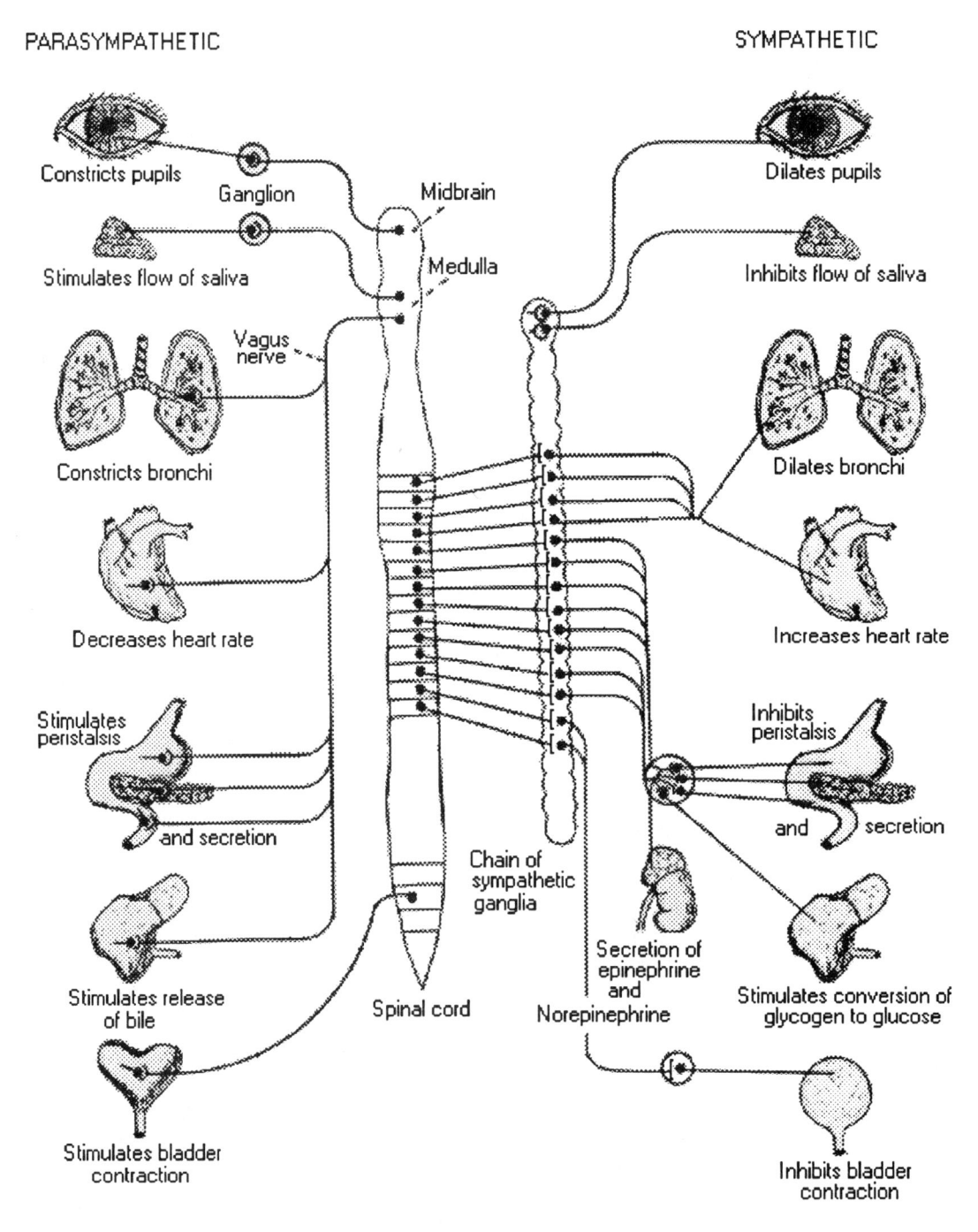

Figure 13-10. A diagrammatic view of the autonomic nervous system, showing the contrasting effects of the sympathetic and parasympathetic divisions.

antagonistic relationship

The parasympathetic division is "wired" differently. It is connected to the brain via cranial nerves (e.g. vagus nerve) and to the lower spinal cord via spinal nerves. The synapse with a second motor neuron occurs near the organs they control, rather than in a chain of ganglia. Figure 13-10 pictures in simple terms the system that coordinates a large number of organs and glands, each having their own crucial contribution to the life of the organism. The opposing effects of the two autonomic divisions provide an antagonistic balance that enables the organism to adjust to changing environmental conditions.

13-H HORMONAL SYSTEMS

How are hormones defined?

Responsiveness and homeostasis in animals depends upon both nervous systems and hormonal systems. Plants rely upon hormones as the principal coordination system. In both plants and animals, **hormones** are defined as chemical messengers that are produced in one part of an organism, and have an effect elsewhere in the organism. Hormones are organic compounds. However, unlike other organic compounds such as sugars that are also transported, hormones produce their effects at very low concentrations. Hormones function in short-term homeostatic regulation, and in growth, differentiation, reproductive cycles, and behavior (in animals).

Hormones have precise targets.

In animals, hormones are produced in small glandular or tissue structures, transported via the circulatory system throughout the organism, and have their effect in specific *target* cells or tissues. In plants, hormones are generally produced in growing tips (meristems), flowers, fruits, leaves, or roots. From there, they are transported cell-to-cell, or by the xylem or phloem.

Hormones act at low concentrations.

The relative structural simplicity of hormonal systems is often a disadvantage to researchers. It is often difficult to identify the precise sites of hormone production, especially in plants, where hormonal glands as such are absent. The difficulty is compounded by the low concentrations of most hormones in tissues which makes detection difficult. A common research technique has been to surgically remove the gland or plant part and observe the effects upon various processes.

plant hormones

The major plant hormones are listed in Table 13-3 along with principal coordinating effects. Most of these plant hormones have a number of physiological effects depending upon the stage of development, plant species, and tissue or organ of the plant receiving the hormone. For instance, auxin can promote stem elongation and inhibit root elongation in the same plant at the same time. Many times one hormone will modify the effect of another, demonstrating that plant hormones do not work individually but "in concert" to coordinate processes within the plant. Several of these antagonistic effects are noted in Table 13-3.

Endocrine glands are ductless.

Animal hormones are generally produced in **endocrine glands** (G. *endo* = inner; within). Endocrine glands have no ducts leading into a body cavity. Instead, hormones are synthesized and secreted from glandular epithelium (Table 13-1) into blood capillaries of the circulatory system for distribution.

animal hormones

Animal hormones vary in chemical structure. Some are modified amino acids (e.g. adrenalin, thyroxin), some are *steroids* (e.g. estrogen, cortisone, testosterone), and many are proteins. The *protein* hormones include insulin and the gonad-stimulating (*gonadotrophic*) hormones.

Table 13-3.   Major Plant Hormones and Principal Effects.

HORMONE	CELLULAR EFFECTS	EFFECTS ON GROWTH AND DEVELOPMENT
Auxins	Cell wall elasticity, and cell elongation Ethylene formation Activity of meristems	Phototropism Gravitropism Fruit development
Gibberellins	Cell division in meristems Reverses effects of abscisic acid	Stem elongation Breaks dormancy in seeds and buds Fruit enlargement
Cytokinins	Cell divisions Influences nutrient translocation	Leaf development Fruit enlargement Delays senescence
Abscisic Acid	Affects protein synthesis Reverses effects of auxins, cytokinins, gibberellins	Seed and bud dormancy Abscission Stomatal opening
Ethylene	Affects cell membranes and metabolism Radial enlargement of cells	Fruit ripening Promotes leaf senescence Inhibits hypocotyl hook straightening

Figure 13-11 illustrates the principal endocrine glands of the human body. Table 13-4 outlines the principal hormonal secretions of these glands and their functions. In Chapter 14, we shall observe hormones in action as they influence responsiveness and homeostasis.

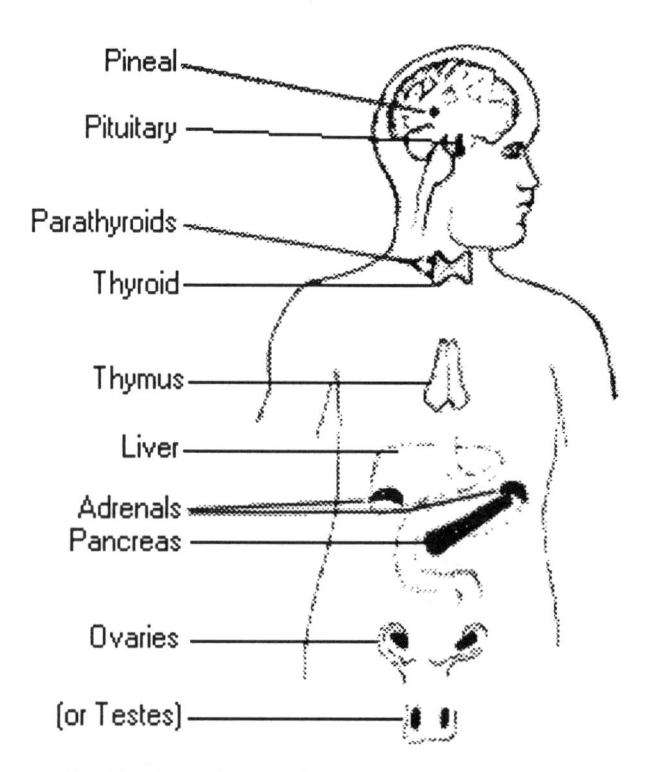

Figure 13-11. Bodily locations of the endocrine glands. The liver and stomach are included for reference.

Table 13-4. Major Endocrine Glands, Hormones, Targets, and Functions.

GLAND	HORMONE	TARGET AND MAJOR FUNCTION
Hypothalamus	Several releasing and inhibiting hormones[1]	Anterior pituitary—regulates secretions
Anterior Pituitary	Growth hormone	Skeleton and muscle—regulates growth
	Gonadotropic hormones (FSH and LH)	Regulate gamete production and sex hormone production
	Thyroid-stimulating hormone (TSH)	Thyroid gland—stimulates secretions
	Corticotropin (ACTH)	Adrenal cortex—stim. adrenal steroids
	Prolactin	Mammary glands—stimulates lactation
Posterior Pituitary[2]	Antidiuretic hormone (ADH)	Kidney nephrons—promotes water reabsorption
	Oxytocin	Uterus; Mammary glands—smooth muscle contract. for birth and milk flow
Thyroid	Thyroxine	Most cells—regulates cell. respiration, growth, development
	Calcitonin	Bone—decreases blood Ca^{++} level by inhibiting release from bone
Parathyroid	Parathyroid hormone	Bone, intestine, kidney—increases blood Ca^{++} by release from bone, and kidney reabsorption
Pancreatic Islet Cells	Insulin / Glucagon	Liver and elsewhere—insulin decreases and glucagon increases blood glucose
Adrenal Medulla	Epinephrine (adrenalin)	Various emergency effects as follow-up to the sympathetic nervous system
Testis	Testosterone	Maintenance of sex organs and male sex characteristics
Ovary	Estrogen	Maintenance of sex organs, female sex characteristics
	Progesterone	Promotes development of uterine lining
Thymus	Thymosins	Lymphocytes—Immune responses
Pineal	Melatonin	Daily rhythms; controls onset of puberty

[1] Hypothalamic neurons secrete *releasing* and *inhibiting hormones* which move via capillaries to secretory cells of the anterior pituitary and promote or inhibit release of each of the anterior pituitary hormones listed above.

[2] Both posterior pituitary hormones, ADH and oxytocin, are synthesized in hypothalamic neurons and transported along axons to the pituitary.

13-I LINKAGE BETWEEN NERVOUS AND HORMONAL SYSTEMS

hypothalamus-pituitary connection

Throughout this chapter, our basic discussion of very complex systems risks conveying that each system operates separate from the others. As we conclude, let's emphasize the close relationship of nervous systems and hormonal systems in responsiveness and coordination. This is most evident in the structure and function of the "hypothalamus-pituitary connection" summarized as follows:

1. *Structurally*, as noted in Section 13-E, hypothalamus and pituitary both contain nervous and circulatory connections to promote information flow between nervous and hormonal systems. Indeed, the *posterior lobe* of the pituitary is largely nervous tissue, and "not a true endocrine gland, but a storage and release center for hormones manufactured entirely in the hypothalamus" (Table 13-4) (8).

2. *Functionally*, the hypothalamus is continually receiving both impulses and hormonal messages from internal body organs via neurons and the bloodstream, respectively. By integrating and responding to this information, the hypothalamus can exercise strategic control over body functions either via hormones or nerve impulses. Note the inseparability of hormonal and nervous system roles in the following hypothalamic activities:

    a. The hypothalamus, ultimately controlled by the brain, in turn controls hormonal secretions of the *pituitary gland*, formerly called the "master gland", because of its broad influence on other endocrine glands. The hypothalamus controls hormonal secretions of the pituitary by producing several *releasing* and *inhibiting hormones*. In addition, two hormones secreted by the posterior pituitary are actually produced by *secretory neurons* in the hypothalamus. See Table 13-4 and footnotes.

    b. The hypothalamus is the seat of the *pleasure center* which controls sensations of hunger, thirst, sexual desire, and other emotional behaviors such as anger and aggression. The mechanism of control involves synaptic junctions of *neurons* in the hypothalamus.

    c. The hypothalamus, along with other brain centers, prepares the body at times of challenge or perceived danger by issuing action potentials along *sympathetic neurons*, quickly causing nerve endings to release *norepinephrine*, which triggers arousal as described in Section 13-G. Among glands aroused is the adrenal medulla of the *adrenal gland* which now responds by releasing a more sustained supply of the norepinephrine and epinephrine (adrenalin), through the slower route of the bloodstream, to enable the body to respond appropriately in "flight-or-fight".

marvelous coordination

In summary, there is amazing coordination even between the two coordination systems! This is made possible through nervous and circulatory interconnections, common hormonal secretions, and the tremendous capacity of the nervous system to gather environmental stimuli and rapidly process both nervous and hormonal information. If you can comprehend and explain these processes to the extent that we have explored them, you are ready to proceed into the study of homeostasis and coordination of life processes, in Chapter 14.

"For as the body is one, and hath many members, and all the members of that one body, being many, are one body: so also is Christ".

QUESTIONS AND DISCUSSION TOPICS

1. What is meant by *responsiveness* and *homeostasis*? How are these attributes crucial to being "alive"?

2. What are the common features of the two major types of coordination systems? In what general ways do they differ?

3. How does the nervous system of *Hydra* differ from that of earthworm? How is the human nervous system unique?

4. Explain the structural and functional distinctions among the various parts of the human nervous system (Figure 13-2). Why are clear-cut distinctions difficult to make?

5. Distinguish the three types of neurons in structure and function. What are the most abundant cells in the brain? How do they function?

6. Distinguish the following: impulse, resting potential, action potential, threshold.

7. Where do impulses originate? How are they transmitted from one neuron to another?

8. How does being separated by synapses influence the way neurons process information (input and output)? Discuss several effects of chemical transmitters?

9. Explain why neurons operate on an "all-or-nothing" basis. If this is true, how can one distinguish among types of stimuli (e.g. light from sound)? Intensity of the stimulus?

10. How is the sensorimotor nervous system different from the autonomic nervous system in structure and function?

11. How does the nervous system of humans interface with the endocrine system? Analyze the case for the hypothalamus being at the center of nervous and hormonal control. Is it the ultimate "authority"? Explain.

TOPICS FOR FURTHER STUDY

1. What is the role of neurotransmitters in controlling human behavior? How are these being used a s drugs to modify human behavioral conditions such as psychotic depression? What are the spiritual implications? REFERENCES: #7, #9, #13

2. How do various narcotic drugs affect the human nervous system? REFERENCES: #7 and #10

3. What are the major scriptural teachings and exhortations in regard to the mind of the Christian? What is to be the controlling force within the life of the Christian? See also TOPIC #6.

4. What mechanical engineering designs were introduced in the creation of the human skeletal system? How do animal skeletal systems differ? REFERENCES: #2, #11, and #14

5. Conduct a study of "virtual reality", the up-and-coming, interactive computer technology that creates convincing illusions that one is immersed in a virtually real world that offers "full-body participation" in a choice of "experiences". What are the implications?
REFERENCES: #15 and #17

6. What biological processes are involved in meditation? REFERENCE #18

7. Examine the biology and purpose of pain in the human experience. REFERENCE #20

8. Is homosexuality inborn? Are there identifiable structural and physiological differences that *cause* homosexual behavior, or does the behavior *cause* these differences? If inborn differences exist, does that justify homosexuality as "natural" and appropriate? REFERENCES: #6 and #12.

9. Explore the technology and implications of *artificial intelligence.* REFERENCES: #5 and #19

REFERENCES

1. Andrusko, D. 1992. Adult brain may be capable of self-repair. *Nat. Right to Life News*, April 7, p .16.

2. *Animal Engineering.* 1974. Readings from *Scient. Amer.* W.H. Freeman. San Francisco, CA. [Many classic articles on mechanical and chemical engineering, and related physiology of animal systems.]

3. Behavioral Endocrinology: Special Issue of *Bioscience*, 33(9), Oct., 1983.

4. Bernstein, R. and S. Bernstein. 1982. *Biology: The Study of Life*. Harcourt, Brace, Jovanovich., New York.

5. Collins, H. 1992. Will machines ever think? *New Scient.* 134: 36ff [Other related articles]

6. Dallas, J. 1992. Born gay? *Christianity Today,* Jun. 22, pp 20-23.

7. Fischbach, G.D. 1992. Mind and brain. *Scient. Amer.* 267(3): 48-57. [Special issue with ten other articles on the brain.]

8. Hickman, C.P. Jr., *et.al.* 1990. *Biology of Animals.*, 5th ed., Times Mirror/Mosby College Publ., St. Louis, MO.

9. Johnson, W.C. 1980. Depression: Biochemical abnormality or spiritual backsliding? *J. Amer. Scient. Affiliation* 32(1):18-27.

10. Julien, R.M. 1985. *A Primer of Drug Action*, 4th ed. W.H. Freeman and Co, New York.

11. Kaufmann, D.A. 1981. Mechanical design in the human body. *Creat. Res. Soc. Quart.* 18(3):155-158.

12. LeVay, S. 1991. A difference in hypothalamic structure between heterosexual and homosexual men. *Science* 253: 1034-1037.

13. Littleton, M.R. 1981. Depression: The chemical side. *Moody Monthly*, October, pp.127-129.

14. McMahon, T.A. 1984. *Muscles, Reflexes, and Locomotion.* Princeton Univ. Press. Princeton, NJ.

15. Rheingold, H. 1991. *Virtual Reality.* Summit Books. New York, NY.

16. Salisbury, F.B. and C.W. Ross. 1991. *Plant Physiology*, 4th ed., Wadsworth, Belmont, Cal.

17. Wagner, D. 1992. Cyberspace: The last frontier. *Currents in Sci. & Tech.* 1(2): 12-19. [Two articles, one an interview]

18. Wallace, R.K. and H. Benson, 1972. The physiology of meditation. *Scient. Amer.* 226(2):85.

19. Wallich, P. 1991. Silicon babies. *Scient. Amer.* 265(6): 124-134.

20. Yancey, P. 1977. *Where Is God When It Hurts?* Zondervan, Grand Rapids, MI.

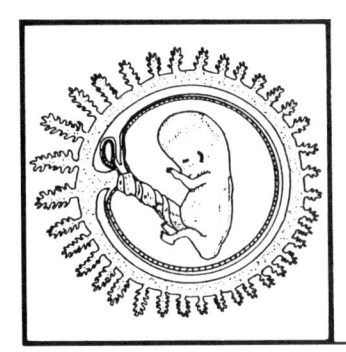

Chapter 14

Homeostasis and Coordination of Life Processes

Living systems are highly organized. Their structural organization is visibly evident in specialized cells that form tissues and organs. These specialized *structures* act as centers wherein specific *life functions* are performed. Structural interconnections permit transport of nutrients, impulses, and hormones between these centers. Precise structural and functional organization permits an amazing array of processes to occur in predictable fashion, all within one living package.

The external environment is variable.

The *internal environment* of living organisms must be precisely and consistently regulated while, at the same time, they face a variable and unpredictable *external environment*. Daily and seasonal fluctuations in light, heat, moisture, mineral nutrients, and food supply are common occurrences. Yet living organisms must be continually exposed to these life-supporting factors. When highly organized, smoothly-functioning, living organisms encounter the choppy, fluctuating environment, "something has to give". Each organism must either regulate its demands and internal cellular environment within tolerable limits in the midst of environmental fluctuations, or it will die. In other words, each organism must maintain **homeostasis** to remain alive.

Regulate within limits or die.

In Chapter 13, we examined the two principal coordination systems—nervous systems and hormonal systems. Now we shall consider the manner in which these systems interact to maintain internal homeostasis.

14-A THERMOREGULATION—HOMEOSTATIC REGULATION OF TEMPERATURE

The metabolic processes of living cells are very sensitive to temperature changes. Therefore, **thermoregulation** is a prime example of the encounter between living organisms and the choppy, fluctuating environment.

adaptations to temperature

All species have certain major structural, functional, and behavioral features that suit them to a particular environment. These **adaptations** (Section 8-A) are traits which each individual inherits as a member of a population. For example, the snowshoe hare and balsam fir tree of the taiga biome are taxonomically and ecologically distinct from the jackrabbit and sagebrush of the desert biome. The ears of the jackrabbit are much larger than those of the snowshoe hare. Large ears are an adaptation to promote radiation of heat in order to cool the jackrabbit's body during hot days on the desert. The needle-like leaves of the balsam fir are designed to withstand freezing temperatures and dry air during the taiga winter. These plant and animal

natural selection—regulation over generations

adaptations were programmed into each created kind of organism. In Chapter 15, we shall examine *natural selection* which "fine tunes" each population within genetic limits by favoring those individuals that are suited to the changing environment.

regulation in the individual

Whereas natural selection maintains a type of homeostatic relationship between *population and environment* over many generations, our focus in this chapter is upon the homeostatic regulation within *individual organisms* facing changes in their environment during their lifetimes. Thus, the balsam fir tree and other "evergreens" have needle-like leaves that respond to cooling autumn temperatures by developing a freeze-resistant or freeze-tolerant cytoplasm. Rhododendron leaves roll up in winter to conserve heat and moisture. Other species can orient their leaves in relation to incoming solar radiation in order to regulate leaf temperature (Figure 14-1). Deciduous trees drop their leaves in the autumn instead of maintaining them as frost-resistant organs. Annual plants produce dormant seeds that can survive unfavorable seasons by withstanding temperature extremes and severe draught.

ectotherms rely on external sources of heat

Animals must either allow their body temperature to fluctuate with changes in external temperatures, or possess mechanisms to conserve or generate heat within their bodies. Animals that rely on external sources of heat, instead of metabolically generated heat, to control body temperature are called **ectotherms**. These "cold-blooded animals" include most invertebrates, fish, amphibians, and reptiles. You have no doubt observed the correlation between air temperature and the activity of bees, flies, and other insects. Lizards are noted for their sun-basking and shade-seeking behavior which allows them to maintain body temperature within an appropriate range. Some lizards can supplement external heat sources with internally generated heat resulting from an increased metabolic rate.

endotherms rely on metabolic heat generation

Most mammals and birds are classed as **endotherms** ("warm-blooded") because they use cellular metabolism to maintain homeostatic body temperature at a relatively high level. This feature allows endotherms to be active under a wider range of outdoor temperatures, provided they can obtain the larger quantities of food energy to sustain the higher body temperature through metabolism.

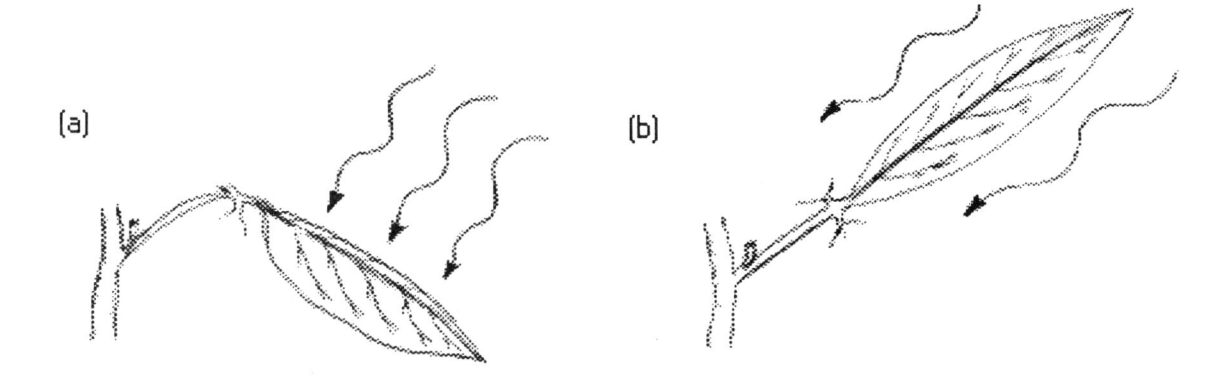

Figure 14-1. Homeostatic regulation of leaf temperature. (a) Normal position when leaf temperature and air temperature are not high. (b) Under high temperatures, the leaf avoids excessive heat absorption by orienting within a plane parallel to the plane of incoming solar radiation.

Humans are endotherms and maintain body temperature within a narrow range from 96° F (35.6° C) to 99.9° F (37.7° C), and the average temperature variably reported from 98.2° F to 98.6° F. Human body temperature regulation is typical of most homeostatic mechanisms within organisms which require the following:

requirements for homeostatic regulation

1. A *parameter* to be controlled; in this case, it is temperature of body fluids and cells (Figure 14-2).

2. A *set point*, the normal or optimal level of the parameter which is genetically and physiologically programmed.

3. *Sensory receptors* to monitor fluctuations or drift of the parameter away from the set point, and to relay *feedback* messages to...

4. ...a *control center* to integrate feedback messages, and initiate corrective measures to oppose "drift" away from the set point.

negative feedback

When a product of a system acts as a feedback message to inhibit its own production, this is called **negative feedback**. Corrective measures brought about by the control center act to hold the parameter within a tolerable range. Figure 14-2 is a model of a typical homeostatic system containing the three major components, as related to thermoregulation.

Human body temperature is generally either above or below the fluctuating environmental temperatures. Therefore, if body temperature is to be maintained within the narrow range near 98.6° F, heat must be continually supplied from within the body and/or radiated to the environment. The homeostatic regulation of body temperature produces a *steady state*, or predictable fluctuation of temperature slightly above and below 98.6° F, not a *static state* of exactly 98.6° F. Indeed, if the parameter were not allowed to "drift" away from the set point as illustrated in Figure 14-2, there would be no internal stimulus to trigger the corrective measures.

***homeostatic* is not "static"**

hypothalamus: control center

The hypothalamus is the control center in human thermoregulation (Sections 13-E, 13-H, 13-I; Figure 13-7). It receives impulses from thermoreceptors in the skin and from its own thermoreceptors that monitor blood temperature. Based upon this information, the hypothalamus will initiate corrective measures as needed to lower body temperature away from the upper limits or to

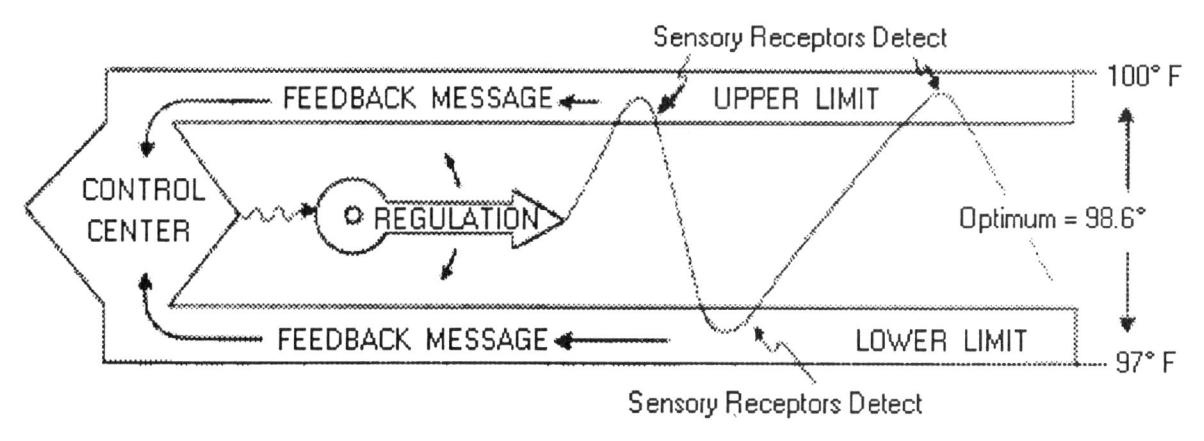

Figure 14-2. A model of the homeostatic regulation of body temperature.

What causes a fever?

raise body temperature from lower limits. Table 14-1 lists the various heat releasing and heat-generating mechanisms that are controlled by the hypothalamus in thermoregulation.

In response to infections, phagocytes (Section 10-H) and other cells of the immune system release chemical signals such as *interleukin-1*. This protein causes the hypothalamus to raise the set point for body temperature. Homeostatic mechanisms (Table 14-1) go into action to raise temperature. The result is a **fever** which, though apparently beneficial to the bodily immune response, can be damaging or fatal beyond 106° F. Aspirin and other drugs can counter the effects of interleukin-1 and lower the set point.

Path of blood regulates heat.

The path of blood flow between the trunk of the body and extremities is a major variable in controlling body temperature. The hypothalamus sends impulses though the autonomic nervous system to smooth muscles in the walls of blood vessels. Vasoconstriction of blood vessels controls the path of blood flow. To conserve internal body heat, blood returning in veins from the cooler arm and leg extremities can be directed to flow alongside the outgoing arterial blood. This is called *countercurrent exchange*, since outgoing heat is exchanged between the two currents. On the other hand, heat loss can be increased when venous blood returns near the skin surface of arms and legs. The surface flow in veins causes the veins to visibly swell above the usual skin profile.

Table 14-1. Thermoregulation in the Human Body Under Two Different Situations.

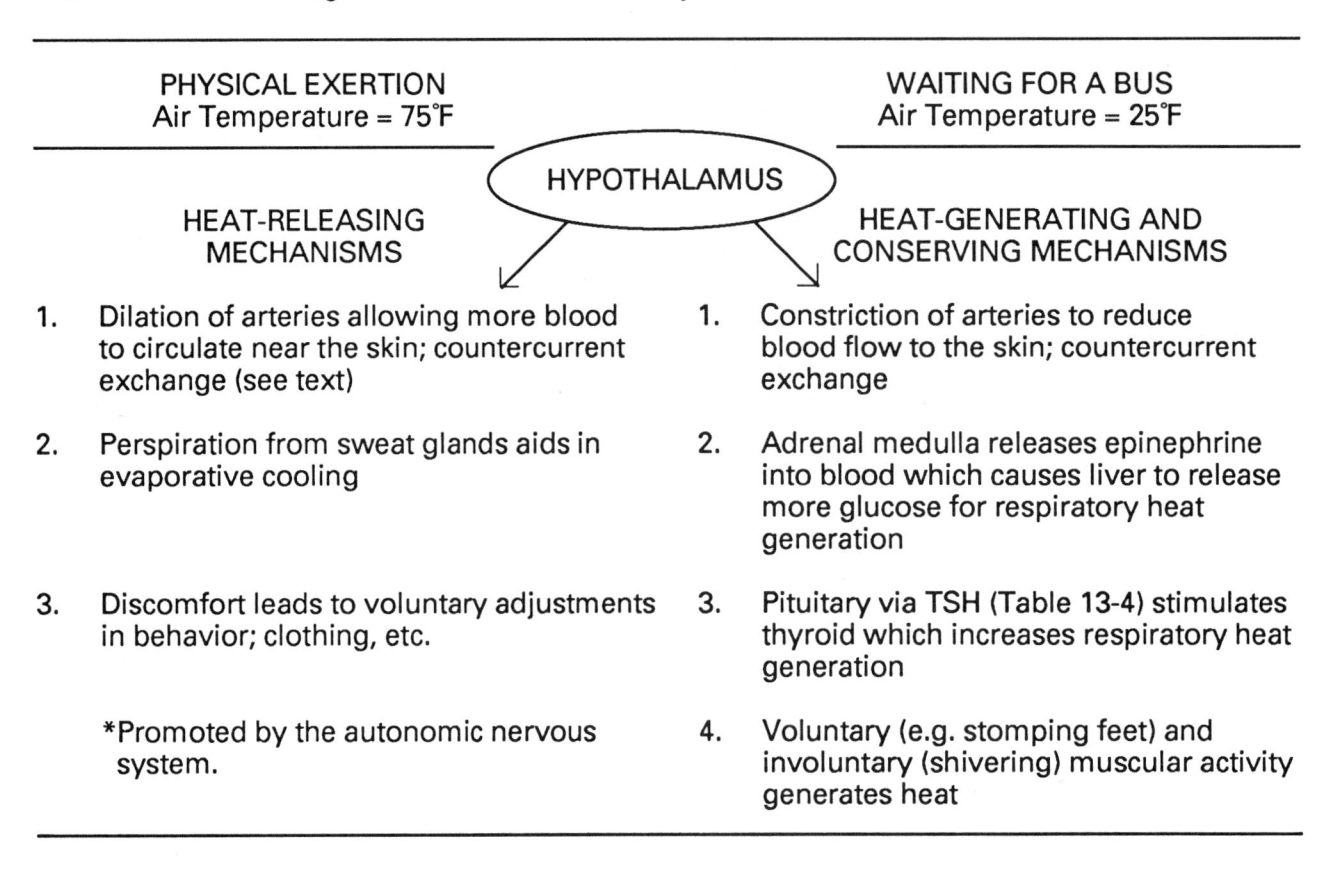

PHYSICAL EXERTION Air Temperature = 75°F HEAT-RELEASING MECHANISMS	WAITING FOR A BUS Air Temperature = 25°F HEAT-GENERATING AND CONSERVING MECHANISMS
1. Dilation of arteries allowing more blood to circulate near the skin; countercurrent exchange (see text)	1. Constriction of arteries to reduce blood flow to the skin; countercurrent exchange
2. Perspiration from sweat glands aids in evaporative cooling	2. Adrenal medulla releases epinephrine into blood which causes liver to release more glucose for respiratory heat generation
3. Discomfort leads to voluntary adjustments in behavior; clothing, etc.	3. Pituitary via TSH (Table 13-4) stimulates thyroid which increases respiratory heat generation
*Promoted by the autonomic nervous system.	4. Voluntary (e.g. stomping feet) and involuntary (shivering) muscular activity generates heat

(HYPOTHALAMUS)

14-B HOMEOSTATIC REGULATION OF BLOOD GLUCOSE

controlling "sugar surges"

Organisms that obtain their nutrition by ingestion and digestion of food are, in effect, bringing a fluctuating variable into their digestive tracts. This is probably best illustrated by the person who ingests a meal high in sugars and starch. Suddenly, a high concentration of sugars such as glucose is accessible to the bloodstream from the villi of the small intestine (Section 10-G). If blood glucose levels were allowed to fluctuate along with the levels in the digestive tract, serious problems would result. Instead, glucose is normally controlled within a certain range (80 to 120 milligrams/100 milliliters of blood). We have already noted that the hypothalamus operates via control over the adrenal gland as one means of regulating blood glucose levels (Table 14-1).

The pancreas is an *exocrine* (duct-containing) gland with internal islands of tissue called the **pancreatic islets** (Section 13-H). The pancreatic islets function as *endocrine* (ductless) glands to control blood glucose level.

absorption and storage

Glucose released from digested carbohydrates is absorbed by blood capillaries in the villi of the small intestine and carried directly to the liver via the *portal vein*. Here, glucose is converted to *glycogen*, a storage polysaccharide (Section 4-G). The liver serves as a sort of fuel tank from which glucose is released as a source of chemical energy.

glucagon

The pancreatic islets function as a homeostatic control center to regulate blood glucose levels. When glucose levels are lower than optimum, certain islet cells release a hormone called **glucagon** into the bloodstream (Figure 14-3). Like all hormones, glucagon travels within the bloodstream and has a target for its regulatory effect, namely the liver. Glucagon stimulates glycogen breakdown into glucose which enters the bloodstream. It also stimulates breakdown of lipid reserves into fatty acids which serve as an energy source.

insulin

disease

If blood glucose levels rise, as they do after meals, the pancreas releases a different hormone, **insulin**, which causes the liver to retain glucose in the form of glycogen. Some of the sugars may be converted to lipids. Figure 14-3 illustrates the manner in which these two antagonistic hormones regulate blood glucose levels. Both insulin and glucagon are small protein molecules representing chains of 51 and 29 amino acids, respectively. Diabetics suffer from a deficiency in insulin production, or from unresponsiveness of target cells to insulin. In some cases the *disease* can be relieved by regular injections of insulin. [How would you define *disease* in relationship to *homeostasis*? [See TOPICS FOR FURTHER STUDY #2.]

14-C REGULATION OF BASAL METABOLIC RATE

"idling" rate

Each person has a certain **basal metabolic rate** (BMR) which represents the rate of respiration when at complete rest. The BMR can be estimated by measuring O_2 consumption by a sleeping person. It represents the energy consumption rate when only the basic metabolic processes are occurring (e.g. involuntary muscle contractions—heartbeat and breathing; active transport across membranes, etc.).

A persons rate of glucose consumption is proportional to his or her BMR. [Can you explain this by the equation for respiration? (Section 14-E)] Recall

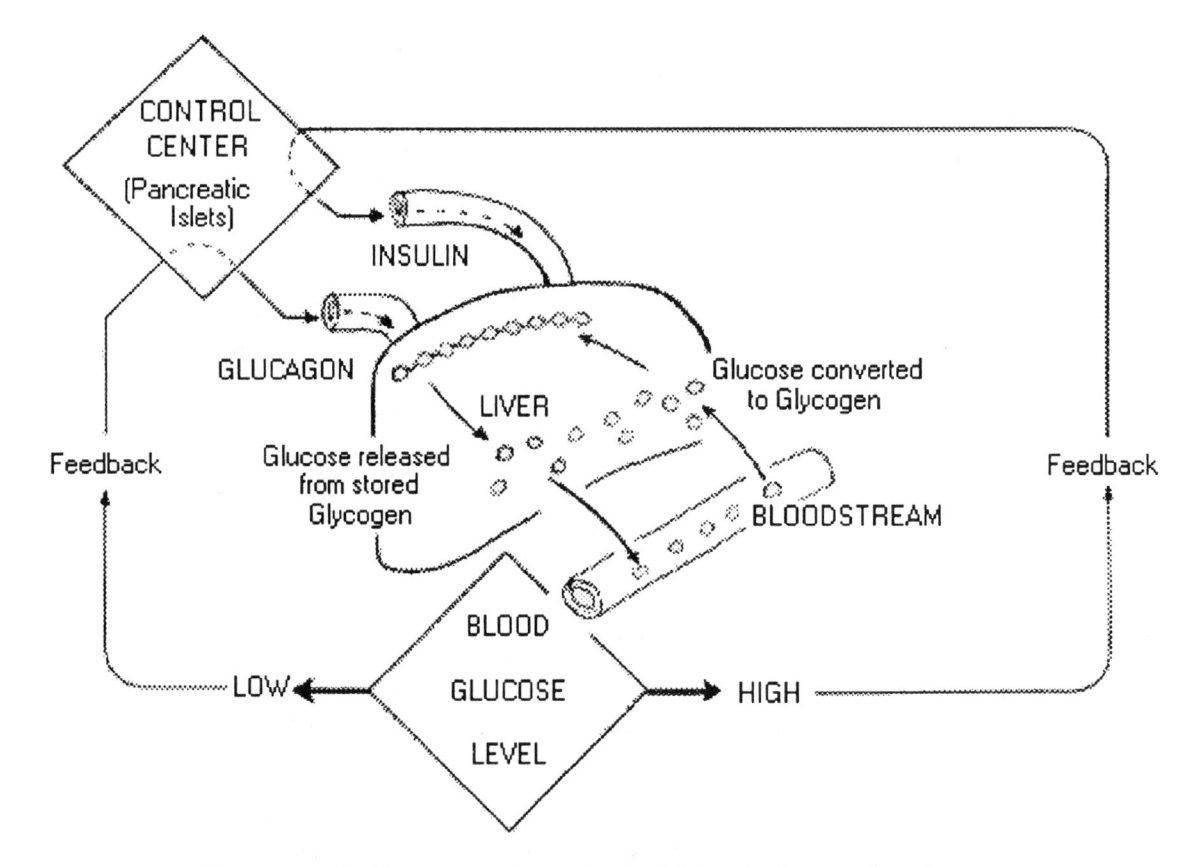

Figure 14-3. Homeostatic control of blood glucose level.

that the pancreas controls blood glucose level. However, BMR is under the control of another gland, the thyroid gland (Table 13-4). **Thyroxine** and other thyroid hormones control the rate of glucose oxidation (respiration) by cells.

BMR imbalances

Hyperthyroidism and *hypothyroidism* are homeostatic imbalances related to the thyroid gland. Hyperthyroidism is due to an overactive thyroid which leads to rapid metabolism, weight loss, and nervousness. Hypothyroidism is caused by a lack of thyroid hormones and is manifested in extreme sluggishness and weight problems. In the long term, thyroid imbalances may also affect human growth and development.

iodine in diet

An enlargement of the thyroid gland, located in the neck region, is called **goiter**, which is often due to a dietary deficiency of iodine, a component of thyroxine. Iodine shortage limits thyroxine secretion and this shortage triggers the pituitary to increase secretion of *thyroid stimulating hormone*, which causes thyroid enlargement (Table 13-4).

14-D KIDNEYS AND OSMOREGULATION

Water is the basic medium within which all metabolic processes occur. All of the nutrients, vital gases, sugars, enzymes, and organelles of living cells are suspended in this marvelous medium. Because water readily diffuses in and out of cells, each cell can tolerate only a relatively narrow range of water concentrations in the medium surrounding its membranes. Thus, bacteria and protista, as unicellular organisms, are active only under certain aqueous conditions. Plants wilt and cease to grow when soil water is insufficient to maintain turgor pressure. Multicellular animals have elaborate mechanisms for circulating fluids through their bodies. You will recall the role of the blood

osmoregulation

plasma in bathing body cells with the proper concentrations of glucose, amino acids, and salts (e.g. Na⁺, Cl⁻, K⁺), while flushing away the waste products. **Osmoregulation** is the homeostatic regulation of the concentrations of water and dissolved substances within the body fluids. In vertebrate animals, the **kidneys** are mainly responsible for osmoregulation. Kidneys also handle the excretion of excess water, salts, and dissolved waste products.

freshwater life

The quality of water available differs among freshwater, saltwater, and terrestrial habitats. Each of these habitats presents its own challenges. Freshwater animals live in relatively pure water with a low concentration of salts. The water tends to diffuse into the bodies of these animals from its higher concentration (purity) outside the body. Salts needed by the animal also diffuse in slowly. For instance, water and small amounts of dissolved salts continually enter the body of freshwater fish through the gills. Their kidneys serve to excrete excess water and retain salt for bodily use. Freshwater fish "drink" very little water.

saltwater life

Saltwater animals must deal with the relatively low concentration (purity) of water in their environment relative to their body fluids. Because of this relationship, water of greater concentration in the body fluids tends to be lost by diffusion into the saltwater exterior. On the other hand, salts tend to diffuse into the animal cells from their higher concentration in the saltwater. Therefore, the saltwater animal must retain water and excrete salts. Saltwater fish do drink saltwater!

What determines direction of diffusion?

Note that both fresh and saltwater animals must meet water on its own terms. Water can readily diffuse across cell membranes by a process called **osmosis**. However, the direction of water diffusion across a membrane depends upon which region has the highest water concentration (purity) or the lowest concentration of dissolved salts. The higher the salt concentration, the more water molecules that are needed to form *spheres of hydration* (Figure 4-6). These molecules are "tied up" and not free to diffuse. Figure 14-4 illustrates the osmotic behavior of water in regard to cells in freshwater and saltwater.

terrestrial life

Terrestrial environments present a third type of challenge to osmoregulation. Terrestrial animals must conserve water, which is often in scarce supply; and, they must maintain the proper salt balance in their body fluids. Although freshwater, saltwater, and terrestrial animals have quite different osmoregulatory problems, their kidneys are remarkably similar in structure and function. In each case, osmoregulation is accomplished by the excretion of varying amounts of salts, nitrogenous wastes (e.g. urea), and water. Thus, the kidneys and associated organs are viewed as an **excretory system**.

overview— excretory system

The human excretory system consists of a pair of *kidneys*, each of which drain *urine* via *ureters* into the *urinary bladder* (Figure 14-5a). Urine leaves the body from the bladder via the *urethra* (Figures 11-17, 11-18). Blood enters each kidney through a **renal artery** and exits via the **renal vein** (Figure 14-5b). However, between these two points, the kidneys produce a major change in the composition of the blood. Excreted salts, urea, and water are removed from the blood in precise amounts and are carried via the ureter to the bladder.

kidney structure

A close-up view of the kidney in cross-section reveals tiny blood-filtering units called **nephrons** (Figure 14-5c). Blood entering each kidney is distributed among the approximately one million nephrons through repeated branching of the renal artery into arterioles. At each nephron, an arteriole leads to a dense ball of capillaries known as a **glomerulus** (L. *glomer* = ball of yarn).

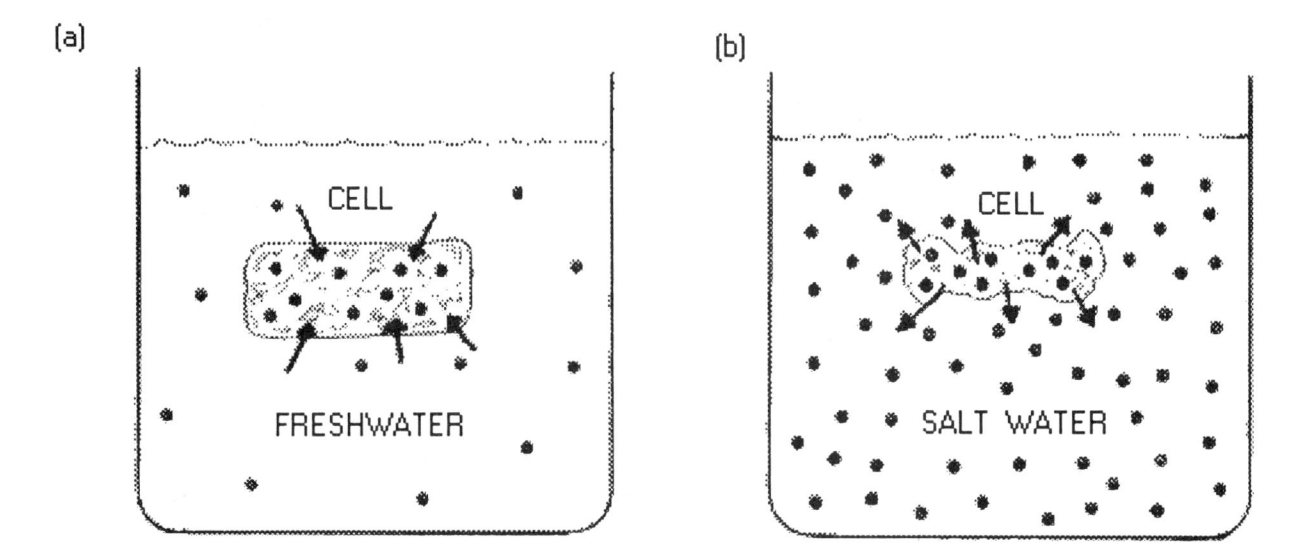

Figure 14-4. The direction of osmosis (—>) depends upon the relative water concentration on either side of the cell membrane. (a) Cells of freshwater organisms are usually surrounded by purer water (higher concentration) which means water tends to diffuse in across the cell membrane. Freshwater organisms must actively excrete water.
(b) Saltwater organisms maintain purer water within their cells than that of the salty water surrounding them. Water tends to diffuse outward. Saltwater organisms must retain water and excrete salts. Dissolved salt ions are represented by darkened circles.

Beyond the glomerulus is an extensive network of capillaries, the **capillary net**. The capillary net leads to a branch of the renal vein through which blood is returned to the heart. Thus, the circulatory vessels include artery, glomerulus, capillary net, and vein (Figure 14-5).

parts of a nephron

Each nephron begins at a **Bowman's capsule**, a hollow cup surrounding each glomerulus (Figure 14-5c). From Bowman's capsule, a nephron has the following parts: proximal tubule, loop of Henle, distal tubule, and collecting duct. The collecting duct empties into the ureter which leads to the bladder.

Having considered the path of blood vessels and excretory tubules, the processes of osmoregulation and excretion occur in the following phases:

1. **FILTRATION** occurs when blood enters the glomerulus under pressure high enough to filter the dissolved salts and small molecules such as glucose, amino acids, and urea through the capillary walls into Bowman's capsule. The blood cells and plasma proteins remain in the bloodstream as it moves into the capillary net. About two-thirds of the blood fluid volume is filtered into these tubules every hour. To avoid dehydration, it is crucial that about 99% of the water, as well as glucose, amino acids, and a portion of salt ions be reabsorbed into the bloodstream. Without this "recycling" of water you would need to consume (and urinate) almost the equivalent of one 50-gallon drum of water daily!

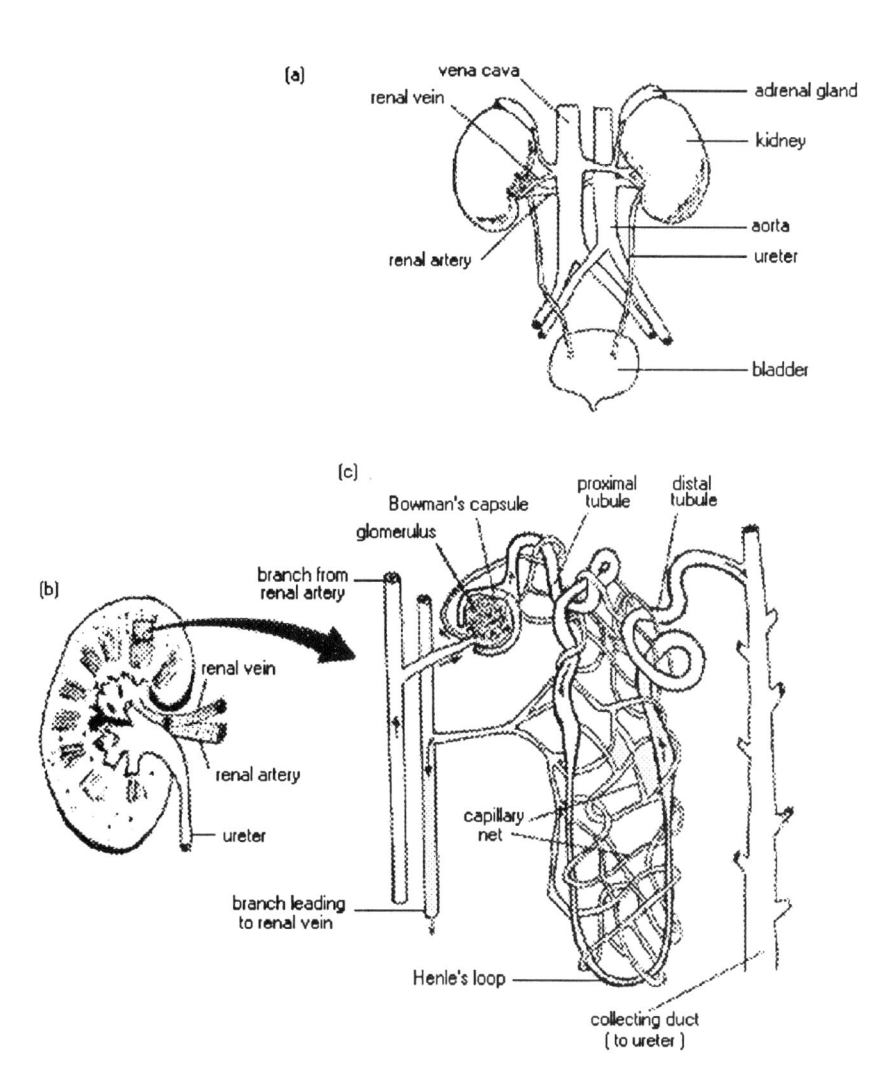

Figure 14-5. The human excretory system: (a) General anatomy (b) Cross-section of kidney (c) Diagram of a nephron

Active transport

promotes osmosis...

2. **ACTIVE TRANSPORT** occurs where the **proximal tubule** forms an intimate contact with the capillary net. Here, energy-driven membrane pumps (Section 13-D) transport glucose, amino acids, and salts through the thin tubule walls to the surrounding fluids. These substances reenter the blood by diffusion into the capillary net. Normally, the capillary net reabsorbs all the glucose, amino acids, and about 80% of the water and salts. This enriched blood has a very low water concentration. This low concentration causes about 80% of the water lost in filtration to return to the capillaries by osmosis.

but not all water reenters.

Urea and other toxic wastes, some salts, and about 20% of the water remain in the tubule leading to the bladder. To this waste is added unfiltered urea, salts, and certain drugs and toxins which are actively secreted from the capillary net. Essential ingredients of the bloodstream have been separated from the wastes remaining in the tubule. However, body cannot afford to loose 20% of the water in body fluids.

Henle's loop, a U-shaped region between the proximal tubules and the **distal tubules** (Figure 14-5), is the site of additional water recovery from the tubules into the capillary net. Notice the lengthened pathway made possible by the

rescuing more water

descending tubule and its "hairpin turn" before ascending to the distal tubule, then a return descent via the **collecting duct** leading to the bladder. The liquid medium around Henle's loop has a high concentration of salts, causing water in the descending arm of the loop to diffuse out of the tubule and become accessible to reabsorption by the capillary net. Salts are actively transported from the ascending arm, leaving a dilute fluid to enter the collecting duct (Figure 14-5c).

hormonal control of blood water

3. **HOMEOSTATIC ADJUSTMENT** of the water balance of the bloodstream is under the control of the hypothalamus as shown in Figure 14-6. Osmoreceptor cells of the hypothalamus monitor the water concentration of the blood. If these receptors detect a low amount of water in the blood, they cause the pituitary gland to release more **antidiuretic hormone**, or **ADH**, into the bloodstream. The ADH increases the permeability of the collecting duct walls to water. More water escapes from the urine in the duct to be reabsorbed into the bloodstream. When the hypothalamus detects normal or high blood water levels, it causes the pituitary to secrete less hormone, causing lowered tubule wall permeability and less water reabsorption from the urine. Blood volume is under similar control through cardiac receptors and hormonal control of sodium balance. Hormones from the adrenal glands (Figure 14-5a) regulate the rate of active transport of sodium from the distal tubule into the bloodstream.

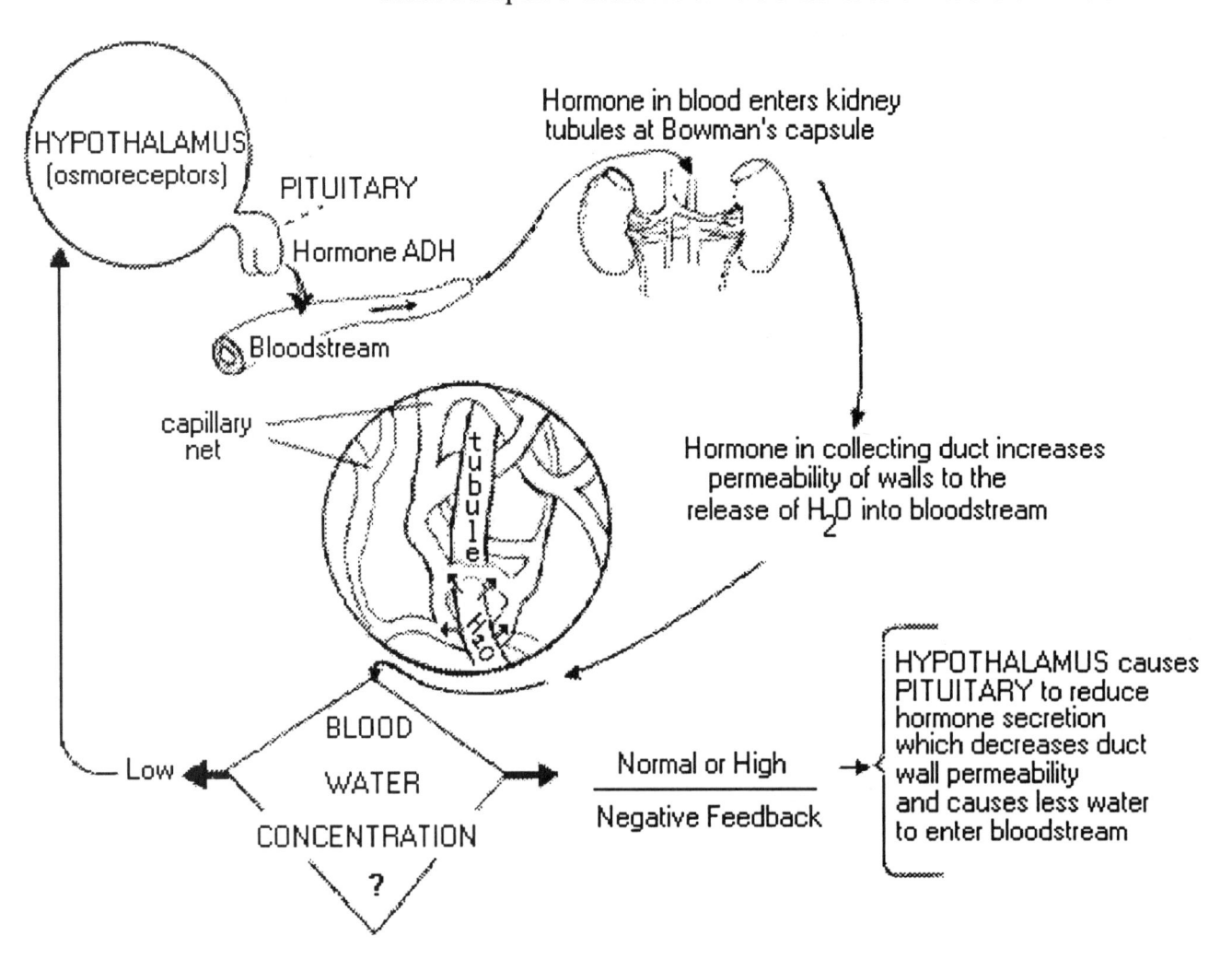

Figure 14-6. Homeostatic control of blood water content.

Summary:

In summary, the kidneys function as blood-filtering devices under the influence of blood pressure. Following this pressure-filtering process at the glomerulus, active transport in the tubules restores necessary nutrients to the blood, and removes wastes and toxic substances. Blood water concentration and blood volume are homeostatically controlled at the collecting duct through hormones.

kidney problems

Ingestion of certain chemicals in our diet affects osmoregulation. Alcohol is considered a *diuretic* because it inhibits secretion of ADH by the pituitary. [Use Figure 14-6 and the above discussion to explain increased urine production by a person who consumes an alcoholic beverage.] The so-called "hangover" is partly the result of body tissue dehydration. Caffeine also acts as a diuretic by increasing the filtration rate at the glomeruli.

Certain kidney diseases disrupt osmoregulation, resulting in excess water loss from the body, or accumulation of urea and other wastes in the blood. In the latter case, an artificial *dialysis machine*, which operates on the principle of osmosis, removes toxic wastes from the blood. [See TOPICS FOR FURTHER STUDY #3.]

14-E HOMEOSTATIC CONTROL OF BLOOD GASES

All living cells require a continuous supply of usable energy (Section 5-B). For most organisms, *cellular respiration* converts glucose sugar into usable energy as summarized in the following equation:

$$C_6H_{12}O_6 + 6\,O_2 \longrightarrow 6\,CO_2 + 6\,H_2O + \text{usable energy}$$

Glucose ($C_6H_{12}O_6$) is obtained via photosynthesis in autotrophs, whereas heterotrophs absorb or ingest food as a source of the sugar and other organic compounds.

All organisms that perform respiration according to the above equation must continually acquire O_2 gas. Oxygen acts as the oxidizing agent to degrade glucose molecules so that usable energy is released. Thus, oxygen is vital to life because usable energy is vital to life.

O_2-absorbing structures:

aquatic life

Structures designed to absorb oxygen are remarkably similar among the great diversity of life forms. First, the absorptive surfaces must be kept moist so that gaseous absorption and diffusion can occur. For aquatic organisms, this is easily accomplished because they are surrounded by water. However, terrestrial animals such as amphibians and earthworms that absorb O_2 through their moist skin are restricted to moist environments. Animals that are able to face the rigors of the dry land environment have their moist O_2-absorbing surfaces enclosed within body structures such as lungs.

moist surfaces

Leaves of land plants exchange CO_2 and O_2 through the stomatal openings in the leaf epidermis (Section 10-C). The epidermis of leaves is designed to reduce water loss by enclosing the moist surfaces of the spongy mesophyll cells within the leaf.

large surface area

Oxygen-absorbing structures must not only be moist, but secondly, they must expose a large surface area of very thin absorptive tissue. This design is evident in the gas-absorbing tissues of gills, lungs, and leaf spongy mesophyll. The large surface area is a design feature that we have encountered in other

vertebrate lungs

animal and plant organs. For example, the villi of the small intestine, and root hairs of plant roots. A large surface area of thin tissues exposes a maximum area to the atmosphere and to other cells, or to a circulatory system that can carry absorbed oxygen throughout the body.

The lungs of vertebrates have a large internal surface area composed of tiny **alveoli** (Figure 14-7). Each alveolus is a thin-walled air sac lined with blood capillaries where O_2 is absorbed from the inhaled atmospheric gases, and respiratory CO_2 is expelled. This represents a diffusion of O_2 from a high concentration in freshly inhaled air to a low concentration in the bloodstream. Carbon dioxide diffuses out of the blood into the alveolar air spaces because of its higher concentration in the blood than in the air spaces. Air is alternately inhaled and exhaled by the respective lowering and raising of the muscular **diaphragm**, and by muscular movements in the rib cage. This rhythmic action produces alternating suction and pressure in the chest cavity around the lungs. The lungs expand and shrink in response to this alternating pressure change around them.

O_2 transport

As our respiration equation indicates, for every 6 molecules of O_2 required to degrade one glucose molecule, 6 molecules of CO_2 are released. Body cells must therefore continually absorb O_2 from blood plasma, and release CO_2 into the plasma as a waste product. The red blood cell protein, *hemoglobin*, has a high attraction for O_2 in the alveoli, and a lower attraction for O_2 within respiring tissues. Consequently, blood is a more effective carrier of O_2 than one would expect strictly from its solubility in water. Although CO_2 also binds to hemoglobin, it is carried mostly in the plasma where it dissolves more readily than O_2 does.

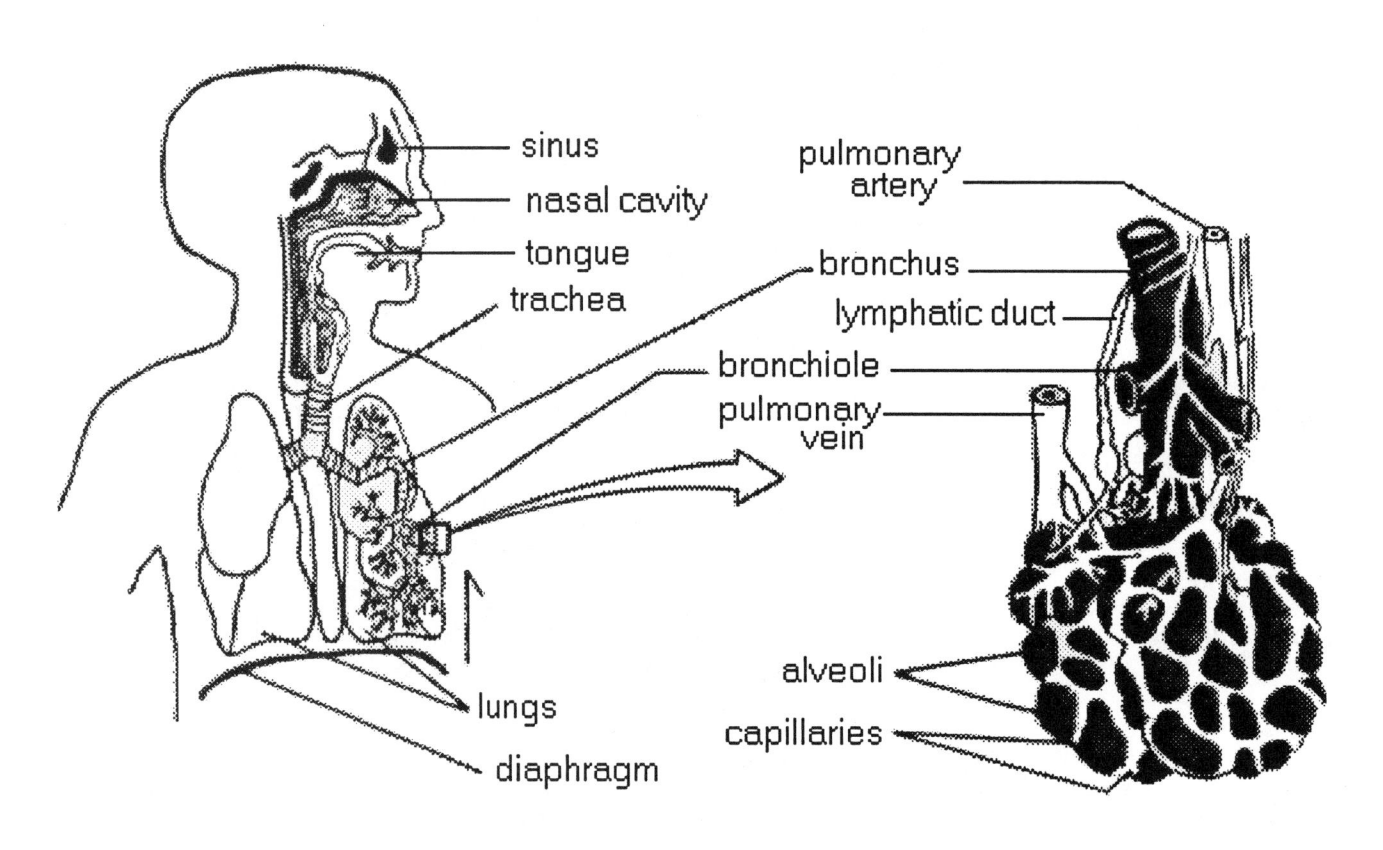

Figure 14-7. Major organs of the human respiratory system.

homeostatic control

Any increase in energy demand by cells increases cellular respiration. More O_2 and glucose will be consumed, and more CO_2 released. There must be a homeostatic mechanism to control blood O_2 supply and CO_2 removal under all physiological conditions. The body maintains appropriate levels of these gases by homeostatic control of breathing rate, heart rate, and blood flow distribution. [How would faster heart rate provide increased O_2 supply if accompanied by an increased breathing rate?]

Carbon dioxide levels in the blood are monitored by sensory receptors located in the lower brain. When blood CO_2 levels rise, impulses are sent to the brain, which in turn sends separate impulses to the heart and diaphragm, causing increased heart rate and breathing rates. Increased breathing rate brings larger volumes of air in contact with the bloodstream in the alveoli, and more rapidly sweeps away carbon dioxide that is diffusing from the bloodstream. Faster heart rate moves the two gases within the body more rapidly to match the higher cellular respiration rate. In addition, vasoconstriction of blood vessels directs blood flow to the more active body parts. Receptors in the aorta and carotid arteries detect decreases in blood O_2 and display regulatory input, as well. [Are heart rate and breathing rate under voluntary or involuntary control, or both? Explain. See Figure 13-10.]

14-F HORMONAL INTERACTIONS — THE MENSTRUAL CYCLE

Reproduction: coordination between mates

So far in this chapter, we have considered coordination of processes within the individual organism. Sexual reproduction requires coordination of reproductive processes in *two separate individuals* so that male and female gametes can unite in fertilization at a precise point in time. In most animals, fertilization involves mating behavior which is also under neurological and hormonal control.

The basic biology of human reproduction and sexual behavior was introduced in Section 11-J. A closer focus upon the human female reproductive cycle, or **menstrual cycle**, illustrates the interaction of several hormones that are produced at different bodily locations as they coordinate female reproductive developments. Although male sexual processes are also controlled by sex hormones, they will not be considered here.

hormonally controlled cycle

The menstrual cycle is normally a 28-day cycle involving (a) the partial maturation and release of an oocyte from an ovary, and (b) the thickening of the uterine lining (**endometrium**) in preparation for nourishment and development of the fertilized egg. The pituitary gland represents a third body location directly involved in this cycle.

hypothalamus and pituitary

In girls 10 to 12 years of age, the menstrual cycle is activated by hypothalamus-releasing hormones (Table 13-4) which stimulate the pituitary gland. Hormones from the pituitary set the menstrual cycle in motion. The cycle is usually dated from the beginning of **menstruation**, the only easily detectable event. Menstruation involves expulsion of the uterine lining from the body.

ovarian follicle

After menstruation, the pituitary releases **follicle-stimulating hormone (FSH)** into the bloodstream. The target sites of FSH are the **follicles**, tiny cellular clusters within each of the two ovaries (Figure 14-8). Each follicle contains a primary oocyte (Figure 11-13), a diploid cell waiting to complete meiosis I (Section 11-J). Usually, the two ovaries will alternate in producing

one per month

estrogen has two targets

LH triggers ovulation

one egg each month; up to 500 will mature during a normal lifetime. The oocyte undergoes the first meiotic division in response to FSH, and the second division when a sperm penetrates the egg cytoplasm.

Follicle-stimulating hormone stimulates a follicle to enlarge and develop around the oocyte. After several days, the cells of the follicle begin to release another hormone, **estrogen**, which has two major targets. First, estrogen causes the endometrium to thicken in preparation for possible implantation of an embryo. Second, estrogen causes the pituitary to gradually decrease the FSH supply. Specifically, estrogen affects the pituitary through negative feedback inhibition of the hypothalamus which controls the pituitary. As FSH secretion declines, the pituitary begins to secrete **luteinizing hormone (LH)** (L. *luti* = yellow). The LH causes the follicle to mature and rupture, releasing the egg from the ovary. The event, called **ovulation,** occurs on about day 14 of the cycle (Figure 14-8). The released egg enters the oviduct where fertilization can occur if sperm are released into the vagina (Section 11-J).

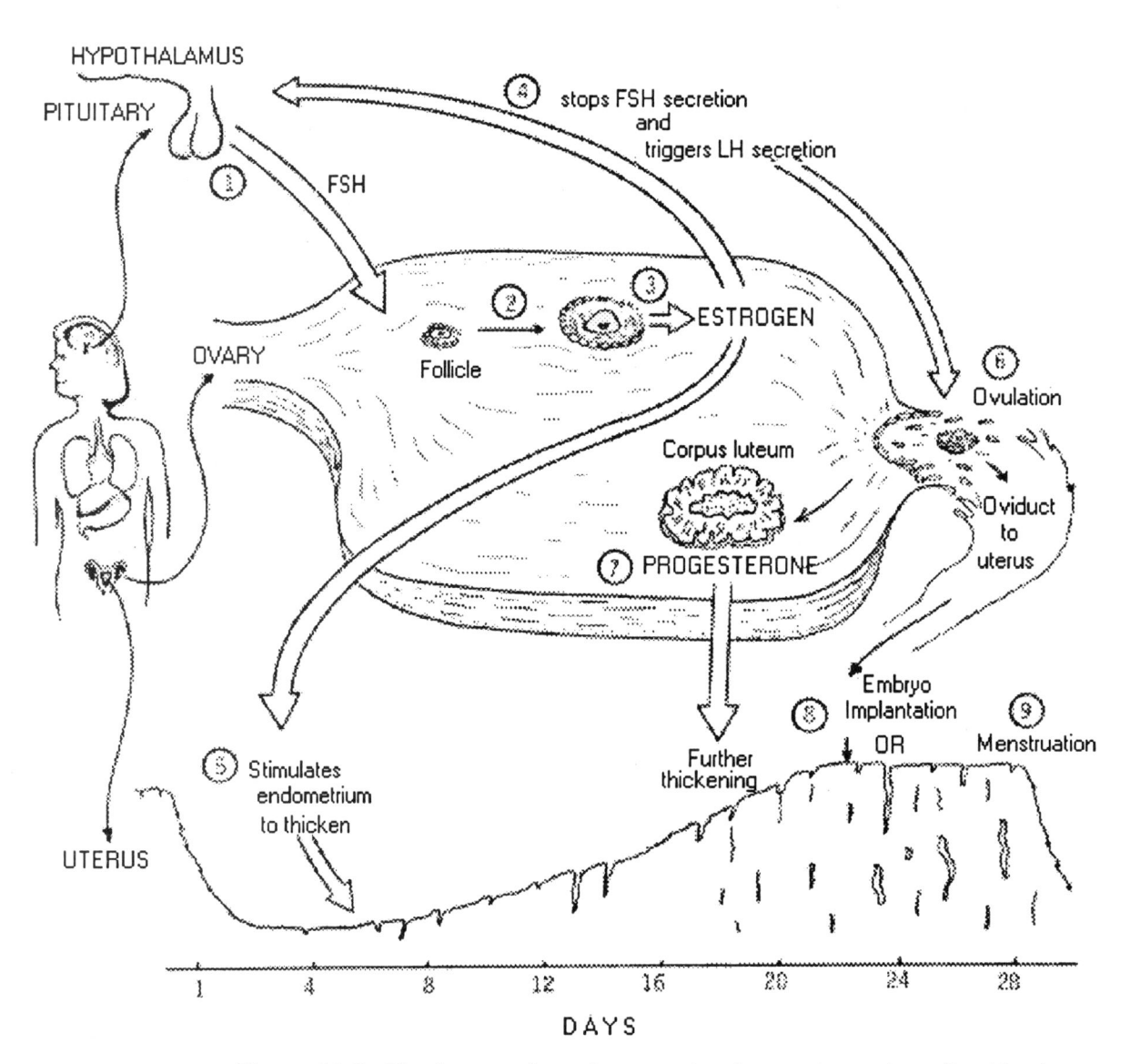

Figure 14-8. The human female reproductive cycle as described in the text. Double arrows indicate secretion and transport of hormones in the bloodstream.

Progesterone is sustained after conception...

Following ovulation, the ruptured follicle is transformed into a yellowish glandular mass called the **corpus luteum**. The corpus luteum produces still another hormone, **progesterone**, which stimulates final development of the endometrium. If the egg is fertilized within the oviduct, it moves on to the uterus and is implanted within the nutritive tissues of the endometrium. The implanted embryo releases **chorionic gonadotropin** which sustains the corpus luteum. The corpus luteum can therefore continue to secrete progesterone needed to sustain the endometrium. After the second trimester (second three months of pregnancy), the endometrium produces enough of its own hormones. Chorionic gonadotropin is detectable in the urine and is an early indicator of pregnancy.

...otherwise, menstruation occurs.

If fertilization does not occur, there is no supply of chorionic gonadotropin to sustain the corpus luteum. Degeneration of the corpus luteum causes a decline in the supply of progesterone. Without sufficient progesterone, the endometrium is expelled from the uterus in menstruation. Menstruation signals the beginning of the next 28-day cycle.

How does "the pill" work?

The *birth control pill* is an oral contraceptive that contains estrogen and a progesterone-like substance. When taken daily from about day 5 until day 26, an artificially induced concentration of these hormones in the bloodstream causes a prolonged feedback inhibition of the hypothalamus so that the pituitary gland is not stimulated to secrete FSH or LH. As a result ovulation does not occur. The hormonal formulation of the pill does allow endometrium development and menstruation.

"abortion pill"

More recent approval of the use of synthetic progesterone administered in the form of *skin implants* claim to provide a 99% effectiveness in preventing conception for up to five years. The formulation inhibits FSH and LH secretion by the pituitary. The more controversial *RU-486*, or "morning after" pill, inhibits progesterone production after conception and induces uterine contractions that lead to abortion. The formulation has also produced serious side-affects.

14-G HUMAN DEVELOPMENT

The development of a human being, or any creature, from a single cell is truly one of the outstanding marvels of creation. Complex coordination of embryonic development by hormones from both the mother and the developing child guide the unfolding of the genetic blueprint inherited through the zygote. Certainly, it is a complex process that we can only briefly address here. [See TOPICS FOR FURTHER STUDY #4.]

What is "development"?

The term **development** is typically used to describe the processes involved in transforming a single-celled zygote into all of the specialized cells, tissues, and organs of the adult stage. Most obviously, this requires *cell division* which involves DNA duplication, mitosis, and cytoplasmic division. Cell division or *cleavage* increases the number of cells, and is essential for **growth** (enlargement). However, growth alone is not development. There must also be an orderly program of **differentiation** in which cells become specialized into various tissues and organs. A brief survey of human **embryology** will serve to illustrate these concepts and point up certain basic questions of interest to us.

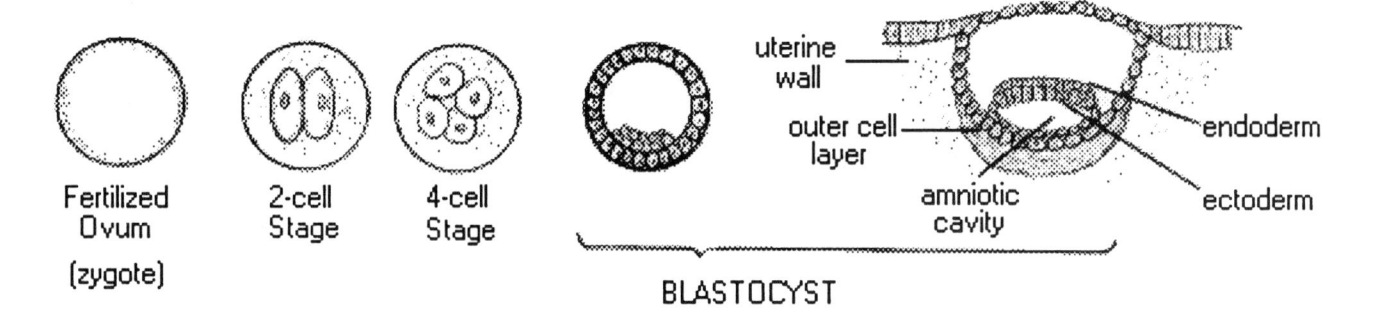

Figure 14-9. Early cleavage stages of the human zygote within the first four or five days after fertilization. Cleavage stages end and differentiation begins as the blastocyst is implanted in the uterine tissues.

embryo to the uterus

The early stages of cleavage and differentiation (gastrulation) in human development are shown in Figure 14-9. Cleavage occurs during the four-day trip from the oviduct to the uterus where implantation of the **blastocyst** will occur. The blastocyst is a tiny fluid-filled ball of cells. By the time of implantation, a mass of cells, called the *inner cell mass*, appears on the inside of the sphere. The blastocyst digests its way into the uterine tissue, absorbing nourishment as it goes.

development in the endometrium

placenta

The outer layer of the blastocyst becomes the **chorion**. The chorion secretes chorionic gonadotropin which acts through the corpus luteum and progesterone to prevent menstrual loss of the uterine lining (Section 14-F). Meanwhile, small branch-like projections of the chorion called **villi** penetrate into the endometrium, and the intimate association of the two tissues eventually forms the **placenta** (Figure 14-10). The placenta becomes distinct by the end of the first trimester (first three months of pregnancy) and nourishes the baby until birth. The mother's blood bathes the maternal side of the villi, but does not flow directly into the child's body. The capillary beds of the villi absorb nutrients, oxygen, hormones, antibodies, and unfortunately certain drugs and other toxic chemicals from the mother's blood. Excretory wastes are released from the blood of the embryo and handled by the mother's excretory system.

amniotic sac

While the above events occur through development of the outer layer of the blastocyst, the inner cell mass forms several cell layers and a fluid-filled cavity called the **amniotic cavity** (Figure 14-9). The cells nearest the outer cell layer soon differentiate into the **amnion** which, along with the fluid, constitutes the **amniotic sac**. The amniotic sac enlarges and completely surrounds the developing embryo providing a fluid cushion to protect it during development (Figure 14-10).

three layers

The remainder of the inner cell mass differentiates into two layers of cells, the **ectoderm** and the **endoderm** (Figure 14-9). A third layer, the **mesoderm**, later arises between the ectoderm and endoderm. Each of these three layers can be traced as it gives rise to specific organs of the developing embryo (Table 14-2).

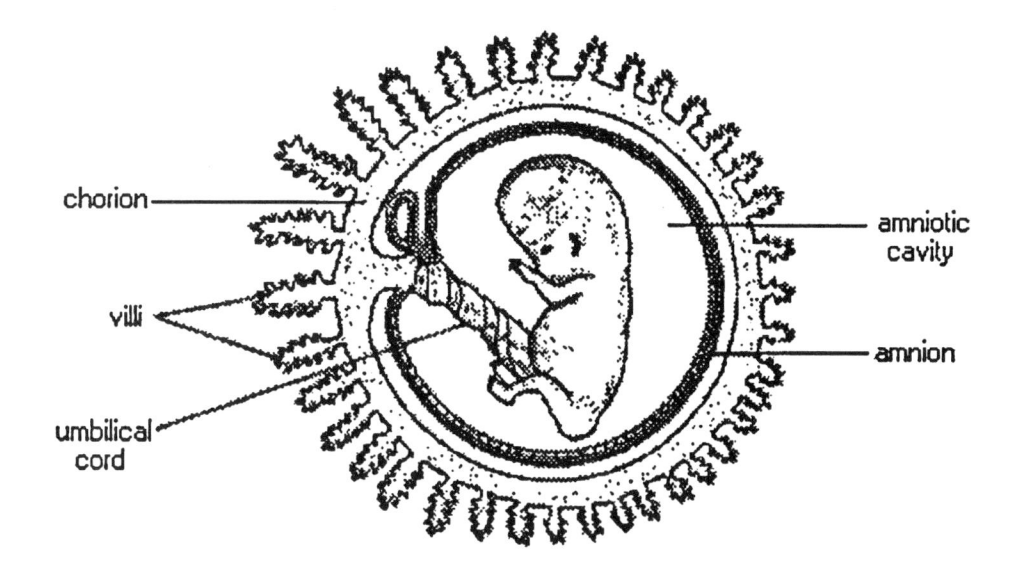

Figure 14-10. A developing human embryo within the amniotic cavity. Nourishment for the embryo is drawn in from the villi which are bathed in a flow of maternal blood.

Table 14-2. Major Body Organs and Tissues Derived from the Ectoderm, Mesoderm, and Endoderm.

ECTODERM	MESODERM	ENDODERM
Epidermis of skin	Skeleton	Inner lining of digestive tract
Hair and nails	Muscles; connective tissue	
Lining of nasal and oral cavities	Lining of body cavity (coelom)	Respiratory system
Tooth enamel	Circulatory system	Liver
Central nervous system	Reproductive system (part)	Pancreas
Peripheral nervous system	Excretory system (part)	Other glands
Sensory epithelium		

By the end of the first trimester, all organ systems are present. However, the developing individual is only about three inches in length, and still dependent upon the mother for survival. Table 14-3 outlines the timetable of major early developmental events. By the end of the third month the baby can move his or her arms and legs and the mother may become aware of some movements within. The face acquires the ability to squint or frown, and sucking and swallowing are possible.

It is estimated that between 10 and 50 percent of all conceptions fail to result in live births. There are a variety of reasons including failure to implant, implantation in the oviduct or body cavity; or, various genetic, nutritional, or environmental factors that interfere after implantation in the uterus. In recent

years, genetic counseling and amniocentesis (Section 12-G) have provided prenatal information to parents regarding possible genetic disorders in the unborn child. There is also increasing awareness of the detrimental effects of such agents as drugs (e.g. thalidomide, painkillers, antibiotics), alcohol, caffeine, X-rays, and pathogens (e.g. viruses causing German measles, flu, polio, etc.) upon the physical and mental development of the baby.
[See TOPICS FOR FURTHER STUDY #5.]

Table 14-3. Major Events in Early Human Development.

DAY	DEVELOPMENTAL EVENT
1	Fertilization of ovum in oviduct
5	Blastocyst stage enters uterus
6 - 9	Implantation of blastocyst in uterine lining
8 - 15	Differentiation of embryonic germ layers
16 - 23	Chorion and maternal tissues form the placenta
	Neural groove complete; skeletal development begins
	Developing heart begins beating
	Mother has missed only one menstruation since conception
28	Arm and leg buds appear; embryo is pea-sized
	Heart actively pumps blood
35	Differentiation of cerebrum and sense organs
42	Brain waves detectable; mouth and lips are present
60	Embryo now called a *fetus* is just over 1 inch long; all organs are established and body features clearly defined; some movement
63 - 70	Fingernails develop; sucks thumb, swallows, squints, jumps
77	Baby begins to breathe amniotic fluid
84	Sex organs distinguishable
168	Some chance of survival outside womb with aid of modern medical technology

Since the U.S. Supreme Court decision that legalized abortion in 1973, over 1 million legal abortions have been performed each year. A continuing controversy exists among various sectors of our society. [For more detail see TOPICS FOR FURTHER STUDY #6.]

**Abortion—
So few realize.**

Although abortions have been performed for centuries, relatively few individuals have experienced or witnessed the act. Few have even seen photographs. It is our purpose here to emphasize briefly the most common techniques of abortion and to focus attention upon the already highly developed human being that is the target for destruction in the process.

Most abortions are now performed during the first trimester. However, in 1988, an estimated 187,714 abortions were performed in the second and third trimesters; 15,908, *after* 20 weeks (3). Embryonic development will normally

Life is present before detection.

abortion methods

have been occurring for about 28 days before the mother suspects that she is pregnant. By this time, the being within her already has organ systems that are partially developed. For example, the heart is already pumping its own blood (Table 14-3). If this mother should decide to have the developing life within her destroyed, there are several common methods of abortion available.

The *D & C* (dilation and curettage) consists of the use of a surgical tool (curette) to scrape out the contents of the uterus through the dilated cervix. A *suction method* employs a hollow plastic tube attached to a vacuum source to draw out the unborn child and uterine lining. For abortions during the second trimester (4 to 6 months), the *saline method* is often used. A needle is inserted through the mother's abdominal wall and into the amniotic sac. A salt (saline) solution is then injected into the environment around the unborn child. The altered amniotic fluid, so amazingly formulated to support life, becomes a cause of convulsions, suffering, and death. Within 24 hours, the "mother" gives birth to a dead baby. In a fourth method, the baby may be removed by Caesarian section in which the mother's abdomen is opened surgically. The baby is removed alive and "put aside" to die.

A final method of abortion worth noting requires a brief introduction. In 1981, a Nobel Prize was awarded to the discoverers of a new class of hormones, the *prostaglandins*. Different prostaglandins are produced throughout the body tissues rather than by glands. They are secreted in response to other hormones or irritations of body tissues. Some prostaglandins produce warning symptoms of inflammation and pain (aspirin inhibits pain by inhibiting prostaglandin synthesis). Others promote blood clotting and uterine contractions. The latter function has led to the use of prostaglandin suppositories to induce abortions. The suppository causes strong uterine contractions that are supposed to suffocate the baby and result in a stillbirth. However, babies reportedly survive this ordeal. Consequently, their death is a final step in the technique.

Is there any question?

Science through medicine has provided the techniques of abortion that allow minimal threat to the life of the mother, particularly if administered during the first three months. In addition, developmental biology is making major progress in understanding the physiological and morphological events from conception to birth. But, can biology tell us when an individual life begins? Genetically speaking, a new generation begins at conception when a new blueprint forms. Embryologically, there is a smooth, uninterrupted unfolding of the genetic blueprint from the first cells to birth. Yet, with all of this knowledge, some would argue that this developing being is simply an unwanted "growth" with no rights to life.

value of life

We must realize that one's position on abortion must be rooted more deeply than science alone can allow. The position must stem from ones worldview. When does a person come into existence? A Christian, theistic worldview would place great *value* on a life because God does, even before birth (Jeremiah 1:5). Humans are not simply a product of the evolution of matter and culture; and, therefore accountable to no other being? We are accountable to the Creator God who has revealed His standards of truth and has spoken these words:

Before I formed thee in the belly I knew thee...Thou didst form my inward parts; Thou didst weave me in my mother's womb.
Jeremiah 1:5a; Psalm 139:13

14-H PLANT GROWTH AND DEVELOPMENT

Coordination of cell division and differentiation is as essential in plant growth and development as in animals. In both instances, development begins with the zygote. For our purposes, we shall consider flowering plant development as occurring in three major phases—seed development, seed germination, and seedling development.

SEED DEVELOPMENT

origin of seeds

A brief reference back to Section 11-I should help as we consider the development of the embryo following fertilization in the flower. The focus of attention is the female sex organ, the **ovary** (Figure 11-16). The ovary contains one or more **ovules**, each bearing an **embryo sac** containing an egg (Figure 14-11). Each ovule develops from the interior tissue of the ovary and remains attached to this tissue, which is known as the **placenta**. When fertilization is complete, an embryo develops within the ovule, and eventually each ovule becomes a mature seed. Figure 14-11 illustrates this relationship in the garden pea.

double fertilization

Double fertilization occurs when the pollen tube penetrates the ovary and delivers the two sperm nuclei to the embryo sac. One sperm nucleus fuses with the egg to form a zygote which will develop into an embryo. The other sperm fuses with one or more other nuclei of the embryo sac to form, after several mitotic divisions, a nutritive tissue called **endosperm**. The endosperm tissue nourishes the embryo as it develops from the zygote.

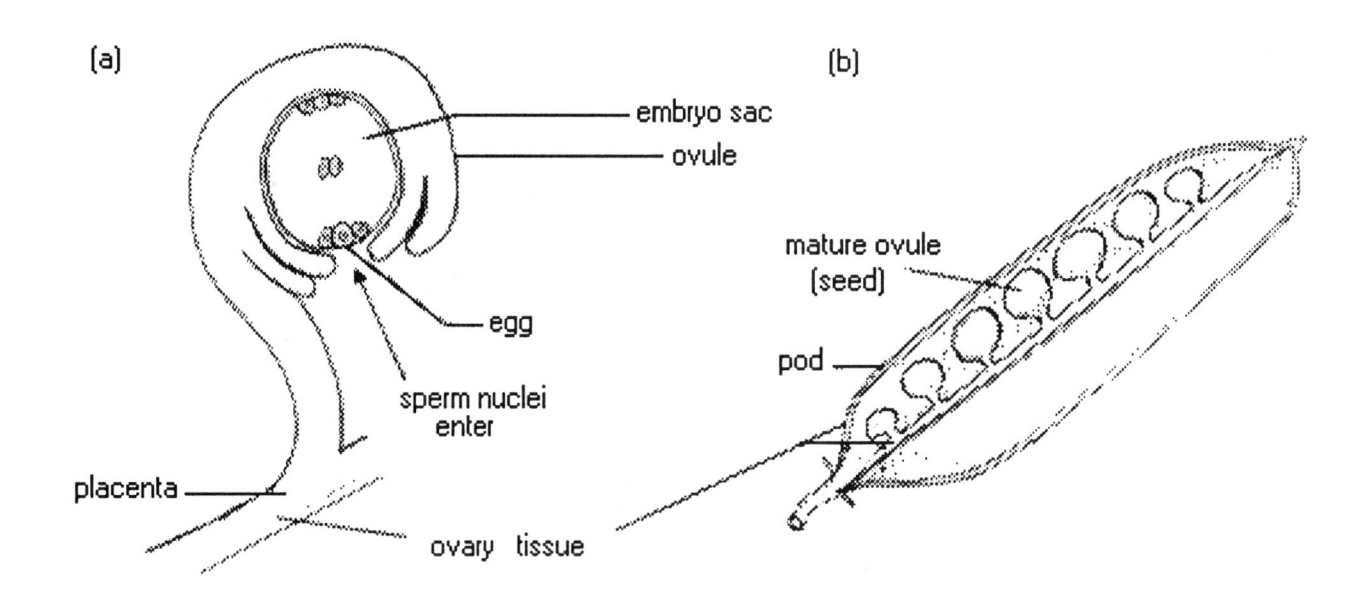

Figure 14-11. Garden pea ovules and related structures (a) just prior to fertilization, and (b) when pea seeds are nearly ready for shelling out of the pods.

zygote to embryo...

...with three tissue regions

apical meristems...

...and cambium

The zygote divides unevenly to produce a small *terminal cell* and a larger *basal cell* (Figure 14-12a,b). A few more mitotic divisions of the basal cell produce a stalk-like *suspensor* which gives support to the globular embryo as it forms from divisions of the terminal cell. Cell differentiation within the globular embryo occurs, producing **protoderm**, an outer layer of cells which will form epidermal tissue (Figure 14-12 d). The innermost cells, the **procambium**, will give rise to vascular tissues of the plant, the xylem and phloem. Between protoderm and procambium is the **ground meristem** which will produce leaf mesophyll, cortex, and pith. Additional cell divisions and enlargement produces two mounds that will form the two **cotyledons**, or embryonic leaves (Figure 14-12e,f). The portion of the embryo below the point of attachment of the cotyledons is the **hypocotyl**. Note that both shoot and root **apical meristems** are visible and will be the major growing points for the shoot and roots when germination occurs. Like *stem cells* in animals (Section 10-H), meristematic cells such as the **cambium** in plants remain embryonic and can produce new cells throughout the life of the organism.

The *seed coat* forms a protective layer around the embryo and its reserves of chemical energy and nutrients. This package constitutes a mature dicotyledon seed. The monocotyledon seed develops similarly, but has only one cotyledon (Figure 14-13).

ovary becomes fruit

Most seeds will remain enclosed within the maternal tissue of the ovary which has grown and matured along with the seed. The mature ovary, or **fruit**, often serves to promote seed dispersal by animals, wind, or water (Figure 11-17). The mature seed may germinate immediately after leaving the parent plant, or in many cases, remain **dormant** (L. *dormit* = sleep) for a period of time.

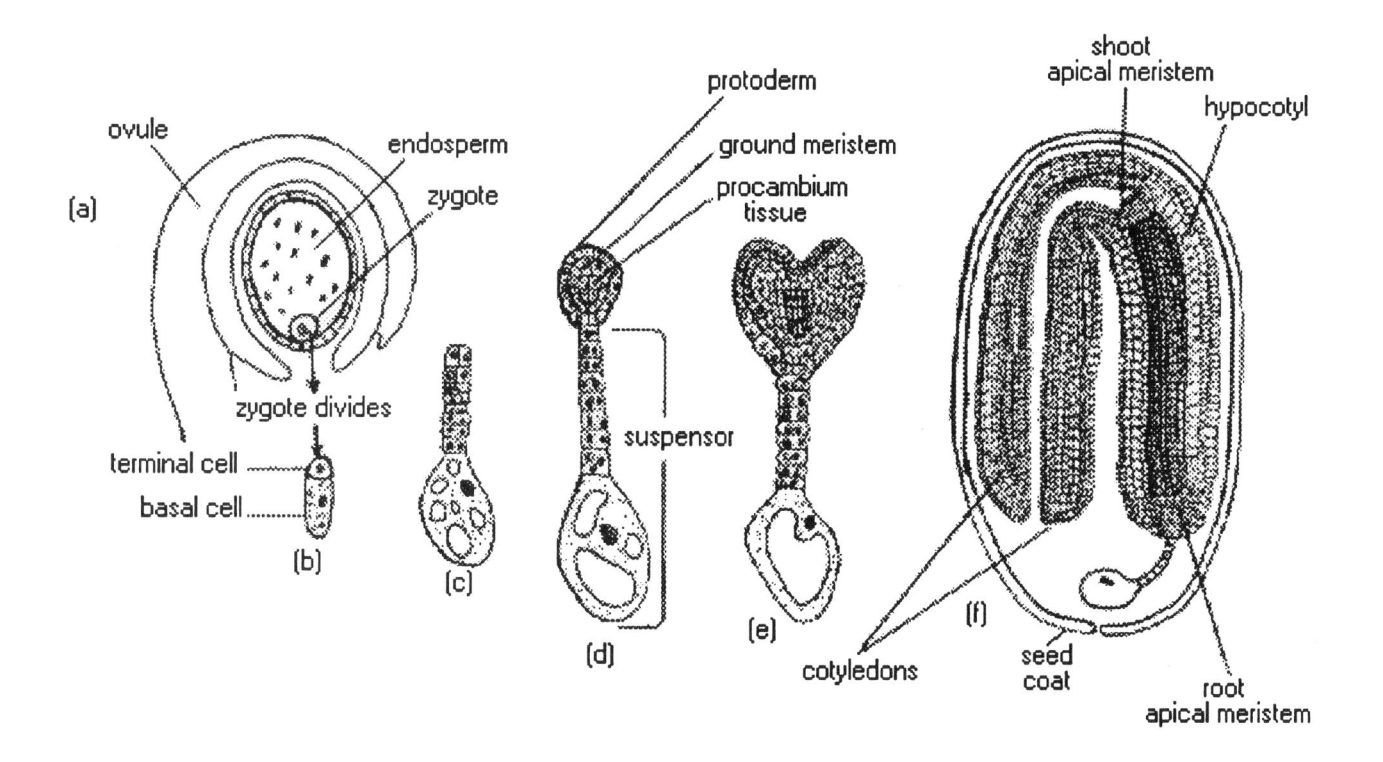

Figure 14-12. Embryo development in the dicot *Capsella bursa-pastoris*, the shepherds purse.

CORN   **BEAN**

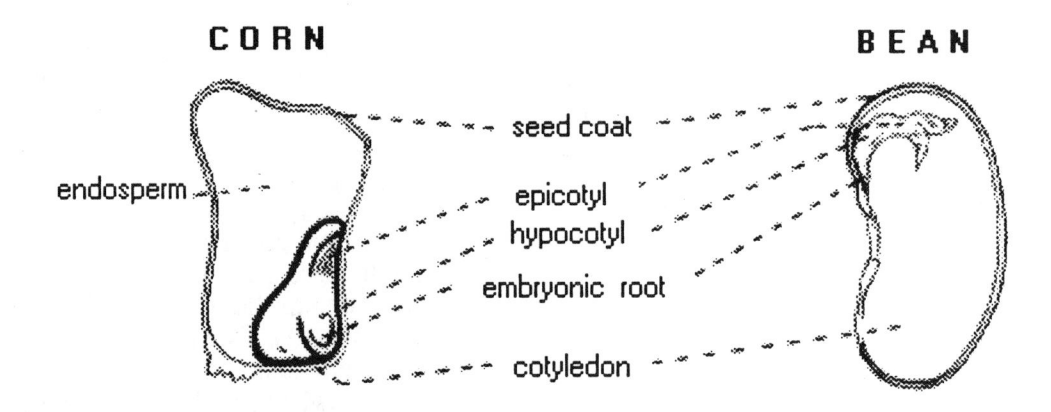

- endosperm
- seed coat
- epicotyl
- hypocotyl
- embryonic root
- cotyledon

Figure 14-13. Cross-sectional diagram of the seeds of a monocot (corn), and dicot (bean, with one cotyledon removed). The corn seed has energy and nutrient reserves in the form of endosperm, whereas bean has these materials stored in the cotyledons.

dormancy

Dormancy is a condition of very low metabolic activity of the embryo. The dormant seed can survive harsh environmental conditions because it requires little or nothing from the environment. Humans have been able to harvest, transport, store, and eat dormant seeds and seed products because of the capacity of seeds to withstand all kinds of environmental conditions.

SEED GERMINATION

mechanisms to assure suitable conditions

The dormant seed is generally the only motile stage in the flowering plant life cycle, and therefore, the only means by which plants can colonize other habitats. Thus, the "mission" of seeds is to reach a new suitable environment for germination, and to germinate, perhaps following an unfavorable season or temporary stop-off at another unfavorable site. If this mission is to be accomplished, the dormant seed must have some mechanism to detect when conditions "out there" are favorable for growth. Otherwise the dormancy will have been in vain. Seeds accomplish this mission because they have been metabolically programmed to respond to reliable environmental stimuli.

large seeds

Large-seeded plants such as corn and bean will germinate under warm temperatures if moisture is available. Thus, water acts as a stimulus as it diffuses through pores in the seed coat. The cytoplasm of embryonic cells (Figure 14-12) becomes *hydrated* as water moves across cell membranes by osmosis. Hydration activates cellular respiration and other metabolic processes which utilize the starch, protein, and mineral nutrient reserves stored in the cotyledons or endosperm (Figure 14-13). Root apical meristem cells (Figure 14-12) begin to divide and elongate in response to turgor pressure (Section 10-C). The swelling and enlargement of the embryo and endosperm in response to the massive absorption of water causes the seed coat to burst. Germination is evidenced by the emergence of the hypocotyl from the seed coat (Figure 14-14). The embryo must rely upon the energy and nutrient reserves of cotyledons or endosperm until the roots are established, and the growing shoot can emerge from the soil and become a photosynthetic autotroph.

**small seeds
need light**

Corn, bean, and other large seeds can germinate and emerge from points deep in the soil because of their large energy reserves. However, lettuce and many annual plants, that are pioneers in ecological succession in bare soil, have insufficient reserves to support germination from deep in the soil. Seeds of many of these species rely upon proper temperatures, moisture, and *light*. Light is taken as the reliable indicator that the seed is near the soil surface. Most soils have a *seed bank*, a reserve of seeds that lay dormant for years until the soil is disturbed and they are brought to the surface where the light stimulus triggers germination. [In view of this, isn't cultivation of "weeds" somewhat self defeating? Explain.]

**other
mechanisms**

There are a host of other requirements for seed germination in different species. These include removal of chemical inhibitors present in the fruit, seed coat, or embryo; a period of cold temperatures; and decay of a thick seed coat by decomposers. Each of these mechanisms is an adaptation that aids in fitting the life cycle of a given species into the environmental conditions of its habitat. [See TOPICS FOR FURTHER STUDY #8.]

SEEDLING GROWTH AND DEVELOPMENT

**hormonal
coordination ...**

... roots first

Seedling growth and development provide yet another illustration of one of the themes of this chapter, hormonal coordination of biological processes. Hormonal coordination of plant embryo growth assures that the hypocotyl will emerge from the seed coat before emergence of the growing shoot and leaves. This timing assures the early establishment of roots needed to anchor the plant, and absorb water and nutrients (Figure 14-14).

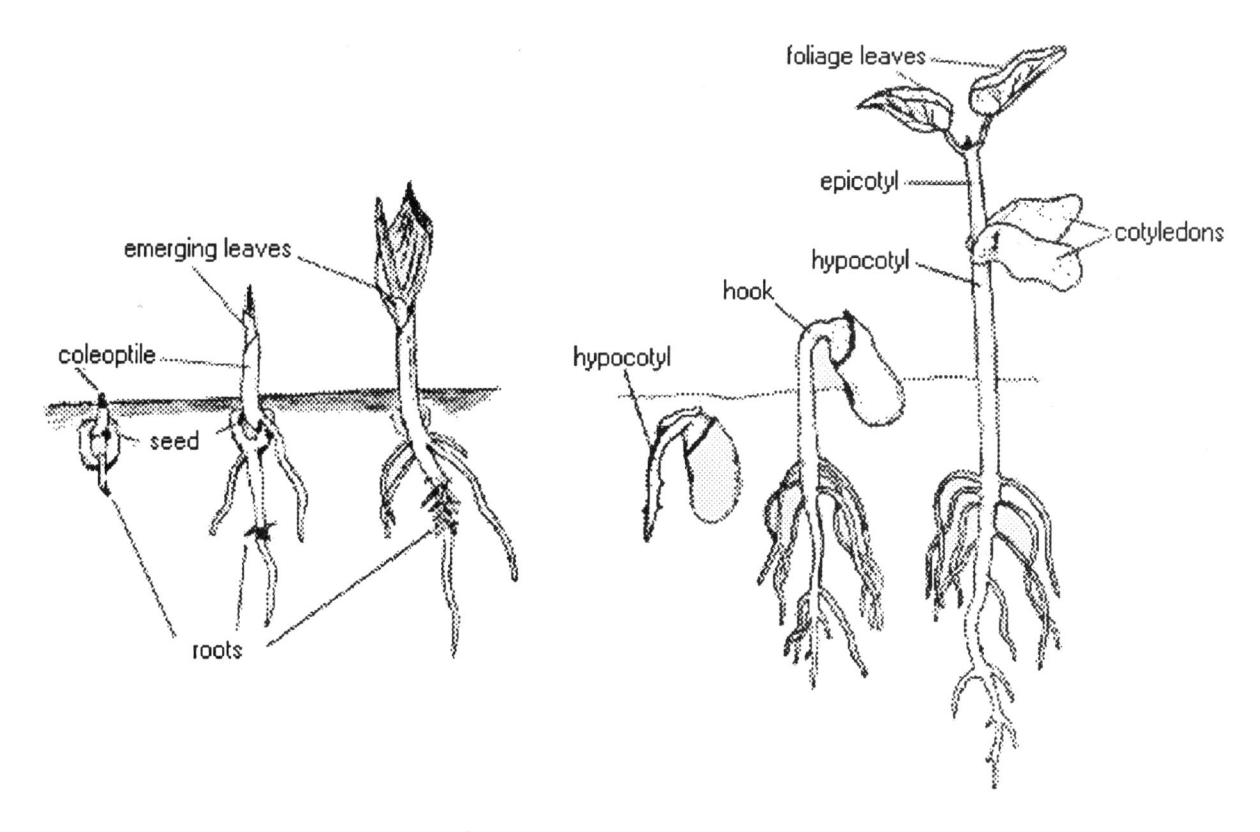

Figure 14-14. Seed germination and seedling development in (a) monocot (corn), and (b) dicot (bean).

**emergence—
cell division
and elongation**

As the root system is developing, the growing shoot prepares to emerge from the dark soil into the light. Seedling emergence also involves hormonal control and responsiveness to light. Cell division and cell elongation force the growing shoot upward through the soil to the surface. The tender shoot apical meristem is the source of all new cells that differentiate into leaves and stems, and must be protected from damage during emergence from the soil. In monocots, a sheath called the *coleoptile* provides the protection (Figure 14-14). The coleoptile elongates faster than the enclosed leaves until it reaches the surface. Here, red light is absorbed by a plant pigment (*phytochrome*) which triggers inhibition of coleoptile cell elongation. Meanwhile, the monocot leaves elongate and push through the top of the coleoptile and unroll in response to light (Figure 14-14).

**Monocots have
a coleoptile.**

**Dicots have a
"hook"...**

**...sensitive
to red light.**

In dicots, a *hypocotyl hook* pushes through the soil until it reaches the surface (Figure 14-14). Here, red light is absorbed by phytochrome which inhibits the synthesis of the gaseous hormone **ethylene** by the hook (Table 13-3). In the absence of ethylene, the hook straightens (Figure 14-14). Red light also promotes leaf expansion and leaf pigment synthesis in the developing seedling.

**responsiveness
to gravity**

Most seeds germinate from some random orientation in the soil. The emerging roots and shoots must be able to sense which direction is "up" and "down" with respect to the center of the earth. Otherwise, subsequent root development would be disoriented with respect to the vertical plane. Gravity is reportedly the stimulus that allows plant roots and shoots to orient their growth properly. Plant growth curvature in response to gravity is called **gravitropism** (or geotropism). Most roots are *positively gravitropic* (grow toward gravitational pull) and shoots, *negatively gravitropic*. Figure 14-15a illustrates gravitropism in a corn seedling. Curvature is due to differential cell elongation between the upper and lower portions of the root and shoot.

**responsiveness
to light**

Once the shoot has emerged from the soil, leaves and stems must be oriented properly with respect to the light source for optimum photosynthesis. **Phototropism** is a growth curvature in response to light. This response is commonly observed when house plants setting on a window sill orient their stems and leaves toward the predominant source of light. Figure 14-15b illustrates a corn coleoptile phototropically responding to unidirectional light.

**an invitation
to inquiry**

Tropic responses all involve curvature which is caused by differential cell elongation. However, this does not explain all of the cause-effect relationships between the stimulus (gravity or light) and the differential cell elongation. There continues to be some uncertainty as to the nature of the plant receptors of light and gravitational stimuli. These receptors somehow initiate curvature through effects on hormone distribution within the tissue. Given certain additional information, a possible mechanism or mechanisms can be proposed through analytical reasoning in a laboratory setting. Perhaps your laboratory instructor will provide such an opportunity as this author has been prone to do.

Summary:

We have touched upon just a few of the plant developmental processes leading to the production of roots, stems, and leaves. The growing plant has a tremendous capacity to detect and respond to daily and seasonal changes. These processes are essential to the survival and reproduction of organisms that must complete their entire life in one location in a changing environment.

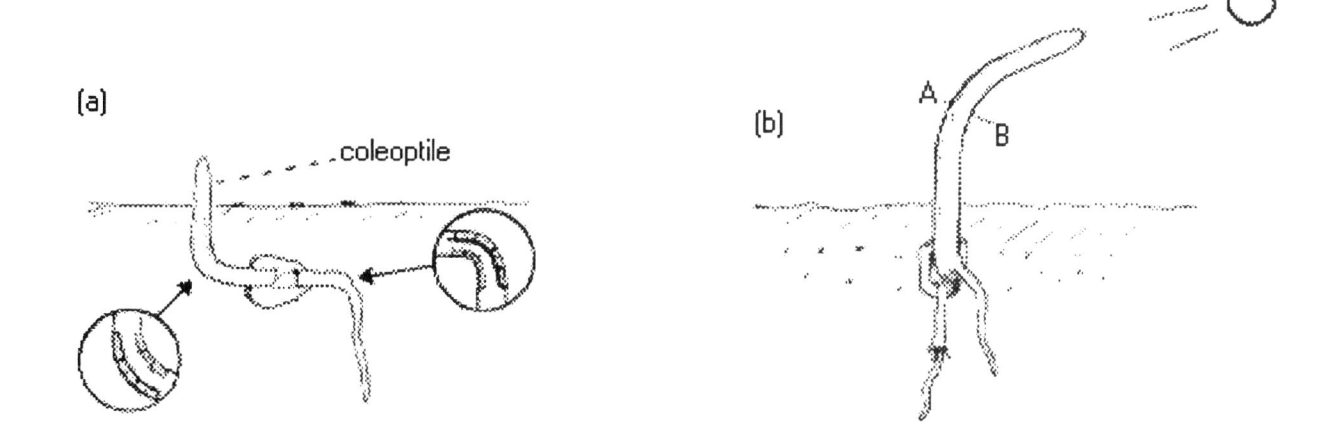

Figure 14-15. (a) Gravitropism in a corn seedling. Curvature is produced by differential cell elongation as shown in the insets. (b) Phototropism, a growth curvature in response to light. [Where would you expect the greatest cell elongation, on side A or B?]

14-I COORDINATION OF CELLULAR DIFFERENTIATION

In this chapter, we have discussed some examples of homeostatic mechanisms that regulate or coordinate life processes. Some of the examples such as regulation of temperature, glucose, oxygen, and other substances in the blood, involve *nutritional* processes. Others, such as the hormonal regulation of the human female reproductive cycle, involve *reproduction*. Finally, the coordination of embryonic *development* of humans and flowering plants involves hormonal control.

How does an adult develop from a zygote?

Coordination of embryonic development centers on the production and differentiation of cells into distinct tissues and organs. But how does a mature adult with billions of cells with vastly different structure and function arise from a one-celled zygote? In spite of much research, complete answers to this question continue to elude researchers. There are some hints, however, as to mechanisms involved.

"hormonal environment"

As soon as a zygote divides several times, cell differentiation begins to occur (Figures 14-9 and 14-12), and one can locate specific cell groups that will later give rise to whole tissues and organs. The *position* of each of these first cells and their *hormonal environment* appear to influence their destiny. In animals, cells migrate within the embryo as they differentiate, and some cells and tissues may be degraded. In plants the plane of cell division following mitosis influences the orientation of cell elongation and resultant shaping of the embryo (Figure 14-12). Hormonal gradients appear to influence these patterns of cell division as well as enlargement.

tissue culture

Figure 14-16 illustrates the influence of the ratio of two plant hormones upon cell differentiation. Such *tissue culture* experiments offer major clues as to how plant cell differentiation is controlled. Notice that pith cells of stems are hormonally induced to become embryonic and differentiate into roots, stems, or leaves depending upon the ratio of auxin to cytokinin. Whole plants can be

cloning
hormones "in concert"

produced *asexually*, or **cloned** from tissue or cell cultures in this manner. This experiment suggests at least two things worthy of your consideration: First, rather than an "all-or-nothing" effect, hormones appear to act "in concert" (like the blending of different orchestral instruments) to produce a particular differentiation outcome, depending upon the *ratio* of concentrations. Second, if a few cells of pith can be induced to function as embryonic cells and generate all of the parts of a complete plant, then hormones must influence the "unfolding" of the genetic blueprint in these cells. How do genes and hormones control differentiation? We will examine this exciting aspect of biology in Chapter 17.

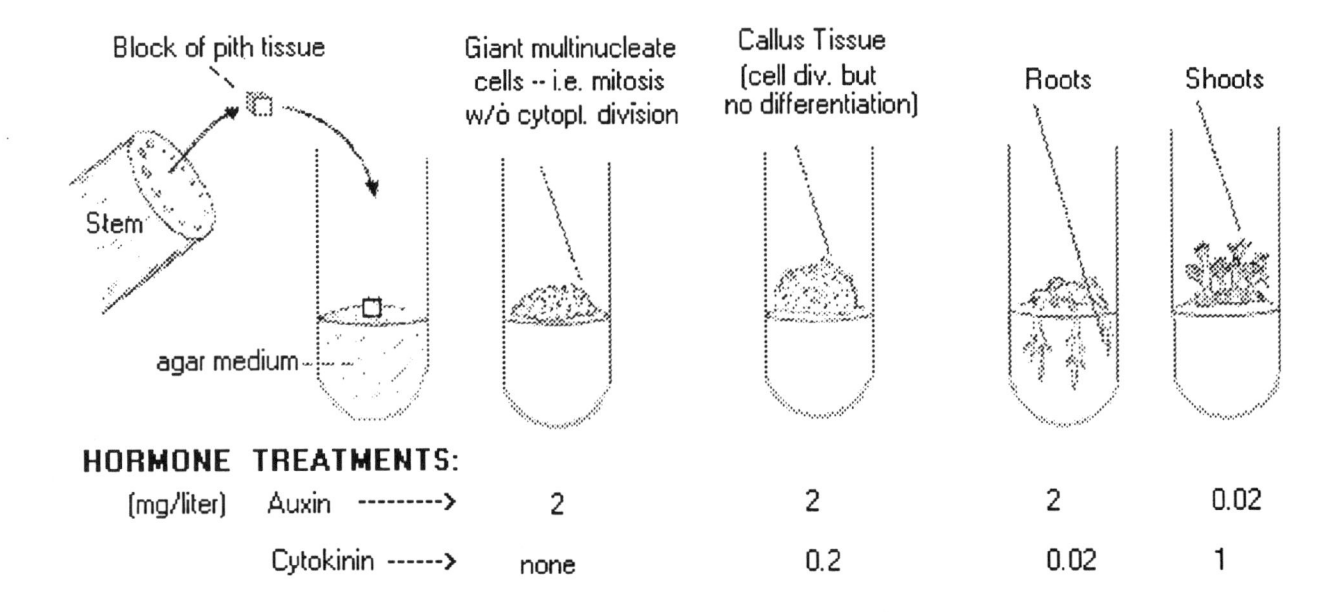

HORMONE TREATMENTS: (mg/liter)	Giant multinucleate cells -- i.e. mitosis w/o cytopl. division	Callus Tissue (cell div. but no differentiation)	Roots	Shoots
Auxin --------->	2	2	2	0.02
Cytokinin ------>	none	0.2	0.02	1

Figure 14-16. Plant hormonal coordination of cell differentiation. Each test tube originally appeared as the one on the left; the outcome of each was determined by the ratio of auxin and cytokinin in the agar medium.

QUESTIONS AND DISCUSSION TOPICS

1. Discuss several means by which plants and animals adjust to unfavorable temperatures.

2. How is the hypothalamus like a thermostat? What other body organs and systems are involved? Use the thermostat analogy to explain fever?

3. Describe the homeostatic control of blood glucose levels. How does control of bodily energy use by the thyroid differ from that of the islet cells?

4. Describe the path of each of the following upon entering the kidney: red blood cell, water molecule, glucose molecule, urea molecule, sodium ion.

5. How is active transport different from diffusion? How does active transport of glucose and other blood components at the proximal tubule cause osmosis?

6. Why would you expect the loop of Henle of water-conserving desert rodents to be relatively longer than most mammals of humid biomes?

7. Explain the menstrual cycle as an example of hormonal coordination of bodily processes and feedback mechanisms?

8. Outline the major events in human embryological development. What aspects demonstrate the systematic nature of development?

9. What is a seed? What are the major roles of seeds (and fruits) in the plant life cycle? What mechanisms allow these roles to be accomplished?

10. To what kinds of stimuli are plants responsive? Describe the tropic mechanisms which cause young seedlings to emerge from the soil and become properly oriented.

TOPICS FOR FURTHER STUDY

1. How do organisms control their body temperature amid temperature extremes in harsh environments? Which have a greater advantage in winter, cold-blooded (ectotherms), or warm-blooded animals (endotherms)? REFERENCES: #15 and #16

2. Research the causes and innovative medical solutions to diseases such as *diabetes mellitus*, hypothyroidism, etc. REFERENCES: #7 and #11

3. Develop a biblical position on family planning, including an understanding of the options and implications of conception control. REFERENCES: #9, #10

4. Explore further one of the marvels of creation, the development of the human from conception to birth, through such REFERENCES as the cited books (#1 and #13), video (#18) and film (#4).

5. What are major causes of miscarriages and birth defects which can be prevented by proper health and nutrition of the mother? REFERENCE: #19

6. What are the statistics, methods, and the moral, ethical, economic, biological, and psychological considerations surrounding abortion? REFERENCES: #3, #6, and #8.

7. What can the developing baby sense, feel, and perform at various stages of development within the mother's womb? REFERENCES: #12 and #13

8. Study the moral, ethical, and legal implications of manipulations of human embryos and fetal tissue. REFERENCES: #2 and #5

9. If you have a "germinating interest" in seeds, some interesting topics include the following: (a) nutritional value of seeds, (b) seed dispersal mechanisms, (c) plant-animal relationships in pollination and seed dispersal, (d) how seeds "know" when to germinate. Or, you may like to "dig deeper" into *gravitropism*, or "shed more light" on *phototropism*, or "get a better feel" for *thigmotropism*? REFERENCES #14 and #17 provide a wealth of background and additional references into plant science literature.

REFERENCES

1. Ankerberg, J. and J. Weldon. 1990. *When Does Life Begin? And 39 Other Tough Questions about Abortion.* Wolgemuth and Hyatt. Brentwood, TN.

2. Arvant, P.S. 1990. Brave new harvest. *Christianity Today* (Nov. 19): 24-28.

3. *Back to Basics.* Fact Sheet. 1992. National Right to Life. Washington, DC.

4. *Everyday Miracle: Birth.* [Film, Intrauterine Photography] 1980. Films Incorporated, British Broadcasting Company.

5. Frankel, M.S. 1990. Freezing human embryos—Value dilemmas when biology and medicine converge. *Bioscience* 40: 40-43.

6. Gish, D.T. and C. Wilson. 1981. *Manipulating Life: Where Does It Stop?* Part IV, "Abortion", pp. 189-220. Master Books, San Diego, CA.

7. Hadley, M. 1988. *Endocrinology*, 2nd ed. Prentice-Hall. Englewood Cliffs, NJ.

8. *Information Please Almanac.* 46th ed. 1993. Houghton Mifflin Co. Boston, MA.

9. *Is Birth Control Christian?* 1991. Christianity Today Institute, Series of Articles in *Christianity Today*, Nov. 11 issue.

10. Lang, B.R. 1991. RU-486 and abortion politics. *Bioscience* 41(9): 594, 654.

11. Notkins, A. 1979. The causes of diabetes. *Scient. Amer.* 241(5): 62-73.

12. Palmer, J.S. 1987. Sensing in the womb. *Amer. Biol. Teach.* 49(7): 411-417.

13. Parker, G.E. 1987. *Life Before Birth: A Christian Family Book.* Master Books. El Cajon, CA.

14. Salisbury, F.B. and C.W. Ross. 1991. *Plant Physiology*, 4th ed. Wadsworth, Belmont, CA.

15. Schmidt-Nielson, K. 1972. *How Animals Work.* Cambridge Univ. Press. New York, NY.

16. Schmidt-Nelson, K. 1978. *Animal Physiology*, 2nd ed. Cambridge Univ. New York.

17. Stern, K.R. 1991. *Introductory Plant Biology*, 5th ed., W.C. Brown, Dubuque, IA.

18. *The Silent Scream* [Candid Video on Abortion]. 1984. Amer. Portrait Films. Anaheim, CA.

19. Winick, M. 1985. *Feeding the Mother and Infant.* Wiley-Interscience. New York, NY.

# Chapter 15

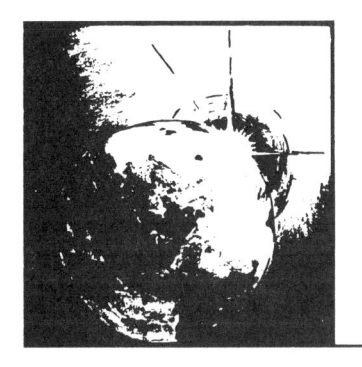

Genetic Variation and Selection

adaptations:
— morphological
— physiological
— behavioral

Previous chapters have surveyed the great variation and biodiversity that exists *among* species (Chapter 9) and *within* each species (Chapter 12). Now, in this chapter, we will consider the processes whereby populations and species adapt to their environment. **Adaptations** are the morphological, physiological, and behavioral features that enable organisms (and populations) to survive and reproduce in their environment. *Life characteristic #5* (Section 2-C) indicates that adaptations are common in all species.

"fine tuned" by natural selection

Adaptations are "fine tuned" by **natural selection**, a result of environmental factors acting upon populations and favoring individuals with phenotypes that better "fit" them to survive/reproduce in that environment (Section 12-I). Because natural selection causes *genetic changes* in populations, two controversial questions must be addressed:

controversial questions

1. What is the **origin of life**?

2. What, if any, are the natural **limits to biological change**?

Before beginning our discussion, we need to emphasize the importance of beginning with the right understanding.

15-A ORIGIN OF LIFE — IT MATTERS WHERE WE BEGIN !

Science probes the *present* ...

Genetic variation is an observable *fact* of life. According to *genetic theory*, nucleic acids are the hereditary material responsible for genetic variation. *Facts* and *theories* concerning life are accepted on the basis of repeated *observations* and *controlled experiments*. Using this method, the natural sciences aim to learn all there is to know about life and the physical universe. Smallness of *size* does not deter molecular biologists from decoding the human genome. Great *distance* does not stop the launching of humans to the moon. Even the *time* required to accomplish tasks is reduced by computers, while video equipment captures past events for "instant replay".

but what about *prehistoric* processes?

The same scientific determination that exposes secrets of the DNA molecule and of outer space also seeks to expose the secrets of *prehistoric time*. But here, there are no "instant replays". The sights, sounds, chemicals, and organisms of the prehistoric past are gone forever, leaving only marks and deposits on the landscape of Earth. Without "instant replay", such marks and deposits can be *observed* as part of a scientific inquiry, or as part of an FBI or Perry Mason investigation; but, marks and deposits do not "tell their own story". They can only be *interpreted* and used to *reconstruct* past events by someone's *speculation*. Therefore, "origins hypotheses" may be plausible, but are not testable. As Popper (36) defines science, hypotheses and theories

origins hypotheses ...

plausible but not testable

are scientific only if they are falsifiable by repeated testing through observation and experimentation.

Should "origins hypotheses" be discarded as "nonscience"? No, according to both scientists and theologians (15, 40). Instead, they emphasize that origin science must be distinguished from operations science. **Origin science** seeks to understand events and processes that occurred before scientific tests and records could be made. On the other hand, **operation science** examines recurring processes to explain the operation of the physical universe. Both origins and operations science seek answers to "scientific questions." However, failure to recognize the limitations of origins science has elevated unfalsifiable hypotheses which are at best, plausible, to the same level as "operations hypotheses" that have empirical support. For example, *prebiotic simulation experiments*, conducted under *conditions as they may have been*, are inconclusive because no one can know *how it really was*. Likewise, "flood theories" attempt to explain the geologic layers of Earth's crust as plausible, but they too are unfalsifiable.

origin science

operation science

Worldview influences origins hypotheses.

Because of the strong role of personal speculation, "origins hypotheses" are naturally influenced by their author's **worldview** (Section 1-A). Worldviews are based upon belief systems which help us deal with questions like "What is the origin of life?", and "How have living organisms changed over time?". Belief systems are sets of *faith presuppositions* (Section 1-C) about higher powers, the universe, and mankind. To illustrate, two persons may view the same **fossil**, the remains or evidence of a once-living organisms. From a theistic worldview, one person proposes a supernatural flood to account for burial and preservation of the fossil. A person with an atheistic worldview rejects this "flood hypothesis" or any supernatural intervention. Neither hypothesis can be refuted or proven. Nor will there be meaningful discourse until presuppositions are recognized.

contrasting philosophies:

Consider the following philosophies, and notice the differing "origins hypotheses" or inferences from each philosophy:

1. **NATURALISM** stems from an atheistic worldview, and sees the natural world as the only reality, controlled by natural laws operating over time; denies existence of a "supernatural being" or "intelligent designer"; therefore, there is no purpose in nature (ateleological) (Table 1-4, Section 1-C).

hypotheses rooted in naturalism

ORIGINS HYPOTHESES (MODELS): (to explain phenomena):

a. EVOLUTION, both *chemical* and *biological evolution*, are changes in matter over time according to natural laws, and are responsible for development of today's complex molecules and living systems.

b. UNIFORMITARIANISM: Geological and biological processes of the past occurred as they would today; therefore, past processes may be explained by extrapolation from present processes. Earth must be billions of years old, based upon these extrapolations. This model is held by some theists as well.

hypotheses rooted in supernaturalism

2. **SUPERNATURALISM** stems from a theistic worldview, and allows that, whereas "natural laws" govern processes in the natural world, "supernatural" events may occur. Indeed, even "natural" laws are manifestations of a "supernatural being" who acts out of purpose.

ORIGINS HYPOTHESES (MODELS):

a. CREATION BY INTELLIGENT DESIGN: A purposeful, "intelligent designer" acted to create the present physical forms, living organisms, and their processes; then, ceased from creating at some time in the past, allowing continuing changes in physical and biological systems within limits.

b. CATASTROPHISM: Geological and biological processes of the past have been periodically interrupted by periods of uncertain length during which unusual conditions and events (perhaps "supernatural") prevailed; therefore, interpretation of past processes by extrapolation from present processes must be done with caution. Earth may be "younger" than uniformitarian estimates.

What is a *creationist*?

The term **creationist** generally applies to theists who have attempted in varying degrees to harmonize scriptural revelation with the natural revelation—*i.e.* the physical and biological world. **Progressive creationists** base their position upon an acceptance of uniformitarian geology and an "old Earth", while believing that God periodically interrupted long periods of natural evolution to create life. The Genesis "days" were actually "day-ages", each of many years. **Scientific creationists** attempt to interpret biological and geological phenomena in view of a literal interpretation of scripture. Genesis "days" were literal 24-hour days; therefore, Earth is quite young. In this chapter, our use of the term, *creationist*, will refer to those who use a literal interpretation of scripture. Theists, who believe God simply used evolution, as presented by evolutionists, to create life, are referred to as **theistic evolutionists**.

Summary:

To summarize, we have noted that natural science cannot falsify hypotheses regarding prehistoric events. However, major questions about the origin of life and the source of biodiversity on Earth involve such nonrecurring events. Therefore, we must distinguish origin science from operation science. Origin science hypotheses are based upon elements of speculation and are consequently influenced by philosophical and worldview bias.

Proceed with caution!!

In view of the uncertainties noted above, I must offer several cautions before you read further. They are as follows:

1. The **scope** of the subject of "origins" is extensive, as is evident from the introduction which touches on geology, theology, and philosophy in addition to biology. Each discipline has a different approach, but there in *one* universe, and *one* right answer to many of the questions of origins. Therefore, *where you begin* and *how you think through the questions* about origins has an important influence on *the answers you obtain*.

2. The **scriptures** present the basis for a theistic worldview. They state clearly *Who* created, but not scientifically *how* He created. Therefore, theists must be careful not to force scripture to "say" what it never intended to say regarding the *how of creation*. Recall the "unfortunate conflicts" of Section 1-C?

3. The **scale** of time for most biological processes we have thus far considered is in seconds to centuries. The origin and genetic variation of species extends at least to many millennia. Therefore, we must put aside arrogance, and humbly recognize the difficulties in thinking according to this time scale.

4. The **space** allotted in this text is limited. Therefore, this chapter is written to help you identify the major considerations regarding origins and variation of life. Many useful references are cited for your further study.

5. Our **strategy** will be to present a brief historical survey, and then examine several physical or biological phenomena as viewed from evolution and creation models. [You may choose any order in which to approach the questions.]

   a. HISTORICAL BACKGROUND — How did contemporary human views of origins develop?

   b. FOSSILS — Are they a record of evolutionary descent?

   c. MUTATIONS — Source of genetic variation or biological entropy"?

   d. BIOLOGICAL CHANGE — Are there limits to variation?

   e. NATURAL SELECTION — Directing force for evolution, or conserver of "kind"?

15-B HISTORICAL BACKGROUND

creationist confrontations

"Creation-evolution" encounters became a focus of attention in our schools, churches, and courts of law twice during this century. In 1925, the Scopes Trial led to a court decision that permitted evolution to be taught along with creationism in public schools. Then, in recent years, creationists have mounted an effective effort to retain creationism in school curricula by supporting a "two-model approach". Having emphasized the importance of worldview and presuppositions in Section 15-A, let us now survey the historical development of the naturalistic evolution model.

complex questions

From the beginning, humans have sought to understand the physical world and their relationship to it. Today, in the information age, we find it hard to imagine living without answers to questions such as the following: "What is the shape of the earth? What does it rest upon? How does it move in relation to the sun and other planets? What is the meaning of the objects that light the sky in the day and night? What is man's relationship to the other creatures? To whom is man responsible? To God? To the gods? To anyone? What is man's purpose and destiny?

curious creatures . . .

The Scriptures indicate that some of these questions were answered for the first man and woman by divine revelation. They knew that they were created by God, responsible to Him, given dominion over the physical and biological world, and challenged to apply God-given rational capacity to study and to understand the creation so that they might exercise wise stewardship. No one knows how many of the questions about the Earth and the starry universe were answered before the fall of man. But rebellious mankind at the fall, in

. . . who prefer the "darkness"

effect snapped off the light, and began to feel his way around in the darkness of creation in an effort to find his place in it. [Read Romans 1:19-22] The Word of God has remained, but history records the attempts of "wise men" to construct their own models of the world around them.

ARISTOTLE (384-322 BC) — FIRST CLASSIFICATION OF LIFE

Western "natural philosophy" was greatly influenced by the Greeks, including Plato and Aristotle, whose wisdom and insight concerning the order of nature were studied and respected for centuries. Aristotle (384-322 BC) developed his own classification system based upon his extensive studies of the natural world around him. His classification, the so-called *Scala Naturae* or "ladder of life", presents a progression of forms of matter from inanimate matter at the bottom, through progressively increasing complexity of plant and animal life. Man, with his unique rational capacity, was at the pinnacle. This is the first theoretical model to be recorded in the history of biology.

ladder of life

a great chain . . .

. . . with "fixed" links

By the time of the Renaissance, Aristotle's "ladder of life" had become ingrained into the worldviews of many prominent scholars. The "ladder" had been expanded into a Great Chain of Being which even included the spiritual beings. Each link in the chain was *fixed*. Therefore, understanding life and the universe was simply a matter of discovering, naming, and describing each link. Each "kind" of creature, or **species** (L. *species* = a kind) was viewed as having a distinct and fixed position in the order of things. Those who acknowledged the authority of the Scriptures were in agreement with this concept of "fixity of kind" as reflecting the created "kinds".

LINNAEUS (1707-1778) — BINOMIAL CLASSIFICATION:

a need for classification

Are the "kinds" *species*?

With the discovery of the New World and voyages of the 16th century, came an awareness of an even greater diversity of creatures. Naturalists were awed by the thousands of new plant and animal species. What's more, each seemed to "fit" into its particular habitat. The existing classification system was swamped and needed revision. Here, Carl von Linne' (Linnaeus) eventually made his great contribution of the binomial nomenclature (Section 3-C). He believed that his system could eventually represent the classification of the original "kinds" created in the garden of Eden. The concept of fixity of kind remained in acceptance, although there is evidence that Linnaeus questioned whether there was a one-to-one relationship between *species* and *created kind*. The Great Chain of Being, even though much expanded with many more "links", nevertheless continued as a unifying principle.

As noted earlier (Section 9-A), taxonomic classification is based primarily upon observed similarities and differences in anatomy. Therefore, as the great variety of species were classified into the hierarchy of genus, family, order, and so on, many began to see the biological realm in a new light. The *diversity (separations between species)* was still evident, but *unity (similarities among taxonomic groups)* also became evident. For example, all vertebrates were observed to have the same basic body plan with vertebral column, two forelimbs and two hind limbs. The limbs of the various species represent **homologous structures**, in that they all had the same bone-like composition and form, although they differed to fit various functions (e.g. wings, flippers, legs) (Figure 15-1).

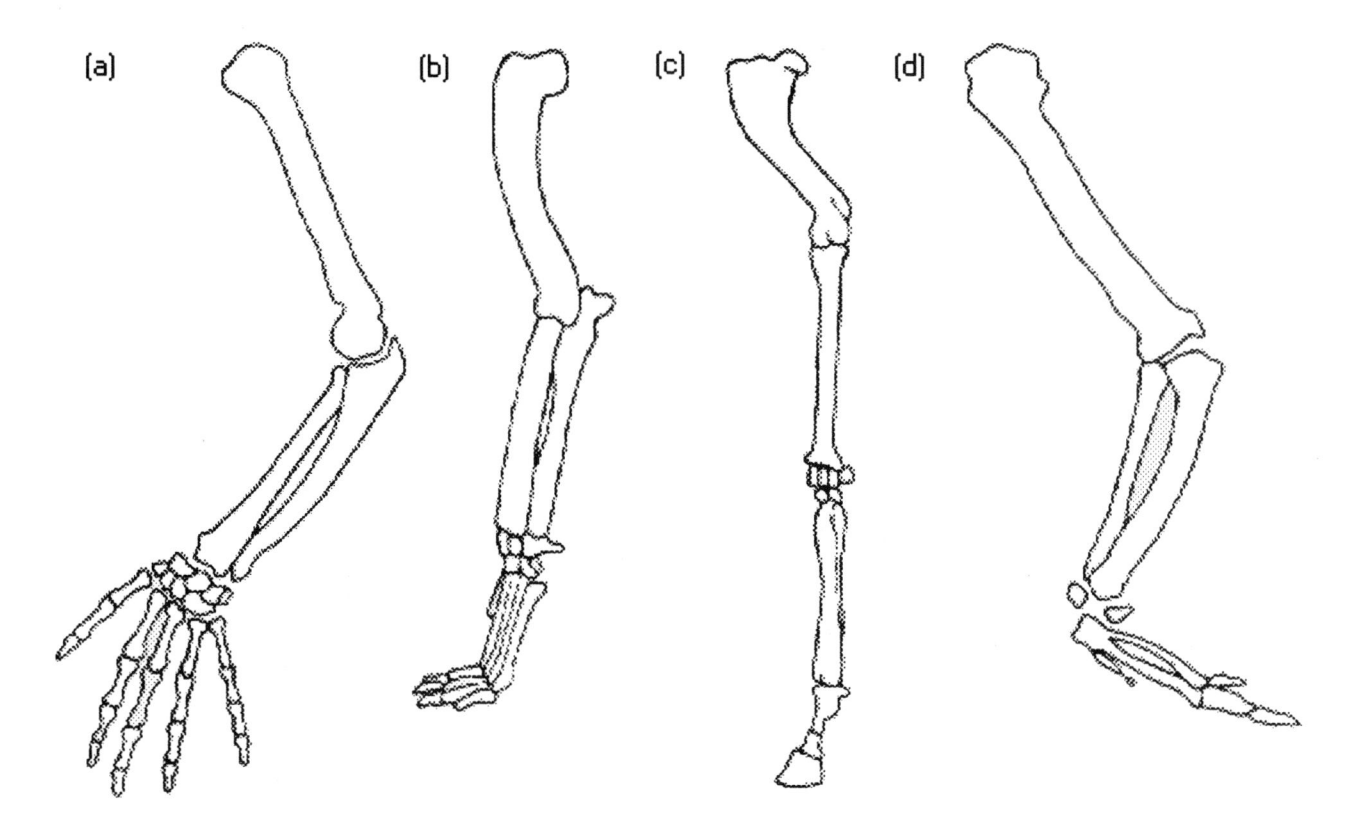

Figure 15-1. Homologous structures as illustrated in the forelimbs of (a) human, (b) dog, (c) horse, and (d) bird.

a creationist contradiction?

Perhaps less obvious than the homologous structures were the **vestigial structures**, body parts that had no known role in body function and often appeared to be a shrunken form of a previously prominent body part. For example, the whales and some snakes were observed to have what appeared to be remnants of pelvic and leg bones; human skeletons showed remnants of a tail (Figure 15-2). If these were created kinds of organisms, supposedly fixed and unchanging, how could vestigial structures be explained?

LAMARCK — INHERITANCE OF ACQUIRED CHARACTERISTICS

Lamarck: species advance up the "chain"

Hypothesis . . .

Jean-Baptiste Lamarck (1744-1829), who coined the term *biosphere* (Section 3-B), also acknowledged the Great Chain of Being. However, Lamarck did not view species as fixed links, but as derivations from similar, created forms that had advanced upward in the chain of being in response to some innate drive. As evidence for his theory, Lamarck cited certain changes in the body structures of animals (and man) in proportion to their use or disuse. Muscle development in athletes and the continual stretching of giraffe's necks to forage in trees caused increase in size. Lamarck believed that these advances in development in each generation were passed on to the next. Likewise, disuse and resultant atrophy of body structures could be reflected in offspring, hence the origin of vestigial structures. Lamarck's proposal of the *inheritance of acquired characteristics* was published in 1809 to account for the observed fitness, or adaptation of each species to its environment.

324

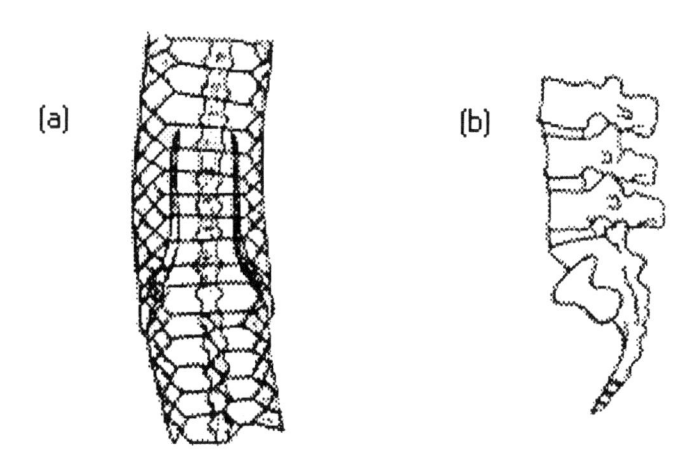

Figure 15-2. Suggested vestigial structures: (a) apparent vestiges of a pelvic girdle in the python, and (b) coccyx (apparent vestige of a tail) in the human.

invites deductions . . .

and testing . . .

Lamarck's proposal was quickly attacked by his contemporaries who saw it as a sweeping assertion with little experimental support. For example, one deduction from his proposal would be as follows: If acquired characteristics are inherited, then weight lifter's *sons* should be distinguishable from banker's *sons*. Even though no clear evidence has even been produced to support Lamarck's proposal, at least it was subject to experimental testing. Modern studies of the effect of environment on gene expression suggest that an *individual's* characteristics can be modified within limits (e.g. weight lifting, skin tanning) (Section 12-H). Thus, we can agree with Lamarck on "acquired characteristics" by use or disuse. However, there is no evidence that these characteristics are *inherited*; that is, passed on in the gametes. Use or disuse of *somatic cells* apparently has no direct influence upon the hereditary material passed on to the offspring in *gametes*.

but no support.

In spite of his somewhat careless extrapolations from limited data, Lamarck did leave his mark. He served to stimulate others to question the fixity of species, and to consider the possibility that evolution might occur.

DARWIN (1809-1882) — DESCENT WITH MODIFICATION

In 1831, young Charles Darwin embarked on an ocean voyage upon the *H.M.S. Beagle* of the British fleet. This voyage changed his life and greatly affected the world. He had failed at medical school, to the dismay of his father who was a noted physician, and the son of another physician (Erasmus Darwin, 1731-1802). Charles had instead trained for the clergy, but became more interested in natural history. Thus, he welcomed the invitation to sail around the world.

Darwin's data

While in South America, Darwin was impressed by the variation in life forms, each seemingly fitting into a particular ecological role, or **niche** (Section 7-A) in its geographic location (habitat). Many of the South American mammals seemed to occupy the same niche as their counter-parts in Europe but were slightly different in anatomy and behavior. These species were obviously different from European mammals; but were they totally unrelated?

His visit to the Galapagos Islands, 600 miles off the west coast of Ecuador, made further impressions. Here on isolated volcanic islands was an incredible array of diversity. He found the bird species especially interesting because they were remarkably similar in some respects, yet different in others (Figure 15-3). Darwin collected specimen's, assuming that they were various species of finches, blackbirds, and wrens. These species differed primarily in their feeding niches. Some were insect eaters, others were seed eaters; some foraged in trees, some on the ground. The bill structure of each species seemed to fit its feeding habits (Figure 15-4).

Upon return to Europe, examination of Darwin's specimens by experts revealed that nearly every one of these species was a finch. Furthermore, Darwin had observed one finch species on the South American mainland, 600 miles from the islands, that resembled each of the island species.

Darwin's hypothesis

Some two years after returning on the *Beagle*, Darwin recorded in his journal his hypothesis that all of the island finches descended from a single pair of finches that had been blown from the mainland thousands of years prior. Having stated this hypothesis, he puzzled as to how one species could have given rise to the many varied species on the islands.

incorporating Malthus' views

Darwin had made some stimulating factual observations. How could these be interpreted? At the beginning of this chapter, we noted the influence that one's presuppositions and worldview can have upon his interpretations of factual data. Here we should note that Charles Darwin was greatly influenced by Thomas Malthus' book, *Essay on the Principle of Population*. In his book, Malthus predicted that each population has a tendency to increase faster than available food supply, leading to a condition of struggle for survival. Darwin had seen the apparent waste in nature brought on by the tendency of each species to overpopulate its habitat. Therefore, he adopted Malthus' ideas of competition and struggle for existence for his theory of evolution.

Darwin's analogy

Whereas there was apparent overpopulation and competition on the one hand, Darwin also remembered the incredible variation among species, each seemingly fitting into a particular niche. He was also familiar with the great diversity that "breeders" could produce by **artificial selection** within species of domesticated plants and animals. The breeder decides upon certain criteria such as body size, growth rate, agricultural yield, or nutritional value, and selects individuals that are the most suitable, generation after generation.

<u>**Cactospiza pallida**</u>

Probing bill; uses cactus spines to probe for insects in cactus

<u>**Camarhynchus psittacula**</u>

Grasping bill with biting tip; carnivorous

<u>**Geospiza magnirostris**</u>

Crushing bill; ground-feeding on seeds, nuts

Figure 15-3. Representative species of Darwin's finches.

Darwin saw as much or more diversity *within* some domesticated species as *between* species. He began to view the *species* as simply a transitional stage in a continual slow progression of change, or evolution. Thus, in addition to *competition*, Darwin added the component of *variation* to his theory.

Competition, variation, and selection . . .

Within the context of competition for survival, Darwin reasoned that those variations that made an individual more suited to compete for scarce resources would enable it to survive and reproduce. The offspring which inherited these traits would tend to appear different from previous generations in that a larger proportion of the population would have the traits of the successful competitors. He called this process **natural selection**. Over long periods of time, new species would originate as new habitats were encountered (e.g. the Galapagos Islands). The pioneer species entering these new areas would begin to diverge in appearance from one another over many generations in response to competition and natural selection. Thus, one species would evolve into two species, and so on.

. . . lead to new species.

In 1859, Darwin published *The Origin of Species*. In it he outlined what he believed to be the cause of the *unity* of common features among living organisms on the one hand, and the *diversity* on the other hand. The cause was **evolution**, or descent with modification through natural selection. The unity, he reasoned, was due to common descent. The diversity of life was the result of modification under the direction of natural selection.

evolution view

In addition to his extensive studies of natural history, Darwin drew upon other observations to support his idea. These included the apparent changes in the fossil record in geologic strata, and the fact that a great variety of past life forms were now extinct. In early embryological stages among widely divergent taxonomic groups of organisms, he saw close similarities in development. He pointed to vestigial structures as evidence of change in body form, or morphology. Darwin saw that the taxonomic classification system was beginning to reflect the phylogenetic descent of complex organisms from simpler forms.

the world responds

The *Origin of Species* was widely circulated and produced various responses. For some, there was an immediate appeal because of its apparent simplicity, and many who knew little about biological processes suddenly began to view the world of life in a new way. There was an intellectual fulfillment in this seemingly comprehensive grip on the biological world and its processes. Darwin had also put a new face on the "gloom-and-doom" prophecy of Malthus. We may look forward with new confidence, said Darwin (9),

> to a secure future of equally inappreciable length. And as natural selection works solely by and for the good of each being, all corporeal and mental endowments will tend to progress towards perfection.

three concerns

In spite of its wide acceptance, Darwin's ideas needed experimental verification. In particular, Darwin identified three areas of major concern. First, what was the source of variation that would provide different degrees of fitness? Second, how could natural selection lead to perfection of complex adaptations such as the eye? After all, the eye is amazingly complex and appeared to have evolved separately in three different taxonomic groups—the arthropods, mollusks, and vertebrates. Darwin's third concern was the fact that the fossil record seemed inconsistent with his idea of descent with modification. Where were the *transitional forms* one would expect if evolution had occurred?

Darwin was unable to gather experimental evidence for his ideas, and relied increasingly upon speculation and personal convictions in his defense against mounting doubt and criticism. Furthermore, while no one could demonstrate the occurrence of evolution, neither could it be clearly denied. W.R. Thompson (41) explains the "*elusive character* of the Darwinian argument" as follows:

> Every characteristic of organisms is maintained in existence because it has survival value. But this value relates to the struggle for existence. Therefore, we are not obliged to commit ourselves in regard to the meaning of differences between individuals or species since the possessor of a particular modification may be, in the race for life, moving up or falling behind. On the other hand, we can commit ourselves if we like, since it is impossible to disprove our statement. The plausibility of the argument eliminates the need for proof and its very nature gives it a kind of immunity to disproof.

Even as Darwin was seeking valid support for his ideas, an Austrian monk was carefully collecting data from plant breeding experiments, and formulating principles of heredity. As you may recall, the monk was Gregor Mendel, and his principles of genetics would cast evolution in a new light.

MENDEL (1822-1884)—PRINCIPLES OF HEREDITY

Mendel's principles of heredity (Section 12-C) had been published in 1866, shortly after the publication of Darwin's ideas. However, Darwin and most others were apparently unaware of the relevance of Mendel's work. Darwin was in fact unable to distinguish clearly between variations induced by environmental factors (e.g. use and disuse of body structures) and those that were hereditary and subject to natural selection. This caused Darwin to admit elements of Lamarck's theory into his thinking.

evolution came before genetics

Both Lamarck and Darwin were greatly handicapped because their efforts predated the unfolding of principles of heredity. Such crucial concepts as *gene*, *chromosome*, *independent assortment*, *mutation*, and *gene pool* were unknown to them. Table 15-1 lists some of the major milestones of genetics in the early 20th century. We should realize two things from a perusal of these names and their contributions. First, Darwin's 'theory' was proposed without the benefit of a knowledge of the most elementary principles governing the inheritance of traits. Second, now that these principles are available, Darwinism must be re-examined for its validity. Indeed, this is already happening in several sectors of modern biology.

Table 15-1. Major Milestones in 20th Century Genetics.

CONTRIBUTOR	MAJOR CONTRIBUTION
Fredrich Miescher (1844-1895)	Discovered nucleic acids (1869), which he named "nuclein"
August Weismann (1834-1914)	Distinguished *germ cells* (gametes) from somatic cells, the former being the carriers of hereditary traits; important step in dispelling Lamarckianism
Hugo de Vries (1848-1935)	Uncovered Mendel's publication in the literature; invoked gene mutation to explain changes in evening primrose plants
E.C. McClung (1870-1946)	First to demonstrate the role of sex chromosomes as studied in grasshoppers (1902)
Walter S. Sutton (1876-1916)	Explained Mendel's principles on the basis of chromosome behavior (1903)
Thomas Hunt Morgan (1866-1945)	First to observe mutation and its inheritance (white eye, in *Drosophila*, the fruit fly); established chromosomal basis for inheritance
G.W. Beadle (1903-) E.L. Tatum (1909-1975)	Discovered that a gene acts by coding for synthesis of an enzyme or other protein that influences metabolism; they studied bread mold, *Neurospora crassa*
J.D. Watson (1928-) F.H. Crick (1916-)	Discovered the molecular configuration of DNA, "the double helix" (1953)
Molecular Biology Research (1970's-)	Recombinant DNA techniques: cutting DNA, transferring and recombining with DNA from other sources

NEO-DARWINISM — EVOLUTION IN LIGHT OF GENETIC THEORY

Charles Darwin died in 1882, having presented biology with a radically different worldview. Natural selection, acting upon organisms with varying characteristics and in competition for limited resources, is producing gradual transformations from primitive to the highest forms of life. Evolution was proposed to occur over millions of years without necessarily any obvious intervention of a supernatural force.

20th century rejection of evolution . . .

The only "fly in the ointment" was Darwin's inability to explain the mechanisms involved; in particular, the source of variation for selection to act upon. This weakness, among others, led to the widespread rejection of Darwinism when Mendelian genetics became known. Indeed, for the first three decades of the 20th century many prominent biologists denied that natural selection could be a major agent for evolutionary change.

...until "modern synthesis"	**Neo-Darwinism** (G. *neo* = new, recent), or the **modern synthetic theory** of evolution represents a revision of Darwinism in light of new understandings in the field of genetics. The modern synthesis received major attention following the 1937 publication of *Genetics and the Origin of Species*, by Theodosius Dobzhansky (1900-1975). Neo-Darwinism is a synthesis of interpretations by Dobzhansky and others of the information from four major areas of research—Mendelian genetics, chromosome theory, mutation theory, and population genetics (see Table 15-1).
something tangible	Mendelian principles and chromosome theory provided tangible, particulate, hereditary "substance" (genes and chromosomes) necessary to provide unity and continuity for an evolutionary progression of life. However, as Darwin's critics questioned, "Where is the source of new hereditary traits?" According to Neo-Darwinism, mutations are the ultimate source of all genetic variation. Mutations produce evolutionary change when they are passed to subsequent generations, and their frequency of occurrence is maintained in a population. How would this happen?
population genetics	The field of population genetics has contributed to the Neo-Darwinian emphasis that *populations*, not individuals, evolve. Each population, or interbreeding group has its own gene pool, and is a unit of evolutionary change. Under certain conditions, the gene (or allele) frequencies in a population may remain constant from generation to generation. This is *genetic equilibrium*, as described by the so-called *Hardy—Weinberg principle* to be discussed in Section (15-G) Evolution is said to occur when allele frequencies change, due to (a) mutations, (b) natural selection, (c) migration, or (d) chance events.
	Neo-Darwinian evolution is thus defined as a series of changes in allele frequencies in a population, or gene pool. When a particular mutation that provides a survival and reproductive advantage appears, that allele becomes the norm of that population. Neo-Darwinian natural selection is actually *differential reproduction*. That is, if changes in allele frequencies are to occur, the bearers of favorable alleles must not only survive, but bear offspring so that favorable alleles and resultant phenotype become more frequent in the succeeding generation.
Remember the cautions...	Having surveyed the history of Neo-Darwinism, we are ready to consider several aspects of the origin and genetic diversity of life from both evolution and creation perspectives. Before preceding, you may wish to review the cautions appearing at the end of Section 15-A regarding *scope, Scriptures, scale, space,* and *strategy*. Our intent is to briefly present a foundation upon which you can build as you consult other sources such as those cited at the end
respect other views, and ...	of the chapter. The ultimate aim should be a better understanding of the biological world, and a recognition and respect for worldviews held by others. In addition, if you are a "creationist", you should aim to develop the mind and
... seek truth.	lifestyle that is consistent with your belief in a Creator who has given us dominion over His creation.

15-C   FOSSILS — RECORD OF EVOLUTIONARY DESCENT?

Geology began with creation worldview.	As discussed in Section 1-C, experimental sciences were launched during the Renaissance and Protestant Reformation. Nicholas Steno ((1638-1687) made careful studies of exposed sedimentary rock layers of the Earth's crust, and his published work in 1669 marked him as one of the fathers of modern *geology* (27). His interpretations may be summarized as follows:

1. **Sedimentary rock strata** (layers) are formed by deposits of sediments laid down upon preexisting layers by water.

2. **Vertical sequences** of strata reflect order of deposition, provided no disruption has occurred since their deposit.

3. **Fossils** are the impressions or remains of once living organisms, and *not* formed by the rocks themselves. Fossil species differ among different strata, and can be used to identify rock layers — now known as *index fossils*.

4. **Genesis flood** caused the catastrophic burial of organisms, some of which are preserved as fossils.

flood and catastrophism

From the influence of Steno and other geologists who believed in a universal Genesis flood, *catastrophism* became a dominant interpretive model of early geology (Section 15-A). To them, rock strata were formed by catastrophic events and forces involving flood waters. While they saw the progression of different fossil species in different rock layers, they did not interpret the arrangements to be a record of life forms deposited over long periods of time while evolution was occurring. Little noted in today's geology texts is the fact that geology was launched by godly men who based their interpretations upon literal interpretation of Scriptures.

How old is the Earth?

In the 1700's, secular and atheistic influences swept into geology. Georges Cuvier (1769-1832) combined his belief in an "old Earth" with catastrophism by proposing that successive strata with different fossils were records of alternating destructions and new creations of life. His work, and that of William "strata" Smith (1769-1839), a canal engineer, led to the development of what is now the system used by geologists for classifying strata. What remained was the question of "time". How long did it take to deposit these layers?

Lyell's uniformitarian view . . .

and Darwin's evolution . . .

Charles Lyell, in the 1830's, brought the *uniformitarian* interpretation of the geologic strata to prominence. In this view, geologic events of the past should be interpreted based upon the presupposition that geologic events and their *rates* in the past were the same as they are in the observable present. Extending his logic, because rates of change in the Earth's crust are happening so slowly in the present, the rock strata must be extremely old. Lyell's uniformitarian view set the stage perfectly for the advent of his friend, Charles Darwin's evolution by natural selection. [How is this true?]

combined to reject the supernatural.

Darwin's model became to biology what Lyell's was to geology—a interpretive model which sought to interpret past events by a logic which excludes "supernatural" interventions or causes. Thus, Neo-Darwinists today interpret the sequence of fossil-bearing strata as an historical account of the progression of evolutionary change over a period of several billion years (Figure 15-4).

uniformitarian interpretations

Uniformitarian assumptions are necessary to determine the ages of rock strata. For example, using uniformitarian logic, **radioactive isotopes** found in mineral and organic deposits are assumed to be "ticking clocks" with constant decay (nuclear change and emission of radiation) rates, which "tell" the age of

radioisotope "dating"

samples. By extrapolating backwards in time using the known rate of decay, the age of strata can be estimated. However, there is good evidence that natural processes can "reset these clocks", or otherwise make questionable the uniformitarian assumptions. Creationists present alternative interpretations (7) which suggest a much "younger" Earth than the estimated ages of geologic strata given in Figure 15-4.

Does evolution account for ...

Regardless of the age of the Earth, if gradual evolution of life was occurring and being recorded periodically by fossil preservation, the following *predictions* should match *observed characteristics* of the fossil record:

TRANSITIONAL FORMS

... the fossil record?

PREDICTION: If evolution produces to new species from ancestral forms over time, then the fossil record should show gradual variations so that phylogenetic lineages could be reconstructed using fossils from older to successively newer rock strata. After all, according to Neo-Darwinism, *all species* are "transitional forms", not "fixed links".

OBSERVATION: Paleontologists have not been able to decipher a phylogenetic line of any taxonomic group of organisms back to simple ancestral forms (18). The transitional forms one would expect if descent with modification were occurring are simply not present. Darwin himself was frustrated by these "missing links" and admitted that this "is the most obvious and serious objection which can be urged against this theory" (9). Instead, one finds the following:

"**Explosions**" — the appearance of 25 to 50 multicellular invertebrate phyla and vertebrates (fish) in the Cambrian strata (Figure 15-4) and the absence of any sequence of simpler ancestral forms in the underlying Precambrian (18, 7). This event is termed the *Cambrian explosion*. Equally impressive, most mammals "suddenly appear" in the Eocene strata, and 50 families of flowering plants in the Cretaceous period (30). Such apparent gaps in the fossil record provide insufficient data for construction of "phylogenetic trees" (Figure 9-1).

EXTINCTIONS

... patterns of extinctions?

PREDICTION: If natural selection operates upon populations to alter gene frequencies, favoring some forms and causing death and extinction of others, then extinction resulting from genetic selection should be gradual and appear throughout the geologic column. Catastrophic events would also occur and cause extinction, but genetic selection could not be expected to depend significantly upon such chance events. [Why?]

OBSERVATION: Continual, gradual extinctions are not evident. Instead, a few mass extinctions apparently occurred in which mass burials of species are associated with possible *catastrophe*. In the Permian Extinction, "half of the families of marine invertebrates and probably more than 90 percent of all species" appear to have become extinct. At the end of the Cretaceous era (Figure 15-4), the dinosaurs were exterminated. These extinctions do not seem to reflect "the sort of thing for which a plausible genetic mechanism exists." (20)

ERA	PERIOD	EVOLUTIONARY EVENT	YEARS, B.P.

CENOZOIC	Quaternary
	  Recent Epoch
	  Pleistocene Epoch
	Tertiary
	  Pliocene Epoch
	  Miocene Epoch
	Oligocene
	Eocene
	Paleocene
		Rise of modern plants,
		animals, and man
		
		Rise of mammals and
		development of the
		highest angiosperms
			25,000
			975,000
			
			12,000,000
			25,000,000
			35,000,000
			60,000,000
			70,000,000

MESOZOIC	Cretaceous	Modernized angiosperms and insects; extinction of dinosaurs, flying reptiles	70,000,000
	Jurassic	Appearance of first birds (reptilian); first of highest forms of insects; first primitive angiosperms	to 230,000,000
	Triassic	Earliest dinosaurs, flying reptiles; primitive mammals; Cycads and conifers common	
PALEOZOIC	Permian	Rise of primitive reptiles, conifers; first modern corals	
	Carboniferous	Earliest known insects; spore-bearing plants; rise of amphibians; peak of crinoids	
	Devonian	First known seed plants and amphibians	230,000,000 to
	Silurian	Earliest known land animals; primitive land plants; Rise of fishes	570,000,000
	Ordovician	Oldest primitive land plants	
	Cambrian	All major phyla of invertebrates; and, vertebrates (fish) represented	
PROTEROZOIC (Precambrian)		Primitive water-dwelling plants and animals Oldest definite fossils Oldest dated rock strata Estimated origin of Earth	1,000,000,000 3,500,000,000 4,000,000,000 5,000,000,000

Figure 15-4. Major divisions and proposed evolutionary events of geologic time based upon radioisotope dating methods. Adapted from Whitcomb and Morris (43). Years BP (before present) and description of events do not reflect the thinking of this author.

... "living fossils"?

In addition to relatively few extinctions, one finds evidence of **stasis**, persistence through successive periods of geologic time with little or no change in morphology. Stasis is best illustrated in the so-called *living fossils*, listed below with their duration in the fossil record, assuming an "old Earth" chronology (44):

> Mosses, ferns, horsetails .... 400,000,000 years
> Starfish .... 500,000,000 years
> Dragonfly .... 170,000,000 years
> Ginkgo tree. (*Ginkgo biloba*). .... 200,000,000 years
> Bats .... 50,000,000 years
> Alligator .... 35,000,000 years

In conclusion, there appears to be no evidence for evolutionary transitions from one species to another, either from studies of living populations, or from the fossil record. As Parker (33) has observed, "Rather than forging links in the hypothetical evolutionary chain, the wealth of fossil data has only served to sharpen the boundaries between created kinds."

Have we found enough fossils?

Evolution proponents who explain the absence of transitional forms by suggesting that not enough fossils have been unearthed are also being challenged. In support of the near completeness of the fossil record, Denton (11) cited data indicating that most *living* taxonomic orders and families of vertebrate animals have now been found as *fossils*—97.7% of orders, and 87.8% of families.

The position of Neo-Darwinism has clearly been weakened. Several individuals including the notable astronomer Sir Fred Hoyle have abandoned evolution, saying in effect that he could no longer deny the obvious. Others have proposed alternate models in place of Neo-Darwinism.

dealing with the "gaps" —

punctuated equilibrium

Stephen Gould and Niles Eldredge are proponents of **punctuated equilibrium** (16). This model claims to address the two puzzling features of the fossil record that we have just noted—the "explosions" and the "living fossils". Punctuated equilibrium proposes that the abrupt appearance of new species is due to the fact that evolution occurs in relatively brief episodes in which *speciation* (evolution of new species) occurs. According to Gould, "In any local area, a species does not arise gradually by the steady transformation of its ancestors (i.e. Darwinism; Neo-Darwinism); it appears all at once and "fully formed" (16). This statement describes the *punctuated* aspect that interrupts long periods of little evolutionary change; that is, stasis or *equilibrium*. The periods of stasis would explain the other Neo-Darwinian dilemma, living fossils.

Where's the mechanism?

Like Neo-Darwinism, punctuated equilibrium is more descriptive than it is explanatory. The proposal fits the fossil record nicely but the mechanisms involved lack clear scientific support. [See TOPICS FOR FURTHER STUDY #5.] Whereas creationists have directed much criticism at the evolution model, they have also proposed alternative explanations for the geologic formations and fossil deposits. The next section addresses this work.

15-D FOSSILS — CREATIONIST INTERPRETATIONS

We have already noted the dominant influence of the theistic worldview in the birth of geology as a science. The belief in the Genesis flood as the catastrophic cause of the fossil-bearing strata was well accepted before the entry of

creationists revive "flood models"

the uniformitarian view in the 1800's. The influence of Lyell and Darwin swayed thinking toward the naturalistic, uniformitarian interpretation in geology and biology, respectively. This view has remained predominant view of the "scientific establishment" to this day. However, since the late 1960's, creationists have increasing attracted attention by proposing "flood models" to account for fossil-bearing deposits. They have promoted their interpretations through lectures, publications (e.g. 32, 43), and professional affiliations (see footnote, REFERENCE # 31).

flood evidence:

Biblical accounts describe the Noachian flood as worldwide, violent, and capable of massive alterations of the landscape (Genesis 7: 19-21; 8:1; II Peter 3:6). The following characteristics of sedimentary deposits have been cited by Coffin (7) as evidence for such a catastrophic geologic event:

1. ORGANIC DEPOSITS: Tremendous numbers of plants and animals now form widespread deposits which often give the appearance of violent, rapid burial. Some massive deposits appear to have been transported great distances. The excellent preservation of the organic deposits would not occur without isolation from decomposers.

2. MASS EXTINCTIONS, noted in Section 15-C, are better explained by catastrophic than by uniformitarian geology.

3. CLIMATE CHANGES occurred as suggested by the fact that a predominance of fossilized plants and animals were adapted to a tropical or subtropical environment.

4. FLOOD LEGENDS have been passed along within human cultures all over the world.

Physical characteristics of the sedimentary deposits are consistent with violent action of water, as Coffin (7) summarizes:

> It is significant that the greatest erosion recorded in the crust occurs between the top of the Precambrian and the bottom of the Cambrian. [Possibly the first surface impacted by flood erosion prior to sediment deposition.] From the lower Cambrian upward we usually observe the sediments passing through a sequence from coarse to fine, a fact that fits in well with the action of rising and deepening water.

"flood model" interpretation of the geologic column

If a catastrophic, worldwide flood is responsible for the geologic strata as reported in Figure 15-4, how do creationists account for the obvious sequence of fossils from the Cambrian upward? The geologic column is seen by many creationists as a reflection of pre-flood ecological distribution of species. Instead of interpreting fossilized species in upper layers as being more evolutionarily advanced, they are seen as an "ecological sequence" reflecting increasing elevations of their respective habitats from ocean depths to upland landscapes. Table 15-2 outlines in very general terms the proposed relationship between "ecological zones" at different elevations and the sequence of fossilized animal species in the geologic column.

Table 15-2. Two Different Interpretations of Fossil Deposits in Geologic Column[1].

INTERPRETIVE MODEL ---> SEQUENCE DETERMINED --->	UNIFORMITARIAN (As Appear in Evolution)	CATASTROPHIC (By Elevation of Habitat)	
REPRESENTATIVE FOSSILS	GEOLOGIC ERA	ELEVATION	HABITAT
Birds, Mammals Reptiles Insects	Lower Cenozoic and Mesozoic	High | | |	Uplands | | Lowlands
Amphibians	Paleozoic	Sea Level |	and Shores |
True Fish Sharks Brachiopods Trilobites	Lower Paleozoic	Shallow | Deep	Open Ocean | Sea Bottom

[1] Modified from Coffin, 1983. *Origin by Design.* Review and Herald Publ.

new respect for catastrophism

In summary, the geologic deposits of the Earth's crust may be interpreted from both evolution and creation perspectives. As noted in Section 15-A, neither are falsifiable. However, the pattern of fossil deposition and the content of the fossil record would seem more consistent with a catastrophic origin. Indeed, many geologists today, regardless of their view of origins, are giving catastrophism a more prominent place in their interpretations. Origin scientists should compare their respective models to see which provides the most plausible explanation for geologic phenomena, past and present. [Why isn't there more free exchange between scientists of all persuasions?]

15-E MUTATIONS — SOURCE OF VARIATION? BIOLOGICAL ENTROPY?

What do Concord seedless grapes, cute little Ancon sheep, and naval oranges have in common? They are the result of natural mutations. Mutations are natural or laboratory-induced alterations in the genetic blueprint. **Gene mutations** and **chromosomal abnormalities** (chromosomal mutations) can produce heritable genetic variation (Section 12-G). Let's analyze their significance.

GENE MUTATIONS

Gene mutations consist of changes in nucleotide sequence of the DNA molecule, either by addition, deletion, or change in a single nucleotide pair (Sections 11-C, 12-G). The altered sequence will code for an altered amino acid sequence in proteins, which in turn, may affect metabolism adversely. Section 12-G discusses a variety of genetic disorders caused by mutations.

Are mutations the source of variation . . .

The discovery in 1910 that gene mutations do occur in natural populations led to what seemed to be the answer to Darwin's search for a source of genetic variation. According to Neo-Darwinism, mutations are the ultimate source of variation as implied in this statement of Dobzhansky, *et.al.* (13)

... from a few original forms?

Changes in the hereditary materials are known as mutations. More than two million species of organisms live on earth, descendants of one or only a few primordial forms of life. Without the occurrence of hereditary changes, life could not have evolved or diversified....

Testing the model ...

In examining the scientific evidence for this proposal, we should first note that mutations can occur spontaneously or may be induced in the laboratory by chemicals or irradiation. They are detectable by a change in phenotype or by analysis of the DNA or chromosomes. Consider the following *prediction* from the evolution model, and actual *observations*:

PREDICTION: If spontaneous mutations over millions of years have produced the genetic variation (new and different alleles) upon which natural selection has operated to produce evolutionary change and new species, then induced mutations in populations under controlled experiments should produce at least some beneficial phenotypic alterations which improve fitness, and thus, are retained in the gene pool.

by induced mutations ...

... gives little support.

OBSERVATIONS: As we have already noted, most mutations are not beneficial and may be so lethal that viable offspring cannot be produced. Such mutated alleles would tend not to be retained in the gene pool. Note the following reports:

1. According to geneticist W.E. Lammerts, many mutant roses showed useful horticultural variation such as increased petal number and more suitable coloration. However, each mutant was physiologically weakened and less fit to survive varying garden conditions (23).

2. The few laboratory-induced, "beneficial mutations" in the fruit fly, *Drosophila*, were generally inferior to the "wild type" alleles. For example, female flies refused to mate with mutant white-eyed males as long as normal males were present (7, 39). Outside the laboratory, natural selection would reduce the frequency of the "white eye, mutant allele". Why? Not because the mutant fly could not *survive*, but because *reproduction*, necessary to pass the allele along, would be less likely. Thus, even so-called "beneficial mutations" may not confer favorable survival value. Pierre Grasse, French zoologist noted that *Drosophila* "seems not to have changed since the remotest of times."

3. Bacteria and viruses have extremely rapid rates of reproduction. Therefore, as expected from the more frequent DNA replications, mutations do occur more often. Gene mutations that confer upon microbes a resistance to antibiotics are frequently reported, but these involve rather simple, metabolic changes (see Section 15-G).

What is origin of gene complexes?

If mutations are to be the source of new hereditary variations, whole complexes of genes must be altered (10). *Gene complexes*, consisting of many genes operating together, are responsible for the development of body organs and even metabolic pathways. How could these gene complexes have appeared by mutations in the first place without natural selection eliminating mutant alleles before all of the favorable genes were accumulated? To answer this concern, some have proposed *macromutations* (or *saltations*) in which a gene complex is systematically introduced or altered all at once, producing a markedly different organism or phenotype. Darwin, and most darwinians, avoid invoking saltations in effect because a naturalistic theory should not need "miracles".

CHROMOSOMAL MUTATIONS

Changes in chromosome number cause variation.

Genetic variation has been produced by alteration of chromosome number. *Individual* chromosomal mutations are so harmful that their importance in evolution is questioned (Section 12-G). However, alteration of chromosome number involving *whole sets* of chromosomes can produce what may be regarded as new species. This condition, known as **polyploidy**, can result when meiosis fails to convert the diploid chromosome number (2n) to haploid gametes (n). Instead, gametes arise with the full diploid number (2n). When diploid sperm and egg fuse, *tetraploid* (4n) individuals are formed, having four sets of chromosomes. Other possibilities can also arise (e.g. triploid (3n), hexaploid (6n), etc.). However, those that produce *fertile* offspring usually have an even number of sets of chromosomes (i.e. 4n, 6n, etc.), as required by homologous pairing in meiosis.

Polyploidy has been most commonly observed in the plant kingdom. Plants do not have *sex chromosomes* that control reproductive physiology as is the case in many animals (Section 12-F). Several important agricultural crops are "natural" polyploids. Others are produced by geneticists using *colchicine*, a drug which is used to control meiotic division to produce gametes with a desired number of chromosome sets. Wheat species are either 2n, 4n, or 6n, but their supposed wild ancestors in Southwest Asia are diploids. The tetraploid potato originated from diploids in South America. Other polyploid crops include alfalfa, cotton, strawberry, and tobacco. Higher growth rates and yield in these species is often associated with the polyploid condition.

unlikely means of evolution

Polyploidy may improve the quality of a species, but does not appear likely as a mechanism for significant evolutionary change (10). There is also some question as to whether the "wild ancestors" of agricultural plants are really ancestors, or degenerate forms originating from superior phenotypes. [See TOPICS FOR FURTHER STUDY #3.]

Few mutations are beneficial.

In conclusion, mutations arise from errors during DNA replication, mitosis, or meiosis. They are exceptional events that have a low probability of occurrence, barring exposure to mutagenic agents. Relatively few mutations and chromosomal aberrations are beneficial, and in many cases even these are not retained in the gene pool unless retained under laboratory conditions. In spite of these results, evolution adherents insist that mutations provide new genetic variation, not only in the present, but as Dobzhansky (12) claimed,

> the thousands of genes now found in the same gamete in most organisms must be the *descendants* of the primordial gene.

Evolution — life arises in primordial sea,

but how could this be?

This statement expresses the broader concept of **organic evolution**. Organic matter is seen as evolving to increasing complexity up the hierarchial levels of organization—atoms to molecules in the "primordial ocean", and then to living cells and organisms. According to J.S. Huxley man represents organic matter having attained a level such that he is capable of directing evolutionary change (41). However, evidence from our discussion so far contradicts such claims. Indeed, mutations may rather be viewed as **entropy** (Section 5-C) as Nobel Prize laureate, Max Planck (35) states:

evolution — one exception to thermodynamics?

The second principle of thermo-dynamics, the principle of the increase of entropy, has frequently been applied...to biological evolution, a singularly unhappy attempt so long as the term evolution is associated with the idea of progress, perfection, or improvement. The principle of entropy is such that it can only deal with probabilities and all that it really says is that a state, improbable in itself, is followed on an average by a more probable state. Biologically interpreted, this principle points towards degeneration rather than improvement: the chaotic, the ordinary, and the common is always more probable than the harmonious, the excellent, or the rare.

creationist view of mutations

Creationists view mutations as evidence of deterioration in the gene pools of populations, and as one of perhaps many biological processes that lead to aging and death of individuals. This view is consistent with the existence of a perfect creation prior to the curse of sin (Genesis 3). Indications of a trend of deteriorating biological and hereditary fitness is seen in the decreasing life spans of men and women as recorded in the Genesis account. Furthermore, children born today from close intermarriages (as was the case in pre-flood times) are very likely to have deformities. This greater frequency of deformity stems from the greater likelihood of recessive lethal genes being expressed in homozygous condition (i.e. Aa x Aa —> aa). [See TOPICS FOR FURTHER STUDY #4]

15-F MACROEVOLUTION *VERSUS* MICROEVOLUTION

We have examined two foundational components of Neo-Darwinian evolution—the fossil record, and mutations as the alleged source of genetic variation. Our analysis of the fossil record has revealed at least four areas which seem inconsistent with gradual evolution of species:

four fossil inconsistencies

1. Absence of transitional forms — Are "kinds" distinct?
2. A few mass extinctions — Where is meaningful selection?
3. "Explosions" of new body plans — Where are ancestors?
4. "Living fossils" — Is "stasis" the predominant mode?

no evidence for *macro*evolution

In addition to the contradictory fossil record, laboratory evidence offers little indication that mutations are the source of the awesome variation in millions of species. Thus, *there appears no mechanism for such extensive variation, and no prehistoric record that extensive variation occurred over time.* That is, there does not appear to be clear evidence for **macroevolution (general theory of evolution)**, which allegedly occurs *with no limits to biological change* (Figure 15-5). Both Neo-Darwinism and punctuated equilibrium claim that there are no limits to genetic variation.

There is evidence for *micro*evolution.

Many creationists suggest that certain limits have separated the created "kinds" since the creation. Thus, while great genetic variation exists within basic kinds such as cat, dog, horse, etc., even intensive artificial selection has not produced variants beyond certain limits. However, **microevolution**, genetic changes *within the limits of what we may call "kinds"* is verifiable from experience. Figure 15-5 and Table 15-3 contrast microevolution and macroevolution.

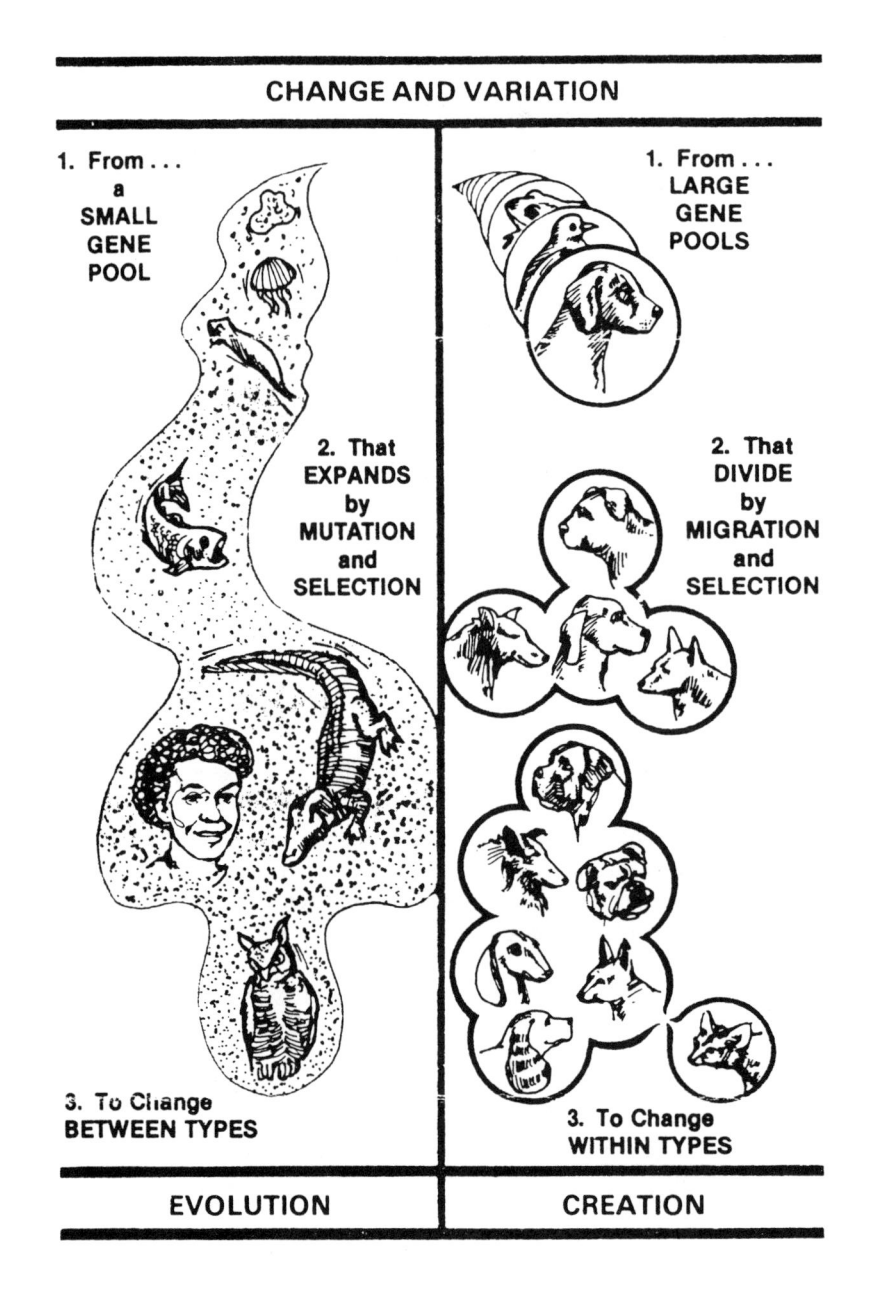

Figure 15-5. *Unlimited* biological change "vertically" from simple forms to complex, as contrasted with *limited* change "horizontally" within "kinds". (Bliss, R.B. 1988. Reprinted with permission of Master Books, El Cajon, CA)

Table 15-3. Two Interpretative Models of Biological Change.

MACROEVOLUTION	MICROEVOLUTION
VARIATION BEYOND "KINDS"	VARIATION WITHIN "KINDS"
BROAD, VERTICAL	NARROW, HORIZONTAL
PREHISTORICAL	HISTORICAL
SUPPOSED	OBSERVED
INFERRED*	TESTABLE

* Macroevolution *presupposes* that life evolved from a common ancestor; then, infers logically that two taxonomic groups (e.g. phyla) with many traits in common must be closely related phylogenetically (Figure 9-1). A "macroevolution explanation" for similarity among taxonomic groups is based on an *inference from* an evolution presupposition, but is not *evidence of* macroevolution.

"Great leaps"...

Creationists who acknowledge the existence of *micro*evolution agree with proponents of *macro*evolution that natural selection does cause "evolutionary change". Indeed, much of the discussion of macroevolution in modern writings relates to *micro*evolution which is discussed accurately from a scientific standpoint, but then is used as evidence for the "great leap" of extrapolation to *macro*evolution. The most extensive "leap" appears when evolution is elevated to *worldview status* as **evolutionism**. Evolutionism combines organic evolution (Section 15-E), biological evolution, and societal evolution into one atheistic, naturalistic framework.

...began with Charles Darwin.

Creationists generally view natural selection, not as *creator* of new kinds, but as a *conserver* of already created kinds. This distinction is one that Darwin failed to make when he rightfully saw how **artificial selection** by geneticists could produce variation *within kinds*, but wrongfully assumed that **natural selection** could move genetic variation *beyond kinds* to produce evolution. Natural selection assigned to this role is misleading because, unlike the breeder, natural selection has no "plan" except in the mind of the evolutionist. In the next section, we will examine the experimentally verifiable roles of natural selection.

15-G NATURAL SELECTION AND POPULATION GENETICS

a summary model of selection:

Natural selection is a process through which the environment influences survival and reproduction of individuals of each population, and in so doing, influences allele frequency in the respective gene pools. Individual organisms which inherit alleles that make them more fit to survive and reproduce will, in turn, pass along more of their alleles to offspring. This relationship in a population may be summarized as follows:

| GENETIC VARIATION (among members) | + | ENVIRONMENTAL RESISTANCE (phenotypes are put to the test) | --> | SURVIVAL/REPRODUCTION OF ORGANISMS EXPRESSING SUITABLE PHENOTYPES (*NATURAL SELECTION*) |

SOURCES OF GENETIC VARIATION IN A GENE POOL:

1. Gene mutation

2. Migration of individuals to/from other gene pools

3. Meiosis: Independent assortment and crossing over

CONDITIONS NEEDED TO STABILIZE GENE FREQUENCY IN GENE POOL:

1. No mutations occur at the chromosome *locus* of that allele

2. No immigration or emigration — *i.e.* isolated population

3. No natural selection — All individuals survive/reproduce

4. Population is large — minimizes effect of "chance events" that would impede random mating throughout, or isolate a small group of individuals

Hardy-Weinberg Principle

The model also lists *sources* of new alleles in the gene pool and *conditions* necessary to stabilize its frequency. Such conditions produce a **genetic equilibrium**, meaning that allele frequencies would remain constant from generation to generation, and no microevolution would occur. This is a statement of the **Hardy-Weinberg Principle** as illustrated in Figures 15-6 and 15-7. [Explain how each of the *conditions* above would oppose changes in allele frequency.]

application of selection model

Conversely, when one or more of the conditions for genetic equilibrium does not exist, allele frequencies will change, and microevolution will occur. To illustrate, consider the alleles for eye color in the fruit fly, *Drosophila*. As noted in Section 15-E, female flies prefer to mate with normal red-eyed males rather than white-eyed males. The allele for red eyes (R) is dominant to the mutated allele (r). Note how the frequency of the r allele will change when each of our four conditions that promote genetic equilibrium are altered:

1. Mutations occur — increases frequency of r

2. Immigration/emigration — r alleles enter/leave gene pool

3. Natural selection — r-carrying males less prone to mate

4. Population is small, or barriers prevent random mating

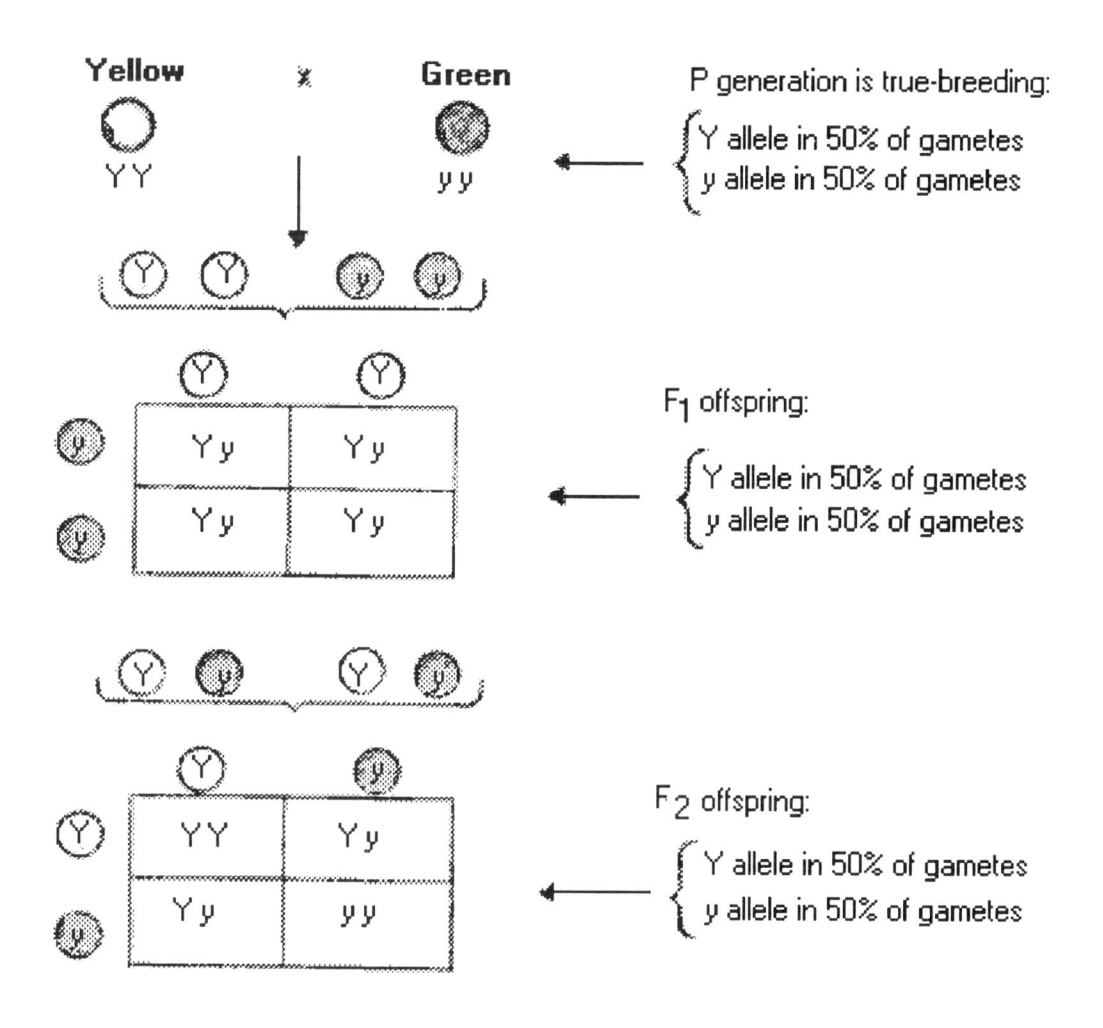

Figure 15-6. Illustration of the Hardy-Weinberg principle beginning with true-breeding pea plants. (See Sections 12-C and 12-D.)

NOTE: Each generation has a 50/50 ratio of Y/y in genotypes of the offspring (i.e. $4Y/4y$ in the total of possible genotypes). You can test the Hardy-Weinberg prediction that the F_3 generation will also be 50/50 by using Punnett squares to show crosses between each of the F_2 genotypes (i.e. $YY \times yy$; $YY \times Yy$; $Yy \times Yy$; etc.) and totalling the alleles.

Remember, you are assuming that each individual has an equal chance of mating with others regardless of genotype. As long as this and the other conditions listed in the model above is true, theoretically the population would be at *genetic equilibrium* generation after generation, and microevolution would not occur with respect to the frequency of that allele. See Figure 15-7 for a general formula for the principle.

(a)

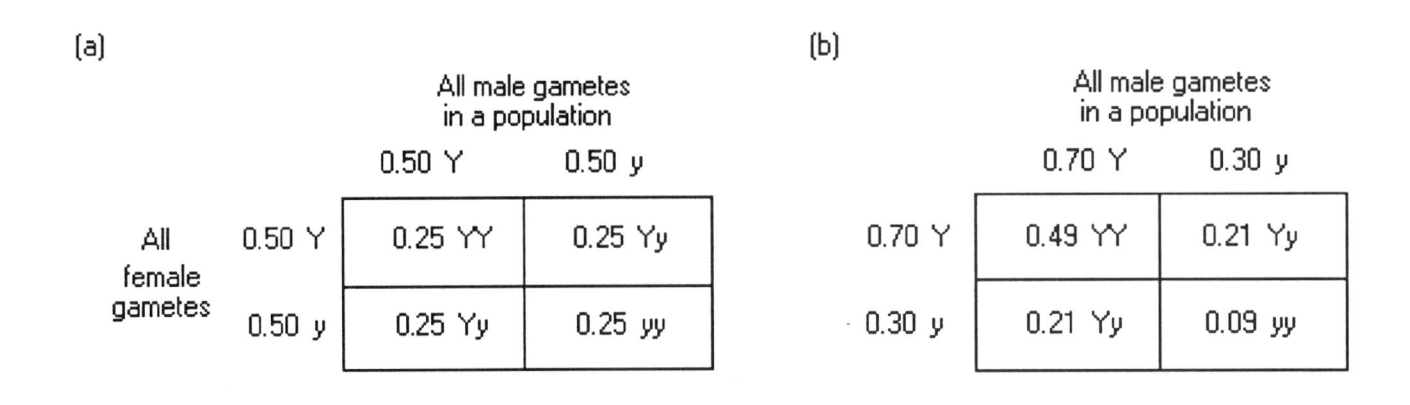

	All male gametes in a population	
	0.50 Y	0.50 y
All female gametes 0.50 Y	0.25 YY	0.25 Yy
0.50 y	0.25 Yy	0.25 yy

(b)

	All male gametes in a population	
	0.70 Y	0.30 y
0.70 Y	0.49 YY	0.21 Yy
0.30 y	0.21 Yy	0.09 yy

(c)

	All male gametes in a population	
	p (Y)	q (y)
All female gametes p (Y)	p^2 (YY)	pq (Yy)
q (y)	pq (Yy)	q^2 (yy)

Algebraic expressions:

$$p + q = 1$$

$$p^2 + 2pq + q^2 = 1$$

Figure 15-7. Applying the Hardy-Weinberg principle to situations where

(a) 50% of gametes in a population have allele Y and 50%, y;

(b) 70% have allele Y and 30% have y;

(c) general algebraic formula that applies to any combination of frequencies. Study the general formula and note its application in parts (a) and (b).

p = frequency of or proportion of gametes with dominant allele

q = frequency of or proportion of gametes with recessive allele

$2pq$ = frequency of heterozygous genotype in resulting offspring

p^2 and q^2 = frequency of homozygous dominant and recessive genotypes, respectively, in resulting offspring

conditions are present for selection

The r allele may be introduced to a given population gene pool either by mutation or by immigrating white-eyed flies. Natural selection will tend to suppress the frequency of r, not because it is deleterious to male health but because of behavioral factors. That is, allele r and the resultant white-eye phenotype lowers male fitness for mating because of "female preferences" for red-eyed males. Therefore, with fewer white-eyed "father flies" and consequently, fewer white-eyed offspring, the frequency of allele r will decrease in the gene pool. The principle here is that phenotypes which influence mate selection positively or negatively will influence allele frequency.

A small population...

Finally, condition #4 would be influential if a small number of fruit flies with a higher frequency of *r* alleles were carried with a fruit container to a separate location. Here a larger percentage of white-eyed males might be accepted because red-eyed males are scarce. Because the population is small, the *r* allele may have a greater chance of increasing in frequency. As this new population increases, red eyes would be observed more frequently than in the original large population "back home". This example illustrates one mechanism of **genetic drift** wherein chance events cause a "drift" in frequency of an allele. In the case of the flies, unless females change their behavioral preferences, the frequency of *r* may drift back to its original level.

...favors genetic drift.

Selection has no "willful plan"...

The fruit fly example should make clear that natural selection can influence gene frequency of alleles in accordance with the fitness, or lack thereof, conferred by the resultant phenotype. Remember, natural selection does not have a "willful plan", it is simply the result of environment acting upon populations and favoring individuals with suitable phenotypes and hence, the alleles responsible. Out of this interaction, natural selection plays three important roles as distinguished by the following outcomes:

...but three roles:

1. Slowing genetic deterioration by limiting frequency of deleterious alleles — *stabilizing selection*

2. Preserving populations in response to a changing environment, allowing adaptation — *directional selection*

3. Allowing new species to form (speciation) when isolating mechanisms are present — *disruptive selection*

STABILIZING SELECTION — Slowing Genetic Deterioration

A "brake" to slow down deterioration

By eliminating deleterious alleles, natural selection acts as a "brake to slow down the decay in genetic quality caused by mutations" (33). In the human population we have discussed hemophilia, caused by a deleterious allele (*h*) inherited as an *X*-linked trait in the same way as allele *r* for white eyes in male fruit flies. Both traits are expressed more frequently in males who always receive the mutated allele from the mother (Section 12-F). [Explain why.] However, unlike white eyes, hemophilia can cause death before marriage and children arrive. Thus, selection suppresses frequency of *h*.

Another hereditary blood disease called **sickle-cell anemia** results from a gene mutation that causes an error in hemoglobin protein (Hb) synthesis. Sickle-cell hemoglobin causes the normally disk-shaped red blood cells to rupture into a sickle-shape. Sickle-cell anemia results from the reduced capacity of sickle-cells to carry blood, and their tendency to clump and clog capillaries.

The disease is caused by the mutant allele Hb*S*. "Normal" persons have genotype Hb*A* Hb*A*, while sickle-cell individuals have Hb*S* Hb*S*. A high percentage of the latter individuals die before adulthood. As you may have guessed from the symbols, Hb*A* and Hb*S* are alleles that display *incomplete dominance* (Section 12-E). Thus, heterozygous individuals (Hb*A* Hb*S*) have a combination of normal and sickle-cell hemoglobin, and normally survive with slight anemic tendencies.

As in the case of hemophilia, we would expect the recessive allele to be held to low frequency because of its lethal effects. However, in tropical regions of Africa and southern Asia the recessive allele has a higher frequency than in most other areas of the world. This is due to the fact that heterozygous individuals have resistance to malaria (Section 9-C), whereas homozygous dominant ("normal") individuals succumb to the infection (Table 15-4).

Table 15-4. Relative Fitness Conferred by Sickle-Cell Genotypes.

GENOTYPE	SICKLE-CELL CONDITION	MALARIAL RESISTANCE
HbA HbA	NORMAL	SUSCEPTIBLE*
HbA HbS	ANEMIC	HIGHLY RESISTANT
HbS HbS	DEATH	DEATH (by s-c. anemia)

*High frequency of sterility and death.

The greater survival and reproductive success of HbA HbS individuals in malarial regions causes higher frequency of the mutant, HbS allele. In nonmalarial regions HbS confers no advantage and is therefore limited more severely by natural selection. Table 15-5 illustrates this pattern.

Table 15-5. Differences in Frequency of the Heterozygous Condition for Sickle-Cell Anemia among Geographic Populations of Humans.

GEOGRAPHIC POPULATION	FREQUENCY OF HbA HbS
African Blacks	25% to 30%
American Blacks	8%
American Whites	<1%

stabilizing selection "screens" mutations

In summary, stabilizing selection screens out mutations when these alleles offer no survival or reproductive advantage. Note that stabilizing selection does not appear to be a vehicle for macroevolution with mutations as the source of new variation. Instead of "creating" new species as Neo-Darwinism suggests, stabilizing selection tends to eliminate mutants, thus opposing the buildup of mutations in the gene pools.

Sickle-cell anemia further emphasizes the limited capacity of natural selection to perfect a population. A third hemoglobin allele, HbC, offers superior resistance to malaria and does not lead to the sickle-cell condition in homozygous individuals (HbC HbC). Yet natural selection has not raised the frequency of this supposedly superior allele above the HbA or HbS alleles. [See "TOPICS" #1.] As Lewontin (25) has stated, "natural selection over the long run does not seem to improve a species' chance of survival but simply enables it to 'track' or keep up with the constantly changing environment". This concept takes us to a second major role of natural selection.

DIRECTIONAL SELECTION — Allowing Adaptation

Currently, there is concern that human activities may be altering the climate of planet Earth (Chapters 5 and 6). Earth's climate is in continuous change as a result of natural patterns and human activities. Species **extinction** occurs when all populations of a given species have reached zero, and the gene pool of that species is terminated. Genetically speaking, living members apparently could not *adapt* fast enough. Even if suitable alleles existed in the gene pools, the combinations and frequencies of these were apparently not expressed as suitable phenotypes in time so that survivors could reproduce. Conversely, those species that still remain on Earth have evidently been successful enough in adapting.

a mechanism to "keep up" with the changing environment

Many creationists, and some evolutionists (e.g. Lewontin quoted above) view natural selection as a conserving mechanism that permits populations to keep up with environmental change. However, creationism uniquely credits the Creator God with having pre-programmed each *kind* with the original structural, physiological, and behavioral adaptations necessary to acquire *nutrition* (Chapter 10), carry on *reproduction* with other members of its kind (Chapter 11), and maintain *responsiveness* and *homeostasis* (Chapters 13,14). Teleologically speaking, these adaptations were "designed" purposely to meet the needs of the *individual* during its lifetime. However, habitat diversity and environmental variability required an additional feature—the capacity for *populations* as "genetic units" (Section 7-A) to respond to change over many generations. In this case, our focus moves from individual to population and the gene pool. Natural selection functions to conserve the kinds in the midst of environmental change by favoring alleles and the resultant *phenotypes* which promote survival and reproduction. Thus, populations can adjust genetically to environmental change instead of being locked into unfavorable phenotypes. [Was there need for natural selection, or could it have functioned before the "fall" of mankind?]

What are limits of variation?

How much change can a population undergo? As much as the genetically programmed limits will allow, and apparently no more, as noted in Section 15-F. Although "boundaries" of *species* and *kinds* are fuzzy, within each kind there is sufficient genetic variation so that populations can "offer" alternative phenotypes to selection as conditions demand.

From your earlier studies of genetics, you will recall the source of much of the genetic variation. For instance, many traits are controlled by *multiple alleles*, and others, by *multiple genes* (Section 12-E). Complex traits such as skeletal size, height, skin color, and seed weight are controlled by multiple genes. For such traits, the phenotypes within a population will show *continuous variation*, and follow a normal distribution as shown in Figure 12-5.

example of *directional selection*

If we were to study the mean (average) root depth in a plant population as a hypothetical trait, we might observe the outcome shown in Figure 15-8. Let's suppose that the environment in the region were to become gradually drier over a period of many years, causing decreased soil moisture. As the mean (average) depth of the water table increases, plants with alleles favoring deeper root systems are favored for survival and reproduction over plants with shorter roots. As a result, gene frequencies shift in favor of deeper roots. This shift is reflected in the shift of the normal distribution of phenotypes (Figure 15-8b). Because the population shifted in the direction of better fitness or

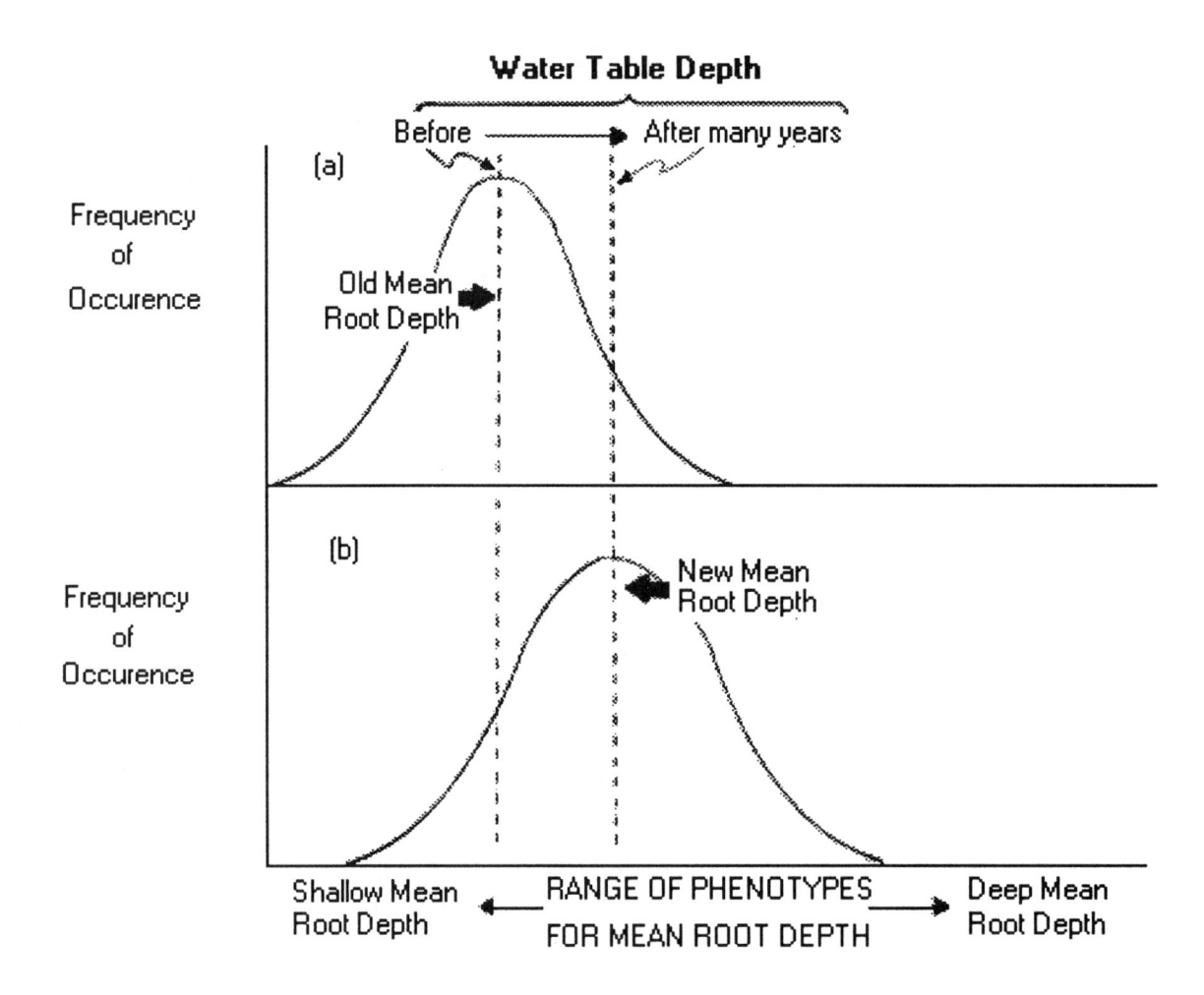

Figure 15-8. Frequency distribution of phenotypes for mean root depth in a hypothetical plant population (a) before, and (b) after a long period of decreasing soil moisture.

adaptation, this process is called **directional selection**. [Compared to Figure 15-8, how would a curve of normal distribution for a population with more narrow (less) variability be shown? Which may have the greater chance of surviving environmental change over time?]

DISRUPTIVE SELECTION — Forming New Species

So far, we have seen natural selection manifested in two roles, as stabilizing selection and as directional selection. The emphasis is on providing continual adjustments within gene pools in the face of mutations and environmental change, respectively. But natural selection can also appear to have disruptive effects upon a gene pool. This manifestation of natural selection is called disruptive selection.

Do you recall our earlier discussion in which female fruit flies preferred red-eyed males? We followed a fruit crate of predominantly white-eyed males and females on a trip to a distant location. Here, with fewer red-eyed males,

New species may result from *disruptive selection* . . .

. . . when populations are isolated . . .

geographically

morphologically

or chemically.

evidence for speciation

Can adaptation keep pace with environmental deterioration?

keeping up with the microbes . . .

females were less reluctant to mate with white-eyed males. Now, let's suppose that, during this time, female behavior changes and begins to favor white-eyed males (white is "in"!). More white-eyed offspring means an increased frequency of r in the gene pool. If *behavioral change* (mate preference) as well as *morphological change* (eye color) are both heritable, future generations may reflect a genetic drift away from the red-eyed flies "back home". Over time, a distinct species may develop because of different *morphology* (different eye color) and different *behavior* (female mating preference). Production of new species, or **speciation**, occurs when individuals become isolated and no longer interbreed with other members of the population. Isolation splits gene pools and allows genetic variation to follow separate paths in microevolution. As a result, many creationists believe that speciation has produced many species within the original created "kinds" (Figure 15-5).

Isolating mechanisms or barriers to "gene flow" may be geographic, morphological, chemical, or chromosomal (incompatibilities during meiosis). *Geographically*, the Galapagos finches (Figure 15-3) microevolved after being isolated from the mainland. Squirrels separated by Grand Canyon formed separate "Albert" and "Kaibab" populations (4). *Morphological* differences, such as fruit fly eye color, may develop and isolate potential mates from each other. A mutation in an allele for flower color may create an isolated gene pool if the flower attracts a different pollinator from the "wild type" allele. The *chemical* makeup of flower stigmas may also restrict fertilization by allowing pollen from only certain male plants to germinate.

Taxonomists who face the challenge of distinguishing closely related gene pools are well aware of the existence of speciation. Hence, they may classify groups as *varieties, races*, or *subspecies*. Much speciation occurs as a result of human actions. For example, microevolution of five distinguishable species of Hawaiian moths has occurred in response to the introduction of banana plants onto the islands from Tahiti some 1,000 years ago. They are distinguished by the separate traits which enable each to feed on banana leaves. Human activities and natural selection is the subject of several other examples in the following paragraphs.

NATURAL SELECTION AND HUMAN ECOLOGY

Unlike the rather quick physiological adjustments in individual bodily functions, genetic adjustments occur over generations of time and are slower. Indeed, selection may be occurring on a scale too slow to preserve some species in the face of deteriorating environmental quality (e.g. mutagens), global climate changes, and habitat destruction. A few examples may illustrate the human dilemma in today's changing world.

ANTIBIOTIC RESISTANCE. You are probably aware of the ongoing efforts of pharmaceutical companies to produce new antibiotics for treatment of bacterial infections. Any given antibiotic prescription such as streptomycin imposes an unchanging environmental "challenge" to pathogenic bacterial populations. Though the antibiotic does not change, the bacterial populations can change. Favorable mutations are highly probable in rapidly reproducing bacteria. As soon as one individual cell develops a favorable allele which affords resistance to the antibiotic, this cell gains a survival-reproductive advantage in the environment of the antibiotic. The appearance of the mutation would be noticed when that antibiotic no longer controls the infection

... and pests.

Selection in moths ...

... in a changing environment

under the dosage administered. Increased dosage or a switch to another antibiotic may be necessary.

Similar mechanisms operate as farmers battle insect and weed pest populations with pesticides to maintain the *monoculture* (Section 5-G). Thus, natural selection operates to conserve even those kinds of organisms that are considered enemies to man.

INDUSTRIAL MELANISM. Perhaps the most commonly cited example of natural selection is *industrial melanism* in moths, discovered in Great Britain (6). Before the mid-19th century, the predominant form of the peppered moth (*Biston betularia*) in Great Britain was a gray-winged phenotype, speckled with tiny pepper-like black spots (Figure 15-9). This wing coloration made an ideal camouflage of the moth against a background of gray lichens (Section 7-F) that normally grow in patches on tree trunks. The camouflage apparently provides protection for the moths against predatory birds.

With the arrival of soot and pollution from the industrial revolution, many lichen populations were darkened or destroyed, leaving only the darker surface of the tree bark; and, leaving the gray-winged moth without camouflage. As a result, gray-winged moths were subjected to heavy predation by birds. During the same period, a melanic (dark-winged) phenotype of the same moth species became more abundant. Previously, the dark-winged phenotype had been a "sitting duck" for predatory birds. Now, with darkened surfaces "in" and gray surfaces "out", the favorable phenotype shifted from gray-winged (recessive allele; *dd*) to the dark-winged (melanic) form (dominant; *DD* or *Dd*). Thus, the predatory removal of gray-winged moths in industrial areas caused a change in allele frequencies in favor of the dark-winged allele.

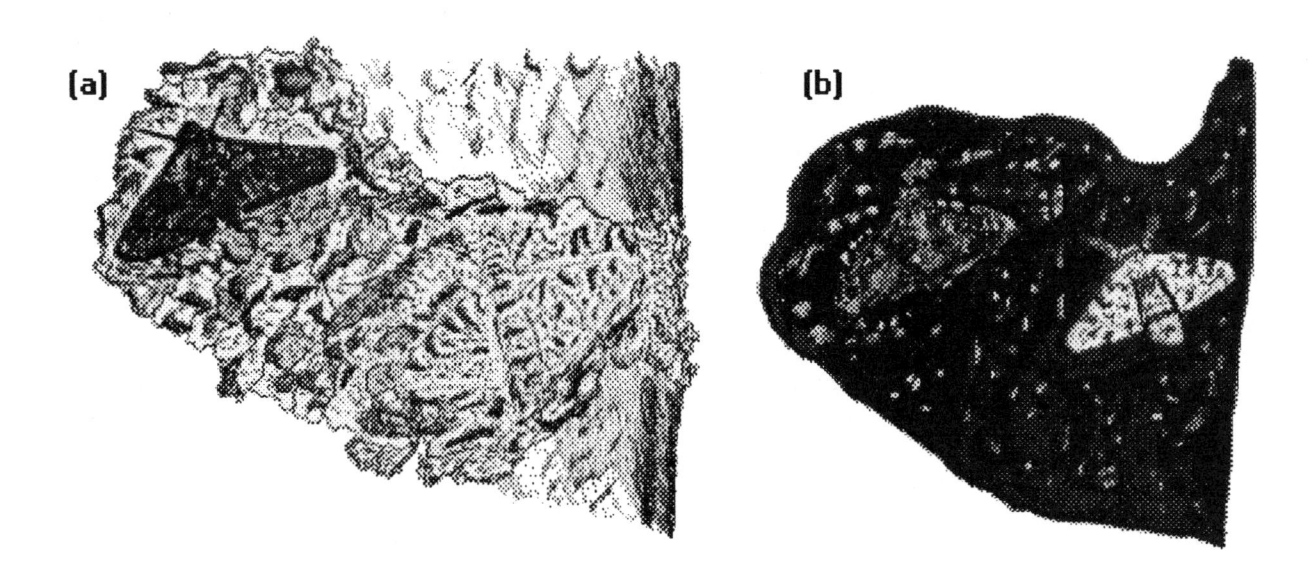

Figure 15-9. Gray-winged and dark-winged variants of the same moth species, *Biston betularia*. The sketches show the two variants as they would appear (a) on lichen-covered surfaces in pollution-free areas, and (b) on a darkened, soot-covered surface.

experimental verification:

It is noteworthy that natural selection in peppered moths can be verified by experimentation as follows:

HYPOTHESIS: Capture of moths by birds depends upon wing color, and natural selection favors the allele (and color) which provides camouflage.

DEDUCTION: If wing color affects capture of moths, then out of known sample populations, more moths with wing color that provides camouflage in a given environment should be recovered as survivors after a period of predation. [Recovery by "light traps" at night.]

EXPERIMENT: Capture and mark a known number of moths of each phenotype and release the two phenotypes into industrial and countryside environments. Estimate percentage that avoided predation by numbers recaptured (the *mark-recapture method*).

RESULTS: Data from Kettlewell (1955) *Heredity* 9: 323-342.

Table 15-6. Effect of Moth Wing Color and Environment on Bird Predation.

ENVIRONMENT	PERCENT OF MOTHS RECOVERED AFTER PREDATION	
	GRAY-WINGED	DARK-WINGED
Industrial	19	40
Countryside	12.5	6

In recent years, pollution control efforts in Britain have apparently caused a return of conditions favoring gray-winged moths (7). [Which type of natural selection is demonstrated by the moth population's response to environmental change? How could the Hardy-Weinberg principle be applied to estimate the frequency p and q of allele D and d, respectively? (Figure 15-7).]

an alternative hypothesis

While it appears that industrial melanism is the ideal textbook example, there may be more involved than meets the eye. R.E. Byers (6) has emphasized that there are several possible chemical agents in the industrial atmosphere that could cause mutation or alter gene expression, resulting in increased production of the brown melanin pigment. Interestingly, some dark-winged forms are darker than those collected a century ago. He proposes several possible hypotheses that could be tested. Could it be that, in addition to natural selection, there is a direct effect of environment on gene expression (Section 12-H)? [See "TOPICS" #2.]

HUMAN LANGUAGE AND GENETIC DIVERSITY.

one of the dark sides to evolution

The origin of human languages and different tribes and nations is an intriguing topic of biology and the social sciences. One's view concerning the biological origin of the human species has obvious influence upon his or her position on issues in the social sciences and ethics. For example, the Darwinian emphasis upon improved fitness through struggle while ignoring biblical principles has led some to justify a "racial struggle" as an inevitable part of evolution. [See also Chapter 12, "TOPICS" #5]

isolation of human populations

The creation model views all humans as one *kind* having descended from common ancestors (32). As with other kinds of creatures, the human gene pool originated with a great store of genetic variation. However, population genetics proposes that there had to have been reproductive barriers to separate the human population into smaller, interbreeding groups. Otherwise, the differences in gene frequencies observed between different racial groups today (e.g. Table 12-2; Table 15-5) could not have developed. Geographic isolation would be the most likely cause, but what would cause people to move apart geographically? The answer probably lies in the existence of different *languages*.

the power of language

Unlike all other creatures, humans were given the capacity to reason abstractly and to articulate ideas using words and other symbols. Assuming that descendants of Adam used one language, they would have been drawn together into a common gene pool as long as this language was universal. Today we can witness the power of common language and culture to unite, or different language and culture to separate people. The confusion of the languages by supernatural intervention at the Tower of Babel is revealed in Genesis 11 as the cause of dispersion of humankind into separate tribes and nations, each with a distinctive language (Figure 15-10). Morris has published a more detailed discussion of the development of human society and culture, from a creationist perspective (32).

Figure 15-10. Genetic variation among representative tribes and nationalities of species, *Homo sapiens*.

We conclude this chapter with the same theme that was introduced at the beginning, and emphasized throughout. A statement by Lester and Bohlin (24) says it quite well:

**Worldview—
it matters
where we begin.**

World view colors our thoughts in everything we do. Science is no exception, and the concept of origins is an integral part of one's worldview. As a result, a scientific investigation into the subject of origins will be greatly affected by our world view, particularly in light of the difficulty in finding adequate tests, as we have seen. A knowledge of how we got here and why we are the way we are is crucial to understanding our world and how we react to each other.

QUESTIONS AND DISCUSSION TOPICS

1. Why is it helpful to distinguish *origin* science from *operation* science? Are they both "natural sciences"? How do they differ?

2. Using the topic of origins, react to the following statement: "Two scientists may *view* the same phenomenon, but they will not have the same interpretation if their *worldviews* are different." What does this say about *observation* in science? About the importance of identifying worldview differences in "origins" discussions?

3. Chart the progression of western thought from Aristotle to Darwin, noting the contributions of key individuals in relation to the trend toward a naturalistic evolution model of origins? At what point was there an apparent divergence of opinion between Christianity and natural scientists with regard to taxonomic classification?

4. What is Darwinian evolution? What were Darwin's supporting arguments, and what evidence was lacking?

5. Make a case for (or against) the position that Darwinian evolution theory was a necessary part of the discoveries from Mendel through those outlined in Table 15-1. How did the discoveries of genetics affect Darwinism?

6. Is there a legitimate place for uniformitarian thinking in interpreting past geologic events? If so, explain to what extent.

7. Where was Darwin (and followers today) wrong about natural selection?

8. Distinguish between creation and evolution interpretations of the following phenomena: fossil record, mutations, limits of evolution, natural selection.

9. From your understanding of stabilizing selection, how is it improbable that bodily organs, which develop under control of many alleles at different chromosomal loci could have arisen from chance mutations of alleles?

10. Based upon your understanding of directional selection, what factors determine whether a species will be able to adapt to environmental change? Illustrate by specific examples from the text. In view of this, why would it be unlikely that the mass extinctions recorded in the fossil record were a consequence of the "normal" operation of natural selection?

11. Are new species evolving today? How do human activities affect speciation? Support your answers and illustrate by examples.

TOPICS FOR FURTHER STUDY

1. How does the sickle-cell condition provide resistance to malaria? Is this an example of evolution by natural selection? REFERENCES: #1, #14, #28

2. Is industrial melanism in the peppered moth an instance of natural selection involving camouflage against predatory birds, or a case of modification of gene expression by a toxic agent in the environment (Sections 12-H, 15-G)? REFERENCES: #3 and #6

3. Is there evidence that the "wild, weedy, ancestors" of modern crop plants are actually degenerate forms within originally superior created kinds? REFERENCE: #19

4. Is there evidence that the human gene pool is deteriorating since Adam instead of evolving to a more sophisticated level? Consider such aspects as intermarriage of Adam's offspring, decreasing life span, and effect of medical advances which allow individuals to bear offspring that couldn't do so previously. REFERENCES: #34, #37, #42

5. What is punctuated equilibrium? Is it any more valid than Neo-Darwinism? See REFERENCES: #24 and #16

6. Was a "day" a day in the Genesis 1 account of creation? REFERENCES: #21, #31, #45

7. Is the creationist "intelligent design" hypothesis compatible with the naturalistic "endosymbiont hypothesis" if one invokes the "gene theme" concept— *i.e.* Creator used the design pattern of the prokaryotic cell to equip eukaryotic cells with mitochondria and chloroplasts. REFERENCE #5.

8. Students of education may wish to critique the way in which evolution is presented in high school or college biology textbooks. See Section 15-A and REFERENCE: #29

9. Look closer at Darwin's Finches, and an alternative theory to explain their existence on the Galapagos Islands. REFERENCE: #17.

REFERENCES

1. Barman, C.R., *et. al.* 1985. Sickle-cell anemia: "Interesting Pathology" and "Rarely Told Stories". *Amer. Biol. Teach.* 47(3): 183-187.

2. Bergman, J. 1980. Evolution, race, and equality of intelligence. *Creat. Res. Soc. Quart.* 17(2): 127-134.

3. Bishop, J.A. and L.M. Cook. 1975. Moths, melanism, and clean air. *Scient. Amer.* 232(1): 90.

4. Bliss, R.B. 1988. *Origins: Creation or Evolution.* Master Books. El Cajon, CA.

5. Brown, C. 1991. The origin of euglena. *Creat. Res. Soc. Quart.* 28: 112-113.

6. Byers, R.E. 1984. Industrial melanism: Proof of evolution or toxicological response? *Origins Res.* 7(2): 15.

7. Coffin, H.G. 1983. *Origin by Design.* Review and Herald Publ. Assn. Washington, DC.

8. Cook, L.M., *et.al.* 1986. Postindustrial melanism in the peppered moth. *Science* 231: 611-613.

9. Darwin, C. 1956. *The Origin of Species.* E.P. Dutton, and Co. New York.

10. Davie, L.R. 1980. A re-examination of the role of mutations and natural selection in the evolution model of origins. *Origins Res.* 3(1): 6-8.

11. Denton, M. 1986. *Evolution: A Theory in Crisis.* Adler & Adler. Bathesda, MD.

12. Dobzhansky, T. 1970. *Genetics of the Evolution Process.* Columbia Univ. Press, New York.

13. Dobzhansky, T., *et. al.* 1977. *Evolution.* W.H. Freeman Co., San Francisco, CA.

14. Friedman, M.J. and W. Trager. 1981. The biochemistry of resistance to malaria. *Scient. Amer.* 244(3): 154.

15. Geisler, N.L. and J.K. Anderson. 1987. *Origin Science*. Baker. Grand Rapids, MI.

16. Gould, S.J. 1980. *The Panda's Thumb*. Norton, New York.

17. Harper, G.H. 1981. Speciation or irruption: the significance of the Darwin finches. *Creat. Res. Societ. Quart.* 18(3):171-175.

18. Heinze, T.F. 1973. *Creation vs. Evolution*. Baker, Grand Rapids, MI.

19. Howe, G.F. and W.E. Lammerts, 1980. Biogeography from a creationist perspective: II. The origin and distribution of cultivated plants. *Creat. Res. Societ. Quart.* 17(June):4-18.

20. Johnson, P.E. 1991. *Darwin on Trial*. InterVarsity Press. Downers Grove, IL.

21. Key, T. 1984. How long were the days of Genesis? *Jour. of Amer. Scient. Affil.* 36(3):159-61.

22. Klotz, J. 1970. *Genes, Genesis, and Evolution*. Concordia Publ. House, St. Louis.

23. Lammerts, W.E. 1965. Planned induction of commercially desirable variation in roses by neutron radiation. *Creat. Res. Soc. Quart.* 2: 39-48.

24. Lester, L.P. and R.G. Bohlin. 1984. *The Natural Limits to Biological Change* Probe Ministries International, Dallas, TX.

25. Lewontin, R.C. 1978. Adaptation. *Scient. Amer.* 239(3): 212-230.

26. Marsh, F.L. 1983. Genetic variation, limitless or limited? *Creat. Res. Soc. Quart.* 19(4): 204-206.

27. McIver, T. 1988. *Anti-Evolution: An Annotated Bibliography*. McFarland and Co., Inc. Jefferson, NC.

28. McCleod, K.C. 1982. The sickle-cell trait. *Creat. Res. Society. Quart.* 19(1): 19-26.

29. Mills, G.C., *et.al.* 1993. Origin of life & evolution in biology textbooks — A critique. *Amer. Biol. Teach.* 55(2): 78-83.

30. Moore, J.N. 1976. Documentation of the absence of transitional forms. *Creat. Res. Soc. Quart.* 13: 110-111.

31. *Moore, J.N. 1983. *How to Teach Origins (Without ACLU Interference)*. Mott Media, Milford, MI.

32. Morris, H.M. 1974. *Scientific Creationism*. CLP Publishers, San Diego.

33. Parker, G.E. 1980. Creation, selection, and variation. *Impact* No. 88 (October), p. 3.

34. Parker, G. 1980. *Creation: The Facts of Life*. CLP Publishers, San Diego.

35. Planck, M. 1936. *The Philosophy of Physics*. W.W. Norton & Co., New York, NY.

36. Popper, K. 1963. *Conjecture and Refutations*. Harper Torchbooks. New York, NY.

37. Rifken, J. 1983. *Algeny*. Penguin Books, New York.

38. Salisbury, F.B. 1971. Doubts about the modern synthetic theory of evolution. *Amer. Biol. Teach.* 32: 335-338, 354.

39. Savage, J.M. 1963. *Evolution.* Holt, Rinehart, Winston. New York, NY.

40. Thaxton, C.B., W.L. Bradley, R.L. Olsen. 1984. *The Mystery of Life's Origin: Reassessing Current Theories.* Philosophical Library. New York, NY.

41. Thompson, W.R. 1960. A critique of evolution. *Jour. of Amer. Scient. Affil.* 12(Mar.): 2-9.

42. Underwood, J.H. 1979. *Human Variation and Human Microevolution.* Prentice-Hall. Englewood Cliffs, NJ.

43. Whitcomb, J.C. and H.M. Morris. 1967. *The Genesis Flood.* Baker Book House, Grand Rapids, MI.

44. Wysong, R.L. 1976. *The Creation-Evolution Controversy.* Inquiry Press, Midland, MI.

45. Young, D. 1982. Genesis: Neither more nor less. *Eternity.* May pp. 14-21.

* J.N. Moore's book (REFERENCE #31 above) has an Appendix that lists societies focusing on creation issues and research; also, publishers of literature and audio-visual titles are listed. I am personally familiar with the following:
Access Research Network, P.O. Box 38069, Colorado Springs, CO 80937
Creation Research Society, P.O. Box 28473, Kansas City, MO 64118
Institute for Creation Research, P.O. Box 2667, El Cajon, CA 92021

Part 4

Cells — Functional Units of Life

Recap of Part 1

We have been studying life from several vantage points. **Part 1** presented a panoramic view of life and the science of life. In Chapter 1, we discussed the *nature of science* as a means of gathering knowledge of the natural world, and its relationship to other ways of knowing. We emphasized the influence that a person's worldview can have on his or her position on life science-related issues such as environment, abortion, and origin of life. In Chapters 2 and 3, we examined the *unifying characteristics* of life, and the *classifications* that make biology a systematized body of knowledge. Chapter 4 concluded Part 1 with a study of *atoms and molecules*, the basic levels of organization of life.

Part 2

Part 2 presented a view of *life in the environment*. We emphasized that living organisms are intimately linked to the environment and to one another through a continual flow of energy and nutrients among the populations of a biotic community. Ecology reveals the nature of mankind's *relationships* to the other creatures. The Scriptures teach us our *responsibilities* as stewards of the creation.

Part 3

As we moved from ecosystems and populations to the *organism* level of organization in **Part 3**, we were confronted with the great diversity among different species. In order to survey this diversity of life, we followed the logical groupings made possible by *taxonomic classification* (Chapter 9), and studied the *basic life functions* common to each taxonomic group—namely, *nutrition* (Chapter 10), *reproduction* (Chapter 11 and Chapter 12), and *responsiveness* and *homeostasis* (Chapters 13 and 14). Finally, in Chapter 15, we studied the basic life function of *adaptation*, through which populations as genetic units can maintain responsiveness to environmental change through *natural selection*. The limits of biological change were examined in the context of the challenge to understand the origin of life on Earth.

Preview of Part 4

We now come to our final vantage point from which to view life—the *cellular level* of organization. **Part 4** opens with a study of the basics of cell structure and function, in Chapter 16. Chapter 17 explores the function of nucleic acids in the control of cellular metabolism, and Chapter 18 addresses energy capture and conversion processes in cells.

Chapter 16

Cell Structure and Function

What occurs in cells?

All life processes and relationships ultimately depend upon cellular processes. We have already drawn back the curtain to expose the activities occurring at the cell level. For example, in our study of the *unicellular* microbes and protista, we learned that a single cell can perform nutritional intake, processing, excretion, reproduction, and locomotion. Thus, for unicellular organisms, one cell "does it all". We have also exposed certain processes of the specialized cells of multicellular plants and animals. These processes include absorption of light, water, or nutrients; and, intracellular digestion, stinging, mitosis, meiosis, fertilization, secretion, impulse transmission, contraction, differentiation, and osmotic processes. Having already encountered cells in the midst of these activities, this chapter will now address some logical questions such as the following: How is each cell structured so that it can perform these functions? How are the activities of each cell guided to make it a "servant" of the whole organism?

16-A METABOLISM — A VIEW FROM DIFFERENT LEVELS

What goes on inside cells?

...cellular metabolism:

e.g. **synthesis and respiration**

If we accept the principle that all life processes are ultimately cellular processes, we are confronted with a logical question; namely, what really goes on inside cells, and how? The term **cellular metabolism** is used collectively to refer to all of the chemical and physical processes that occur within cells. Examples of cellular metabolism include nucleic acid and protein synthesis (Chapter 17), photosynthesis of glucose (Chapter 18), and cellular respiration (Chapter 18). Each of these processes involve metabolic reactions that are performed within cells. Cellular respiration nicely illustrates this concept. As we have noted repeatedly, *every living cell must have a continuous supply of energy* (Section 2-C). Cellular respiration provides usable energy according to the following equation:

$$C_6H_{12}O_6 + 6\,O_2 \longrightarrow 6\,CO_2 + 6\,H_2O + \text{usable energy}$$

As shown in Figure 16-1, glucose and oxygen enter living cells, and cellular respiration converts the chemical energy of glucose into usable energy. Carbon dioxide and water are released as waste products. The unicellular heterotrophs (e.g. bacterial decomposers) perform this process when each cell encounters a source of dissolved glucose. [What must they do if they encounter only complex carbohydrates?]

unicellular *vs.* multicellular

Unlike unicellular organisms, each cell of multicellular organisms is dependent upon other cells to supply glucose and oxygen for cellular respiration. For example, a muscle cell deep within your thigh has no direct access to either glucose or oxygen. Therefore, your digestive, respiratory, and circulatory systems must provide these molecules. In addition, the liver and hormonal system must function to homeostatically regulate glucose levels in the

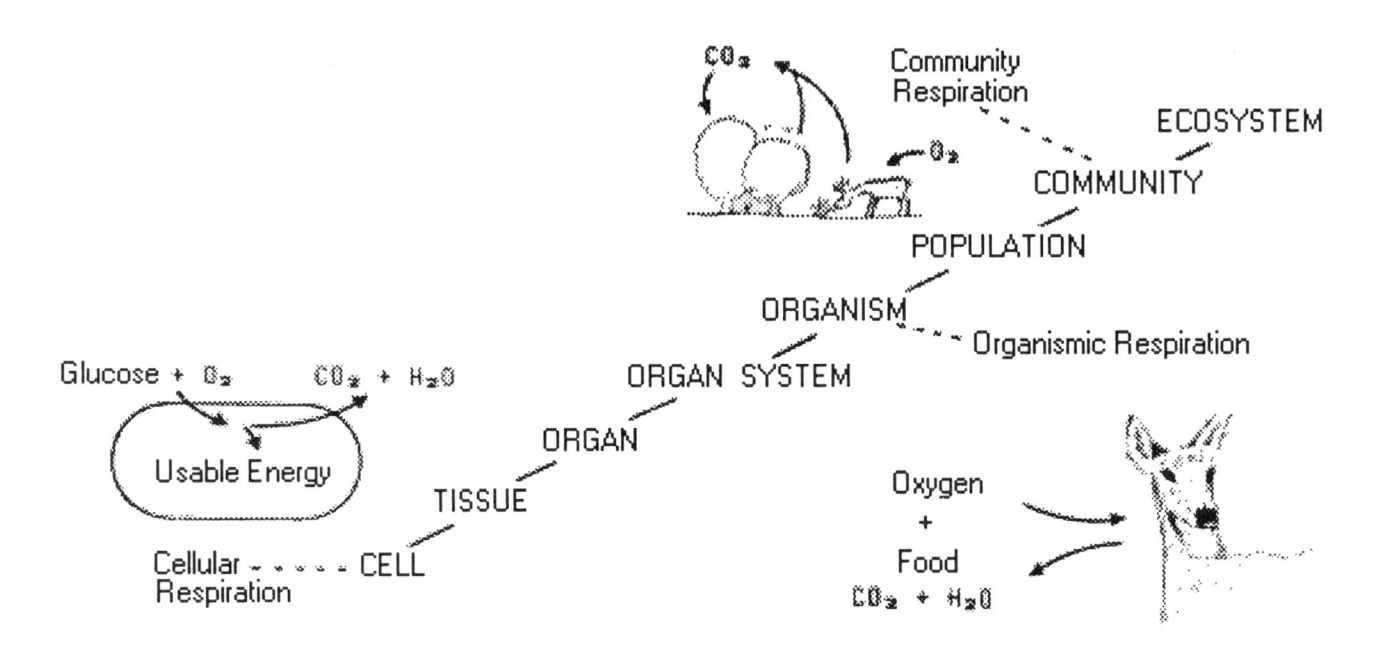

Figure 16-1. Respiration is an example of cellular metabolism that is manifested in different ways as it is viewed at increasingly higher levels of organization (See also Figure 2-1).

organismic metabolism— cells depend on one another

body fluids around each cell. Thus, every cell of a multicellular organism is dependent upon the metabolism of other cells which collectively represent **organismic metabolism.** Organismic metabolism involves groups of specialized cells, each submissively functioning as a part of a higher level of organization, through tissues, organs, and organ systems (Figure 16-1). Finally, each organism depends upon other organisms in the biotic community for a source of energy and nutrients (including oxygen). Here we see respiration as a part of **community metabolism.** Recall that the energy income of each biotic community via autotrophs (i.e. gross primary productivity) must be greater than its energy expenditures via respiration if its populations are to survive within a given ecosystem (Section 5-D).

Community metabolism...

...is cellular metabolism in a broader context.

As we have seen from Figure 16-1 and our discussion thus far, even though *cellular metabolism* occurs within cells, the significance of these processes extends beyond the single cell to the whole organism and its environment. Therefore, it is this author's belief that prior studies of "community" and "organismic respiration", for example, should provide a broad and more concrete experience that will make "cellular respiration" more meaningful. Others prefer to introduce cellular biology near the beginning of a biology course. Regardless of what sequence is used, this chapter is intended to enhance your ability to integrate life processes across the hierarchy from cells to ecosystems. To the extent that you are successful in integrating in this way, you will be drawing closer to the essence of what is meant by *life*.

16-B   CELLULAR METABOLISM — WHAT CONTROLS IT?

Cells are controlled...

Right now, each of the billions of cells in your body are humming away absorbing raw materials, respiring glucose, synthesizing molecules, pumping ions, and doing all sorts of other things that keep you alive. What controls these activities of cellular metabolism? Specifically, what causes each cell out of billions in the body to function as a "servant" for the benefit of the whole?

from within the cell . . .

. . . by nucleus

Cellular metabolism is controlled both from *within* each cell, and from outside each cell. Control from within is mediated through the nucleic acids, the hereditary blueprint. We have already emphasized the influence of the genes upon the physical appearance and behavior of an organism. In Chapter 17, we shall examine the cellular processes through which nuclear genes exert their influence upon the cytoplasm and upon processes beyond the cell. Interestingly, as we shall see, the cytoplasm has some controlling influence over the nucleus.

. . . and from without . . .

Cellular metabolism is also controlled by factors outside the cell. This is essential if there is to be coordination and regulation of each cell for the benefit of the whole. We already have encountered some of these external controls by the *physical environment* and by *hormonal signals*. Environmental factors such as light and temperature can affect the expression of genes for pigment formation. Thus, genes for chlorophyll synthesis are only expressed when plants receive light. The expression of genes for hair color in siamese cats depends upon body temperature (Section 12-H).

. . . by hormones

and impulses.

Hormonal effects upon target cells represent another form of control form outside of cells. For example, thyroxin from the thyroid gland regulates the cellular respiration rate of each body cell. Auxin influences plant cell elongation and final cell shape. Other activities of cells that are imposed from outside the cell include impulse transmission, muscle cell contraction, and cell differentiation during growth.

Summary:

In summary, cellular metabolism is controlled by factors both within and outside of each cell. The relationship between these two levels of control is not unlike the relationship between state and federal government. Each of the fifty states has jurisdiction over certain activities involving its citizens. However, certain other affairs of each citizen are controlled from outside state boundaries by the federal government. The unity of our nation depends upon the proper exercise of these laws and the submission of each citizen to them.

16-C   CELL SIZE — WHY SO SMALL?

Small size requires "bridges" . . .

While peering through the lenses of a microscope, we may observe the frenzied activity of tiny microbes or the cytoplasm of human cells. Cells are alive and real, but they may seem removed from us by their smallness and because of the indirectness of our observations. These handicaps have challenged cell biologists to try to build bridges of understanding across this gulf to the microscopic world of the cell. The same challenge should be taken up by the student who wishes to understand cells.

. . . including measurement . . .

We rely upon linear measurement to quantify objects in the macroscopic world. Similar units can be used to span the gulf between the macroscopic and microscopic worlds. Of course, one must have a working knowledge of the units of measurement; for example, knowing and being able to visualize the relative lengths of a meter and millimeter.

Table 16-1 illustrates a range of decreasing sizes of different organisms and components of organisms. To represent this great range of sizes, four linear metric units are included in the table. The **meter**, represented by the length of one meter stick, is the largest unit. The other three metric units differ in length by a factor of one thousand. Check the metric side of a ruler, and be able to visualize in your "mind's eye" the size of a *centimeter* and *millimeter*. [Based upon Table 16-1, about how many human red blood cells would have to be placed in a row to reach across a distance of 1 millimeter?]

...and reference to other objects.

Most cells are invisible to the unaided eye. Notice from Table 16-1 that the human ovum falls just below the lower limit of objects that are visible to the unaided eye. Yet the human ovum is much larger than bacterial cells. [About how many times larger, based on Table 16-1?]

Table 16-1. Comparative Sizes of Biological Structures.

STRUCTURE	meter (m)	millimeter (mm)	micrometer (μm)	nanometer (nm)	MODE OF OBSERVATION
Organism: Human	10 m				
	1 m	1,000 mm			Unaided Eye
	0.1 m	100 mm			
Mouse	0.01 m	10 mm			
Ant	0.001 m	1 mm	1,000 μm		
CELLS/ORGANELLES:					
Human ovum		0.1 mm	100 μm		
Paramecium					
Human hair					
Human r. blood cell		0.01 mm	10 μm		Compound Microscope
Chloroplast					
Mitochondrion		0.001 mm	1 μm	1,000 nm	
Bacterial cell					
Large virus			0.1 μm	100 nm	
MOLECULES:					
Gene			0.01 μm	10 nm	Electron Microscope
Ribosome					
Cell membrane (thickness)				1 nm	
DNA helix (diameter)					
Amino acid				0.1 nm	Submicroscopic

three reasons for smallness

There are at least three explanations for the smallness of cells. First, the small size aids in exchange of substances across the membrane. The smaller the cell the greater the *surface-to-volume ratio* between membrane surface and cytoplasmic volume. Recall that human body cells are continually bathed in fluids which supply nutrients and carry away waste products. Because these cells are so small, no part of the internal cytoplasm is far from the membrane surface where exchange can occur. This relationship helps explain the second benefit of the smallness—as an aid to communication within the cytoplasm. Metabolic regulation of events in the cytoplasm is issued from two major points, the nucleus and from outside the membrane (Section 16-B). Smaller cell size shortens the distances over which regulatory substances and signals must travel within the cytoplasm. There are undoubtedly other reasons why cells are so small. Such teleological ("why") questions (Section 1-D) may have other answers that we have yet to learn.

16-D CELL STRUCTURE AND COMPOSITION

Draw a quick cell on paper.

Every line is a membrane!

The major chemical elements that compose living matter are listed in Table 4-1. How are the atoms of these elements assembled so as to form cellular structures that contribute to life processes? To illustrate the incredible organization of cells, draw the outline of a typical animal cell and include any internal structures, such as the nucleus, that you may recall. [Are you finished? Don't label it.] Every line you have drawn, both the outline of the cell and its internal organelles, represents a *membrane*! In fact, cells may be described as complex networks of membranes which have at least three major structural roles as illustrated in Figure 16-2. Let's examine each of these roles.

Cell Membrane -- a membranous boundary

Organelles -- membranous compartments

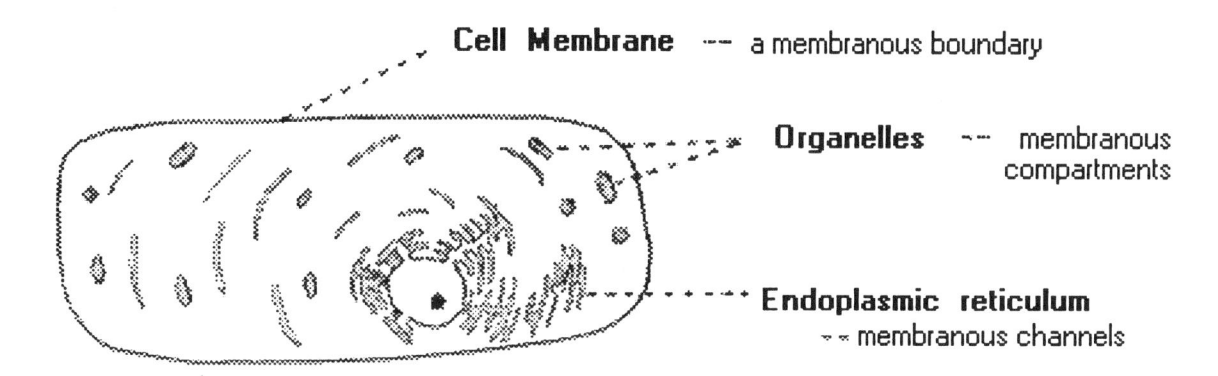

Endoplasmic reticulum -- membranous channels

Figure 16-2. Each cell is a complex network of membranes that serve three major structural roles as shown.

MEMBRANOUS BOUNDARY

traffic control

The **cell membrane** (or plasma membrane) forms the *membranous boundary* around the cell. The cell membrane holds the contents of the cell and regulates the movement of substances in and out of the cytoplasm. The term **selective permeability** is used to describe this property of membranes that actively control which substances can pass through them.

There are two major cellular regions within the volume defined by the cell membrane—the nucleus and the cytoplasm. The **cytoplasm** includes all of the cellular contents and organelles inside the cell membrane but outside the nuclear membrane.

MEMBRANOUS COMPARTMENTS

Organelles separate processes.

The nucleus is representative of a second major structural role of membranes, the formation of *membranous compartments*, called **organelles**. The cytoplasm is not simply a solution in which substances are dissolved in water. Rather it is a complex *suspension* in which organelles and macromolecules (e.g. large proteins) are suspended in a viscous, aqueous medium. Organelles are tiny compartments that allow diverse metabolic processes to be carried out separately within the same cell.

"generalized cell"

Figures 16-3 and 16-4 illustrate generalized animal and plant cells, respectively. The "generalized cell" is actually a creation of biology instructors and textbook authors. In real life, cells are so diverse and specialized that many different cells would have to be shown to represent all of the possible structures. The generalized cell and other models of cellular structures must be interpreted in light of their limitations. This caution is but another aspect of the challenge we face as we seek to comprehend the nature and function of cells.

Both of the generalized cells, illustrated in Figures 16-3 and 16-4, represent **eukaryotic** cells because they have membrane-enclosed organelles. **Prokaryotic** cells of the kingdom Monera lack membrane-enclosed organelles.

Generalized plant and animal cells differ in four principal ways:

separations

1. BOUNDARIES AND JUNCTIONS: **Cell walls**, porous meshworks of **cellulose** fibers, surround the outer face of the cell membrane. Fiber cells, which provide strength and rigidity, may develop thicker walls with cement-like *lignin* added. Following mitosis, a *cell plate* divides the cytoplasm and determines the position of new cell wall formation (Section 11-E). Then, each daughter cell secretes substances for its own part of the new wall, thus separating adjacent cell membranes with cell wall.

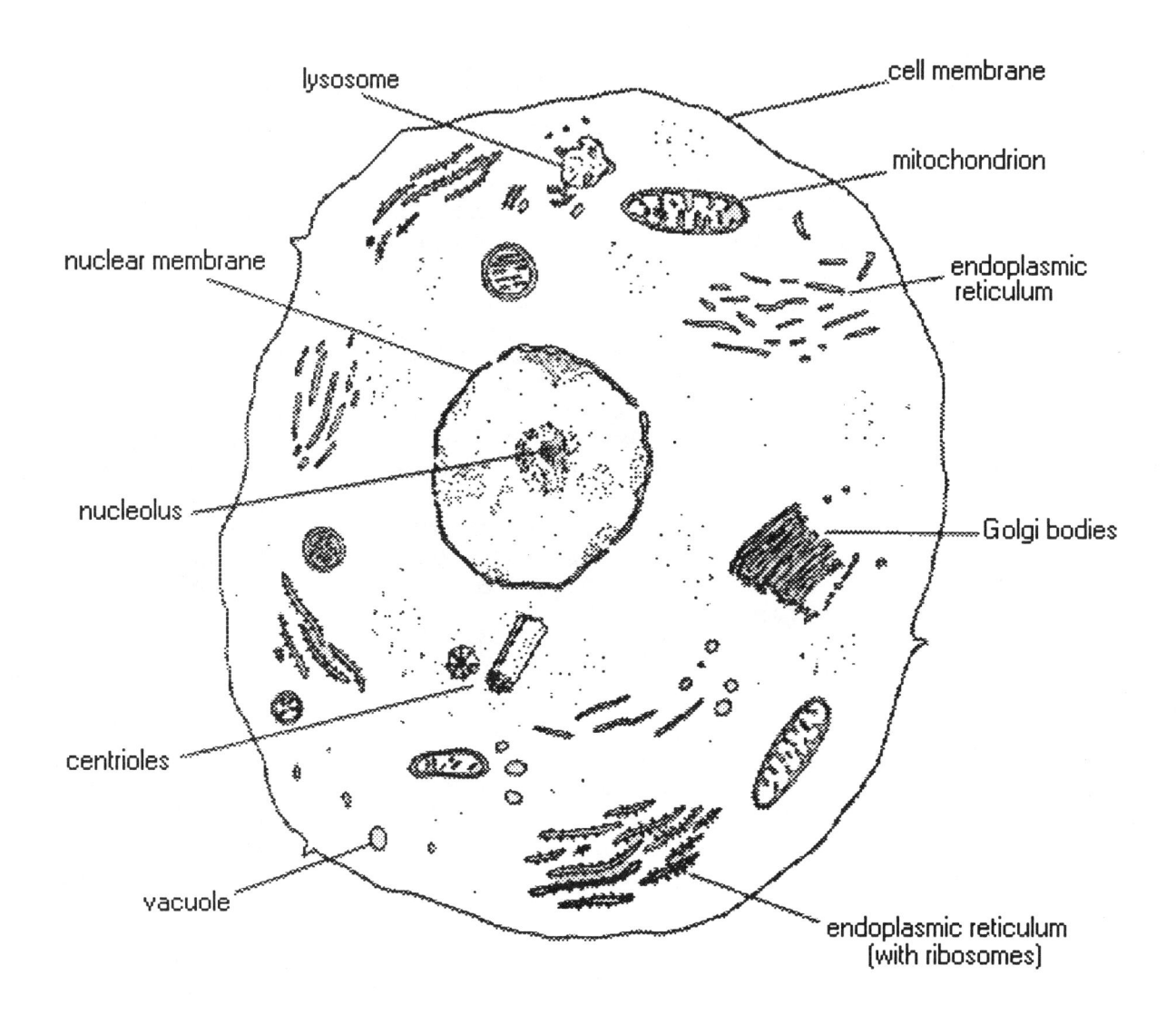

Figure 16-3. A generalized animal cell.

366

```
                cell wall
   cell membrane
                                    vacuole
                                    Golgi bodies
                                    nucleolus
                                    nuclear membrane
                                    endoplasmic reticulum
                                    (with ribosomes)
   chloroplasts

   mitochondria
                                    plasmodesmata
```

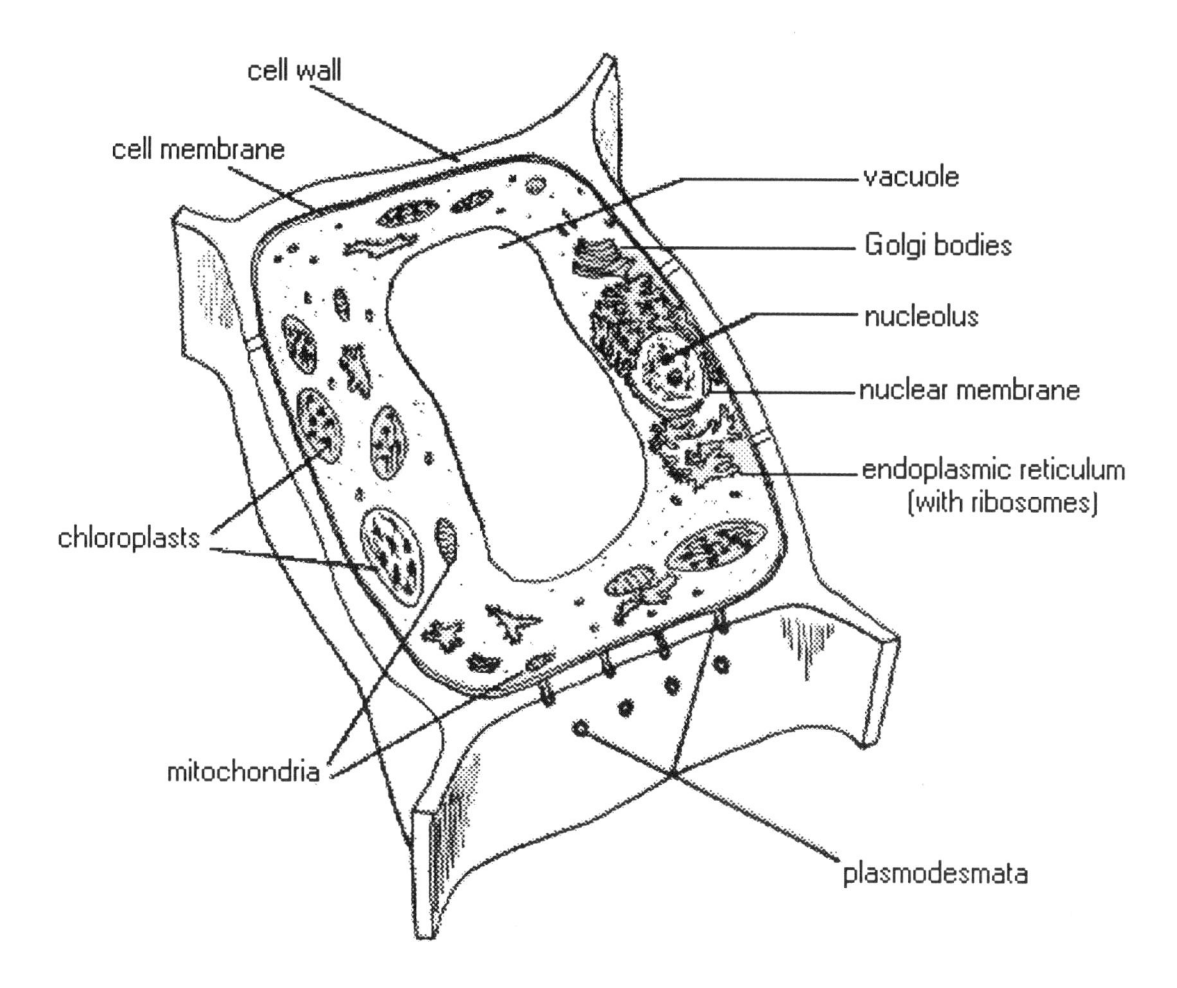

Figure 16-4. A generalized plant cell.

passageways	2.	PLASMODESMATA are junctions between adjacent cells consisting of continuous strands of cell membrane and cytoplasm that extend through the cell wall (Figure 16-4). Substances can pass readily *through* plasmodesmata between adjacent cells, thus apparently compensating for any hindrance of the cell walls. Animal cells, whose membranes are not separated by a cell wall, have *cell junctions* at points along adjacent membranes, but no plasmodesmata. [Are plant cells really separated?]
photosynthesis	3.	CHLOROPLASTS, organelles of photosynthesis, contain green and orange pigments, **chlorophyll** and **carotenoids**, respectively. These light-absorbing pigments are found only in cells of photosynthetic tissues. [Name several types of plant cells that lack mature chloroplasts?]
disposal	4.	LYSOSOMES are found in animal cells, but not plant cells. Lysosomes are membranous sacs containing powerful digestive enzymes that can digest macromolecules, other organelles, whole bacterial cells (e.g. engulfed by white blood cells), and even the cell itself in cases of embryonic development. [Could this explain the disappearance of a tadpole's tail? See TOPIC #1.]

other organelles

Most of the remaining organelles are present in at least some types of both animal and plant cells. These include the **mitochondria** (singular, mitochondrion), the sites of cellular respiration; **Golgi bodies** which collect and transport substances that are synthesized within the cell; **ribosomes**, sites of protein synthesis; and **vacuoles** which have a variety of functions.

MEMBRANOUS CHANNELS

The third structural role of membranes is to form *membranous channels* throughout the cytoplasm. These channels are called **endoplasmic reticulum**, or ER (Figures 16-3 and 16-4). The ER forms a complex, interconnected, membranous network which is associated with the cell membrane, nuclear membrane, and Golgi bodies. The ribosomes are attached to the surfaces of some regions of the ER, giving it a rough appearance. Proteins synthesized at the ribosomes are transported in the ER and in conjunction with the Golgi bodies.

Is the cell theory correct?

As noted earlier, plant cells commonly have interconnections through adjacent cell walls, the plasmodesmata (Figure 16-4). In fact, even the membranous channels of ER of one cell are connected with ER in adjacent cells via the plasmodesmata. Research in this area is leading some plant biologists to suggest that the *cell theory* with its emphasis on cells as distinct units may be a misleading paradigm in attempts to understand organismic biology of plants. [See TOPICS #4.]

Summary:

In summary, a cell is a complex network of membranes. Membranes perform three structural and functional roles. They form the membranous boundary called the cell membrane, the membranous compartments called organelles, and the membranous channels of the endoplasmic reticulum. These membranous structures are suspended within a medium of water. Together, water and membranes account for most of the marvelous substance we call "living matter". A closer examination of water and its properties will give clues to the structure and function of membranes.

16-E   WATER — FLUID OF LIFE

vital "medium" of life

Water is the most abundant molecule in living matter. Approximately two-thirds of the average person's body weight is water. [About how many pounds of H_2O are you carrying?] Someone has said that the "chemistry of life is dominated by the chemistry of water". All known life processes require the presence of this vital medium.

We have discussed several of the specific chemical and physical properties of water in Section 4-E. The stable and slightly polar covalent bonds between one atom of oxygen and two hydrogen atoms accounts for the tendency of water molecules to remain chemically and physically stable (i.e. resistant to change).

Can you list valuable properties?

The hydrogen bonding among water molecules (Figure 4-1) is responsible for the high specific heat of this liquid. That is, water can absorb large amounts of heat energy with little change in temperature or physical state. The resulting temperature stability is crucial for body temperature regulation in the midst of fluctuating environmental temperatures (Section 14-A). Otherwise, cells would tend to "boil over" as a result of heat absorption from the surroundings or from metabolically generated heat within. In addition, organisms have evaporative cooling mechanisms (e.g. sweat glands) that rely upon the high specific heat of water to draw heat energy away from the body during evaporation. A large oak tree may release 300 gallons of water as vapor into the atmosphere per day. This evaporative loss of water and heat serves to cool the leaves and even the air around the tree, as shade tree-lovers know.

As water approaches freezing, its density decreases, causing the coldest water to rise to the surface of lakes. Thus, ice forms on the surface first. If the water froze from lake bottom upward, aquatic organisms would be trapped between freezing water and chilling air with little chance of surviving.

The aquatic food web as we know it could not exist if water were not transparent to light which is necessary for autotrophic photosynthesis. Vision is possible, because light entering the eye passes through a transparent aqueous medium before reaching the retina.

The fluid properties of water ideally suit it for providing physical support, or turgidity, in plant cells. Water is also an important component of the fluids that lubricate joints, and body organs as they rub across one another.

The hydrogen bonding property of water molecules makes water an excellent **solvent**, meaning that it can dissolve a great variety of **solutes** (e.g. salts, sugars, amino acids). Water molecules form spheres of hydration around solute particles and hold them in solution (Figure 4-2). Molecules that are too large to dissolve are held in suspension. For example, water holds many protein and lipids in suspension. It is this relationship that provides for the existence of membranes.

16-F   MEMBRANE STRUCTURE

Membrane structure is invisible.

The membranes of cells and organelles are so thin that their structure is very difficult to ascertain even with the electron microscope (Table 16-1). Therefore, we must refer to one of several likely models of membrane structure. Membrane models are based upon *in vivo* and *in vitro* studies of membranes, chemical analyses of isolated membranes, and electron microscopy.

chemical components

Chemical extractions of isolated membranes consistently reveal various proportions of lipid and protein molecules. The lipid component of most membranes is predominantly **phospholipids** (Figure 16-5). Cholesterol, a steroid-type lipid, is also an essential component of many membranes. Phospholipids are similar to fat molecules, except that a phosphate group is substituted for one of the three fatty acid chains (Compare Figures 4-8 and 16-5.). The negatively charged phosphate group and some other positively charged groups such as ethanolamine form a charged, "head" portion which tends to dissolve in water. The fatty acid chains composed of neutrally charged hydrocarbons tend to repel water. Therefore the phospholipid molecule has what might be considered as a "split personality". The water-loving, or **hydrophilic**, "head" region [G. (*hydro* = water) + (*philo* = loving)], contrasts with the water-fearing or **hydrophobic** "tails" (G. *phobia* = fear).

phospholipid arrangement

The physical behavior of phospholipids in water gives support to the possible arrangement of these molecules in a real cellular membrane. When a sample of purified phospholipid is placed in a container of water, they tend to form spherical aggregates (Figure 16-6a). The hydrophobic tails repel water and form a waterless gathering of tails pointing toward the inside of the sphere. The hydrophilic heads dissolve in the surrounding water and form the outside

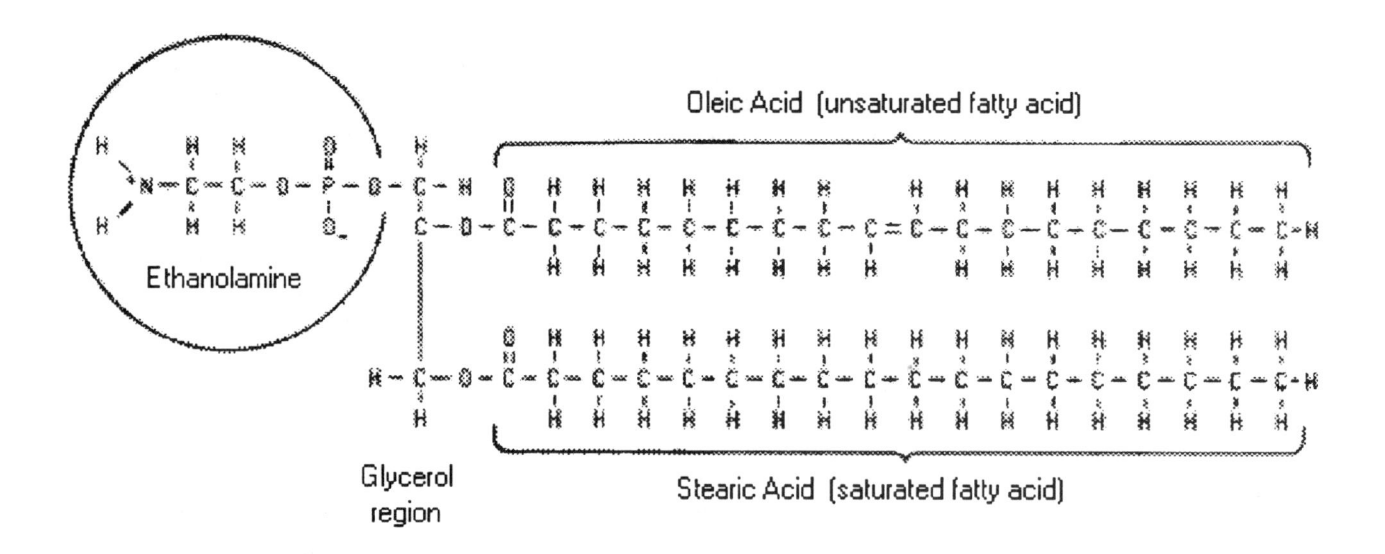

Figure 16-5. Molecular structure of a typical phospholipid, phosphatidyl ethanolamine. The molecule consists of a polar (charged) hydrophilic "head", encircled, and nonpolar (neutral) "tails", in brackets.

of the sphere. Spherical aggregation of phospholipids is a spontaneous process that leads to a more stable configuration with minimum "stress and strain" among the molecules. This behavior is the basis for the grease-sequestering action of many detergents. The hydrocarbon tail portions gather around grease and fats on dishes or clothes while the hydrophilic heads mingle with water, allowing the whole complex to be rinsed away. Egg yolk contains a natural phospholipid called lecithin which can make salad oil dissolve in vinegar. The product is mayonnaise.

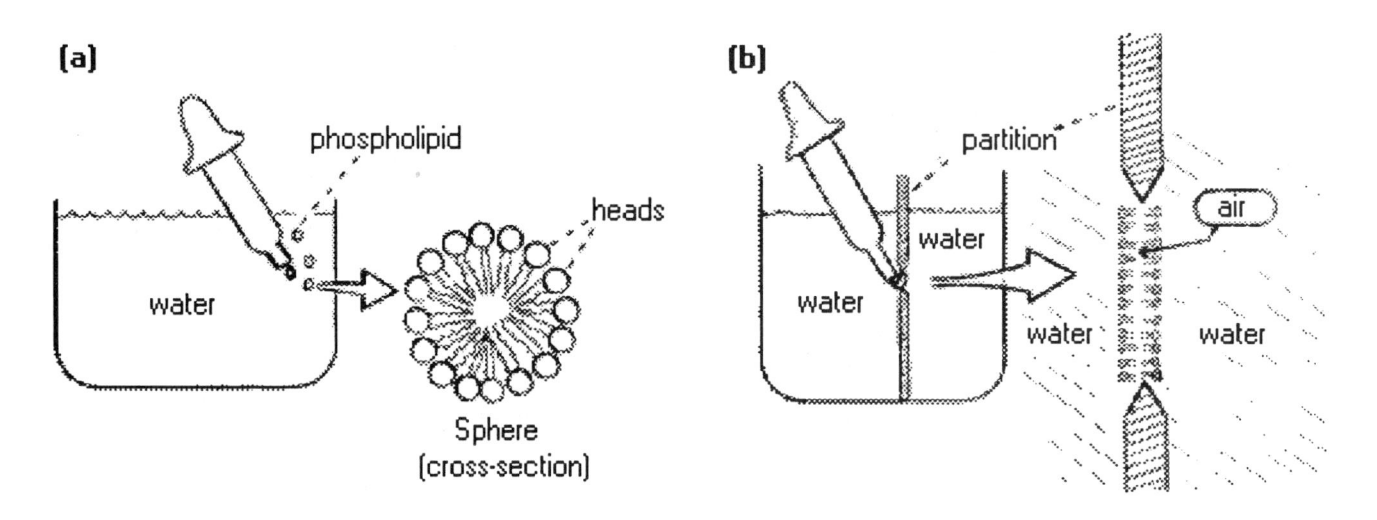

Figure 16-6. Diagrammatic models of the arrangements of phospholipid molecules (a) dispersed in water, and (b) across a small slit with water on both sides. Enlargements are simplified and appear enlarged more than present electron microscopy allows.

Artificial membranes...

When a sample of phospholipid is placed near a small opening in a partition under water (Figure 16-6b), the partition interferes with sphere-formation. Instead, one layer of phospholipid molecules forms on each side of the opening. As expected, the hydrophobic tails exclude water molecules from between the layers and the hydrophilic head dissolve in water to form an outer surface for each layer. The result is an artificial membrane composed of two phospholipid layers. This model strongly suggests the probable arrangement of phospholipids in natural cellular membranes. The model is also consistent with electron microscope views of membranes which reveal a distinct double-layered appearance. However, at least three other observations each suggest

are less complex than biological membranes.

that cellular membranes are more complex than the artificial membrane. First, water can diffuse through cellular membranes, suggesting that the water—excluding phospholipid tails are interrupted at intervals by hydrophilic substances that provide channels for water to diffuse through. (A membrane is said to be **permeable** to those substances that can pass through it.) Second, proteins have been detected in chemical analyses of membranes, as noted earlier. Third, charged substances such as Na^+ and Ca^{2+} move through cell membranes much more slowly than uncharged H_2O molecules.

bimolecular layer model

Figure 16-7 illustrates the most widely accepted model of a "typical" cellular membrane. A bimolecular layer of phospholipids is interspersed with proteins situated within, or upon, the surface of the membrane, like icebergs floating in a sea. These *globular proteins*, unlike *fibrous proteins* of muscle and hair, are composed of chains of amino acids coiled into a three-dimensional configuration (Section 4-I). Certain ones of the 20 different amino acids are more

water influences configuration

hydrophobic than others. Therefore, the actual configuration of each globular protein in the membrane is the result of hydrophobic and hydrophilic interactions among the different amino acids in the phospholipid and aqueous surroundings. Short polysaccharide chains may be attached by covalent bonds to their outer surfaces. Such globular proteins, called *glycoproteins*, function in *recognition* of, or as *receptors* of, extracellular substances or other cells. [Recall post-synaptic receptors (Section 13-D)?] *Cholesterol* is common in animal cell membranes where it promotes fluidity, fluid-like properties of the membranes.

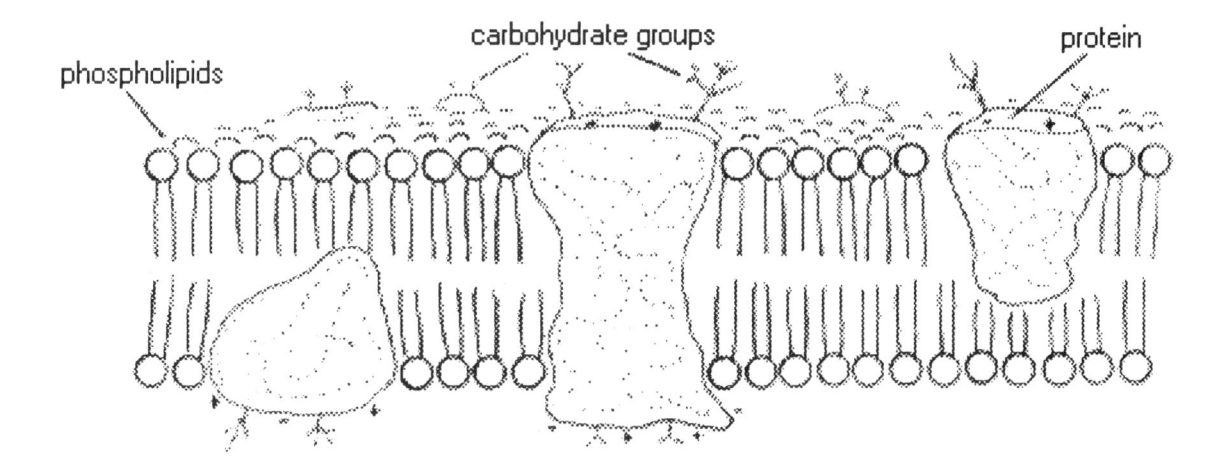

Figure 16-7. The 'fluid mosaic model' of membrane structure suggests that a bimolecular layer of phospholipids form a fluid matrix that is interspersed with various globular protein molecules to form a 'mosaic'. The model was first proposed by Singer and Nicolson, in 1972 (9).

Membranes have protein channels.

Protein molecules that extend completely across the membrane provide hydrophilic *channels* through which water and water-soluble substances can move. Other proteins may be *carriers* which "pump" substances across the membrane by *active transport*. Movement of substances across membranes is the subject of the next section.

Summary:

In summary, membranes cannot be directly observed in sufficient detail to ascertain the molecular arrangements of proteins and phospholipids. However, models such as the one in Figure 16-7 have been developed to account for the known chemical composition and permeability properties of membranes.

16-G MOVEMENT OF SUBSTANCES ACROSS MEMBRANES

Life processes within cells require a continual exchange of energy and nutrients with the surroundings. There are several important mechanisms by which substances move across cellular membranes. These are diffusion, osmosis, active transport, and pinocytosis.

DIFFUSION AND OSMOSIS

The concepts of *diffusion* and *osmosis* account for the movement of water and many water-soluble substances in and out of cells. **Diffusion** is the movement of ions or molecules from a region of high kinetic energy to a region of lower kinetic energy.

kinetic theory

According to the *Kinetic Theory of Matter*, all molecules and ions are in constant motion, and thus, they possess *kinetic energy* (Section 5-B). *Temperature* is a measure of kinetic energy. The higher the temperature of water or some other substance, the faster the molecules are moving and bombarding against one another. Our touch and pain receptors can detect differences in the amount of kinetic energy of matter. [Consider how kinetic energy can do the "work" of lifting mercury vertically inside a thermometer!]

concentration and kinetic energy

The kinetic energy of molecules is manifested not only in temperature differences, but in diffusion. When a high concentration of a certain substance is placed in one location within a gaseous or liquid medium, the molecules will eventually diffuse throughout the medium. This occurs by the random motion of each molecule which has a high probability of moving away from the region of its highest concentration (Figure 16-8). Thus, fragrant molecules of perfume or cologne move off the skin and through the air, partly by diffusion. [And, partly by convection currents which move the molecules in bulk. Does cytoplasm have a means of producing *bulk flow*? Keep this question in mind.]

We have already encountered several examples of diffusion in living systems. For example, CO_2 and O_2 both must diffuse into the air spaces of absorptive tissues such as spongy mesophyll cells of leaves, and the alveoli of lungs.

osmosis defined

When randomly moving molecules encounter a membrane that is permeable to that particular substance, the molecules will tend to move through the membrane. Water, which is the most abundant molecule in cells and body fluids, is continually diffusing across cellular membranes. The movement of water molecules across a *differentially permeable* membrane as a result of a difference in kinetic energy is called **osmosis**. Note that the concept of osmosis is focused on *water* movement, specifically movement of water

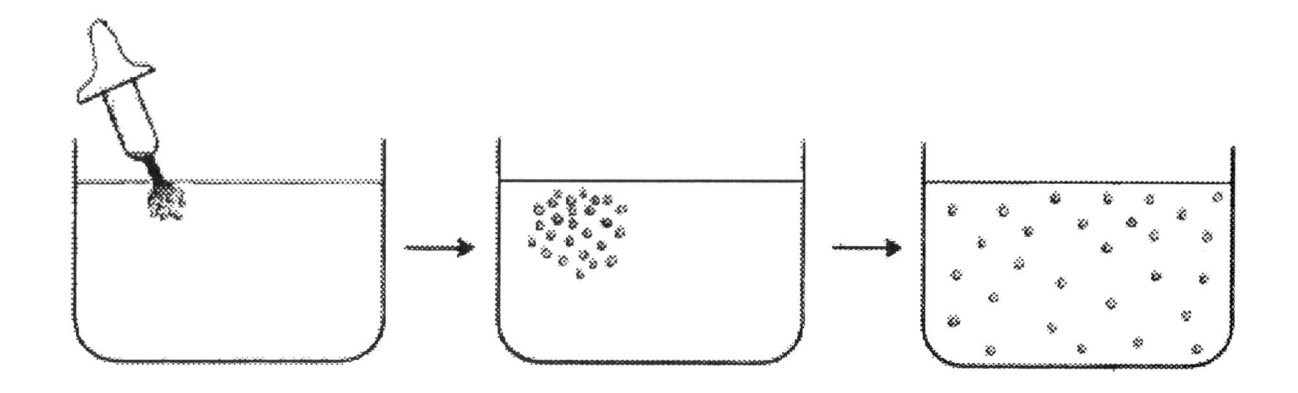

Figure 16-8. A diagrammatic representation of diffusion. Arrows between the containers indicate elapsed time.

Membranes are differentially permeable.

across a membrane that is more permeable to water than to other particles such as ions and large molecules; hence the name, **differentially permeable membrane (DPM)**. Kinetic energy of water may be increased by increasing temperature and/or pressure, or decreased by adding **solutes**, ions or molecules which cause water to expend its kinetic energy to dissolve them into solution. Diffusion and osmosis are manifestations of the *second law of thermodynamics*, which predicts that reactions occur in a direction so as to decrease *order* (or purity) and increase *entropy* (Section 5-C).

osmosis demonstrated

Osmosis can be demonstrated by the model in Figure 16-9. The DPM separating side A from side B is made of cellophane. When the concentration of water on one side of the membrane is lowered due to the addition of solutes, water molecules become less free to diffuse. Their "freedom", or measure of their kinetic energy, is now devoted to dissolving the solute by forming *spheres of hydration* around solute particles (Figure 4-2). Side A, which has a lower solute concentration and the higher water concentration will have more water molecules "free" to collide with the membrane and pass through to side B. Indeed, these "free" water molecules may even be "put to work" dissolving solutes when they arrive at side B. As that happens, the probability of their return to side A is decreased, and more water will accumulate on side B. Thus, osmosis is the (a) movement of water molecules (b) across a differentially permeable membrane (c) from a region of higher water concentration (or higher kinetic energy, purity, freedom to diffuse) to a region of lower water concentration (higher solute concentration).

It takes "work" to dissolve.

Osmosis can cause pressure.

In reality, osmosis itself cannot lead to a condition of equal concentrations of water on both sides of a membrane. The solute particles cannot readily cross the differentially permeable membrane. Therefore, side B will always contain at least a very dilute concentration of solute, and therefore have a lower water concentration. Does this mean that water will continue to move from side A to side B, indefinitely? The answer is, no. The net movement of water by osmosis from side A to B occurs only until the pressure of the rising liquid in B reaches a certain point. This pressure is due to the mass of the column of solution in side B being lifted against the pull of gravity; and is a manifestation of the kinetic energy of the water in side A. The kinetic energy of the water drives water molecules across the membrane from side A to side B, and causes the solution to rise. The pressure against the membrane at equilibrium is called the **osmotic pressure**.

osmotic pressure

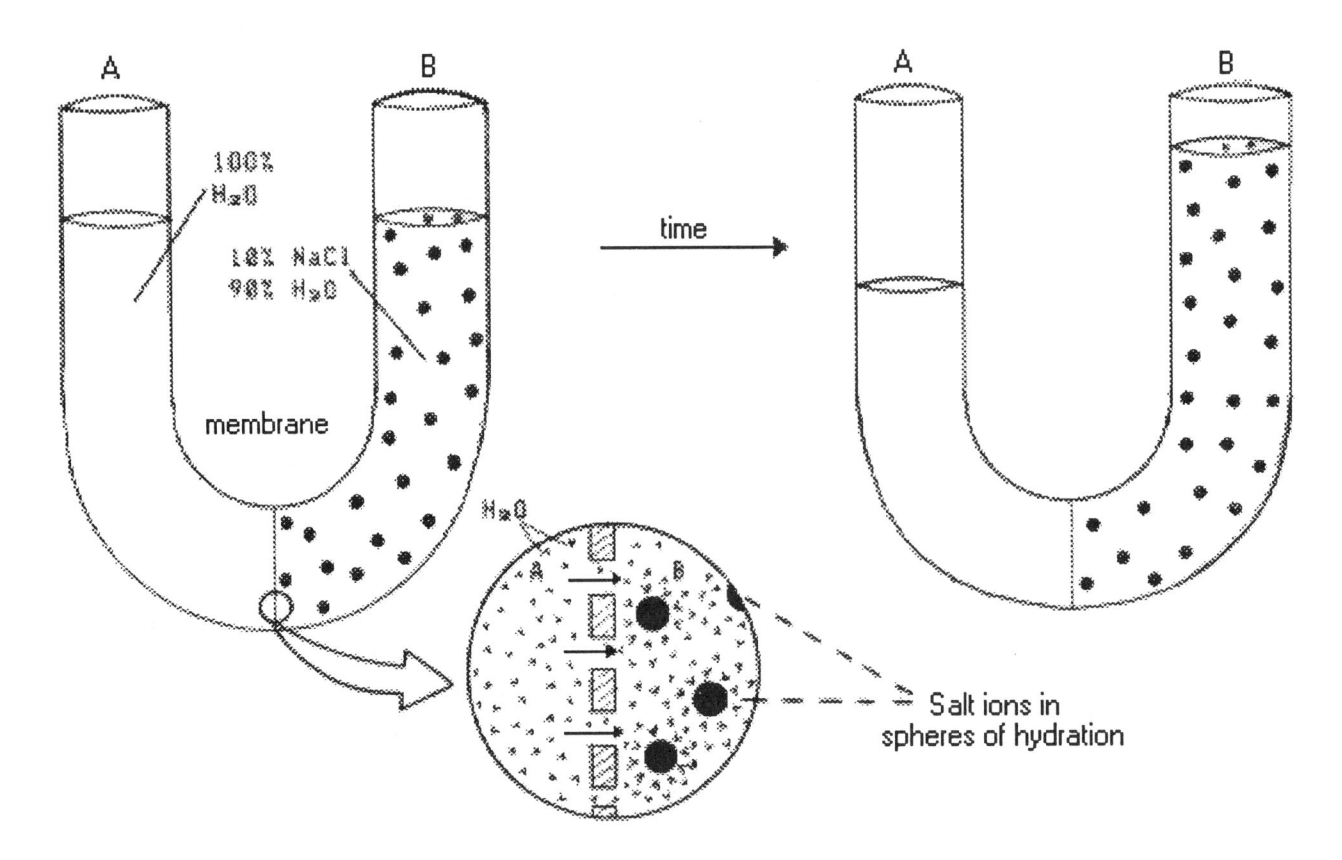

Figure 16-9. A U-tube with side *A* containing pure water is separated from side *B*, containing 10% NaCl (solute) and 90% water, by a differentially permeable membrane (DPM). The DPM theoretically allows water but not salt ions to cross the membrane. Water molecules on side *A* are not involved in dissolving solute. Those on side *B* expend kinetic energy to form spheres of hydration (see enlarged view). Therefore, water molecules tend to move from this region of higher water concentration and kinetic energy (side *A*) to side *B*. Net movement of water from side *A* to side *B* occurs until the pressure of the rising column of solution on side *B* equalizes the kinetic energy of water on side *A*.

model of a cell

Later, we will emphasize the difference between a laboratory U-tube system and real cells, but for now the U-tube system can at least help us understand some basic relationships. The cytoplasm of plant cells contains a variety of solutes and macromolecules dissolved or suspended in water. Therefore, water outside most plant cells is in higher concentration, and tends to move into the cells by osmosis (Figure 16-10a). The rigid cell wall prevents cells from bursting as a result of water uptake. When internal pressure of the cell membrane against the wall reaches a certain point, osmotic equilibrium is reached just as in our U-tube model (Figure 16-9). Plant cells are said to be *turgid* when the cell membranes are extended outward against the inner face of the cell wall by water pressure within.

Turgor pressure supports plants.

The **turgor pressure** of leaf and stem cells is responsible for holding these organs erect. When soils dry to the point where water cannot enter root cells, the entire plant may wilt because plant cells loose more water by osmosis than they gain. This condition can be roughly simulated by placing plant cells in a highly concentrated salt solution (Figure 16-10 b). Water concentration within the cytoplasm is higher than in the surrounding solution, and the cells

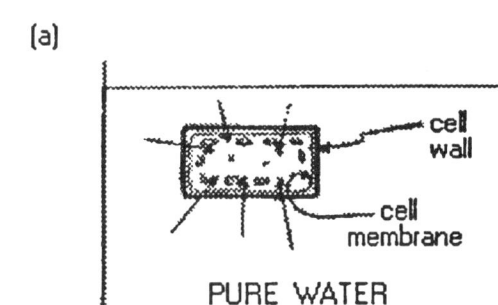

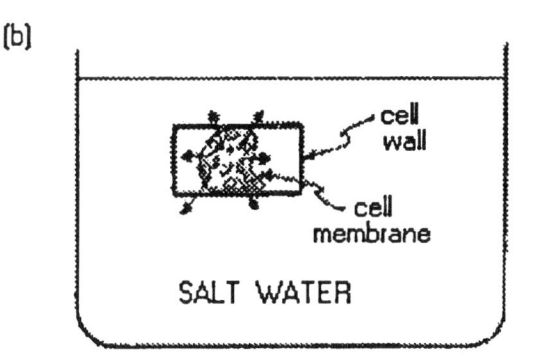

Figure 16-10. The turgidity of plant cells depends upon the concentration of water inside the cells relative to that on the outside. Arrows indicate primary direction of osmosis that led to the condition shown. See text for explanation.

Why do plants wilt?

lose water by osmosis. Loss of turgor pressure is evidenced as the cell membrane collapses around the dehydrating cytoplasm. This condition is called **plasmolysis**. Major metabolic disruptions occur when this condition is reached. [Can you suggest why, based upon Section 16-D and 16-E?]

role of kidneys

The osmotic properties of body fluids have a major influence upon the metabolism of animals. Section 14-D discusses the important role of the kidneys in osmoregulation. Each body cell must be continually bathed in fluid with the proper osmotic properties to maintain its cytoplasm and cellular metabolism.

ACTIVE TRANSPORT

As promised earlier, we must admit that U-tubes and cellophane membranes do fall short as a model of cellular membranes. Cellophane membranes are differentially permeable because they are like "screens" with fixed pore sizes. Thus, the cellophane is permeable to small molecules but not to large molecules like starch or proteins. Likewise, cell membranes are permeable to *nonpolar* (neutral) substances such as O_2, CO_2, ether, and ethanol which diffuse through the lipid matrix. However, simple diffusion will not account for the movement of other equally small molecules such as water, sugars, ions, and other *polar* substances. Indeed, cells can actually accumulate potassium ions in higher concentration than the surrounding fluids, and sodium is partially excluded.

countering diffusion

Active transport gives selective permeability to membranes...

Clearly, biological membranes are more than fixed "screens". Instead, membrane proteins function as "carriers" or "channels" to give the membrane a **selective permeability**. When substances move across membranes through protein channels from a high to a low concentration without an expenditure of energy, *passive transport* is said to occur. The process of **active transport** requires energy expenditure to activate membrane proteins as "carriers" or "pumps" which function to accumulate ions and molecules against concentration or pressure gradients (Figure 16-11).

...and regulate composition of cytoplasm.

Active transport produces a relatively greater concentration or purity at the expense of high quality chemical energy supplied by cellular respiration. On the other hand, diffusion is a random, spontaneous process that is driven only by the low quality, kinetic energy of the particles involved. Both processes are essential to cellular metabolism. However, active transport can actively regulate the chemical composition of the cytoplasm and surrounding fluids.

OUTSIDE   MEMBRANE   INSIDE
(x-section)

(a) Diffusion

OUTSIDE   MEMBRANE   INSIDE
(x-section)

ENERGY INPUT

(b) Active Transport

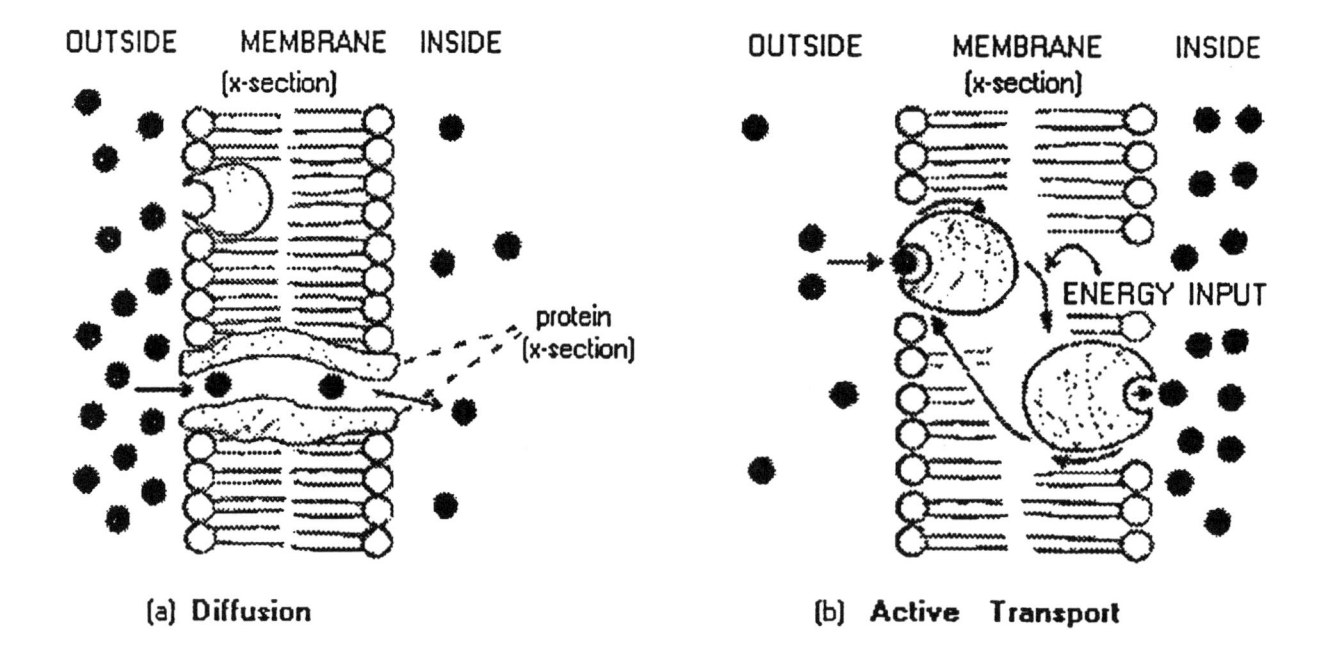

Figure 16-11. Comparison between (a) diffusion of a substance across a membrane from a high concentration outside to a lower concentration inside the cell; and (b) active transport which involves expenditure of energy to operate a membrane protein "pump". The pump is shown transporting a substance against a concentration gradient.

membrane "pumps"

Sodium-potassium pumps are complex membrane proteins that actively transport Na^+ and K^+ to form a concentration gradient across cell membranes of neurons in readiness for impulse transmission (Section 13-D). Osmoregulation and excretion processes of the kidneys require active transport of glucose, amino acids and salts by the kidney tubules (Section 14-D). Therefore, active transport is a fundamental part of the living condition of cells because of its orderly regulation of cellular import and export. Cessation of active transport would immediately cause important gradients to disappear into equilibrium with the cellular environment —a condition associated with cellular death.

ENDOCYTOSIS AND EXOCYTOSIS

How do "big molecules" cross?

Diffusion, osmosis, passive transport, and active transport all involve the movement of dissolved ions and small molecules across cellular membranes. Endocytosis and exocytosis occur when substances too large to pass though the lipid bilayer are transported by membrane-bound spheres or **vesicles** (Figure 16-12). Large molecules or particles are taken into the cell by **endocytosis**. Here, the cell membrane forms a depression in which the substances collect, after which the membrane pinches off to form the vesicle, surrounded now by cytoplasm (Figure 16-12). **Exocytosis** occurs when cells synthesize large molecules in the cytoplasm and "wrap" them in a transport vesicle, which fuses with the cell membrane and expels (*secretes*) the substances to the outside. A special category of endocytosis, in which liquids are taken in, is called *pinocytosis*. [(*pino*, drink) + (*cyto*, cell) = "cell drinking"].

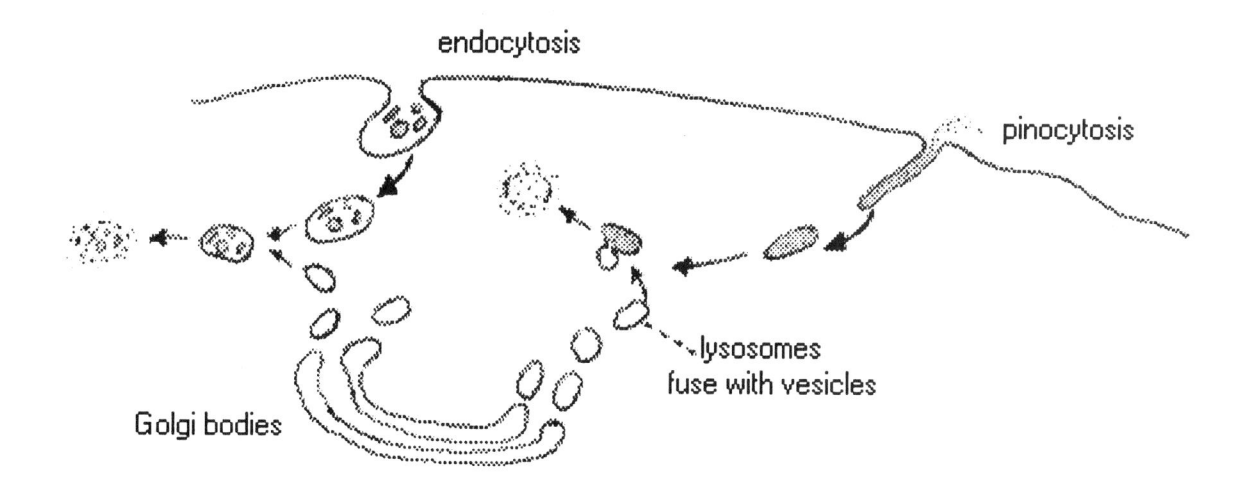

Figure 16-12. Pinocytosis and phagocytosis. Lysosomes supply digestive enzymes to break down the engulfed material within the vesicles.

The intake of food particles by digestive cells in sponges and in the gastrovascular cavities of cnidarians and flatworms involves endocytosis. The well-known *Amoeba* provides a classic example of feeding by endocytosis, in this case termed *phagocytosis* (Figure 9-4). White blood cells, called *phagocytes*, provide defense against pathogenic bacteria that enter the human body by engulfing and digesting them (Section 17-E).

Digestion of the large molecules or cells, taken in by endocytosis or pinocytosis, occurs within membrane vesicles. Lysosomes released from the Golgi bodies fuse with the vesicle and release digestive enzymes into contact with the engulfed material (Figure 16-12). The digestion products can diffuse outward across the vesicle membrane and be used as raw material in cellular metabolism.

16-H SUMMARY

We have completed our general survey of the cell as the basic unit of living matter. All life processes are ultimately cellular processes. Cells of unicellular organisms are unspecialized and can perform all of the basic life processes. However, cells of multicellular organisms are specialized, and perform only certain specific functions. These functions have their basis in cellular metabolism. Cellular metabolism is regulated from within each cell. However, cellular metabolism is also controlled from sites outside the cell membrane through hormonal secretions, nerve impulses, or environmental stimuli.

Each cell is a complex network of membranes. The cell membrane is a membranous boundary which is selectively permeable. Depending upon their size and chemical properties, various substances are allowed to pass through the membrane by such processes as diffusion, osmosis, active transport, and pinocytosis.

Within the eukaryotic cell are a host of different organelles, membranous compartments within which different metabolic processes can occur. The membranous channels of the endoplasmic reticulum serve as a transport and communication network between the nucleus and other parts of the cytoplasm. In the next chapter, we shall examine the role of the nucleus and the DNA blueprint in controlling cellular metabolism.

QUESTIONS AND DISCUSSION TOPICS

1. Explain the basis for the sequence of topics as they are presented in Units II, III, and IV of this text. How has this order, or that chosen by your professor, helped (or hindered) your understanding of cellular metabolism.

2. Use the approach of Figure 16-1 to illustrate how DNA metabolism is viewed at successively higher levels of organization, starting with the cell. Repeat for the concept of *homeostasis*.

3. To understand how cell size affects surface-volume ratio, make the following computations of area (A in cm^2) and volume (V in cm^3):
   a. Compute A, V, and area-to-volume ratio (A/V) of a 10-cm cube.
   b. Compute A, V, and area-to-volume ratio of a 1-cm cube.
   c. Which has the larger A/V, the 10-cm cube or the 1-cm cube?
   d. Compare the A/V's of 10-cm and 1-cm cubes with that of a 0.1-cm cube. Summarize the relationship between volume and surface area, and explain how smallness aids cells in exchange of substances between cytoplasm and surrounding environment.

4. What is a scientific model? What cautions must be observed in using models such as the "generalized cell"?

5. What is *cellular metabolism?* What controls it? [Suggestion: Use examples of cells, the function of which we have studied — e.g. muscle cells, ovarian follicle cells, cells in a growing embryo?

6. Describe the molecular structure of cellular membranes. What if, instead of lipids and proteins, membranes were made of salts and sugars?

7. Explain the three general structural and functional roles of membranes in living cells. What specific examples fall within each category?

8. Without looking in the text, be able to draw a "generalized cell" and label the different organelles, and identify the function of each.

9. Analyze the claim that "the chemistry of life is dominated by the chemistry of water", from the following philosophies: (a) mechanistic, (b) vitalistic, and (c) biblical theism. See Introduction to Chapter 4.

10. What is osmosis and why are osmotic processes so important in the maintenance of cellular structure and metabolism? How do plants and animals perform osmoregulation?

11. Distinguish active transport from diffusion. How is active transport different from endocytosis and pinocytosis? How are they similar?

12. Why would inhibitors of cellular respiration inhibit nerve impulse transmission? How would such inhibitors affect the turgidity of plant cells, all other factors being in abundant supply?

13. To you, what cell structure or process determines if a cell is *living* or *dead*?

TOPICS FOR FURTHER STUDY

1. Why have lysosomes been called "suicide bags"? What is their role in embryonic development? REFERENCE #3.

2. What is it about the structure and function of membranes that makes living matter "alive"? What techniques are used to study structures so small as membranes and organelles? REFERENCES: #1 and #8

3. Have a closer study of Golgi bodies and their relationship to the "suicide bags" (see TOPIC # 1, above). REFERENCES: #6 and #7

4. Does the cell theory really account for morphology and development of plants? Are plants even truly *multicellular*? REFERENCE #5

REFERENCES

1. Bretscher, M. 1985. The molecules of the cell membrane. *Scient. Amer.* 253(4): 100-108.

2. Darnell, J., *et.al.* 1990. *Molecular Cell Biology*, 2nd ed. Freeman and Co. New York, NY.

3. de Duve, C. 1983. Microbodies in the living cell. *Scient. Amer.* 248(5): 74-84.

4. deDuve, C. 1985. *A Guided Tour of the Living Cell*. Freeman and Co. New York, NY.

5. Kaplan, D.R. and W. Hageman. 1991. The relationship of cell and organism in vascular plants. *Bioscience* 41(10): 693-703.

6. Rothman, J.E. 1981. The Golgi apparatus: Two organelles in tandem. *Science* 213: 1212-1219.

7. Rothman, J.E. 1985. The compartmental organization of the Golgi apparatus. *Scient. Amer.* 253(3): 74-89.

8. Scientific American Book. 1980. *Molecules to Living Cells*. W.H. Freeman, San Francisco.

9. Singer, S.J. and G.L. Nicholson. 1972. The fluid mosaic model of the structure of cell membranes. *Science* 175: 720-731.

10. Starr, C. and R. Taggart. 1992. *Biology: The Unity and Diversity of Life*, 6th ed. Wadsworth, Belmont, CA.

11. Stillinger, F.H. 1980. Water revisited. *Science* 209: 451-457.

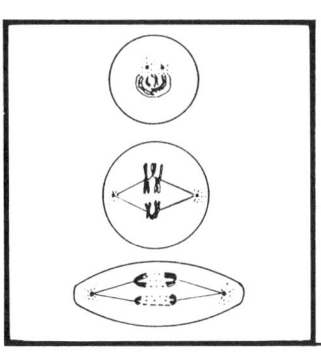

Chapter 17

Genetic Control of Cellular Metabolism

DNA, deoxyribonucleic acid, "genetic blueprint", "genetic code", hereditary material—these are the most common names applied to one of the most intriguing parts of living cells. In this chapter we shall explore a few of the mysteries of DNA and the way in which genes control cellular metabolism. Understanding these processes will aid our understanding of viruses, the immune system, and the technology of *genetic engineering*, discussed later in the chapter.

17-A GENES AND GENETICS — REVIEWING PREVIOUS ENCOUNTERS

A REVIEW: from *population* to *molecule*

If your biology "course" has followed the path of the chapter sequence of this text, you have become acquainted with DNA on several earlier occasions. Let's review them using the hierarchy of levels of organization as we did for the concept of *respiration*, illustrated in Figure 16-1. The following major principles provide a progression from the *population* to the *molecule* level:

1. A *population* is a group of interbreeding organisms of the same kind, occupying a particular space (Section 7-A).

2. Individual *organisms* are temporal carriers of genes which can be shared only with other members of the same population via mating and production of offspring. The shared hereditary material of each population is referred to as a *gene pool* (Section 7-A).

3. The frequency of each allele in a gene pool depends in large part upon its contribution to the overall fitness of individual organisms as they compete for energy and nutrients, acquire mates, and bear offspring. Natural selection tends to suppress the frequency of unfavorable alleles (Section 7-D and 15-G).

4. Sexual reproduction is a basic life process in which, through gamete production, fertilization, and nurture of offspring, the genetic material is passed on from one generation to another (Figure 17-1). These processes are accomplished within and among specialized *tissues* and *organs* associated with the distinction of male and female (Chapter 11).

5. Genetics is the branch of biology concerned with the study of the genetic material and the manner in which genes are passed from one generation to another (Chapter 12).

6. With some exceptions, every *cell* contains at least one set of genes which are borne on a specific number of chromosomes (Chapter 11).

7. Each new generation, beginning with the zygote, receives an identical copy of the genetic material in every cell. This is made possible by *mitosis*, or nuclear division (Chapter 12).

8. Growth of an individual from a zygote involves not only mitosis and cytoplasmic division, but also *differentiation* (Figure 17-1). The genetic material somehow controls the pattern of differentiation via an influence on cellular metabolism. The result is the development of a mature individual as specified in the genetic blueprint (Sections 14-G, 14-H, and 14-I).

9. DNA *molecules* are long sequences of nucleotides linked together to form a polymer. DNA is believed to be the actual genetic material (Section 4-J, 11-C, and 11-D).

10. Our present though incomplete understanding of the function of the genetic material is the result of the work of many individuals spanning the past 100 years or more. Table 15-1 highlights the major milestones.

Genetic code is fundamental.

It should be evident from the above principles that the genetic code is fundamental to our understanding of all of the basic life processes (i.e. nutrition, reproduction, responsiveness, and homeostatic regulation) as viewed at every level of organization from ecosystem to molecule. But, how can these mere *molecules* exert such influence upon complex life processes? This question is only partly answered by research in *molecular biology*. Let us begin our study by addressing one of the early questions of molecular biology, namely, "How can we be certain that DNA really is the actual hereditary material of living cells?"

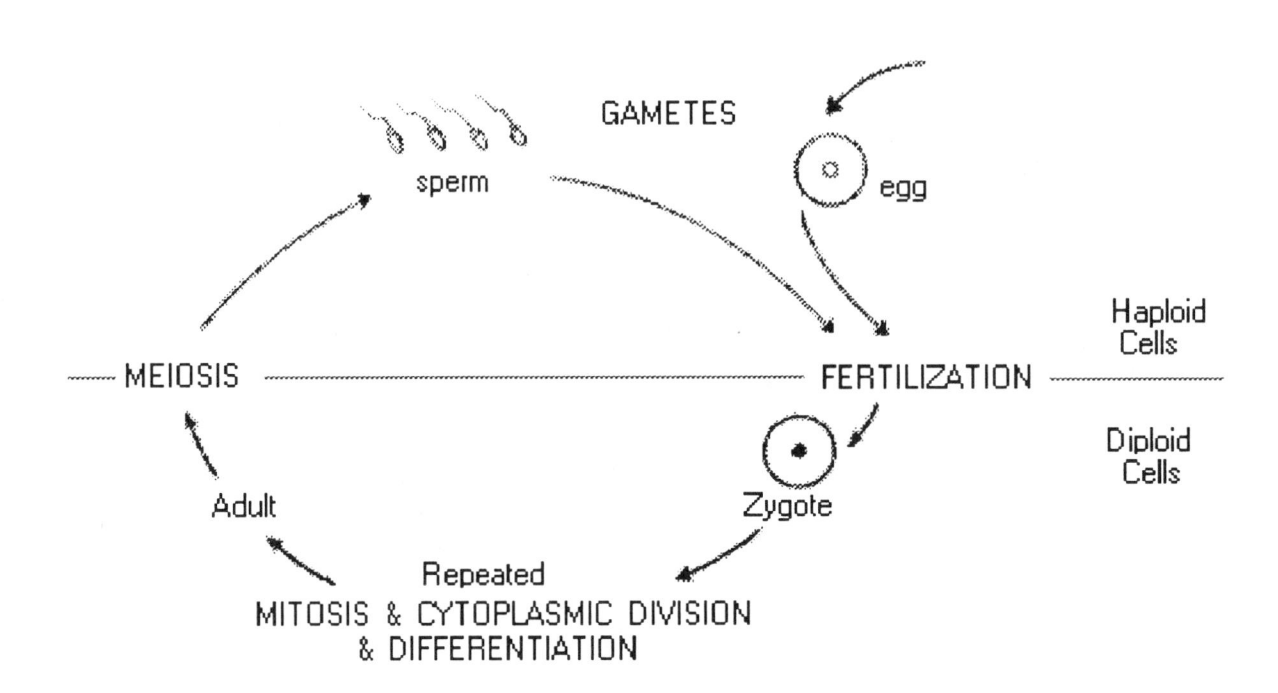

Figure 17-1. A generalized life cycle. By means of sexual reproduction, the genetic material (DNA) is passed on from one generation to another.

17-B DNA — IS IT REALLY THE HEREDITARY MATERIAL?

Mendel fostered hypotheses:

In perhaps the most significant biological discovery of the 19th century, Gregor Mendel demonstrated that hereditary traits are controlled by discrete factors or particles that are passed unchanged from one generation to another (Section 12-C). In light of this principle, the following *hypotheses* were made by early 20th century geneticists:

a "code"

1. The hereditary material must be capable of carrying a precise "code" for the repeatable production of each of the many traits in organisms.

its duplication

2. There must be a mechanism for the exact duplication of the genetic material to account for its passage from generation to generation unchanged.

its control

3. There must be a mechanism through which the genetic material controls the activities of cells.

These hypotheses were to serve as valuable springboards for scientific progress leading up to recent times. A few brief references to important discoveries are in order.

DNA from pus?

In 1867, unaware of Mendel's published work dated three years earlier, Friedrich Miescher, a German physiologist, isolated a substance from the nuclei of pus cells. He called it *nuclein*, and studied it for many years without determining function of this nuclear substance. In the late 19th century, several biologists proposed that nuclein was responsible for the transmission of hereditary traits, but were unable to gather supporting evidence.

Stains reveal chromosomes.

chemical makeup of DNA

finally, the "spiral helix"

Evidence shows DNA can carry a "code" ...

In 1914, another German named Robert Feulgen discovered that the red dye, fuchsin, would combine with the nuclear material of cells. The Feulgen staining technique was later used to demonstrate that nucleic acids were present in all cells; and, more specifically, in the chromosomes. Later experiments demonstrated that all cells of a given organism have the same amount of DNA except the gametes which have only half as much. Chemical analyses by P.A. Levene in the 1920's revealed the existence of four nitrogen bases, adenine, guanine, thymine, and cytosine. He proposed that the nitrogen bases were linked into polymeric form by intermittent sugar molecules and phosphate groups (Figure 4-12). In 1953, James Watson and Francis Crick (16) proposed the now-accepted *double-stranded spiral helix* model for DNA molecules (Figure 11-3). Based upon these findings, DNA would appear to be capable of coding for an almost endless variety of genes, as demonstrated in Section 11-C. This evidence supports *HYPOTHESIS #1* above — *i.e.* the genetic material must be capable of carrying a "code".

... DNA can be duplicated ...

The presence of DNA in the chromosomes and the discovery of the roles of mitosis and meiosis suggest that these processes permit the precise transfer of DNA during cell division and gamete production. These data would be consistent with *HYPOTHESIS #2* — *i.e.* must be a mechanism for duplication of the genetic material.

HYPOTHESIS #3 above predicts that the actual genetic material must be able to influence cellular activities and expression of traits. In the 1940's, A.T. Avery and co-workers demonstrated that a substance obtained from a heat-killed, virulent (disease-causing) strain of bacteria could transform a nonvirulent strain into a virulent form. They suggested that a non-living

...and DNA influences cell metabolism.

chemical substance could act as a "transforming factor" to alter an observable characteristic of an organism. The purified "transforming factor" was found to be DNA. These experiments set apart DNA from proteins, lipids, carbohydrates, and other cellular components as being *the* hereditary material, capable of influencing cellular metabolism.

17-C   HOW GENES ARE EXPRESSED

Thus far we have seen evidence that DNA is indeed the hereditary material or genetic code of life. In Chapter 11, we noted that the zygote is the first cell of a new generation following fertilization (Figure 17-1). Maturation to adulthood involves the duplication of DNA and repeated mitotic divisions that accompany cell division. Mitosis assures that every new somatic cell has an identical copy of the DNA. But, not every cell of a developing individual is identical in structure and function. How can this be, if DNA is really the hereditary material? Current theory answers by stating that *all diploid cells of an organism have the same DNA blueprint, and hence, the same genes, but different genes are expressed in different types of cells*. To understand the significance of this relationship, the following two questions must be addressed:

How can cells with same genes look different?

1. How are genes expressed within cells?

2. What determines whether a gene is expressed or not?

It will be helpful if we address these two questions separately, beginning with the first.

OVERVIEW OF GENE EXPRESSION.

The DNA of the cell nucleus (nuclear DNA) remains in the nucleus, yet it exerts control over nearly every aspect of cellular metabolism. This control, which is exerted through **gene expression**, is mediated through a "chain of command" involving two other types of polymeric molecules, namely **RNA (ribonucleic acid)** and **proteins**. The chain of command may be illustrated as follows:

"chain of command"

```
         TRANSCRIPTION     TRANSLATION
   DNA --------> RNA --------> PROTEIN ---> METABOLIC ROLE
nucleotide    nucleotide     amino acid    e.g. enzyme, hormone,
 sequence      sequence       sequence      or membrane protein
```

DNA extends its "authority" by *transcription*.

You should recall from our study of mitosis (Section 11-E) that, when a cell is in interphase and preparing for mitosis, **DNA duplication** must occur so that the daughter cells resulting from mitosis will each have their own copies of DNA. However, cells perform many other metabolic processes besides dividing. During interphase, or following the last mitotic division in a mature cell, DNA is continually directing these metabolic processes via the chain of command noted above. It does so by **transcription**, coding for the synthesis of RNA, a complementary copy of the DNA. The RNA moves out of the nucleus, and carries out the commands of a specific portion of the DNA code by directing the synthesis of a specific protein. Protein synthesis under the direction of RNA is called **translation**. The protein then becomes the active agent to influence metabolism, either as an enzyme, membrane protein, or hormone. Thus, gene expression involves the synthesis of proteins which exercise control over cellular metabolism. Let us examine the process in detail, beginning with transcription.

TRANSCRIPTION — RNA Synthesis

Transcription—RNA synthesis

If gene expression involves DNA coding for the synthesis of proteins, and if there are hundreds of different proteins at work in each cell, then it is apparent that the single copy of DNA is a focal point in cellular metabolism. Transcription, or RNA synthesis, makes it possible for the single copy of DNA to remain in the nucleus while simultaneously forwarding coding instructions for many different proteins via translation, in the cytoplasm.

Figure 17-2 illustrates the transcription of RNA from a portion of the DNA double helix. The two strands of the DNA double helix separate so as to expose one strand to the enzyme, **RNA polymerase**. This enzyme catalyzes the synthesis of a single-stranded RNA molecule alongside the exposed DNA strand. The RNA strand is made by joining individual nucleotides, synthesized prior to this time, in a sequence so that each nucleotide is paired in *complementary* fashion with an adjacent nucleotide on the DNA strand.

a complementary copy

Several points should be noted from Figure 17-2. First, the RNA differs from DNA in the following ways:

RNA differs from DNA

1. RNA is single- instead of double-stranded.
2. RNA contains **uracil (U)** in place of thymine (T).
3. RNA contains ribose sugar in place of deoxyribose.
4. RNA can move out of the nucleus into the cytoplasm.

A new view of the *gene*.

Second, we should emphasize that RNA "transcripts" made from one segment of the DNA will be different from an RNA transcript from another DNA segment. There is a starting point and a stopping point to each of these *transcription units* which are roughly equivalent to a **gene**. The newly formed RNA from the transcription unit of DNA contains **exons**, portions necessary for later translation; and, **introns**, intervening segments which will be "clipped out" by enzymes, digested, and the resultant nucleotides reused in making other RNA's.

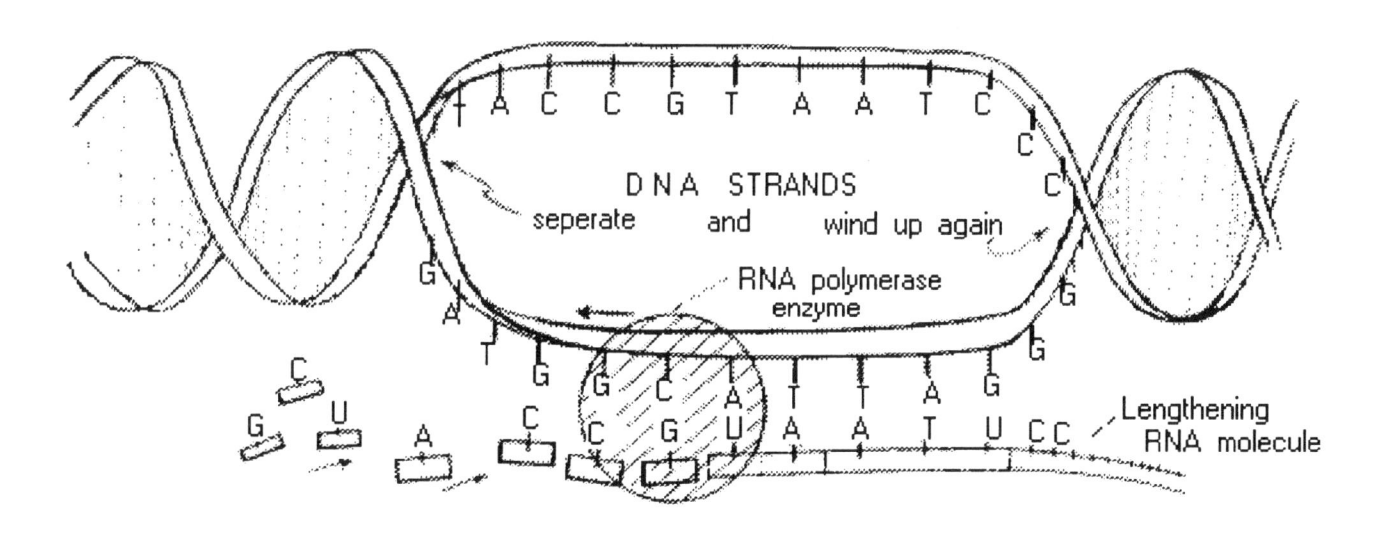

Figure 17-2. Transcription — the synthesis of RNA as a single-stranded, complementary copy of a segment of DNA.

Three kinds of RNA . . .

Finally, before leaving transcription, we should note that there are three different kinds of RNA, each formed under the direction of a different RNA polymerase enzyme. Perhaps the most noted RNA is **messenger RNA,** or **mRNA,** because its nucleotide sequence will determine the amino acid sequence in the protein molecule. **Transfer RNA,** or **tRNA,** functions like an "adaptor" that makes a plug compatible with a wall outlet. The tRNA is involved in transferring amino acids into proper sequence at the **ribosome,** the site of protein synthesis. Ribosomes are composed of proteins bound to a third type of RNA, **ribosomal RNA,** or **rRNA.** Ribosomes act like tiny "stations" on the endoplasmic reticulum where mRNA and tRNA, with attached amino acids, meet under the direction of enzymes to form proteins. Each of the three RNA's are synthesized by transcription in the nucleus, and released through the nuclear membrane into the cytoplasm where they participate in *translation*, or protein synthesis.

are made in nucleus . . . move to cytoplasm.

TRANSLATION: PROTEIN SYNTHESIS.

Once in the cytoplasm, mRNA attaches to one or more ribosomes associated with the endoplasmic reticulum (Figure 16-3, 16-4, and 17-3). Here, its nucleotide sequence serves as a pattern, or *template*, upon which a protein polymer can be assembled from individual amino acids (Section 4-I). Protein synthesis, or **translation,** involves a "change in language" from nucleotides to amino acids—that is, a change from a nucleotide sequence contained in mRNA to an amino acid sequence contained in the eventual protein molecule.

Translation— assembling amino acids into protein

mRNA codons

The position of each amino acid in a protein molecule is determined by a three-nucleotide segment, or **codon,** on the mRNA. Each mRNA codon specifies one out of the 20 common amino acids, as shown in Table 17-1. Transfer RNA, serving like the "plug adaptor", has two "ends". One end,

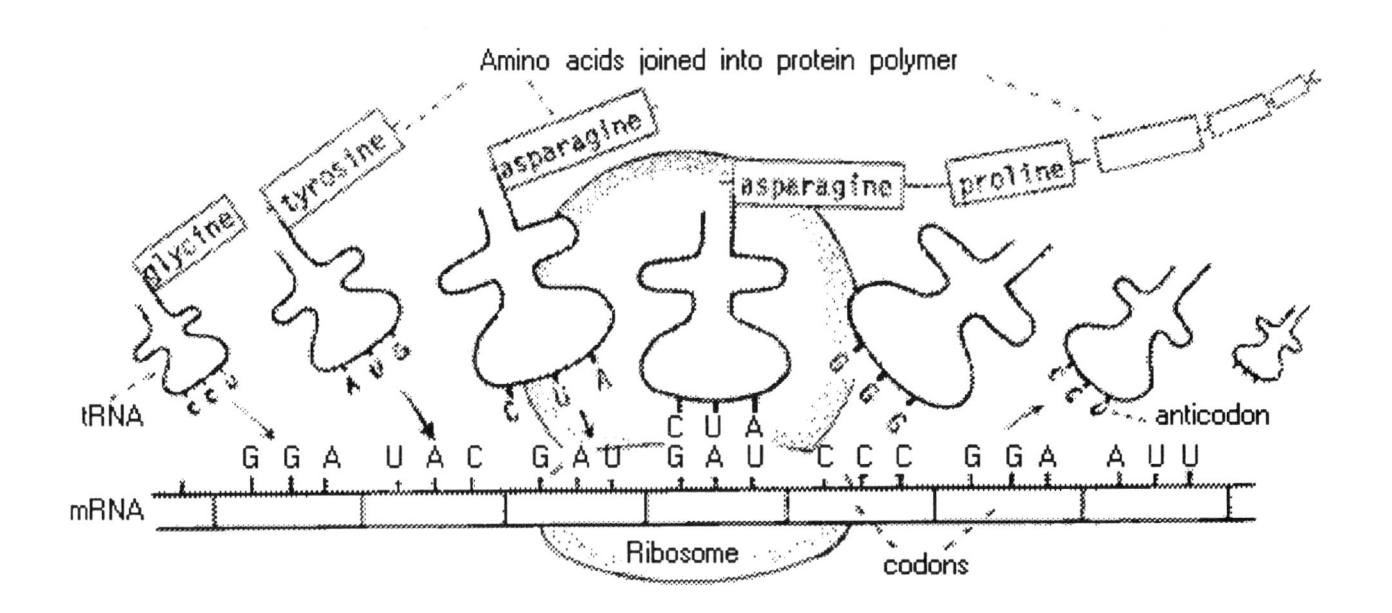

Figure 17-3. Translation — the synthesis of a protein polymer representing an amino acid sequence as specified by the sequence of codons on an mRNA molecule.

anticodons match codons

the **anticodon**, has a "three-prong" structure which "fits" a *complementary* mRNA codon. To the other end is attached the designated amino acid matched according to the anticodon of this particular tRNA. Thus, each mRNA codon will accept a specific tRNA molecule *if* its *anticodon* matches the complementary pairing on the mRNA codon. The tRNA, in turn, carries only a certain one of the 20 amino acids. Thus each position on the growing protein molecule is assured of receiving only the amino acid specified by the codon (Table 17-1), and carried by the matching, incoming tRNA. Transfer RNA's are "reused" after they release a given amino acid at the ribosome.

examples:

To illustrate translation, note in Figure 17-3, that the next amino acid to be added to the protein polymer is asparagine, being transferred by a tRNA molecule with anticodon C-U-A which matches the complementary codon G-A-U on the mRNA. A quick reference to Table 17-1 shows that this is a correct translation; G-A-U "means" asparagine. [Use Table 17-1 to determine the name of the amino acid that is just to the right of proline in the growing protein polymer.]

Table 17-1. The Genetic Code Based Upon Messenger RNA Codons, Three-Nucleotide Units That Code for a Specific Amino Acid.

FIRST Letter of Codon	SECOND Letter of Codon: U	C	A	G	THIRD Letter of Codon
U	phenylalanine	serine	tyrosine	cysteine	U
	phenylalanine	serine	tyrosine	cysteine	C
	leucine	serine	STOP	STOP	A
	leucine	serine	STOP	tryptophan	G
C	leucine	proline	histidine	arginine	U
	leucine	proline	histidine	arginine	C
	leucine	proline	glutamine	arginine	A
	leucine	proline	glutamine	arginine	G
A	isoleucine	threonine	asparagine	serine	U
	isoleucine	threonine	asparagine	serine	C
	isoleucine	threonine	lysine	arginine	A
	START	threonine	lysine	arginine	G
G	valine	alanine	asparagine	glycine	U
	valine	alanine	asparagine	glycine	C
	valine	alanine	glutamate	glycine	A
	valine	alanine	glutamate	glycine	G

START... protein synthesis... and STOP

As mRNA is coding for the amino acid sequence, it moves through the ribosome like a train pulling through a station. As each codon passes through the "station", it codes for the attachment of an amino acid to the growing protein polymer by a peptide bond. When the translation has progressed so far that the end of the mRNA appears, a special "STOP codon" will indicate that the end has been reached. Consult Table 17-1 for several "STOP codons" and "START codons".

Sequence is important...

in determining protein shape and function.

Protein molecules that result from translation vary greatly in size from the relatively small polypeptides such as insulin to the large globular proteins such as hemoglobin with hundreds of amino acid monomers. Regardless of size, the final protein product must have precisely the right *sequence* of amino acids. Otherwise the metabolic role of the protein may be jeopardized. This is because amino acid sequence determines the final three dimensional configuration or shape of the molecule. The insulin molecule in Figure 4-10 gives some indication of this coiling and twisting configuration. Some amino acids are more hydrophobic than others and tend to join other hydrophobic portions of the polymer as they locate within the molecules, and away from water. Hydrogen bonds and disulfide bonds also form from one site to another (Figures 4-1, 4-10). The result is a molecular configuration that is always the same micro-precision shape, provided there were no mutations in the genetic code or errors in transcription or translation to cause an improper amino acid sequence.

RNAases control "lifetime" of mRNA.

Both mRNA's and tRNA's are reusable. However, the "life" of a given mRNA molecule is often relatively short. Enzymes known as RNAases degrade mRNA into individual nucleotides. This controls the duration that an mRNA molecule can exercise its authority. Individual nucleotides from mRNA breakdown can be reassembled into new mRNA sequences in another transcription.

Summary:

In summary, we have established that DNA controls cellular metabolism through gene expression. Gene expression occurs in two major steps. First, mRNA is transcribed from the DNA in the nucleus. Second, the mRNA codes for the synthesis of a protein polymer in the cytoplasm. The resultant protein represents the influence of DNA upon metabolism while it functions as an enzyme, membrane protein or hormone.

Case in point— the hormone, insulin

The hormone, insulin, shown in Figure 4-10, illustrates the principle of genetic control through the chain of command of DNA —> mRNA —> protein. Insulin is a small protein (polypeptide) with 51 amino acids, each occupying a precise position in the sequence as assured in translation. Properly synthesized and secreted into the bloodstream by cells of the *pancreatic islets*, insulin carries tremendous metabolic influence as it participates in the regulation of blood glucose level (Section 14-B).

Gene expression is known to occur only through the synthesis of RNA and proteins. Smaller protein molecules may arise entirely from a single gene, whereas larger proteins such as hemoglobin may consist of several polypeptide chains, each coded for by separate transcription units on the DNA.

17-D HOW GENE EXPRESSION IS CONTROLLED

Differentiation results from controlled gene expression.

It is estimated that, in each human cell, each set of 23 chromosomes carries about 100,000 pairs of genes (Section 11-D). Obviously, not all of these genes, nor the same genes, are expressed in every cell. The distinctions in structure and function among different kinds of cells depends upon the expression of only certain genes at the proper time. In fact, **cell differentiation** (Section 14-I) may be defined as the development of cells according to the controlled expression of genes.

The mechanisms that control gene expression in eukaryotic cells are not well understood. However, the prokaryotic bacteria, having a much simpler genetic control system and fewer genes per cell (ca. 4,000-5,000), have yielded valuable information concerning the control of gene expression. *Escherichia coli*, a harmless member of the biotic community within the mammalian large intestine, has been extensively used in molecular biology research.

operon theory

In 1961, following extensive studies of gene expression in *E. coli*, Francois Jacob and Jacques Monod published their theory of genetic control. They proposed a control system called the **operon**, consisting of a set of three or more genes and associated regulatory components. Control of gene expression is accomplished by the control of transcription, they claimed. Thus, the operon would contain an ON/OFF switch to control the transcribing of RNA from DNA, an essential step in gene expression.

What is an operon?

The basic components of an operon are illustrated in Figure 17-4. First, it has one or more **structural genes**, genes that code for specific proteins. The structural gene or genes are those over which control is to be exercised. Second, there is the **repressor gene**, a key part of the control system. This gene codes for the transcription of mRNA which codes for the synthesis of a **repressor protein**. Once synthesized, the repressor protein binds like a magnet to the **operator gene**, a relatively short segment of DNA. In order for gene expression to occur, RNA polymerase (Figure 17-1) must be able to bind to the DNA near the operator (Figure 17-4) to initiate transcription of the structural gene.

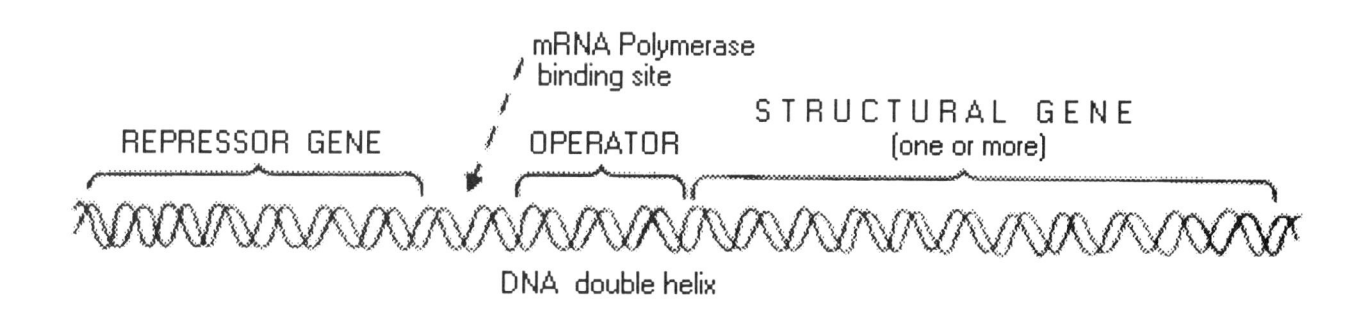

Figure 17-4. The basic form of an operon as proposed by Jacob and Monod.

an ON and OFF "switch"

As long as the repressor protein is bound to the operator, the RNA polymerase binding site is blocked and transcription is inhibited (See Figure 17-5a). Hence, the control of gene expression would seem to depend upon whether or not the repressor protein binds to the operator. This leaves the question of what determines whether or not the repressor protein can bind to the operator.

Jacob and Monod observed that *E. coli* could be cultured on either glucose or the disaccharide sugar, lactose. However, when grown exclusively on lactose, *E. coli* was induced to synthesize three different enzymes for the metabolism of lactose. These are called **inducible enzymes**. That is, the bacterial cells do not synthesize the enzyme proteins unless they are needed—a very fitting mechanism considering the high "cost" in nutrients, energy, and space within a microscopic cell to produce such complex molecules as mRNA and protein.

An operon is "responsive" to environment.

Figure 17-5 illustrates the repressed and induced modes of the lactose operon. When lactose is present in the culture medium, it is absorbed by the bacterial cell. The repressor protein has an even greater affinity for lactose than for the operator. Therefore, lactose binds to the repressor protein molecule and causes it to release from the operator. RNA polymerase is now free to bind to the DNA and catalyze transcription of mRNA from the structural genes. As a result, and three lactose-metabolizing enzymes are formed by translation under the direction of the mRNA. (Here, one mRNA segment codes for three proteins.) When lactose is no longer present, the repressor protein is free to reattach to the operator and the repressed mode resumes.

genetic feedback

The lactose operon and others like it illustrate the principle of **genetic feedback control**. That is, control of cellular metabolism (e.g. lactose utilization for energy and growth) by a genetic system (operon) which is itself responsive to an environmental condition (presence or absence of lactose). Genetic feedback control thus allows *E. coli* to maintain its metabolism in a steady state, or homeostasis, during a period of time in which its source of energy is shifted from glucose to lactose, and back to glucose.

Genetic control systems are known to operate in eukaryotic cells as well. However, they are much more complex. For example, translation of proteins can be blocked even after transcription of RNA has occurred. Nevertheless, the same principles seem to apply and the same ends are in view.

gene expression and its control

Figure 17-6 presents a general model that incorporates much of what we have discussed thus far. The following relationships are evident from the model and our earlier discussion:

1. DNA directs the synthesis of proteins.

2. Proteins regulate cellular metabolism.

3. Metabolism determines the structure and function of each cell (and hence, organism) as it differentiates into a mature form— i.e. its phenotype.

4. Metabolic products of one cell may influence the environment and hence, metabolism of other cells.

(a) Lactose Absent --- Repressed State:

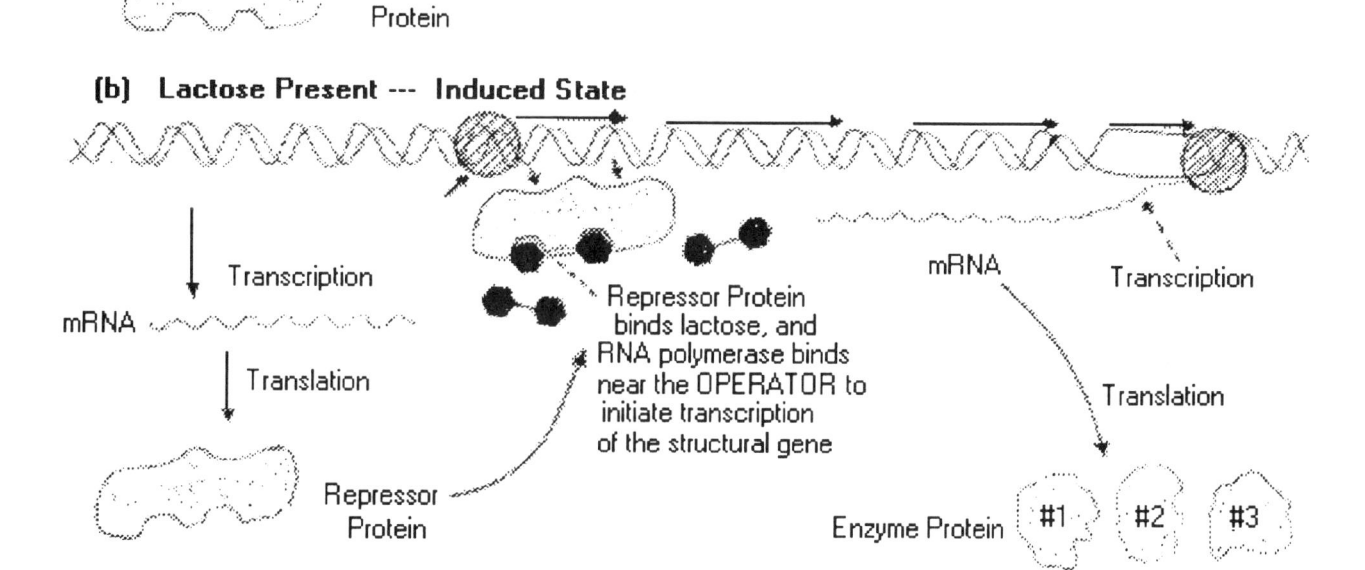

Figure 17-5. The function of the lactose operon. Unless lactose is present in the bacterial culture, the operon remains in the repressed state (a), and unneeded enzymes are not synthesized. When lactose is added (b), it binds to the repressor protein and induces protein synthesis.

5. Protein molecules in the cytoplasm and within membranes and organelles are the "molecular receptors" of environmental stimuli, and mediators of genetic feedback control. This system allows homeostatic regulation of gene expression in responsiveness to environmental conditions.

6. The phenotype of an organism (i.e. sum total of all of its structural, functional, and behavioral features) is the result of environmental conditions operating through the genetic feedback control system.

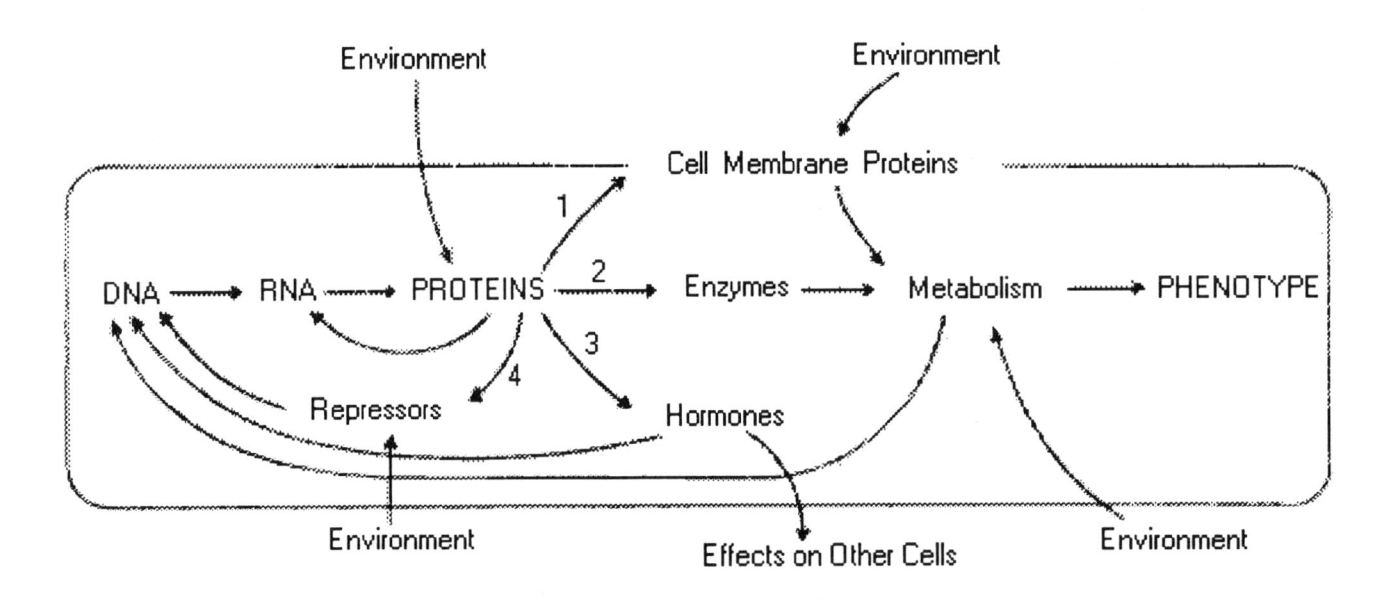

Figure 17-6. A genetic feedback control model as explained in the text. Reprinted with the permission of Macmillan Publishing Company from PLANTS AND LIFE by Alan W. Haney and John Noell. Copyright © 1978 by Macmillan Publishing Company.

Which is most influential—genes or environment?

Having studied the genetic feedback model and the relationships outlined above, you should be reminded in yet another way of the intimate association between organism and environment. You may wish to consult Section 12-H for a discussion of the relationship between heredity and environment, and Section 15-G for a consideration of the influence of environment on phenotype via natural selection. The genetic feedback model (Figure 17-6) should provide a new perspective from the cell and molecule level with which to consider these earlier relationships at the organism and population levels of organization.

17-E   CELLULAR DEFENSE SYSTEMS

Our study thus far has portrayed cellular metabolism as a highly complex array of chemical reactions involving molecules and ions. Hundreds of proteins and other molecules are precisely formed and arrayed within a single microscopic cell and its organelles. As shown in Figure 17-6, these systems must be poised and *responsive* to environmental *stimuli* in the form of various ions, organic molecules, or other disturbances (e.g. light, electrical, or mechanical energy changes). But how can cells distinguish "legitimate" environmental stimuli from toxic substances and pathogens that disrupt normal metabolism and cause disease? This is the task of the human immune system and related bodily processes.

distinguishing the "good guys"

two reasons for studying the immune system

A brief discussion of the human immune system will be relevant for two reasons. First, cellular defense mechanisms are pertinent to those concerned about personal health, hygiene, and environmental health-related problems. Second, cellular defense mechanisms must be understood in order to appreciate the challenges faced by genetic engineers as they seek to introduce "foreign DNA" into cells.

ENTRY OF FOREIGN AGENTS

Our bodies are continually exposed to dust, pollen, spores, bacteria, viruses, and toxic chemicals from the environment. Dust, pollen, and spores have surface proteins and carbohydrates that often trigger allergic reactions. These are considered non-pathogenic in contrast to pathogenic bacteria, fungi, and viruses which enter cells and spread by reproduction.

pathogens

Some common bacterial diseases include tetanus, strep throat, and venereal disease. Fungi cause athletes foot and urinary tract infections. Colds, flu, measles, polio, and AIDS (Acquired Immune Deficiency Syndrome) are caused by viruses. Common avenues of entry into the human body include the mouth, eyes, nasal passages, sex organs, anus, insect bites, and wounds.

avenues of entry

defenses are spiritual . . .

Bodily defenses are of two principle types—spiritual and biological. Professionals in medicine and counselling are aware of the importance of "peace of mind" as it contributes to the biological health and well-being of an individual. Mental events such as depression, anger, anxiety, and stress can weaken the body and render it more susceptible to disease. The Scriptures emphasize the importance of "peace with God" as the source of hope and joy which is fundamental to human health and well-being. [See TOPIC #4.]

and biological.

FIRST LINE OF DEFENSE — Physical Barriers

skin and mucous membranes

The skin is the most obvious physical barrier and initial biological defense against the entry of foreign substances. However, cells of mucous membranes and the lining of the stomach secrete powerful enzymes that can destroy microbes that are inhaled or ingested. Ciliated cells of the respiratory tract "sweep out" foreign particles and pathogens.

SECOND LINE OF DEFENSE — General Responses

general defense

If foreign material enters the body fluids beneath the dermal tissues, a second line of defense involves the *complement proteins, inflammation,* and *phagocytes*. The effectiveness of these responses depends upon their capacity to accomplish several objectives that are somewhat analogous to those of a successful national defense. These objectives are as follows:

Complement system warns and directs defenses.

1. RECOGNITION of the presence of "enemy" cells and toxic substances is possible, in part, through the **complement system** which consists of an array of plasma proteins that act as a "radar system" by detecting invasion. The complement system then activates other proteins to create chemical gradients pointing toward the site of invasion. Others of these *complement proteins* bind to pathogens "marking them" as targets for destruction. They may also begin destroying microbial invaders.

histamines help mobilization

2. MOBILIZATION of the defense arsenal of complement proteins and phagocytes to the site of invasion is facilitated by the *inflammatory response*. In response to complement proteins, *histamines* are released from infected white blood cells (*e.g.* basophils, Figure 17-7) and cause dilation of capillary walls. This allows leakage of complement proteins and the escape of amoeboid phagocytes so both can move more freely to the invasion site. The site will appear red and swollen until the invasion is quelled. The inflammatory response also eventually aids in restoration of infected tissue.

phagocytes destroy

3. DESTRUCTION of "enemy" invaders occurs by **phagocytes**. Phagocytes, or "eating cells", consist primarily of *neutrophils* (Figure 17-7). These amoeboid cells squeeze through capillary walls and follow gradients in complement proteins to the infection site. There, they engulf microbes and debris by phagocytosis (Section 16-G). A specific immune response may not be necessary if phagocytes destroy foreign invaders.

THIRD LINE OF DEFENSE — Immune System Responses

immune system: specific responses

If the invasion is not stemmed by the "general response" noted above, the body's **immune system** is activated. The immune response is *specific* in that specific proteins and white blood cells are activated by the invading agents. Proteins called **antibodies** are specifically made in response to the presence of **antigens**, foreign proteins or carbohydrates on the surfaces of foreign cells. Antibodies are produced by white blood cells, called **lymphocytes**. Lymphocytes are produced by divisions of *stem cells* in the bone marrow. Two major classes of lymphocytes are the **T cells** which mature and differentiate in the *thymus* and **B cells** which mature in the bone marrow. Below, we present a brief introduction to the immune system, with objectives stated in military terms:

Lymphocytes produce antibodies.

distinguishing "self" *versus* "nonself"

1. RECOGNITION of "enemy" cells and substances must occur by distinguishing *self* (normal body components) from *nonself*. This is possible because each of your body cells has membrane proteins called **MHC markers** which are coded from a complex of genes (major histocompatibility complex). Each cell representing your *self* has this unique "self marker" or "ID card" which are molecular, "don't destroy me" signals to your patrolling lymphocytes. Bacterial or viral invaders which do not have your MHC markers (*i.e.* nonself) are prime targets for destruction.

The cellular and molecular "battle" rages.

2. MOBILIZATION of the immune system is triggered if neutrophils fail to quell the invasion during the inflammatory response. Larger white blood cells, the **macrophages** ("big eaters"), begin phagocytically attacking the microbial invaders. They digest microbes and foreign debris but do not digest certain microbial proteins. Instead your macrophages expose these *antigens* on their surfaces along with your MHC markers. The *antigen-MHC marker" complexes* make contact with special lymphocytes, the **helper T cells**. Helper T cells, like "field commanders", trigger cell divisions of other lymphocytes, including the B cells and other T cells.

antibodies aid macrophages

3. DESTRUCTION of invaders, any cell or substance which bears *antigens*, is normally accomplished by the B cells and T cells, as well as by phagocytes. *Clones* (genetically identical populations) of each of several types of B cells and T cells are formed by mitosis and cell division as commissioned by the helper T cells. By virtue of their large numbers and movement, these lymphocytes become a potent defense. The B cells synthesize and release *antibodies*, protein molecules "taylor made" to "lock onto" the antigens of foreign cells, marking them for destruction by macrophages (Figure 17-8). The T cells move about like tanks with antigen-specific antibodies projecting from their membranes. They directly bind to antigens and cause destruction with the aid of macrophages attracted to the scene.

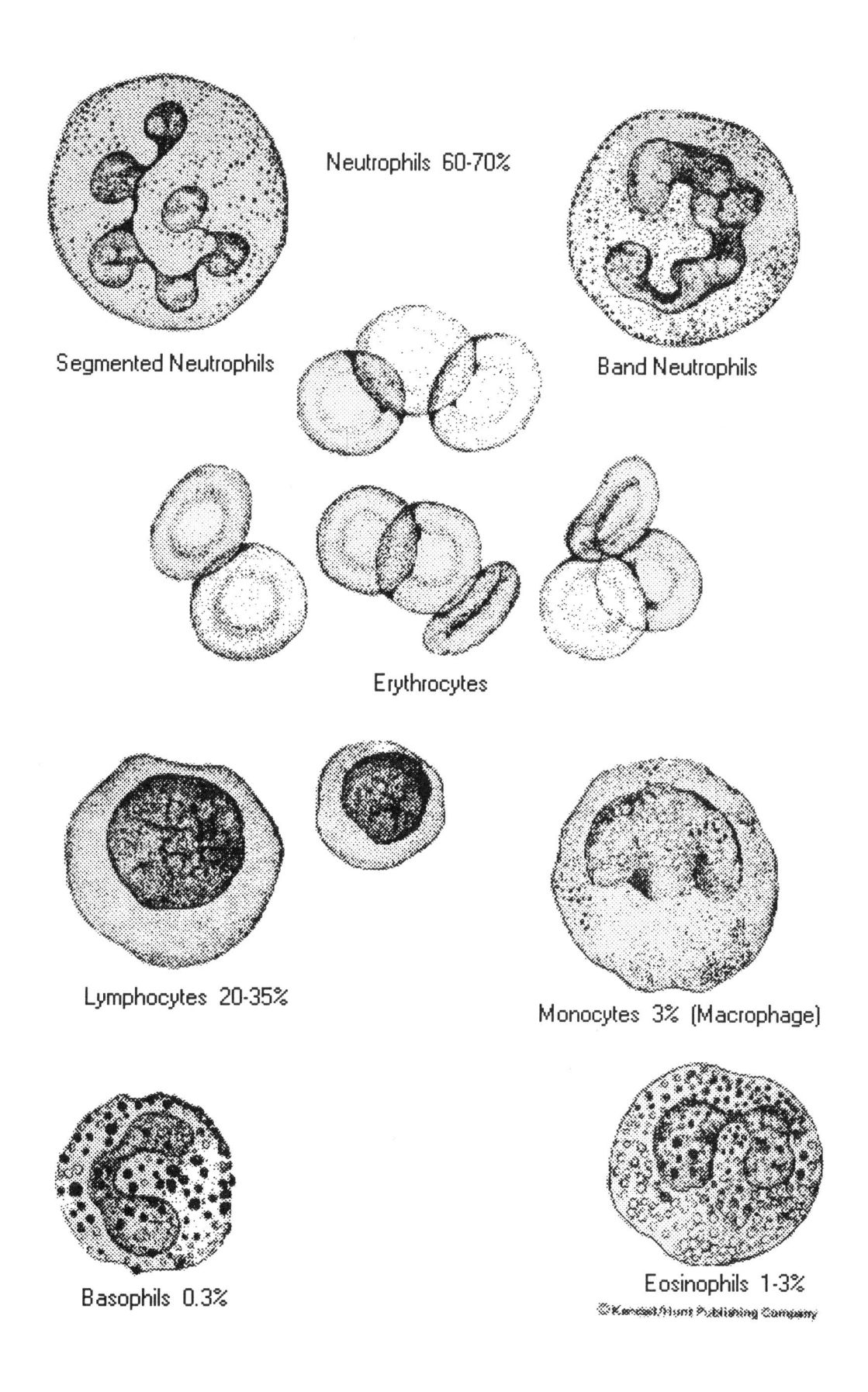

Figure 17-7. Types of blood cells and relative abundance in normal blood.

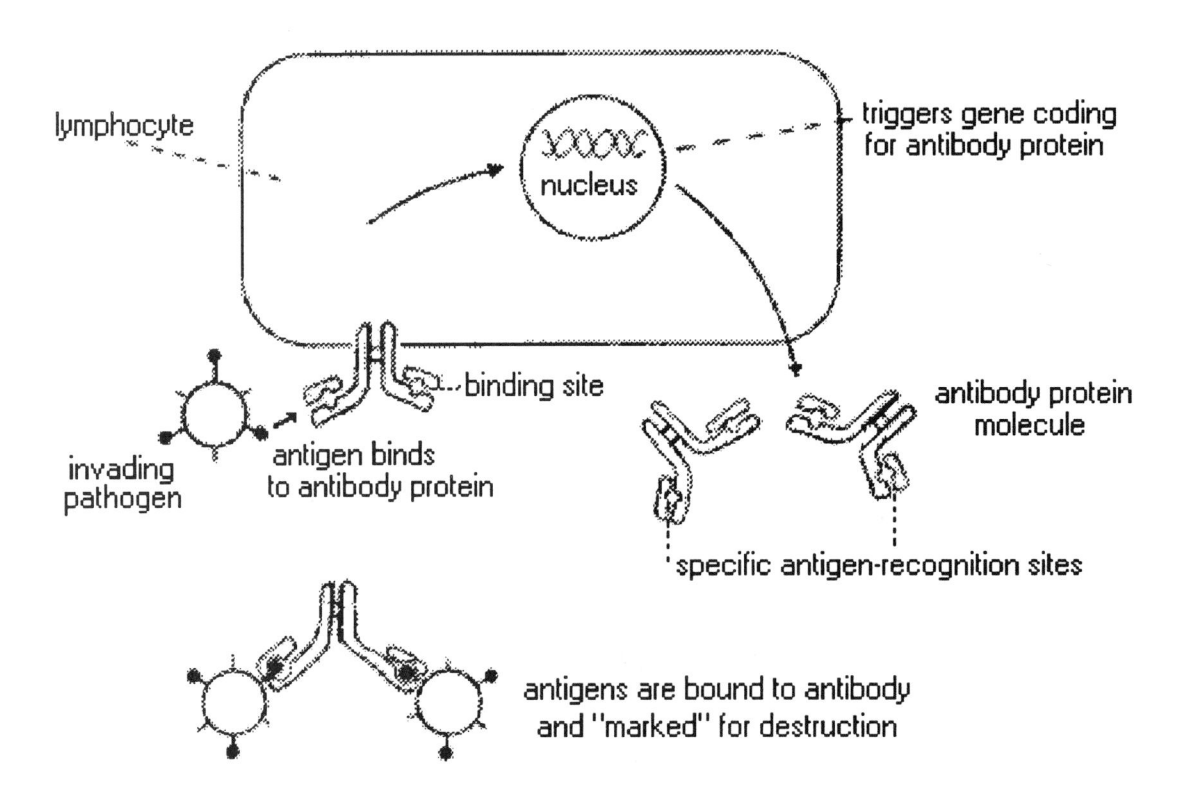

Figure 17-8. Antibody production by a lymphocyte cell in response to the presence of an antigen. Bound antigens are targeted for destruction by macrophages.

4. REDUCTION OF FORCES must occur when the "battle" has been won. As more and more antigens are bound and destroyed, fewer B cells are induced to produce antibodies. *Suppressor T cells* chemically call the attack to a halt.

5. FOLLOW-UP AND PEACE-KEEPING When the "battle" is over, the lymphocyte population in the body will decline, but certain *memory B cells and T cells* remain in readiness for any secondary attack by that antigen. This is an *acquired immunity*. *Passive immunity* provides a quick, temporary defense against an antigen via the injection of antibodies from another source.

acquired immunity

Cancer is often stopped by immune system.

Cancer cells may arise from some disruption in the genetic control system, either by viral activity (Section 17-F) or by mutagens such as chemicals or ionizing radiation. Usually, abnormal membrane proteins, perceived as antigens by T cells, will signal their early destruction before the disease is ever humanly detected. So-called *immune therapy* is being researched for its potential in fighting cancer by mobilizing specific T cells to destroy tumor cells.

What problems may arise?

In spite of the apparent effectiveness of the immune system as a defense against foreign invaders, and as a search-and-destroy system against cancerous cells in the body, problems may arise. Certain allergies are caused by an apparent overreaction against harmless antigens such as those on pollen grains. *Antihistamines* may relieve allergic reactions by countering histimine production during the inflammatory response. The immune system acts in opposition to modern medicine when T cells attack the cells of an organ transplant. In cases of rheumatoid arthritis, rheumatic fever, and other autoimmune diseases, the body's immune system attacks its own body tissues.

What is AIDS?

AIDS (acquired immune deficiency syndrome) is a deadly disease that results from an attack of the human immunodeficiency virus (HIV) on the helper T cells, thus strategically interfering with antibody production. The AIDS victim gradually loses the ability to fight off other infectious diseases and cancer. The disease is apparently transmitted by direct contact with body fluids such as saliva, blood, and semen. Because of these modes of transmission, AIDS is most likely encountered by promiscuous sexual or homosexual contact, by intravenous drug use with unsterilized needles, or by receiving a transfusion of blood donated by an AIDS victim.

How can AIDS be avoided?

While extensive, costly research is underway to develop reliable blood tests and a vaccine against AIDS, there seems to be an overall unwillingness on the part of society to recognize that the disease is largely a consequence of immoral living. Instead of encouraging abstinence as a logical, let alone moral, alternative to controlling AIDS, condoms are being distributed. According to Arnold, *et al* (2), researchers found "channels of 5 micrometers (Table 16-1) that penetrated the entire thickness of" surgical gloves made of the same type of latex used to make condoms. The HIV virus, measuring between 0.1 and 0.3 micrometers, could easily pass through such channels. This concern, plus the high failure rate of condoms in preventing pregnancy during only the few fertile days per month (15.7% among married couples), make the prescription of condoms both illogical and unethical. The biology of the HIV virus will be discussed in the next section.

17-F VIRUSES — THE ULTIMATE PARASITES

From our previous mention of such familiar viral diseases as influenza, polio, and AIDS, it should be evident that viruses are quite capable of invading the body. Indeed, a major cellular target of the viruses is the DNA and the transcription-translation machinery of the cell. These characteristics make viruses a likely tool in the hands of genetic engineers seeking to alter the genetic code of cells. Before we consider this particular application, a brief discussion of viruses is in order.

Viruses attack DNA and RNA.

viruses described

Viruses are parasites that invade cells and assume control of cellular metabolism so as to reproduce more virus particles at the expense of the host cell. Viruses come in all sizes and shapes as shown in Figure 17-9. Outside of the host, they consist of a protein "coat" surrounding a "core" of nucleic acid, either DNA or RNA. The infective form of a virus cannot carry on metabolism. The virus has no cytoplasm, and can be stored, like a chemical reagent, in crystallized form. Thus, they "travel light", and have nothing to offer a host cell, and everything to gain. The origin of viruses is unknown. Some suggest that they may represent aberrations of portions of genetic material of normal cells.

virus reproduction

Viruses have various means of reproducing themselves. Some, such as the bacteriophages ("bacteria eaters"), assume control of the cell's nucleic acid metabolism and use it to generate new protein coats and nucleic acid cores. The completed virus particles eventually burst out of the doomed host cell and infect other cells (Figure 17-9a).

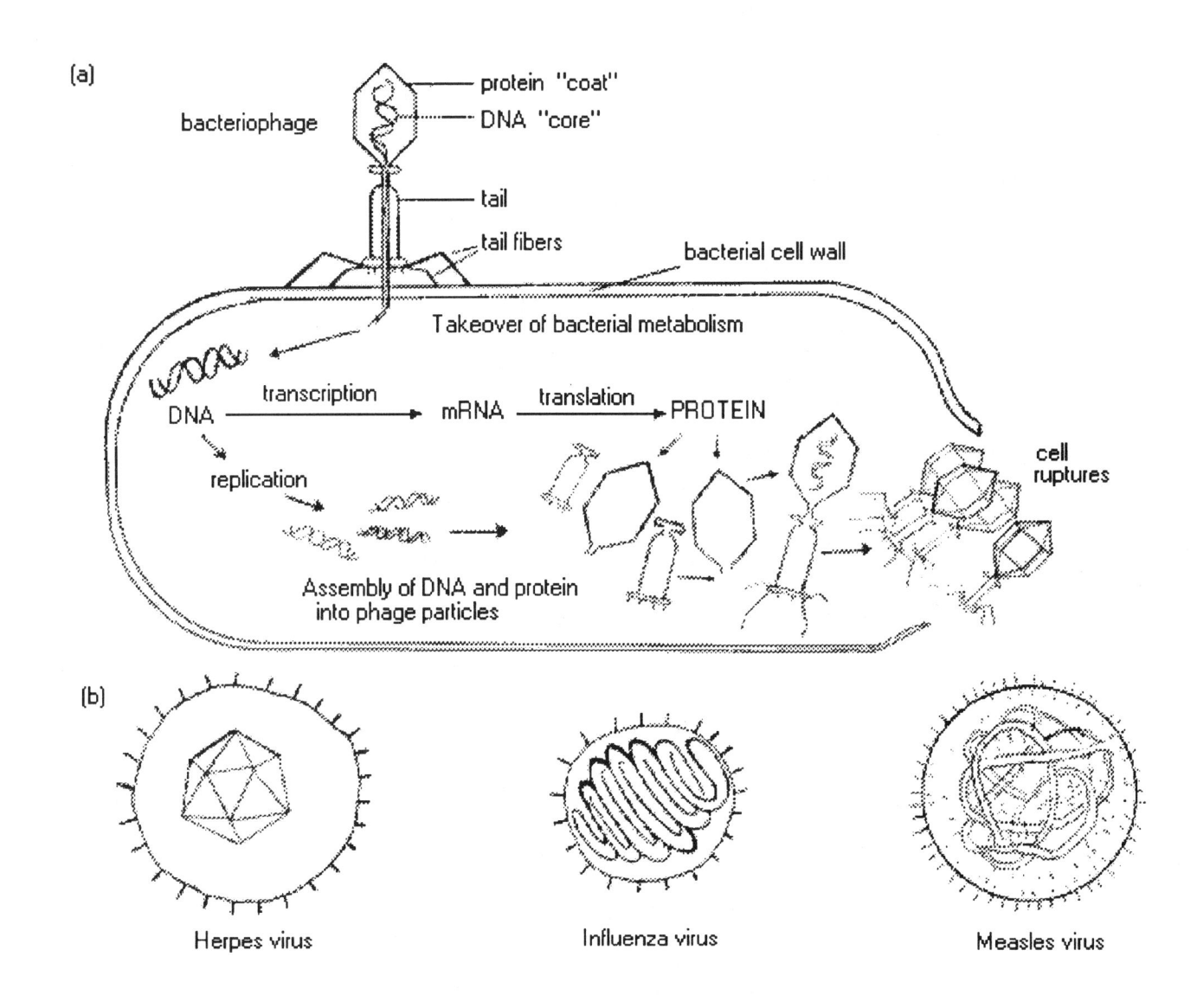

Figure 17-9. Viruses. (a) Reproduction of a bacteriophage virus. (b) Various other viruses.

HIV virus: a *retrovirus*

The human immunodeficiency virus (HIV) which causes AIDS infects helper T cells of the immune system (Section 17-E). The HIV is classed as a **retrovirus**, meaning that it has the capacity to convert RNA into DNA by reversing transcription (Figure 17-10). To do this, it uses its enzyme, *reverse transcriptase*. The viral DNA is then inserted into a host cell chromosome. Later, the viral DNA may be clipped from the host chromosome and transcribed into viral RNA. The RNA codes for viral proteins which are assembled around copies of viral RNA. Soon, new HIV particles burst from the T cell host to infect other T cells. Some retroviruses have been known to speed up their proliferation by *transforming* host cells into cancer cells.

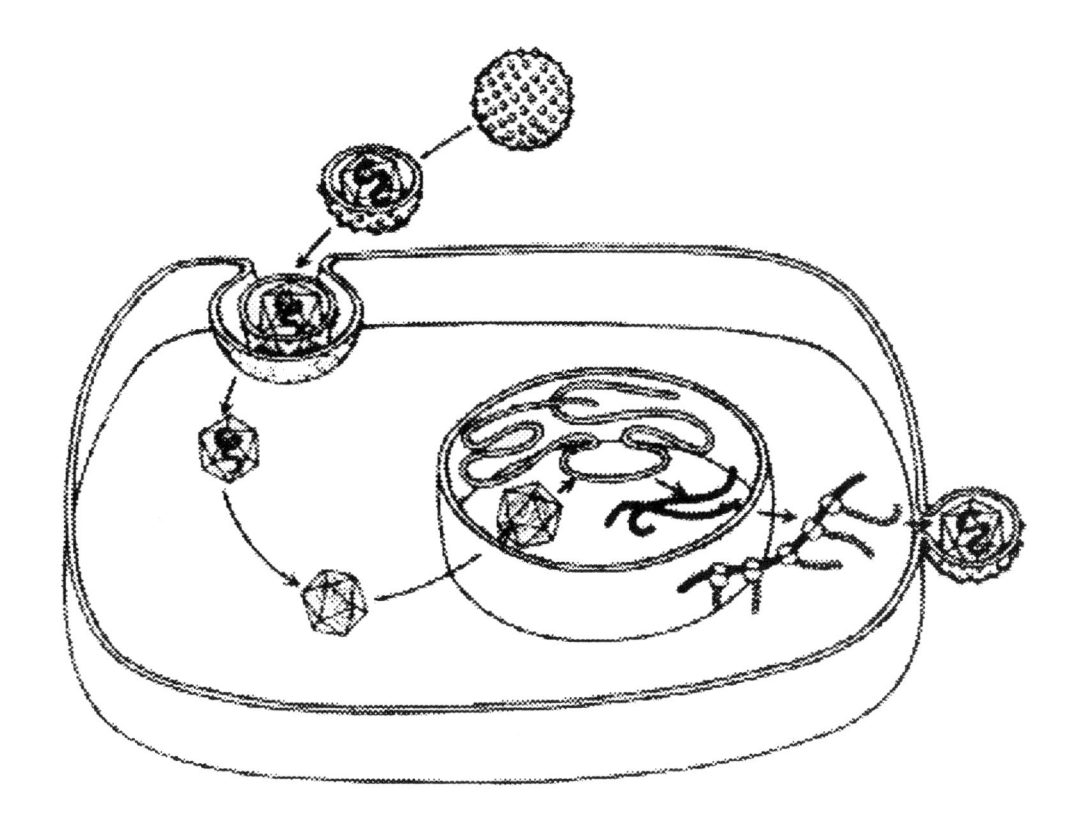

Figure 17-10. Diagrammatic view of human immunodeficiency virus entering a host cell by endocytosis, shedding the membrane derived from the previous host, and interacting with the host cell DNA as described in the text. Modified with permission: H. Varmus. 1988. *Science* 240: 1429; used by permission. Copyright © 1988, AAAS.

explaining AIDS symptoms

There is often an apparent "inactive period" between the time that the HIV is discovered in the blood and the time when immunodeficiency symptoms become pronounced. In 1993, AIDS researchers (14) discovered that, during this time the HIV is being filtered by the lymph nodes. However, because the virus is not being destroyed there, pockets of the increasing viral population eventually spill out into the bloodstream, ending what was thought to be an "inactive period". Effective elimination of HIV will have to begin immediately upon detection, and before the accumulation in lymph nodes.

viruses microevolve

As we have seen, the human immune system is an effective defense against most viruses. Your body has an arsenal of numerous antibodies, each produced as you encounter various genetic strains of viruses. New strains continually arise, apparently by mutations. The continual appearance of new types of influenza and the several strains of HIV are apparently the result of mutations in previous viral strains.

17-G   RECOMBINANT DNA TECHNOLOGY

The capacity of DNA molecules to separate and recombine should not surprise you. *Crossing over* (Section 11-H) is a natural process in which a portion of the DNA of one chromosome separates and exchanges with a portion from the homologous partner during meiosis. Transposons (Section 12-G) involve genes or fragments of DNA that "jump" from one location to another. These separations and recombinations of DNA are major causes of genetic variation. Whereas these processes have been known for decades, deliberate removal of DNA from one organism or species and splicing it into the genome of another has only been possible since the 1970's. This **recombinant DNA** technology, or *biotechnology* has mushroomed out of molecular biology, and it promises to bring many benefits, and potential challenges.

transposons—"jumping genes"

Biotechnology . . .

. . . has many applications.

Crop plants are being genetically modified for improved resistance to insect pests (6). Tobacco and tomato are among plants now genetically engineered to become "workhorses" to synthesize an array of materials including human albumin and interferon. *Gene therapy* is offering promises of correcting genetic diseases such as cystic fibrosis and Huntington disease. The number of new techniques and applications in medicine and agriculture is multiplying rapidly.

PROBLEM: how to penetrate the immune system . . .

The immune systems of humans and other mammals presents a major obstacle to gene transfer. It is no small challenge to prepare precise segments of DNA and introduce them into cells in such a way that they can escape destruction by cellular defense mechanisms. Then, the DNA must be spliced into chromosomes in such a way that it will be transcribed to RNA and expressed through protein synthesis. Several techniques employing viruses and bacteria have been developed to accomplish these feats.

. . . using virus

In the previous section we saw that many viruses can produce recombinant DNA when they enter a host cell and insert segments of their own DNA into sequence with host genes. This property of viruses makes them useful tools for biotechnology's efforts to modify the genetic blueprints of organisms. So far, bacteria such as *Escherichia coli* have been the focus of much of the research with recombinant DNA. For example, the mouse gene that codes for the synthesis of the protein, insulin, has been inserted into *E. coli*. The resultant strain of the bacteria is then cultured as tiny "insulin factories", providing this much needed hormone for treatment of diabetics. All sorts of enzymes, hormones, and other chemicals are now being produced in this way.

. . . or bacteria

use plasmids to transfer genes

Before discussing recombinant DNA techniques using *E. coli* and other bacteria, a brief explanation of the genetics of these microscopic monerans (Section 9-B) is in order. Being prokaryotic, *E. coli* has no nucleus or other organelles. The DNA is found in the cytoplasm in a single circular chromosome (Figure 17-11). These bacteria can take up smaller rings of DNA, called **plasmids**, from cells of the same or different species. The plasmid has only a few genes, usually related to self-replication and transfer. [Does this sound familiar? See Section 17-F.] One plasmid has a gene that transforms a bacterial cell into a plasmid donor ("male") by directing the formation of a *conjugation tube* (Figure 17-11). The plasmid is then duplicated as one copy of the plasmid DNA in linear form moves through the tube leading to a recipient ("female") bacterial cell. Once the transfer is complete the plasmid is enzymatically closed into a ring and the bacterium has officially been "infected" with maleness. Plasmids may also join with the bacterial chromosome and be replicated with the "host" DNA and hence, be passed throughout the bacterial progeny. Exchange of parts of the chromosome of one cell with that of another through a conjugation tube may also be mediated through a plasmid.

Because plasmids function as agents of gene transfer from one bacterial cell to another, and their genes can be replicated throughout a bacterial culture, they are very useful in genetic engineering. We shall briefly describe the general approach to gene splicing, using plasmids as the agents of transfer. Isolated plasmids are exposed to a specific **restriction enzyme** extracted from bacteria. This enzyme acts like a very specific "molecular scissors" to clip open the plasmid rings at a particular *cleavage* site. The enzyme clips the DNA only where a nucleotide sequence of A-A-T-T is located, as shown in Figure 17-12a.

This specific cleavage creates two "sticky ends", consisting of a precise sequence of four unpaired nucleotides— T-T-A-A and A-A-T-T. Now suppose the technician wishes to recombine DNA from a eukaryotic cell; for example, a gene for insulin synthesis in the mouse. Mouse DNA is extracted and treated with restriction enzyme to cleave the genetic material into segments. Then, the cleaved plasmids and mouse DNA segments are combined at their respective sticky ends when exposed together with a *ligase enzyme* (L. *liga* = bound). As shown in Figure 17-12b, these plasmids, complete with genetic code for the synthesis of insulin, are taken up by *E. coli* and the recombinant DNA is replicated along with bacterial DNA as cells divide repeatedly. Insulin produced by these cultures can be extracted and purified for use.

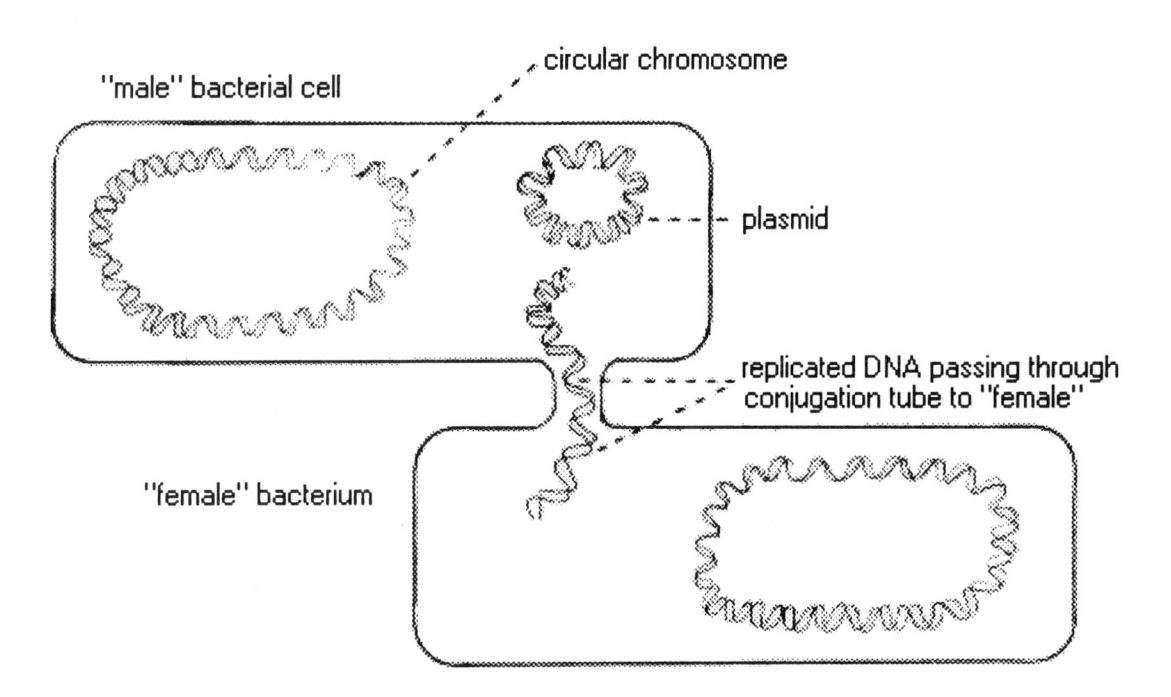

Figure 17-11. Transfer of plasmid DNA from one bacterial cell to another. The plasmids are occasionally inserted into the large circular chromosome by enzymes that recognize compatible sequences of plasmid and chromosome.

(a) Cleavage of a bacterial plasmid:

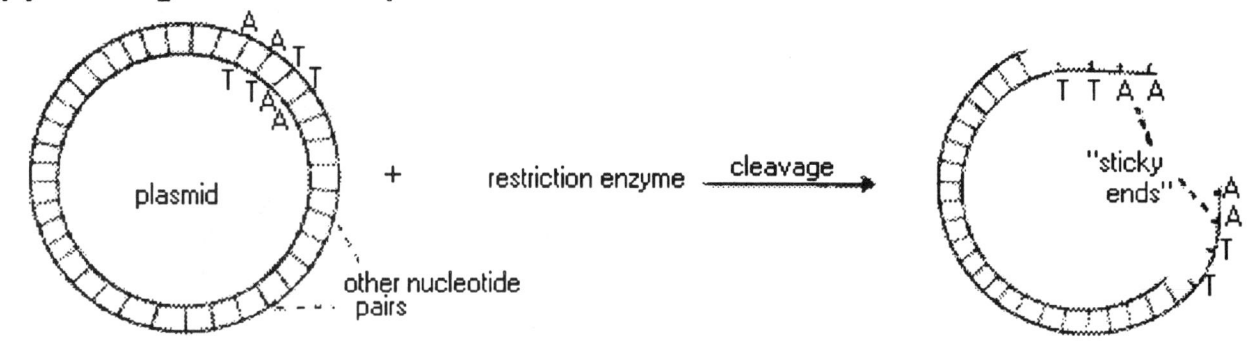

(b) Splicing mouse DNA into cleaved plasmids:

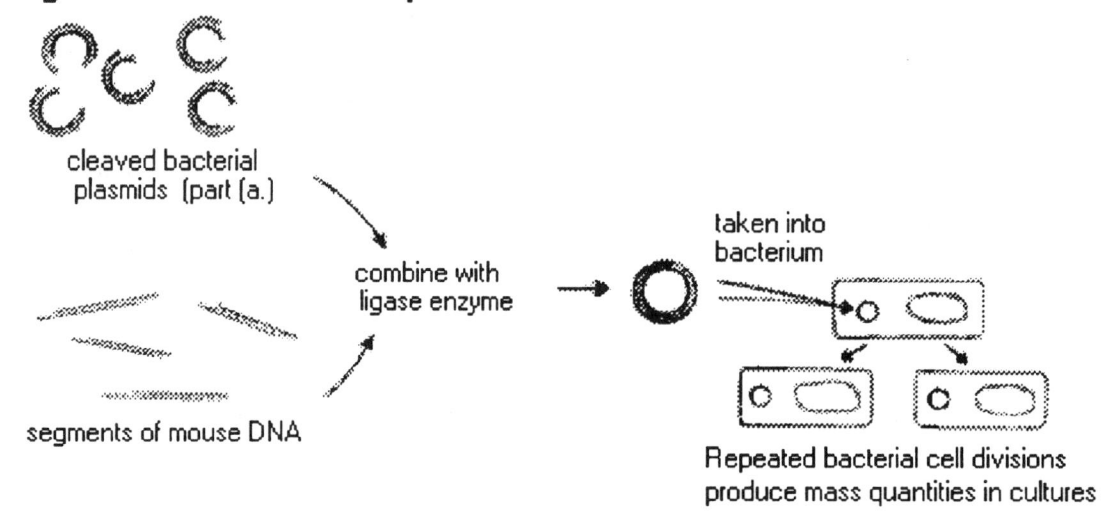

Figure 17-12. A simplified representation of a technique for inserting DNA from one organism into another using bacterial plasmids.

amplifying the DNA supply

Molecular biologists now have a *DNA library*, consisting of various recombinant DNA segments that have been spliced into bacterial plasmids. The quantity of each of these may be increased by *cloning* the plasmids using bacteria or yeasts which reproduce rapidly. Alternately, the *polymerase chain reaction* (PCR) is capable of rapidly replicating DNA sequences in the test tube. The sequence in question is heat denatured, causing the double helix to separate into single strands. Then, DNA polymerase from *Thermus aquaticus*, a bacterial species inhabiting hot springs (and tolerant of temperatures necessary to denature DNA) is added along with free nucleotides. Two complete sequences are formed by base pairing with each single strand as a template. Repetition of denaturation followed by the polymerase reaction yields more and more copies of the sequence.

It should be noted that recombinant DNA techniques are much more complicated and varied than our brief description indicates. It is important that citizens of the "biotech society" be aware of this growing technology, and able to analyze the medical, ecological, and ethical implications of its use in

gene therapy to combat cystic fibrosis

medicine and agriculture. As a case in point, consider research progress on identifying the gene responsible for cystic fibrosis (CF). The CF gene has now been cloned, and it is believed that it codes for a protein that functions as a chloride ion channel in cell membranes (Section 16-G). Teams of researchers representing diverse disciplines such as molecular biology, ion channel physiology, and pharmacology are collaborating in a surprisingly harmonious fashion to bring an end to this devastating disease. Questions they must address, according to Collins (5) are as follows:

biological considerations:

1. What body cells should be given gene therapy?
2. What fraction of cells must be repaired for success?
3. Will overexpression of the CF gene be toxic?
4. How can the CF gene be integrated into respiratory stem cells? Will the immune system intervene?
5. What are chances of inadvertent gene transfer into sex cells, or environmental contamination by "aerosol vectors"?

The CF research, here considered primarily as *somatic cell gene therapy* may raise fewer concerns than *germline gene therapy*. The latter approach involves manipulation of the genome in gametes prior to conception. Aside from concerns about damage to future generations, there are the ethical and theological questions as follows (1):

ethical considerations:

1. Do infants have a right to inherit an unmanipulated genome?
2. Does the concept of informed consent have validity for parents who do not yet exist?
3. At what point do we begin "playing God"?

a call for humility and . . .

. . . reverence.

Once again we need to recognize our position as stewards of a marvelous creation. We are one among millions of species, each equipped with an intricate genetic control system; a control system which we have seen to involve the interaction of minute molecules in precise fashion. Its function can be described with drawings and models, but we understand so little about what makes it work. Should not the "genetic engineer", and those of us who study and benefit from his or her labors, be able to see that all of this points to an intelligent Designer? Scientists, legislators, and citizens need much wisdom and direction for good stewardship in the use of biotechnology. [See TOPIC # 2.]

QUESTIONS AND DISCUSSION TOPICS

1. If *genes* were the topic of a round-table discussion among population biologists, what biological processes and important applications would you hear? Contrast this with another discussion among molecular biologists?

2. What convinces you that DNA is really the hereditary material?

3. Both your cheek epithelial cells and leukocytes originated from the fertilized egg, yet they differ in structure and function. How do you explain this? What is differentiation?

4. What is gene expression? What clues do we have as to what controls gene expression?

5. What controls metabolism—DNA or the environment of an organism? Explain with reference to the lactose operon theory. With reference to the genetic feedback model (Figure 17-6)?

6. Use the genetic feedback model, Figure 17-6, to explain the effects of each of the following upon the metabolism and/or *phenotype* of an organism (See also Sections 12-H and 15-G):
   a. Addition of lactose to *E. coli* bacteria previously fed on glucose.
   b. Exposure to cells to ionizing ultraviolet radiation.
   c. Pancreatic islet cells that are unable to synthesize insulin.
   d. Corn plants grown in darkness are whitish even though they are not albino mutants.

7. How is the concept of homeostasis evident even at the molecule level of organization?

8. Discuss the major lines of defense against foreign invasions in the human body. Distinguish *general responses* from *immune system responses*.

9. Discuss the major internal events that would occur in the event that you received an injection with an infected needle (HIV not included). How would HIV infection affect your immune system?

10. Do you consider viruses as living or non-living systems? Explain.

11. What evidence is there that viruses may have originated from aberrant genetic material of normal cells? Can this proposal be proven? How is this idea consistent with the view that gene pools have deteriorated steadily since the fall of man?

12. Discuss your position with regard to somatic cell gene therapy? Germline gene therapy? How is the introduction of pig DNA into the human cell any different from the injection of pig insulin into the body of a diabetic?

TOPICS FOR FURTHER STUDY

1. Are oncogenes the key to understanding the cause of cancer? REFERENCES: #11 and #14

2. For further study into the rapidly growing field of biotechnology and its applications and implications for medicine and agriculture, check periodic indexes for recent references, or consult REFERENCES #1, #2, and #4.

3. Follow your curiosity about the immune system into an intriguing learning experience. Starting REFERENCES are #3 and #6.

4. Study the AIDS epidemiology and the implications for Christians. REFERENCES: #8 and #10.

5. Plants have some surprising "immune responses", too. Check into the important implications for pest control in gardening and agriculture. REFERENCES #2 and #5

REFERENCES

1. Anderson, W.F. 1992. Human gene therapy. *Science* 256: 808-813.

2. Arnold, S.G. *et.al.* 1988. Latex gloves not enough to exclude viruses. *Nature* 335 (Sept. 1): 19.

3. Biotech special report. 1992. Special issue of *Science*, Vol. 256: May 8.

4. von Boehmer, H. and P. Kisielow. 1991. How the immune system learns about self. *Scient. Amer.* 265(4): 74-82.

5. Collins, F.S. 1992. Cystic fibrosis: Molecular biology and therapeutic implications. *Science* 256: 774-779.

6. Day, S. 1993. A shot in the arm for plants. *Sci. News* Jan. 9: 36-40.

7. Goodenough, U.W. 1991. Deception by pathogens. *Amer. Scient.* 79(4): 344-356.

8. Hoffman, W.W. 1990. *AIDS: Ministry in the Midst of an Epidemic.* Baker Book House. Grand Rapids, MI.

9. Judson, H.F. 1979. *The Eighth Day of Creation: Makers of the Revolution in Biology.* Simon and Schuster, New York.

10. Kaslow, R.A. and D.P. Francis, eds. 1989. *The Epidemiology of AIDS.* Oxford Univ. Press. New York, NY.

11. Kinoshita, J. 1990. The oncogene connection: Proto-oncogenes encode proteins with a neutral role. *Scient. Amer.* 262(2): 24-26.

12. Lehninger, A.L. 1982. *Principles of Biochemistry.* Worth Publ., New York.

13. Lester, L.P. and J.C. Hefley. 1980. *Cloning: Miracle or Manace?* Tyndale. Wheaton, IL.

14. Pantaleo, G., *et.al.* 1993. The immunopathogenesis of human immunodeficiency virus infection. *New Eng. J. Med.* 328(5): 327-336.

15. Vile, R. 1990. Cancer and oncogenes. *New Scient.* 125: S1-S4.

16. Watson, J.D. 1978. *The Double Helix.* Atheneum. New York, NY. Personal account of DNA discovery.

Chapter 18

Energy Capture and Conversion in Cells

Imagine yourself traveling with your family or friends across the wide open plains of the Dakotas. Unfortunately, the gauge indicates that your fuel tank is nearly empty. Fortunately, you are within miles of the next town. Unfortunately, the engine "dies" and you steer the coasting car off the road. Fortunately, your weary body has enough energy left to start out for the nearest town.

After several miles of trudging under the hot prairie sun, you spot a farmhouse. The owner graciously drives you and a can of gasoline back to your stranded car. After emptying the precious liquid into your fuel tank you pay the man for his help and wave good-by. Everyone climbs into the car, anxious to be travelling again and to enjoy the refreshment of air conditioning.

Cells must have energy continuously.

Living cells are like machines in that they must have a continuous supply of energy. This principle is based upon *"Life Characteristic #2"* as discussed in Section 2-C. Since that early discussion of energy, we have repeatedly encountered energy-dependent processes and relationships at each level of organization. Table 18-1 summarizes various encounters with energy in earlier chapters. Section 16-A has already given an "energy overview" from different levels of biological organization.

Table 18-1. Energy Relationships and Life at Different Levels of Organization.

TEXT CHAPTER	LEVEL OF ORGANIZATION	RELATIONSHIP
4	ATOM/MOLECULE	Chemical bond energy — essential to metabolism
5	COMMUNITIES	Food web — energy relationships among populations
10	ORGANISMS	Nutrition — energy intake/processing by organisms
13	ORGANISMS	Responsiveness to stimuli — forms of energy
14	ORGANISMS	Homeostasis — body temperature, sugar (energy), etc.
18	CELL/MOLECULE	Photosynthesis and cellular respiration

Energy is essential to all organisms.

All forms of life depend upon *chemical bond energy* contained in the bonds between atoms. We observed that solar energy is an essential energizing force for all living things. Energy enters the biosphere as radiant energy and eventually is lost as heat. *Autotrophs*, or producers, have the capacity to convert radiant energy to chemical bond energy via photosynthesis. *Heterotrophs* (*viz.* consumers, detritivores, or decomposers) depend upon autotrophs directly or indirectly, thus defining their positions in the trophic structure of a biotic community. (Figure 3-2). The *basic life processes* listed in Table 18-1 all involve energy relationships—acquisition, regulation, and expenditure for growth and reproduction.

Parts 2 and 3 are helpful.

We shall now examine *energy relationships within cells*. Your success in grasping this topic should be enhanced by these encounters with energy and life in previous chapters. If your class has arrived here before studying Parts 2 and 3, it will help to draw upon pre-college encounters with organismic biology and ecology as a helpful context. You may also wish to skim through the chapters noted in Table 18-1, and list the biological terms that suggest energy relationships and processes. An outline of major "energy principles" will help you to integrate the concept of energy as we did for "genes and DNA" in Section 17-A.

Must also have an energy conversion system.

Let's return to the stranded automobile, now supplied with energy, ready to head west. Or is it? Unfortunately, as the key was turned, the only sound is "Rur-r-r, Rur-r-r...." Could it be that the engine wasn't out of gas after all? Fortunately, a brief search beneath the hood revealed that a wire in the ignition system had become detached. There had been sufficient gasoline after all. Sufficient *energy* was available, but the *energy conversion system* was defective. The same is true for cells. Cells must not only have a continuous supply of energy, but also a *metabolic conversion system* to transform the energy supply into a usable form for cellular uses. Cells, like the automobile, may have abundance of energy, but still malfunction unless one or more metabolic conversion systems are functioning to transform the energy into usable form.

All cells have some type of respiration . . .

. . . and enzymes.

Living cells perform many different energy conversions as outlined in Table 18-2. However, photosynthesis and respiration are basic to all of them. All living cells, whether autotrophic or heterotrophic, conduct some type of respiration. In addition, certain cells of autotrophs can convert light energy into a usable form of chemical energy by the process of photosynthesis. Both photosynthesis and respiration involve a series of biochemical reactions that are catalyzed by **enzymes**, or biocatalysts. In this chapter, we shall discuss the nature of enzymes and the two energy conversion processes.

Table 18-2. Various Energy Conversions Performed by Living Cells.

ENERGY MAY BE CONVERTED FROM	TO	BY CELLS WITHIN
Radiant	Chemical	Green Plants (photosynthesis)
Radiant	Electrical	Organs of sight (eyes)
Chemical	Mechanical	Muscles
Chemical	Radiant	Bioluminescent organisms (e.g. firefly)
Chemical	Electrical	Electric fish and eels
Chemical	Heat	All organisms (2nd Law of Thermodynamics)
Sound	Electrical	Organs of hearing (ears)

18-A HOW CELLS USE ENERGY

Before we examine the two metabolic conversion systems, photosynthesis and respiration, it is helpful to understand clearly why all living cells have an immediate and continuous need of energy. [How long can your body cells survive after your oxygen supply, needed for cellular respiration, is cut off? Hopefully, you can estimate how long without experimenting.]

All cells need ATP...

...a molecule with suitable energy content.

First, let's expand our energy principle as follows: *all living cells require a continuous supply of chemical energy in the form of ATP*. **ATP**, or **adenosine triphosphate**, is the basic "energy currency" of cells. ATP is not like other energy-containing organic molecules such as starch, glycogen, glucose, or lipids (Sections 4-G and 4-H) which contain large amounts of energy. Instead, each ATP molecule contains a small quantity of chemical bond energy that is directly usable in many energy-requiring process of cells. In this respect, ATP is like coinage. Several dollars have more monetary value than several dimes, but dimes are more readily useful in a large variety of money-requiring situations—vending machines, pay phones, parking meters, etc. Conversely, when you wish to transport currency on an overseas trip, dimes are much too cumbersome, so you are wise to "convert" your dime collection into dollars or travellers checks. Likewise, cells use ATP for "local" metabolism, but "convert" ATP into sugars for long distance transport and into starch, glycogen, or lipids for energy storage.

How ATP reacts.

Each ATP molecule consists of the nitrogen base **adenine** (Figures 4-12 and 11-2) bonded to a **ribose** sugar (Figure 4-6a) to which three **phosphate** groups (Table 4-3) are attached. ATP can be represented diagrammatically as follows:

How ATP reacts.

When an ATP molecule is utilized in an energy-requiring process, a chemical bond between two phosphates is *hydrolyzed* and one phosphate is released ("water-split", Section 4-E). The ATP is converted to **ADP, adenosine diphosphate**:

The overall chemical reaction is represented as follows:
ATP + H_2O ⎯⎯⎯> ADP + P + usable energy

ATP "drives" endergonic reactions.

However, if this energy-releasing, or *exergonic* reaction is to produce any useful work, it must be *coupled* to an energy-requiring, or *endergonic* reaction. **Coupled reactions** channel the free energy released from a "downhill" (exergonic) reaction into driving an endergonic (uphill) reaction. ATP hydrolysis is coupled to a host of energy-requiring reactions in cells, including those noted in Table 18-2. These may be grouped into three basic types of reactions as illustrated in Figure 18-1. They are synthesis, mechanical work, and active transport.

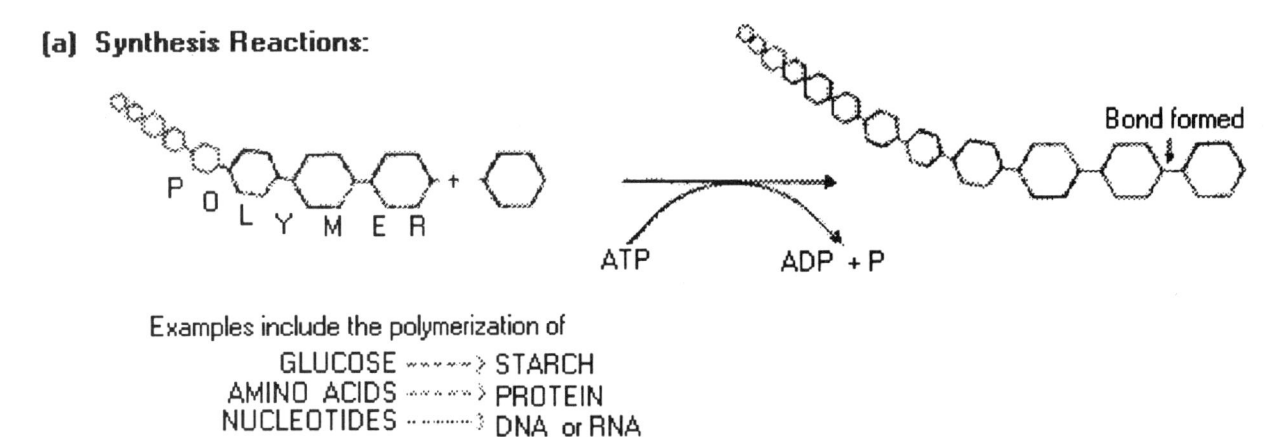

(a) **Synthesis Reactions:**

Examples include the polymerization of
GLUCOSE ·····> STARCH
AMINO ACIDS ······> PROTEIN
NUCLEOTIDES ········> DNA or RNA

(b) **Mechanical Work (Movement):**

General principle: Fibrous protein molecules, empowered by ATP, slide lengthwise over each other in response to electrical impulses.

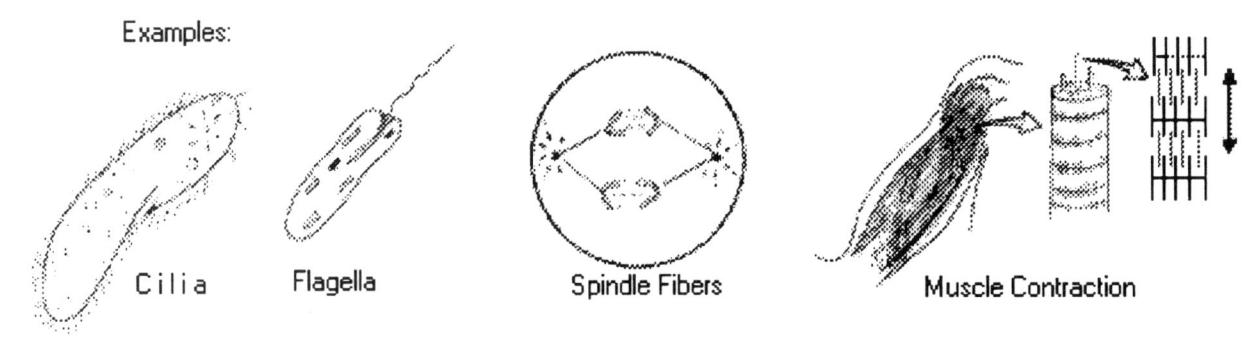

Examples:

Cilia   Flagella   Spindle Fibers   Muscle Contraction

(c) **Active Transport (Membrane "Pumps"):**

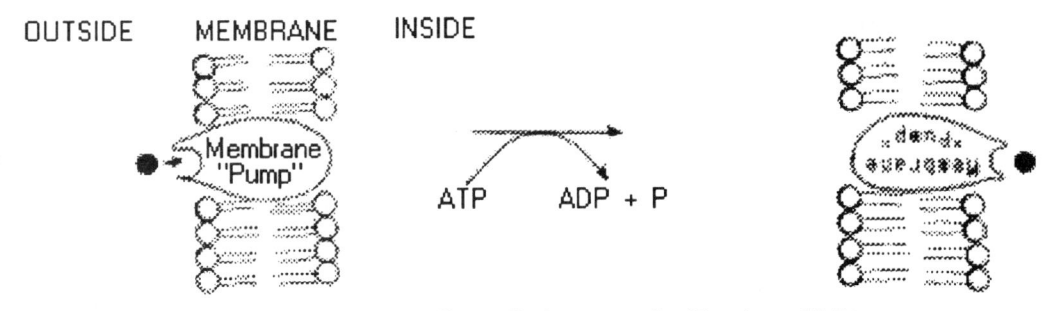

Examples: Sodium/Potassium "pump" of nerve cells (Section 13-D)
Selective permeability of cell membranes of kidney nephrons (Section 14-D)

Figure 18-1. Three basic types of ATP-requiring reactions of living cells.

In each of these energy-requiring reactions, ATP is coupled as represented below:

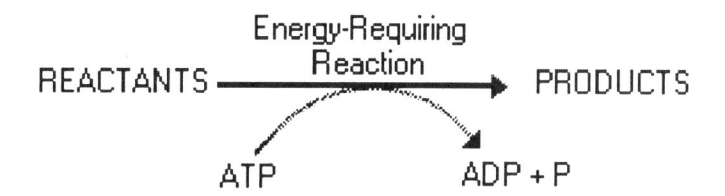

```
                    Energy-Requiring
                        Reaction
   REACTANTS  ─────────────────────▶  PRODUCTS
                  ╲               ╱
                   ATP       ADP + P
```

The product or outcome of the endergonic (uphill) reaction has more energy than the reactant(s) because of the input of energy from ATP. For example, polymers contain more energy than individual monomers because ATP supplies added energy to bond them together (Figure 18-1a and Section 4-G). Even individual monomers such as glucose or adenine are synthesized from simpler molecules (e.g. H_2O and CO_2) with energy supplied by ATP. Polymer synthesis is *growth* as seen at the molecular level as it occurs bond-by-bond. In addition, growth requires movements of chromosomes and cytoplasm as cells divide, each energized by the chemical energy of ATP. Finally, active transport of molecules and ions across cell membranes is "activated" by ATP (Sections 13-D and 16-G).

What endergonic reactions does ATP "drive"?

If ATP-driven processes were suddenly to come to a halt, synthesis and growth would cease. Orderly movements and contractions would be replaced by random motion and stillness. Chemical gradients sustained by active transport would lapse into equilibrium. In short, cells and organisms would die. According to the *second law of thermodynamics* (Section 5-C), systems tend to proceed to a state of maximum disorder or *entropy*. Living systems are no exception. The only way to stay alive is to overcome this tendency, and that requires a continual supply of high quality, light and/or chemical energy to build molecular order, produce motion, and maintain concentration gradients. Death has no energy requirement, but life's energy requirement is continual.

Disorder can be deadly...

...unless high quality energy is available.

In summary, living cells require a continuous chemical energy supply in the form of ATP. The ATP is a versatile energy currency that is essential to a variety of cellular reactions which may be grouped under three major categories—synthesis reactions, mechanical work, and active transport.

18-B HOW CELLS ACQUIRE ENERGY

Having established that cells require ATP, we are ready to consider how cells sustain their supply of this essential molecule. Because cells are continually using ATP, there must be a continual means of replenishing the supply. Otherwise, the ATP level would drop and ADP + P would accumulate in cells.

In reality, each cell generates ATP at a rate sufficient to maintain suitable levels, and like the battery in an automobile, maintains a level of "charge"—sometimes called "energy charge". This relationship between ATP use and ATP production is illustrated in Figure 18-2. From this model we can identify two energy conversion systems that can generate ATP—photosynthesis and respiration. However, photosynthesis only operates in green, autotrophic cells of plants, in the daytime. This is one of three approaches to the production of ATP—i.e. autotrophic cells in light.

ATP synthesis— like keeping a battery charged

...either by photosynthesis

411

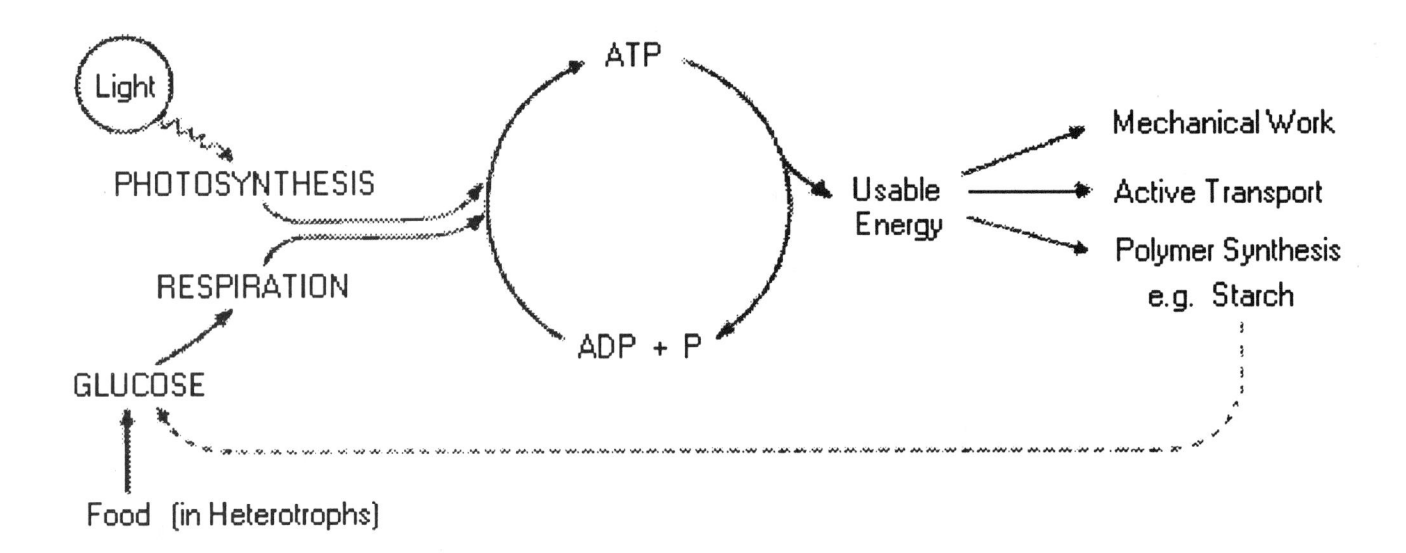

Figure 18-2. Sources of cellular ATP are the energy-rich glucose via respiration (autotrophs and heterotrophs) and light energy via photosynthesis (autotrophs only). When insufficient ATP is generated by photosynthesis in autotrophic cells, respiration of glucose meets the remaining requirement (dotted line).

... or respiration.

A second approach involves autotrophic cells under insufficient light, or in darkness. As sunlight fades in the afternoon, ATP energy charge in green cells begins to decrease, triggering an increase in respiration which "kicks in" to convert glucose into ATP. This means that the autotrophic cell must use some of its ATP generated in light (daytime) to synthesize glucose and starch polymers which are then drawn upon as reserve energy for the night hours (dotted line, Figure 18-2).

Heterotrophs use food energy.

Heterotrophs represent a third approach to ATP synthesis. These cells, including non-green cells of autotrophs and all living cells of heterotrophs, cannot photosynthesize ATP and must rely entirely on respiration of glucose. The glucose supply is obtained either by ingested food (consumers and detritivores), or by absorption of food following extracellular digestion (decomposers). Consult Figure 3-2 for a model of these distinctions from an ecological viewpoint.

Summary:

In summary, all cells require a continuous supply of ATP. This supply is obtained through the energy conversion reactions of photosynthesis and/or respiration depending upon the time of day and upon whether a given cell is autotrophic or heterotrophic.

18-C PHOTOSYNTHESIS— LIGHT ENERGY TO CHEMICAL ENERGY

Photosynthesis: light energy into ATP

Photosynthesis is an energy conversion process in which green, autotrophic cells convert light energy into chemical bond energy. This definition suggests two essential components—a source of light energy and a mechanism to bring about energy conversion.

Electromagnetic radiation reaching Earth from the sun is absorbed by autotrophs. In order to gain a general understanding of photosynthesis, we must consider the nature of electromagnetic radiation and what happens when radiation is *absorbed* by matter.

Different rays affect matter differently.

Electromagnetic radiation is emitted from the thermonuclear activity of the sun (Section 5-B). The radiation travels to earth as waves of varying wavelength as shown in Figure 5-2. The quantity of radiant energy is inversely proportional to the wavelength. Thus, long-wavelength radiation has lower energy than short-wavelength radiation. This difference in energy content determines the effect of the rays as they are absorbed by matter.

Short-wave light expels electrons...

When electromagnetic radiation strikes matter, it displays the properties of "particles", or **photons**. When the low-energy photons of longer wavelengths such as infrared radiation are absorbed by matter they cause vibrations of molecules, but are too weak to cause breakage of covalent bonds. This molecular "tickling" can be sensed as heat by receptors under our skin (Table 13-2). At the opposite end of the spectrum, short-wave radiation such as ultraviolet rays contain sufficient energy to cause electrons to be propelled out of their atomic orbitals. Covalent bonds are often broken by this electronic disruption. Radiation with sufficient energy to expel electrons, break bonds, and create ions is called ionizing radiation. Gene mutations are one manifestation of the bond breaking effects of ionizing radiation (Section 12-G).

and ionizes.

Visible light excites electrons...

Intermediate-wavelength radiation that is detected by the visual receptors of humans is called **visible light** (Figure 5-2). Photons of visible light contain just enough energy to cause electrons to move to a higher energy orbital. The electron in this **excited state** contains the energy absorbed from one photon of visible light and has the potential to do work. Visible light absorption by the pigment molecules of **chlorophyll** and *antenna* (light collecting) *pigments*, the yellow-orange **carotenoids** constitutes the first phase of photosynthesis in green plants.

Thylakoids are membranes that absorb photons.

Figure 18-3 illustrates the basic structural organization of the light energy conversion system of a typical leaf. [How many levels of organization can you identify?] Chlorophyll and other pigment molecules are localized within a complex membranous network within the chloroplasts, called **thylakoids**. Photons of visible light must penetrate the translucent cell walls and cytoplasm to reach the chloroplasts within the palisade and spongy mesophyll cells. Light absorption occurs when the energy of each photon is transferred into the excitation of an electron of a chlorophyll molecule.

Because of its importance as the key energy conversion system that supports life on Earth, countless hours of research have been devoted to the study of photosynthesis by researchers all over the world. Crop physiologists study the agronomic factors that influence photosynthesis by individual plants and stands of plants in the field. The goal of such studies is to increase the rate and efficiency of crop plant photosynthesis to meet the needs of an increasing world population. In the laboratory, fresh leaves are homogenized to break the cell walls and release chloroplasts into an osmotically controlled medium. There, these amazing organelles can be studied. Under suitable conditions, isolated chloroplasts can perform the complete photosynthesis process summarized by the following reaction:

"cell-free" photosynthesis

$$6\,CO_2 + 6\,H_2O + \text{light energy} \longrightarrow C_6H_{12}O_6 + 6\,O_2$$

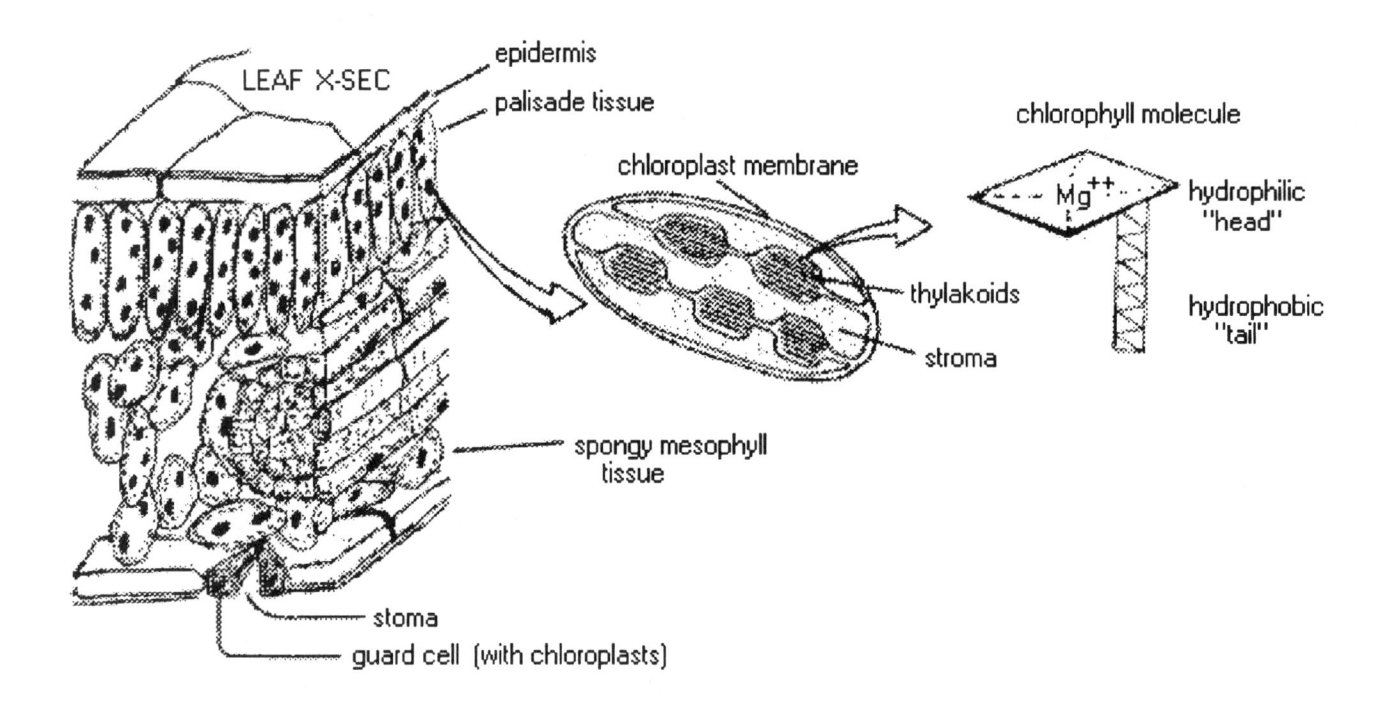

Figure 18-3. Simplified diagram of the arrangement of various components of the photosynthetic system. Chlorophyll molecules are arranged along with carotenoids, phospholipids, and proteins of the thylakoid membranes of chloroplasts.

However, what happens when chlorophyll molecules are isolated from leaves or chloroplasts and exposed to light? When this experiment is performed, as you may do in your laboratory experience, some interesting insights are gained. First, you will see that water is not a suitable solvent for the extraction—even boiling water. Rather, one must use ethanol or some other organic solvent. This demonstrates that chlorophyll molecules have hydrophobic properties (Section 16-F). They are arranged along with phospholipids within the grana membranes. Furthermore, when a tube of beautiful green chlorophyll in ethanol is exposed to a beam of light in a spectrophotometer, the extract is found to absorb visible light, preferentially in the red and blue regions of the spectrum (Figure 18-4). Thus, chlorophyll is more effective in capturing photons of red and blue than of green. Much of the green light is reflected, giving plants the familiar green appearance.

chlorophyll is water-insoluble

red and blue light favored

So far we have observed that photosynthesis is an energy conversion process in which photons of visible light are absorbed by chlorophyll molecules arranged within the thylakoid membranes of chloroplasts. When a photon is absorbed by a chlorophyll molecule, an electron is propelled into an excited state as shown in Figure 18-5. The excited electron (e^-) will either return to its original orbital, or *ground state*, within the atomic structure of the chlorophyll molecule, or it will be picked off in the excited state by another molecule functioning as an **electron acceptor**. If the electron returns to the ground state, the energy absorbed will be emitted as a photon of red light. Light emission from an excited electron as it returns to the ground state is called **fluorescence**. If the excited electron goes to the electron acceptor, then it is ready to cause the generation of usable chemical energy. It does so through a series of reactions called **electron transport** (Figure 18-5). Here, the high energy electrons move from the electron acceptor through a series of intermediate acceptors. As electrons move through the electron transport "chain",

excited electrons carry energy

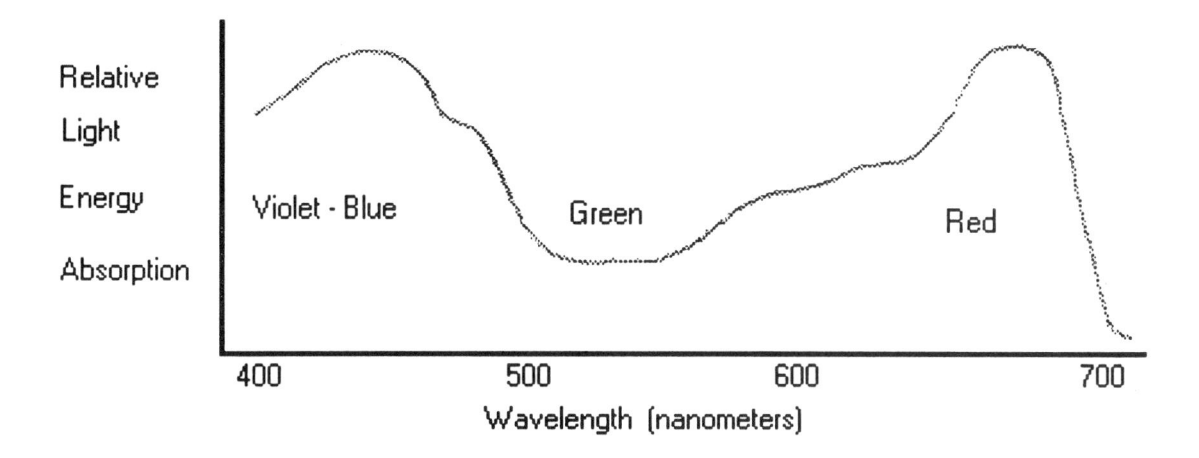

Figure 18-4. Absorption spectrum of chlorophyll molecules extracted in ethanol and measured at different wavelengths in a spectrophotometer (See Figure 5-2).

energy is released in sufficient quantity to convert ADP + P to ATP. This time, ATP is synthesized in an endergonic reaction (uphill) because it is coupled to the exergonic (downhill) reactions of the electron transport chain.

electron transport

When one photon of light is absorbed by a chlorophyll molecule, an electron is excited and may be lost to the electron acceptor (Figure 18-5). This increases the attraction of chlorophyll for an electron to replace the one that was lost. Replacement electrons are attracted from nearby H_2O molecules. Removal of electrons from H_2O causes the breakage of covalent bonds and release of oxygen gas from the "splitting" of the water molecule.

NADPH$_2$ carries hydrogen atoms from water.

Hydrogen ions (H^+), representing H atoms robbed of their electrons, join electrons from electron transport in the conversion of NADP to $NADPH_2$. **NADP**, or nicotinamide adenine dinucleotide phosphate, is an electron and hydrogen carrier that participates along with ATP in the synthesis of glucose. Thus, photosynthesis involves a light-driven flow of high-energy electrons which cause the formation of ATP and $NADPH_2$. These compounds represent a form of chemical energy that is usable in a variety of energy-requiring processes performed by the autotroph in the light. One of these processes is the synthesis of glucose and starch. Figure 18-5 shows that the light-driven synthesis of ATP occurs only in the light. Therefore, glucose and starch synthesis must occur in light to generate a reserve of chemical energy for the synthesis of ATP in the dark via respiration. In addition, glucose and other carbohydrates serve as a source of carbon skeletons for the synthesis of all other constituents for growth of the autotroph (e.g. amino acids, nucleotides, etc.). This relationship is illustrated in Figure 5-3 and Figure 18-5.

Summary:

In summary, we have described photosynthesis as an energy conversion process in which photons of light energy are converted to the chemical energy of ATP and $NADPH_2$. Chlorophyll molecules are the focus of energy capture. Electrons and H atoms of H_2O are transferred to $NADPH_2$, and the oxygen atoms of H_2O become the source of O_2 gas. The ATP and NADP are usable in a variety of energy-requiring processes, especially in the synthesis of glucose and other molecules. These molecules serve as an energy reserve for periods of darkness, and are essential for growth.

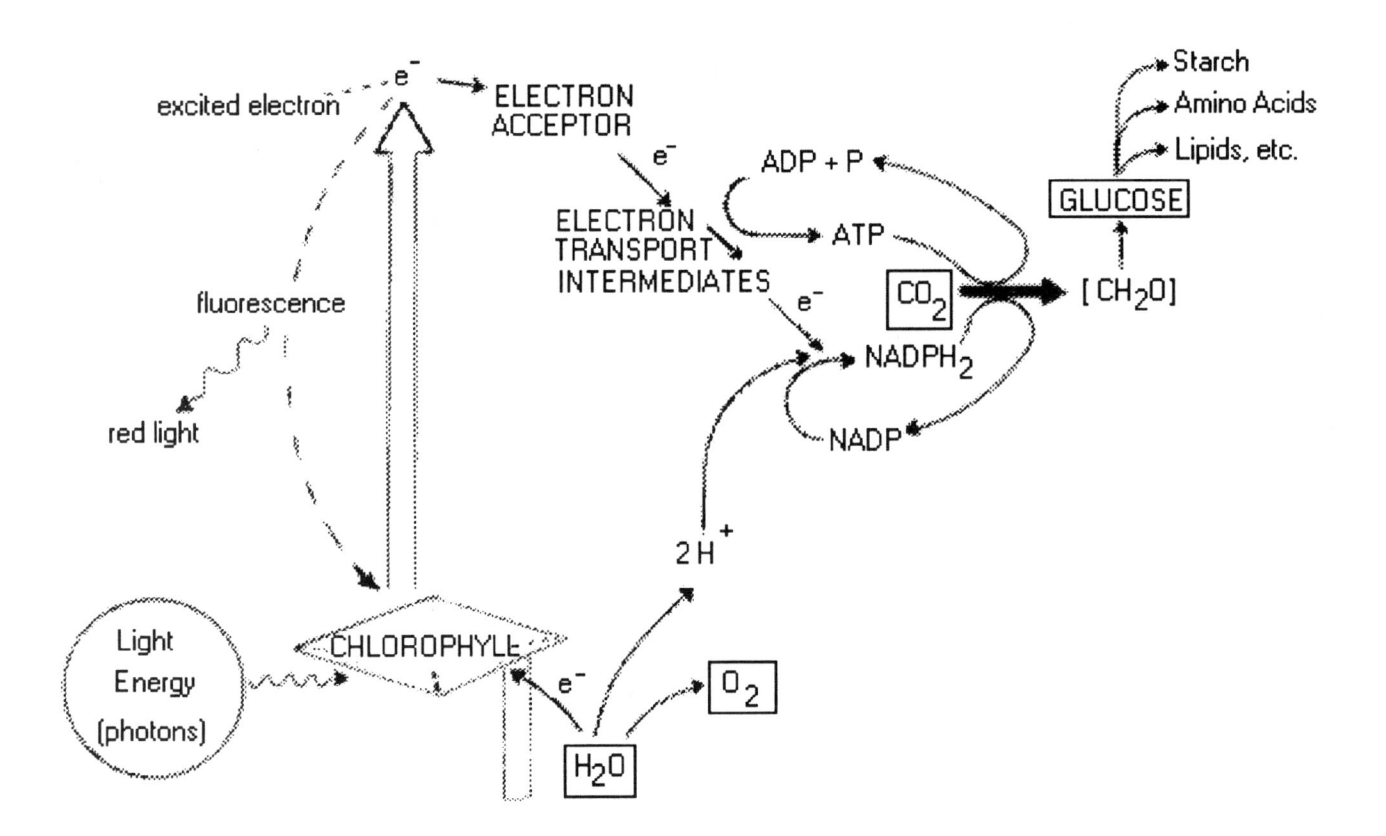

Figure 18-5. Conversion of light energy to chemical bond energy of ATP and NADPH$_2$ which are used in the synthesis of glucose and other growth constituents. The initial reactants and final products are encircled or boxed for emphasis.

18-D CELLULAR RESPIRATION

cells that can't photosynthesize

The green autotrophic cell that is absorbing light on a sunny day has a ready source of ATP via photosynthesis (Figure 18-5). However, what happens when the "lights go out"? Or, what about the non-green cells of autotrophs, and all cells of heterotrophs. For these cases, cellular respiration assumes the dominant means of ATP production, provided glucose or other energy-rich organic molecules are available. Let's consider three characteristics of respiration as follows:

Characteristics of respiration:

1. Respiration is a cellular process—occurs in all cells.
2. Respiration is a combustion process—"slow burn"
3. Respiration is an enzyme-catalyzed process—controlled

aerobic respiration:

First, respiration is a *cellular process* and occurs in some form in all living cells. Most autotrophs and heterotrophs alike perform **aerobic respiration** in which O$_2$ is used to degrade glucose into CO$_2$ and H$_2$O. The multi-step process is summarized as follows:

$$C_6H_{12}O_6 + 6\,O_2 \longrightarrow 6\,CO_2 + 6\,H_2O + \text{usable energy}$$
$$(36 - 38 \text{ ATP})$$

Anaerobic respiration . . .

Many bacteria and fungi can perform **anaerobic respiration** in which glucose is partially degraded in absence of O_2. *Strict anaerobes*, including microbes that live in the mud of marshes, survive only where O_2 is absent. *Facultative anaerobes* will respire aerobically if O_2 is present, but can switch to anaerobic respiration under anaerobic conditions. Yeast and many other microbes are facultative anaerobes. Anaerobic respiration may be summarized by the following equation:

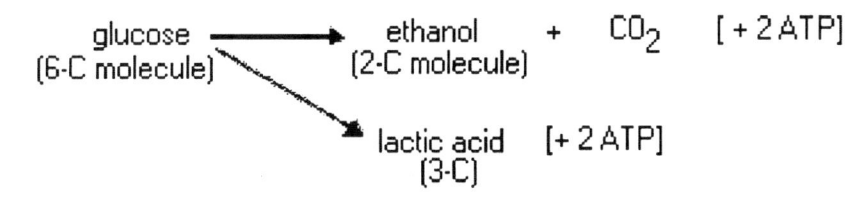

$$\text{glucose (6-C molecule)} \longrightarrow \text{ethanol (2-C molecule)} + CO_2 \quad [+ 2 ATP]$$
$$\searrow \text{lactic acid (3-C)} \quad [+ 2 ATP]$$

is much less efficient than aerobic . . .

Anaerobic respiration, or **fermentation**, is an important part of the brewing and baking industry, as you can see from the products. The "holes" in bread and cheese are produced by the release of CO_2 gas. Note that anaerobic respiration is much less efficient in ATP production than aerobic respiration. Much of the chemical energy is still locked in the bonds of ethanol (or lactic acid). Indeed, ethanol produced by fermentation of agricultural biomass is marketed in engine fuel mixtures.

a type of "combustion"

Respiration is also a *combustion type* process which consumes glucose in order to release energy for the production of ATP. The heat energy that warms our bodies from within is one product of cellular respiration. If one gram of glucose is completely burned in the presence of O_2 in a calorimeter, approximately 3.8 Calories of heat energy is released (Section 5-D). Glucose may be great fuel for a "heat engine" but living cells require more than heat energy. They must convert some of the Calories available in glucose into the usable form of ATP. The 11 Cal./minute required for bicycling, 5 Cal./min. for rapid walking, 7 Cal./min. required for tennis, and even the 1.5 Cal./min. required for relaxing must come largely from ATP. How then does respiration direct some of the energy of glucose into the production of ATP? The answer involves the third characteristic of respiration—it is an *enzymatic process*.

an enzymatic process

From earlier encounters with **enzymes**, you may recall that they are large protein molecules that function as bio-catalysts. A **catalyst** is defined as an agent that promotes a chemical reaction without itself being consumed in the reaction. Chemical reactions between two molecules will only occur if they are brought into contact, or collision, with each other. The frequency of collisions depends upon the number of molecules per unit volume (concentration) and the kinetic energy of the molecules. The latter depends upon the temperature. Addition of heat or actual ignition of combustible material will certainly cause reactions to occur. However, heat and light energy are often the only products.

Energy is captured as ATP and $NADPH_2$.

Respiration is a "slow burn" process because it is under enzymatic control. These catalysts bring about a step-by-step dismantling of glucose molecules at temperatures suitable for other life processes. Energy released from covalent bond breakage is captured in the formation of chemical bonds of ATP and $NADH_2$ which are usable by cells. Heat loss is kept to a much lower level than in a fire. Molecules react at this lower temperature because enzymes draw them into contact with each other to facilitate a chemical reaction.

Enzymes are specific . . .

. . . as explained by *induced fit* hypothesis.

complementarity among molecules

Amazingly, each metabolic reaction is catalyzed by a specific enzyme with an **active site** designed to fit the reactants, called **substrates** (Figure 18-6). The configuration or shape of the active site depends upon the three-dimensional structure of the enzyme protein molecule. This 3-D structure, in turn, depends upon a precise sequence of amino acids formed in the translation process (Section 17-C), and subsequent hydrophobic and hydrophilic interactions which influence the coiling of the protein into the three-dimensional form. The shape of the active site is such that the substrate(s), say molecule A and molecule B, can "fit" into it. Indeed, the "induced fit hypothesis" contends that when substrate molecules enter the active site, the enzyme protein changes shape to fit. Thus, the enzyme-substrate "fit" is apparently even more wondrous than the old "lock-and-key" analogy often used to teach enzyme action. It is also a marvelous example of the complementarity of structure and function at the molecule level of organization.

(a) Enzyme protein synthesis:

STRUCTURAL GENE

transcription ↓

mRNA

translation ↓

Protein molecule coils into a "minimum stress configuration" as determined by the amino acid sequence, and hydrophobic/hydrophylic interactions.

RESULT:
An enzyme molecule with a unique shape, suited to act only on certain substrates (see b. below)

active site

(b) Enzyme Action:

ENZYME + SUBSTRATE ------> ENZYME-SUBSTRATE COMPLEX ------> ENZYME + PRODUCT

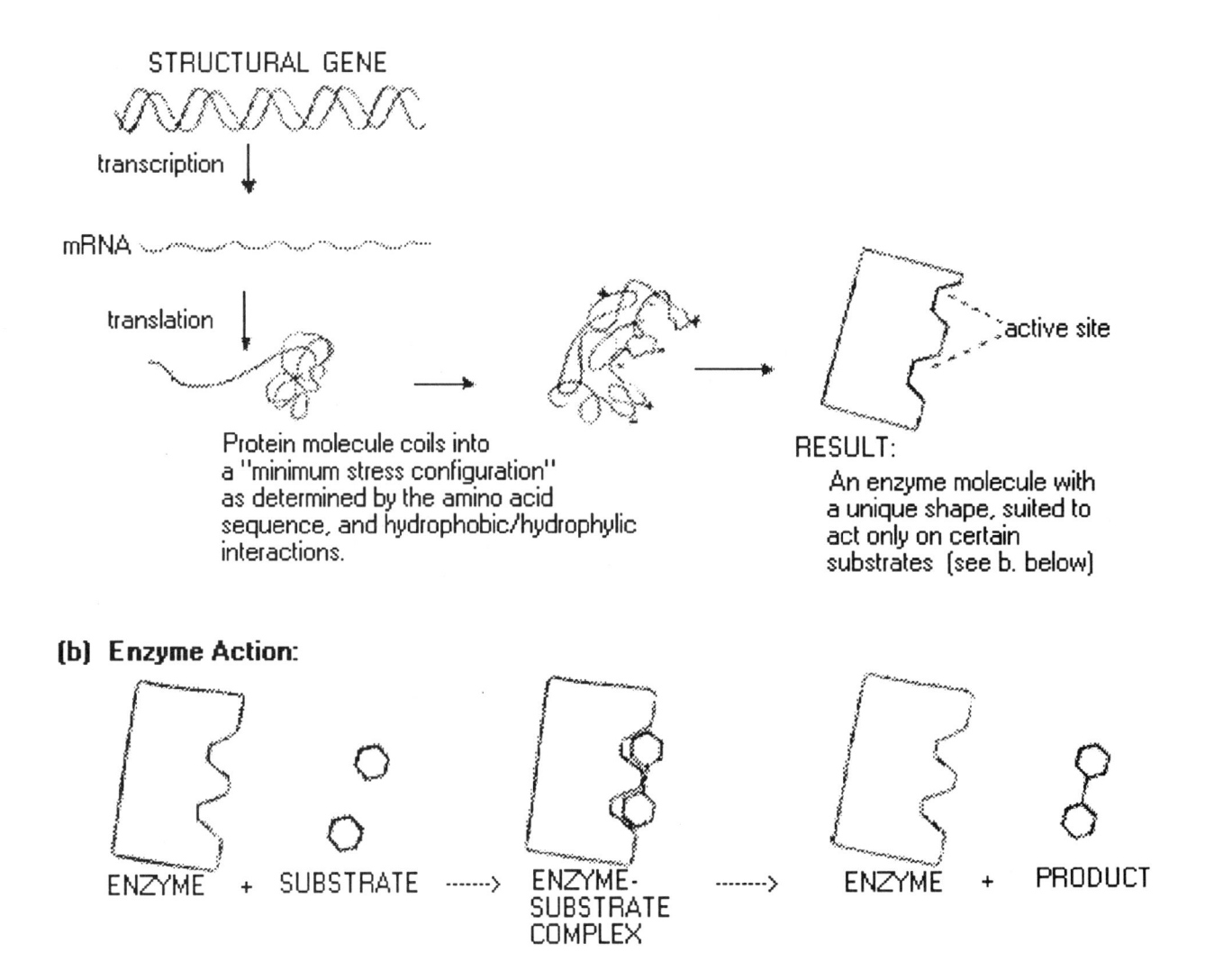

Figure 18-6. Synthesis and metabolic function of enzymes.

How do inhibitors act?

The importance of enzyme molecule "shape" is evidenced in the effect of inhibitors that block the active site (Figure 18-6). For example, cyanide is a deadly inhibitor of respiration because it blocks the active site of one out of the many respiratory enzymes. Other conditions such as extreme temperature and extreme pH will disrupt metabolism by *denaturing* enzymes, destroying their three-dimensional structure.

Aerobic respiration has three phases:

glycolysis

Krebs Cycle

electron transport

Aerobic respiration may be considered as having three phases, each with several enzyme-catalyzed steps. The individual reactions and their respective enzymes will not be addressed in our general discussion. In phase one, called **glycolysis** [G. (*glyco* = sweet) + (*lysis* = splitting)], glucose (6-carbon molecule) is split into two 3-C molecules known as pyruvic acid (pyruvate). Two ATP per glucose molecule are produced in this cytoplasmic phase. The pyruvate moves into the mitochondria where phase two, the **Krebs Cycle** occurs. Here, each pyruvate is stripped of H atoms which are collected by NAD (nicotinamide adenine dinucleotide), a hydrogen carrier. The final phase, **electron transport** involves the reaction of $NADH_2$ with oxygen gas. Electrons move along a series of e⁻ acceptors similar to photosynthetic electron transport (Section 18-C). Three ATP's are usually produced for each $NADH_2$ involved. The summary reaction of electron transport between $NADH_2$ and O_2 is as follows:

$$NADH_2 + 1/2\ O_2 \longrightarrow NAD + H_2O$$
$$3\ ADP + 3P \longrightarrow 3\ ATP$$

total energy converted to ATP

The three phases of aerobic respiration are summarized diagrammatically in Figure 18-7. A total of 36 to 38 ATP's are produced for each glucose molecule, depending upon how hydrogen atoms of $NADH_2$ produced in glycolysis enters the electron transport chain to ATP (19). Each glucose molecule has a total available energy equivalent of about 94 ATP's. Therefore, the efficiency of aerobic respiration is approximately 40% —i.e. 38 ATP / 94 ATP x 100). The remaining 60% of the energy escapes as heat in accordance with the second law of thermodynamics (Section 5-C).

when O_2 isn't available

Obviously, oxygen is an essential requirement for aerobic respiration. During periods of bodily exertion, high ATP demand may exceed the O_2 supply to body cells. This causes the Krebs Cycle and electron transport, which make direct use of O_2, to lag behind glycolysis. The result is a "damming up" effect in a backward direction from the electron transport where O_2 is a substrate to pyruvate. Pyruvate is then converted to lactic acid which may accumulate in muscle cells, causing aches and muscle fatigue. When a person ceases the exertion, O_2 supply is replenished. Adequate O_2 allows the conversion of lactic acid back to pyruvate from which point entry into the Krebs Cycle and electron transport are again possible.

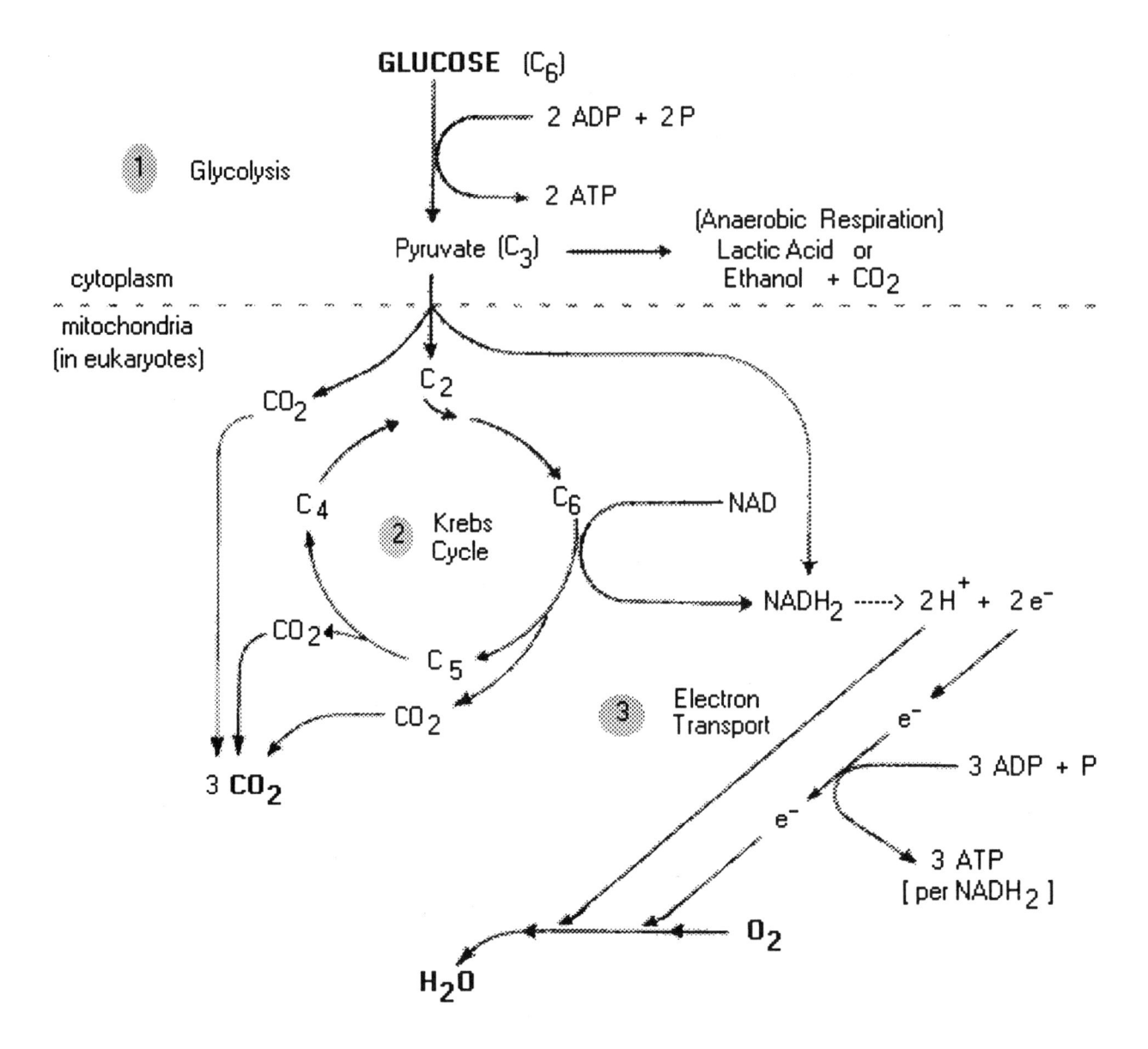

Figure 18-7. Summary diagram of aerobic respiration as discussed in the text. The initial reactants and final products are in boldface for emphasis.

18-E ENERGY METABOLISM—
A RETURN TO THE ECOSYSTEM PERSPECTIVE

putting it all together ...

Chapter 18 has been a challenging chapter to master. However, if you have made a faithful, consistent effort to master the concepts and principles encountered in earlier chapters, now you can more fully appreciate the role of cellular energy conversion as being fundamental to the life of every creature. We have inserted many cross-references to earlier chapters to aid your understanding of photosynthesis and respiration from both an ecosystem perspective and a cellular and organismic perspective. A clear understanding of the two processes from an ecosystem perspective can keep you from sinking into a quagmire of details that seem meaningless.

...ecosystem to molecule

Figure 18-8 is provided as a learning challenge and an aid to help you "put it all together". The model invites you to draw upon your knowledge of life at the *ecosystem* and *community* levels. What is the role and significance of autotrophs and heterotrophs, their relationship to each other as *populations* in the food web, and their interaction with the abiotic environment? The biology of the individual *organism* and its *organs* is included in processes of intake, digestion, and transport. *Cells* and *organelles* provide the fundamental metabolic distinctions between autotroph and heterotroph. Both cell types perform respiration, but only autotrophs can photosynthesize. It is the autotrophs of Earth that support life as we know it by provision of food and oxygen.

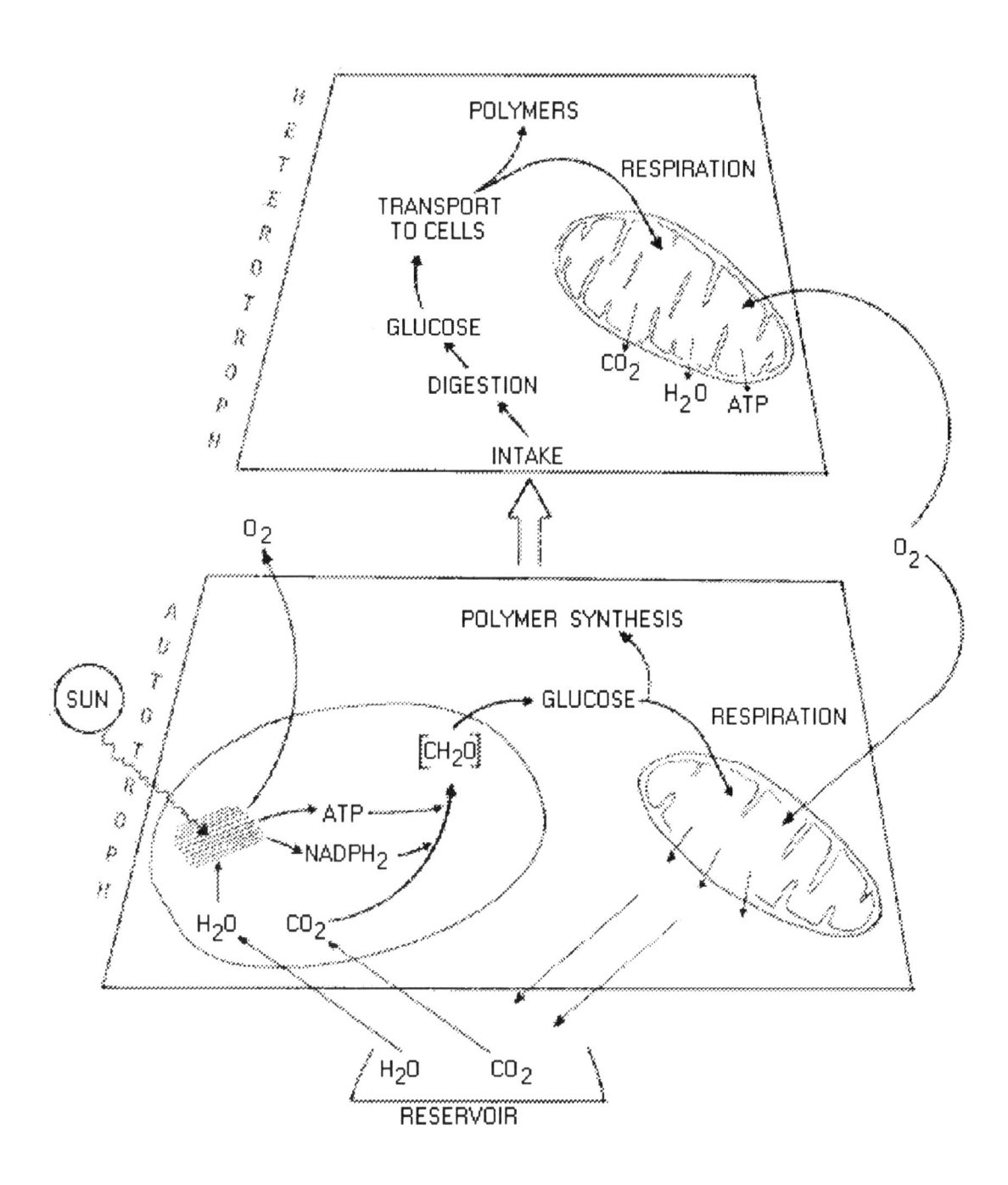

Figure 18-8. Distinctions between autotrophs and heterotrophs—an integrative model featuring both a cellular and ecological perspective.

Challenge—

Keep on learning...

judge carefully...

with God's Word, as a standard...

live as an obedient steward.

We have come a long way since beginning the course of study. It is hoped that you have learned much about the living world, and perhaps more importantly, how to keep on learning about it. A multitude of facts have been omitted. Those we have included have been accompanied by principles and models to provide a framework for your present and future study, and application.

We have observed many areas of biological science that require value judgements and raise moral and ethical questions. Indeed, science may be reaching the point where the question will not be so much, "Can we do it?", as "Should we do it?" Issues of environmental quality, human population control, biomedical ethics, and genetic engineering may enter into decisions you will have to make, personally or professionally. Science provides technique and application, but it cannot provide a standard of right and wrong. We hope you have a personal relationship with the Creator of life, and trust His Word as an unfailing standard of objective truth. If so, we hope this text has helped to expand your vision of God's power, plan, and purpose. Finally, this text will have accomplished its purpose if it has in some small way challenged you to be more alert to the living world, to enjoy the creation, but to recognize the fulfillment that can only come as you honor the Creator by wise stewardship of body, fellow creatures, and environment.

QUESTIONS AND DISCUSSION TOPICS

1. How can an understanding of life at the ecosystem level (Part 2) and organismic level (Part 3) of organization aid in ones understanding of photosynthesis and respiration at the cell level of organization (Part 4)?

2. Assuming an organism exists in an environment where energy is available, what active part must this organism take in utilizing that energy? How is this role different for autotrophs than for heterotrophs?

3. Both glucose and ATP molecules supply chemical energy. Which is more immediately usable in cells, and why? Which contains more energy per molecule?

4. Describe the energy source and energy conversion system that supplies energy to each of the following cells:

    a. a palisade cell in the daytime.
    b. a spongy mesophyll cell at night.
    c. a root hair cell in the daytime—and at night.
    d. a leg muscle cell.

5. From your understanding of organismic biology, explain the role of transport systems (if any) in supplying chemical energy to each cell type in item #4.

6. Suggest cellular processes in your body that would be inhibited if you were exposed to cyanide, an inhibitor of respiratory ATP synthesis. Cite one example representing each of the three major uses of ATP (Section 18-A).

6. How is light energy actually captured by the material substance of a leaf? How are each of the following involved in light energy conversion: electrons, chlorophyll, electron acceptors, water, ATP, NADP, CO_2, and O_2?

7. What are chemical reactions? Why does high temperature generally increase reaction rates? Why doesn't cellular respiration require high temperatures to extract energy from glucose?

8. Explain how enzymes facilitate reactions according to the "induced fit hypothesis". How is this hypothesis consistent with the fact that high fever can be fatal?

9. In what sense are human cells like those of facultative anaerobes?

10. Trace the flow of energy from sunlight to your leg muscle cells as obtained from mashed potatoes. Develop a model similar to Figure 18-8 and include the processes of photosynthesis, phloem transport, digestive processes, transport within the circulatory system, homeostatic control of blood sugar, insulin (protein) synthesis, and cellular respiration within mitochondria of muscle cells.

TOPICS FOR FURTHER STUDY

1. How are muscles like engines? How do muscles and the body adjust during aerobic training? REFERENCES: #10, #11, and #13

2. How do plant leaves "know" how much energy to store as starch in the daytime to supply the energy and growth needs during the night? Examine this interesting area in which your author was privileged to conduct research. REFERENCES: #5, #14, and #15

3. Photosynthesis offers an intriguing study for anyone who wishes to go further than our brief coverage. Consider an aspect of photosynthesis as a topic for a speech or assigned report. For example, how will increasing global CO_2 levels (Section 6-D) affect photosynthesis, which is so essential to all life on Earth? REFERENCES: #2 and #14   See also TOPICS #4 and #5.

4. Ribulose bisphosphate carboxylase (Rubisco), which "fixes" CO_2 into an organic form, is believed to be the most abundant protein in the world. Rubisco is inhibited by oxygen gas, of all things. Certain plants such as sugarcane and corn ("C_4 plants") have an additional enzyme system that avoids this problem. REFERENCES: #9, #12, #14, and #18

5. Animals (e.g. endotherms) can maintain body warmth by heat generation through increased respiration. Read about some plants of the arum family (Araceae) that are "thermogenic". REFERENCES: #14 and #17

6. Proteins are not the only molecules with enzyme properties. You may wish to expand your understanding of the "world of RNA". REFERENCES: #3 and #4.

REFERENCES

1. Baker, J.J.W. and G.E. Allen. 1982. *Matter, Energy, and Life*, 4th ed., Addison-Wesley Publ. Co., Reading, MA.

2. Bazzaz, F.A. and E.D. Fajer. 1992. Plant life in a CO_2-rich world. *Scient. Amer.* 267(1): 68-74.

3. Cech, T. 1986. RNA as an enzyme. *Scient. Amer.* 255(5): 64-75.

4. Celander, D. and T. Cech. 1991. Visualizing the higher order folding of a catalytic RNA molecule. *Science* 251: 401-407.

5. Chatterton, N.J. and J.E. Silvius. 1981. Photosynthate partitioning into starch. II. Irradiance level and daily photosynthetic period duration effects. *Plant Physiol.* 67: 257-260.

6. Darnell, J., N. Barnes, D. Baltimore. 1990. *Molecular and Cell Biology*, 2nd ed. W.H. Freeman and Co.

7. Govindjee and R. Govindjee. 1974. The absorption of light in photosynthesis. *Scient. Amer.* 231(6): 68-82. [A classic article]

8. Hinkle, P.C. and R.E. McCarty. 1978. How cells make ATP. *Scient. Amer.* 238(3): 104-123.

9. Howe, G.F. 1981. Photosynthetic pathways lead to creation design. *Creat. Res. Soc. Quart.* 18(Dec.):176.

10. Lazarides, E. and J.P. Revel. 1979. The molecular basis of cell movement. *Scient. Amer.* 240(5)100.

11. McMahon, T.A. 1984. *Muscles, Reflexes, and Locomotion*. Princeton Univ. Press. Princeton, NJ.

12. Monson, R.K., et.al. 1984. C_3-C_4 Intermediate photosynthesis in plants. *Bioscience* 34(9): 563-574.

13. Nadel, E.R. 1985. Physiological adaptations to aerobic training. *Amer. Scient.* 73(4): 334-343.

14. Salisbury, F.B. and C.W. Ross. 1991. *Plant Physiology*, 4th ed. Wadsworth, Belmont, CA.

15. Silvius, J.E., N.J. Chatterton, D.F. Cremer. 1979. Photosynthate partitioning in soybean leaves at two irradiance levels. *Plant Physiol.* 64: 872-875.

16. Snyder, C.H. 1992. *The Extraordinary Chemistry of Living Things.* John Wiley and Sons, Inc., New York.

17. Solomos, T. 1977. Cyanide-resistant respiration in higher plants. *Ann. Rev. of Plant Physiol.* 28: 279-297.

18. Stern, K.R. 1991. *Introductory Plant Biology*, 5th ed., W.C. Brown, Dubuque, IA.

19. Wivagg, D. 1987. How many ATP's per glucose molecule? *Amer. Biol. Teach.* 49: 113-114.16.

Appendix A

Review of Basic Chemistry

A-1 SUBATOMIC PARTICLES AND ATOMS

subatomic structure

All substances, whether living or nonliving, are composed of minute particles called **atoms**. It was once believed that atoms were the final, irreducible units of matter [Gr. (*a* = not) + (*tomos* = cut)]. However, atoms can be "cut" into one or more positively charged subatomic particles called **protons** (p^+) and an equal number of negatively charged particles called **electrons** (e^-). In addition, all atoms (except for hydrogen atoms) have at least one **neutron** (n^o) with a neutral charge. The protons and neutrons of each atom are tightly clustered together in a mass called the **atomic nucleus**. Over 99 percent of the mass of each atom, and therefore of all matter, is contained in these nuclear particles.

attractions and repulsions

Electrons are negatively charged subatomic particles in constant motion forming a cloud around the positively charged nucleus. Because like charges (e.g. $p^+ p^+$, or $e^- e^-$) repel and unlike charges ($p^+ e^-$) attract, the centrally located nucleus exerts a positive attraction for the negatively charged electrons. Meanwhile, the like-charged electrons find a "comfortable" distance from one another within which to move along at nearly the speed of light (Figure A-1).

Elements differ in the ...

Suppose you were to break down a sample of matter to its individual atoms and then sort the atoms into "piles" according to the **atomic number**—i.e. the number of protons within the nucleus of each atom. Then, by definition, all atoms having the same atomic number represent the same chemical **element**. Atoms of each different element have different atomic numbers. The number of protons equals the number of electrons in each atom, and therefore atoms carry no charge. If you are given the atomic number of an element, you can determine the number of protons and the number of electrons contained within the atoms of that element.

... number of subatomic particles ...

Out of the 92 naturally occurring elements found on Earth, only about 20 are required by the various forms of life. Table A-1 presents atomic structural data for the more common elements with low atomic numbers. With the exception of hydrogen which has 1 proton and 1 electron (Figure A-1), atoms of all known elements in the universe have in common the three subatomic particles—protons, neutrons, and electrons. Furthermore, the difference between elements doesn't reside in differences in the nature of their subatomic particles but in the *number* of subatomic particles. For example, nitrogen atoms differ from carbon atoms in having 7 each of the subatomic particles instead of 6, respectively.

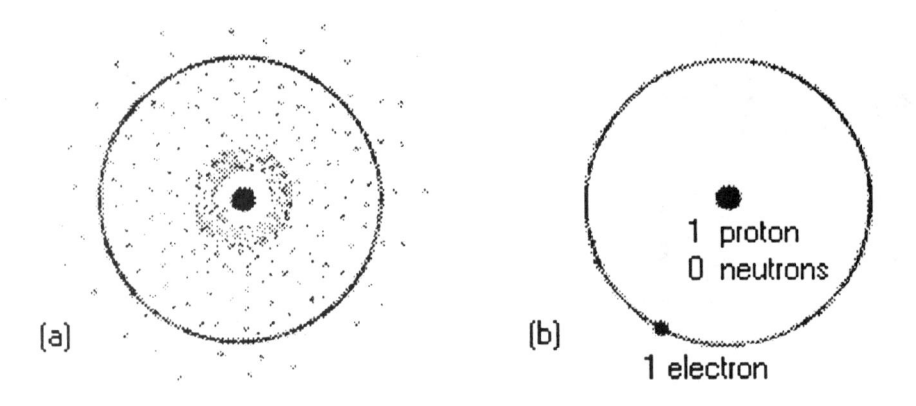

Figure A-1. The hydrogen atom as represented in two different two-dimensional models. Remember that atoms are submicroscopic and have never been observed. Their three-dimensional nature is difficult to represent on paper. The central gray area in (a) is the nucleus and the dots represent the lone electron of the H atom. The greater the density of dots the greater the likelihood of the electron being found in that location at any instant. The circle (a. and b.) outlines the volume of a sphere expected to contain the electron 99% of the time.

...and resultant mass number.

Many elements consist of atoms having a different number of neutrons than of protons and electrons. For example, carbon atoms have 6 protons and neutrons, whereas sodium atoms have 11 or 12, respectively. The **mass number** of an atom is the total number of protons and neutrons (Table A-1).

isotopes

Atoms of the same element (same atomic number) may have different mass numbers due to a variation in the number of neutrons. These are called **isotopes**. For example, the element carbon may exist as carbon-12 ($6p^+ + 6n^o$), carbon-13 ($6p^+ + 7n^o$) or carbon-14 ($6p^+ + 8n^o$). Isotopes exist in nature and can be produced in nuclear reactors. They have several important uses in biological research and in medicine as we shall see.

Different isotopes of the same element all have the same name or chemical symbol accompanied by the distinguishing mass number. For example, the isotopes of carbon we just mentioned are ^{12}C, ^{13}C, and ^{14}C; and those for oxygen include ^{14}O through ^{19}O. For our purposes, it may be said that all isotopes of a given element are chemically indistinguishable. Because nuclei differ in mass among different isotopes of the same element, they can be distinguished physically. [Do you know how?] Thus, as long as the number of *charged* subatomic particles (protons and electrons) is the same among different isotopes, we have the same element and essentially the same chemical behavior. In fact, much of what we now need to understand about chemistry relates to the electrons orbiting around the atomic nucleus.

A-2 WHY ATOMS REACT CHEMICALLY

energy relationships among atoms

Each orbiting electron zips along in a path whose distance from the nucleus depends upon the energy it contains. Energy has been described as the capacity to do work on a system—to produce motion, pressure, change of state, or change in orderliness. Chemists use the concept of *charges* to explain the

428

energy relationships of repulsion and attraction among subatomic particles. Thus, the positively charged nucleus can attract negatively charged electrons. Electrons, in turn possess energy which enables them to "resist" this attraction. Hydrogen atoms have one electron which occupies a spherical *orbital*, or most probable path around the nucleus (Figure A-1). In increasingly larger atoms, the orbitals followed by electrons are arranged in successive layers around the nucleus called **shells**. The first shell is filled by the single spherical orbital mentioned above, and is located closest to the nucleus. Each orbital can be occupied by no more than two electrons. Thus, the first shell can hold only two electrons. The element helium has just enough electrons to fill the first, single-orbital shell (Table A-1). The second and third shells each have four orbitals and hence a capacity for 8 electrons apiece. Study Table A-1 and compare with the diagrams in Figure A-2.

**Table A-1. Atomic Structures of Elements Common in Living Organisms
Arrangement of Electrons in Shells**

ELEMENT	ATOMIC NUMBER	MASS NUMBER*	FIRST SHELL (capacity=2e⁻)	SECOND SHELL (capacity=8e⁻)	THIRD SHELL (capacity=8e⁻)
Hydrogen (H)	1	1	1		
Helium (He)	2	4	2		
Carbon (C)	6	12	2	4	
Nitrogen (N)	7	14	2	5	
Oxygen (O)	8	16	2	6	
Neon (Ne)	10	20	2	8	
Sodium (Na)	11	23	2	8	1
Phosphorus (P)	15	31	2	8	5
Sulfur (S)	16	32	2	8	6
Chlorine (Cl)	17	35	2	8	7
Argon (Ar)	18	40	2	8	8

*Mass number of most common isotope.

Electrons may absorb energy.

An electron can move from one orbital or shell to another only if it absorbs or emits energy. For example, when atoms absorb visible light, an electron may acquire a discrete quantity of energy and change orbitals or shells, moving to a position more distant from the nucleus. This electron is said to exist in an **excited state**. The additional energy absorbed by this electron can be released instantaneously and returned as light energy, or be used to perform work in chemical processes. This is the means by which plants trap the energy of sunlight and make it chemically usable for life processes.

In summary, we have noted that electrons are high-energy, subatomic particles that occupy precise orbitals depending upon their energy status. We are now ready to understand why very few atoms exist individually in nature.

Atoms react and gain stability.

The chemical properties of each atom depend largely upon the extent to which its outermost electron shell is filled. Atoms such as helium, neon, and argon which have full outer shells (Table A-1 and Figure A-2) are said to have a **stable number** of electrons. These elements are essentially inert chemically. The stable condition achieved by these elements is the "goal" to which the atoms of all other unstable elements appear to be "striving". To reach as

CARBON

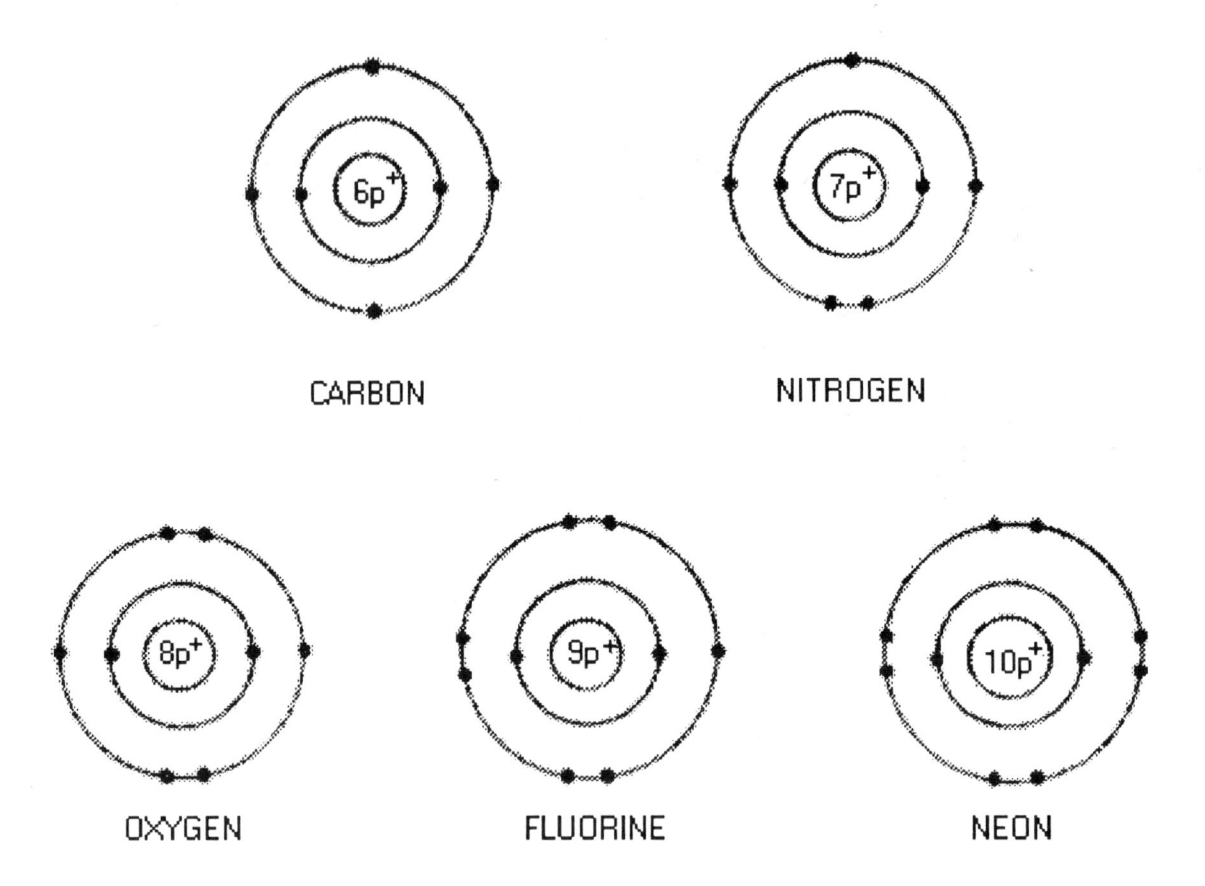

NITROGEN

OXYGEN

FLUORINE

NEON

Figure A-2. Diagrammatic representation of atoms based upon the model proposed by Neils Bohr. Note the progressive filling of orbitals from the innermost shell outward. Compare to Table A-1.

stable of a condition as possible, these atoms will enter chemical reactions either with other atoms of the same element, or with atoms of different elements. In so doing, atoms such as chlorine which have 7 electrons in the second shell (Table A-1) can acquire the one needed electron to fill the "shell" by reacting with an atom such as sodium. Sodium can readily release its single electron of the third shell, leaving a complete second shell (Figure A-3). As a result, sodium and chlorine atoms each achieve a more stable condition approaching that of argon and neon, respectively. Interactions such as these provide the basis for explaining the chemical and physical properties of matter, both living and nonliving.

A-3   CHEMICAL BONDS OF LIVING MATTER

When two atoms react and mutually achieve a stable number of electrons, the electron cloud of the one atom becomes linked to the electron cloud of the other. In other cases, certain atoms already linked to other atoms in a molecule may enter into a weak attraction with atomic portions of another molecule. Regardless of whether linkages are formed through loss, gain, sharing of electrons, or through weak attractions between molecules, the result is called a **chemical bond**.

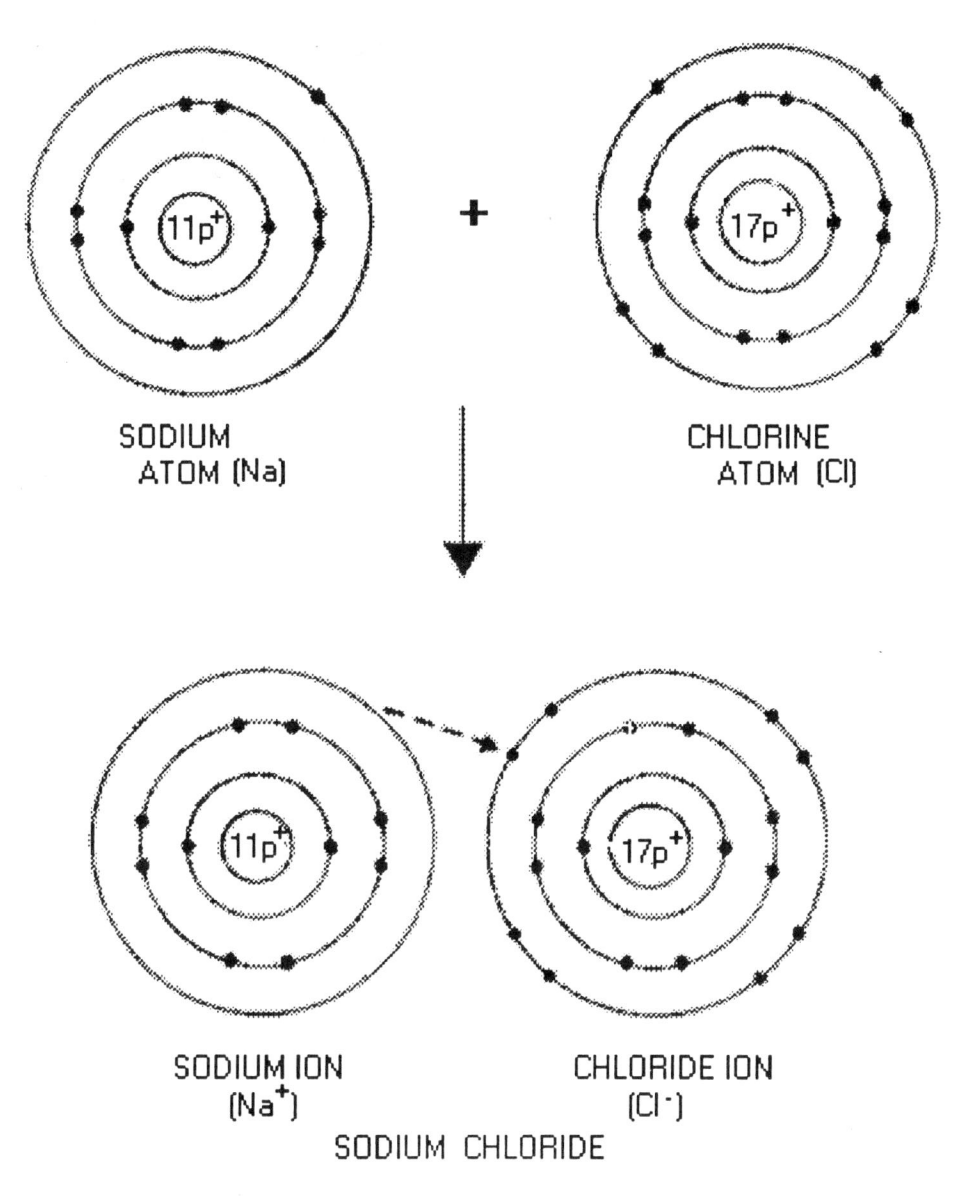

Figure A-3. An ionic bond is formed between a sodium and chlorine atom when the single electron from the outermost shell of sodium is transferred to the outermost shell of chlorine, providing a completed outer shell in both atoms. The resultant atoms, however, have an imbalance between positive charges and are called ions. The net charge is designated beside the chemical symbol for that element.

What are chemical "bonds"?

The word "bond" had a variety of meanings prior to its use in chemistry. For example, the word was used to mean a physical object or social arrangement to restrict freedom. Chemical bonds are not objects even though our models on paper may suggest they are. Instead, they represent *energy relationships* that restrict the freedom of atoms or molecules to varying degrees. Three types of chemical bonds are of major significance in explaining the composition of living matter—covalent bonds, ionic bonds, and hydrogen bonds. The strength and consistency of living matter depends upon the attractive forces of these bonds.

Covalent bonds have shared electrons.

Covalent bonds form when two or more atoms share pairs of electrons and achieve stable numbers of electrons in their outermost electron shells. For example, two hydrogen atoms form a covalent bond representing an *equal sharing* of an electron pair, thus achieving a completed first shell (Figure A-4a). The product is a hydrogen gas molecule.

The reaction in Figure A-4 is often diagrammed using dots or lines to represent electrons or electron pairs as follows:

H. + H. ——> H:H, or H + H ——> H–H

The hydrogen gas molecule is said to form **nonpolar covalent bonds** because the shared electrons orbit in such a manner as to allow equal distribution of negative charge throughout the volume surrounding both of the nuclei (Figure A-4b).

Two atoms of hydrogen combine with one oxygen atom to form covalent bonds of the water molecule (Figure A-5). As a result, oxygen attains the stable number of eight electrons in its second shell and hydrogen again achieves the stable number of two electrons. This reaction may be represented more simply as follows:

2H. + O: ——> H:O:H   or   2H + O ——> O
                                       / \
                                      H   H

Because the oxygen atom has more protons in its nucleus, the resultant greater positive attraction for electrons on the oxygen end of the molecule causes electrons to average more time on the oxygen end. This unequal sharing of negative charge is representative of **polar covalent bonds**.

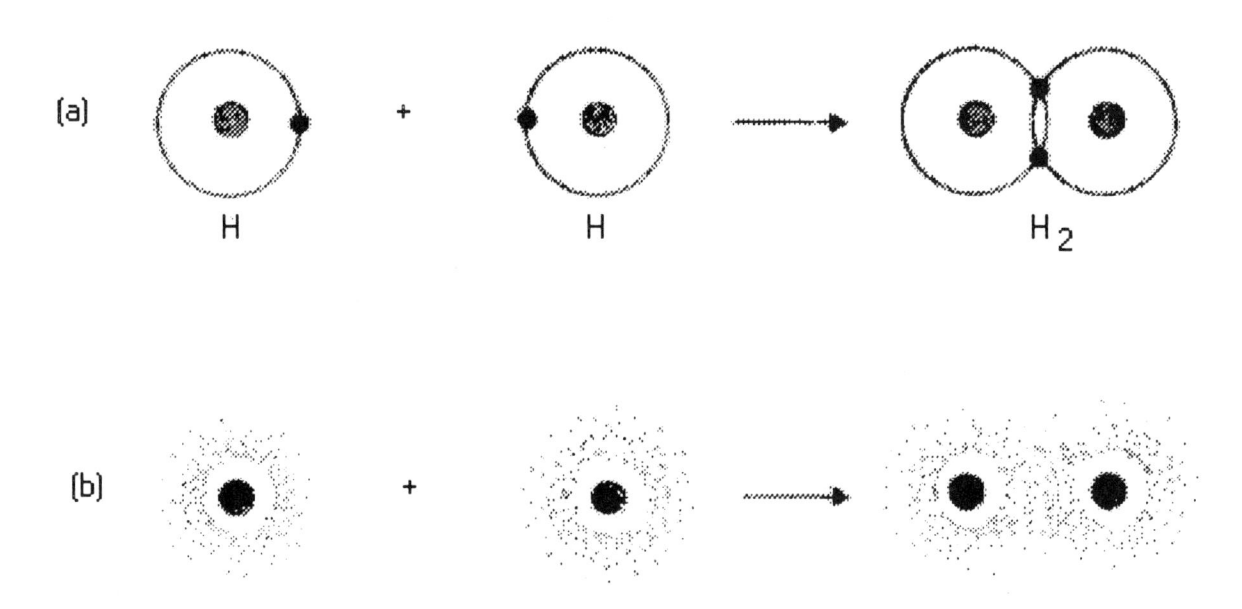

Figure A-4. Formation of a nonpolar covalent bond of the hydrogen gas molecule in which electrons are equally shared between the two hydrogen atoms.

(a)

2H + O → H₂O

(b)

+ −

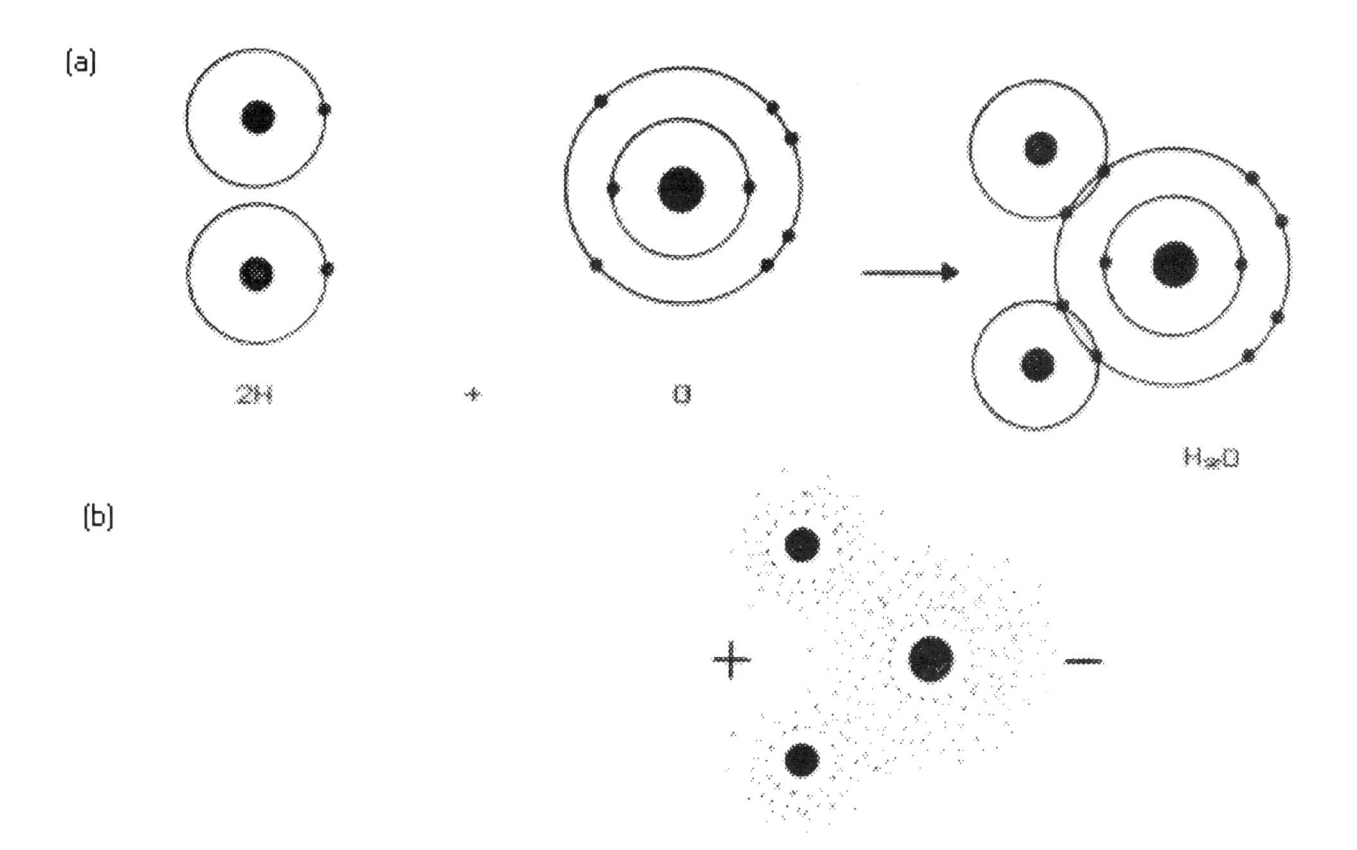

Figure A-5. Formation of polar covalent bonds of the water molecule in which electrons are unequally shared between an oxygen atom and two hydrogen atoms. The greater attraction for the electron cloud by oxygen causes that end of the molecule to be more negative, and the hydrogen end, more positive.

Nonpolar covalent bonds can result from the bonding of atoms of *two different elements* provided their affinities for the shared electrons are similar. Nonpolar methane gas molecules each consist of a carbon atom sharing four electrons of its outer shell, one with each of four hydrogen atoms:

$$H\cdot + \cdot \overset{\cdot\cdot}{\underset{\cdot\cdot}{C}}\cdot \longrightarrow H:\overset{\overset{H}{\cdot\cdot}}{\underset{\underset{H}{\cdot\cdot}}{C}}:H \quad \text{or} \quad H-\overset{\overset{H}{|}}{\underset{\underset{H}{|}}{C}}-H$$

Ionic bonds — donated e⁻'s

Ionic bonds are formed between atoms of different elements when one or more electrons are completely donated from one of the atoms to form a stable number of electrons in the outer shell of the other atom. Figure A-3 illustrates the formation of an ionic bond between a sodium atom and a chlorine atom to form sodium chloride, common table salt. Sodium atoms have a very loose affinity for the single electron in the third shell, whereas chlorine has a very strong affinity for one electron to complete its third shell. The result is a complete transfer of one electron from sodium to the electron cloud of the chlorine atom. Because the resultant sodium atom has one less negative charge than its total of positive charges, the atom is referred to as a sodium **ion**, designated Na^+. Ions are atoms that have an unequal number of electrons and protons. Thus, the chlorine atom with its extra electron is called a chloride ion, (Cl^-).

If you are now able to describe the atomic structure of atoms, explain the chemical differences between atoms of different elements, and discuss the concept of bonds and how they are formed, please move on to Chapter 4. There, you will be introduced to concepts of chemistry necessary to understand introductory biology.

Appendix B

An Abbreviated Taxonomic Classification

KINGDOM MONERA:
    Subkingdom ARCHAEBACTERIA - "ancient" bacteria
    Subkingdom EUBACTERIA - "true" bacteria
    Subkingdom CYANOBACTERIA - "blue-green bacteria

KINGDOM PROTISTA
    PLANT-LIKE DIVISIONS:
        Division PYRROPHYTA - "fire algae", dinoflagellates
        Division CHRYSOPHYTA - golden algae, diatoms
        Division EUGLENOPHYTA - euglenoids, green flagellates
    PROTOZOA:
        Phylum MASTIGOPHORA - flagellated protozoa
        Phylum SARCODINA - amoeboids, radiolarians, foraminiferans
        Phylum SPOROZOA - spore-bearing, parasitic protozoa
        Phylum CILIOPHORA - cilia-bearing protozoa

KINGDOM FUNGI
    Subkingdom MYXOMYCOTINEAE - plasmodial slime molds
    Subkingdom MASTIGOMYCOTINEAE - flagellated fungi
    Subkingdom EUMYCOTINEAE - true fungi
        Division ZYGOMYCOTA - hyphae lack crosswalls
        Division EUMYCOTA - hyphae have crosswalls
            Class ASCOMYCETES - sac, or cup fungi
            Class BASIDIOMYCETES - club fungi
            Class DEUTEROMYCETES - imperfect fungi

KINGDOM PLANTAE:
    ALGAE:
        Division RHODOPHYTA - red algae (*rhodo* = red)
        Division PHAEOPHYTA - brown algae (*phaio* = brown)
        Division CHLOROPHYTA - green algae (*chloro* = green)
    BRYOPHYTES:
        Division BRYOPHYTA - mosses and liverworts
    VASCULAR PLANTS (major divisions):
        Division SPHENOPHYTA - horsetails
        Division PTEROPHYTA - ferns
        Division CYCADOPHYTA - cycads
        Division GINKGOPHYTA - ginkgoes
        Division CONIFEROPHYTA - conifers
        Division ANTHOPHYTA - flowering plants
            Class DICOTYLEDONAE - dicots
            Class MONOCOTYLEDONAE - monocots

KINGDOM ANIMALIA:
- Phylum PORIFERA - sponges
- Phylum CNIDERIA - corals, jellyfish, *Hydra*
- Phylum PLATYHELMINTHES - flatworms, flukes, tapeworms
- Phylum ASCHELMINTHES - round worms (nematodes), rotifers
- Phylum ECHINODERMATA - sea stars, sea urchins, starfish
- Phylum MOLLUSCA - snails, clams, squid, octopus
- Phylum ANNELIDA - segmented worms—earthworms, leeches
- Phylum ARTHROPODA - jointed appendages, exoskeleton
- Phylum CHORDATA - cordates
  - Subphylum UROCHORDATA - Tunicates
  - Subphylum CEPHALOCHORDATA - Lancelets
  - Subphylum VERTEBRATA - Vertebrates (See Table 9-3)
    - Class AGNATHA - Jawless fishes
    - Class CHONDRICHTHYES - Cartilaginous fishes
    - Class OSTEICHTHYES - Bony fishes
    - Class AMPHIBIA - Amphibians
    - Class REPTILIA - Reptiles
    - Class AVES - Birds
    - Class MAMMALIA - Mammals

INDEX

Abiotic factor 138
Abiogenesis 42
ABO blood group 249-250
Abortion 308-309
Acid 55
Acid rain (& deposition) 103
Acquired immune deficiency syndrome (see AIDS)
Action potential 274
Active transport 273, 299, 375
Adaptation 30, 114, 268, 319
Adenine 66, 216
ADH (antidiuretic hormone) 300
ADP (adenosine diphosphate) 409
Adrenal gland 285
Age structure 117
Agriculture 84-88, 91
AIDS 397-399
Air pollution 98-100, 102-106
Albinism 252, 259
Algae 172-173
Allele 245
Aluminum 108-109
Amino acid 63-64
Ammonia 55-56, 101
Amphibian (Amphibia) 180
Amniocentesis 258
Amnion 306
Amniotic cavity 306
Amoeba 169, 193
Anabaena 164, 166
Anaerobe 186-187
Angiosperm 174
Animalia 177-180
Annual plant 190
Annelid (Annelida) 179
Anorexia nervosa 205
Antibiotics 165, 349
Antibody 250, 394ff
Anticodon 387
Antigen 250, 394ff
Applied science 19
Aquifer 96
Archaebacteria 164
Aristotle 323
Arteriole 203
Artery 202ff
Arthropod 179
Artificial selection 326, 341
Asexual reproduction 223
Atherosclerosis 208
Atmosphere 34
ATP (adenosine triphosphate) 409
Atom 28, 427

Atomic number 427
Autonomic nervous system 271, 278-281
Autosome 254
Autotroph 39, 76, 185
Auxin 284
Axon 272

Bacon, F. 9
Bacteria 164-166
Bacteriophage 397-398
Barr body 254
Basal metabolic rate (BMR) 295
Base 55
B cell (B lymphocyte) 394
Bile 201
Binomial nomenclature 36
Biochemistry 27, 57
Biogenesis 42, 213
Biogeochemical cycle 94
Biological magnification 105-106
Biome 36, 140
Biosphere 34-36
Biotic potential 115
Bird (Aves) 180
Birth control pill 305
Birth rate 115
Blastocyst 306
Blood 202
Blood group, human 249-250
Blue-green algae (see Cyanobacteria)
Bond, chemical 430
Brain 270, 276
Bryophyte (Bryophyta) 173-174
Bulemia 205
Calorie (kilocalorie) 79-81
Cambium 190, 311
Cancer 396
Capillary 198, 202
Carbohydrates 59
Carbon cycle 98-100
Carbon dioxide 98-100, 301
Carcinogen 257
Carnivore 39-40
Carrying capacity (K) 119
Catastrophism 321
Categorical complementarity 12-14
Cell division 188
Cell elongation 193
Cell structure 28, 365
   animal 365-367
   plant 365-367
Cell theory 42
Cellular metabolism 362, 387
Cellulose 60-61

Cell wall 163, 191, 366-367
Central nervous system 270, 276
Centriole 221-222
Centromere 220
Chaparral 147-148
Chemical bond 52, 430
Chemical bond energy 94, 431
Chemosynthesis 76, 91
Chlorophyll 187, 413-416
Chloroplast 413-415
Cholesterol 208
Chordate (Chordata) 180
Chorionic gonadotropin 305
Chorionic villus sampling 258
Chromatid 220
Chromosome 218, 252
Circulatory system 198, 202
Classification system 34-36, 161ff
Climax community 123
Clone 224, 316
Clostridium 165
Cniderian (Cnideria) 179
Cocaine 276
Codon 386-387
Coelom 178-179, 197
Commensalism 123
Community, biotic 28, 71
Condom 397
Conifer (See Gymnosperm)
Consumer 39-40
Continuous variation 251
Coral, reefs 148
Corpus luteum 305
Cortex 190, 193
Cotyledon 174
Covalent bond 52, 432
Creationism 321
Crick, F.H.C. 216, 329, 383
Crossing over 227
Cyanobacteria 166
Cytoplasm 365
Cytosine 66, 216

Darwin, C. 38, 241, 325
Darwin's finches 326
Day-age theory 12
Death rate 115
Decarte, R. 10
Deciduous forest 145, 150
Decomposer 39, 76, 88-90
Deforestation 98-99
Dehydration synthesis 54
Deism 10
Dendrite 271-272
Denitrification 102
Density-dependent factor 120

Density-independent factor 120
Deoxyribonucleic acid(DNA) 65, 216, 383
Desert biome 147, 150
Detritivore 39, 88-90
Detritus 39, 88
Development,
  human 305
  plant 315
Diatom 167-168
Dicot (dicotyledon) 174-176, 311-312
Diet 204-209
Differentiation 193, 215, 305, 311, 315, 389
Diffusion 372
Digestion 186, 193
  earthworm 197-198
  human 199
  Hydra 196
  intracellular 193-197
  planaria 197
  protozoa 194-195
  sponge 195
  vertebrates 198-202
Dihybrid cross 246
Diploid cell 226
Disease 122
DNA (deoxyribonucleic acid)
  molecular structure 65-66
  replication 219, 384
  gene expression 384
Dominant allele 244
Down syndrome 257
Drosophila (fruit fly) 337, 342ff
Drugs 276

Earthworm 197-198
Eating disorders 205
Echinodermata 179
Ecosystem 28, 71
Ectoderm 306-307
Ectotherm 292
Egg (see Ovum)
Electromagnetic radiation 74
Electron 427
Electron transport system 414, 419
Embryo 215
  human 305-309
  plant 174, 310
Endocrine gland 199, 283-286
Endocytosis 376-377
Endoderm 306-307
Endoplasmic reticulum 368
Endosperm 232, 310
Endotherm 179, 292
Energy 73, 407ff
Energy and matter 106-108
Energy flow 75ff

Entropy 76, 336
Environmental resistance 120
Environmental stewardship 151-154
Enzyme 63, 417-418
Epidermis
  animal 269
  plant 188ff
Epistemology 5
Erosion 97
Escherichia coli 400
Essential amino acid 86, 208
Essential fatty acid 208
Estrogen 304
Euglena 167-168
Eukaryote 42, 163, 166
Eutrophication 105
Evaporation 96
Evolution 3-4, 320ff
  Darwinian 325-328
  human 345-346, 351
  modern synthetic theory 329-330
  neo-Darwinian 329-330
  organic 338
  punctuated equilibrium 334
Evolutionism 16-17, 320, 341
Excited state 413-416, 429
Excretory system 297
Exocrine gland 199, 283-286
Exon 385
Extinction 139-140, 213, 332

Faith and science 5
Fat 61-63, 208
Fatty acid 61
Feedback control
  in organisms 293
  in populations 119-120
Fern 174-175
Fermentation 165, 417
Fertilization 213, 232
Fibrinogen 202
Fire 98-99, 123
Fission 223
Flagellum 165-168
Flatworm (Platyhelminthes)
  177, 179, 197
Flower 231
Flowering plant (Angiosperm) 174, 231
Fluid mosaic model 371
Fluorescence 414-416
Follicle 228, 303
Follicle-stimulating hormone (FSH) 303
Food chain 39, 76
Food vacuole 193
Food web 82ff
Fossil 320, 330, 334

Fossil fuel 99, 100, 102
Fruit 311
Fungi 170-171, 186

Gaia hypothesis 137
Gallbladder 201
Gamete 40
Gametophyte 230
Gastrovascular cavity 196
Gene 217, 244
Gene expression 384
Gene pool 114, 341ff
Genetic feedback control 390-392
Genetic counseling 258
Genetic engineering 400-403
Genetics 241, 381
Genetic variation 251
Genome Project, Human 252
Geotropism (see gravitropism)
Global warming 99
Glucagon 295
Glucose 59-61
Glycogen 61
Glycolysis 419
Golgi bodies 368, 377
Grassland biome 146, 150
Gravitropism 314
Grazing food web 88
Greenhouse effect 99
Gross primary productivity (GPP) 79
Ground tissue 190
Guanine 66, 216
Guard cell 188
Gymnosperm 174

Habitat 113
Haploid cell 226
Hardy-Weinberg principle 342
Heart 203-205
Hemoglobin 198
Hemophilia 254, 345
Henle's loop 299
Herbaceous plant 190
Herbivore 76
Hermaphrodism 225
Herpes virus 398
Heterotroph 39, 76, 185-186
Heterozygous 246
Histamine 393
Homeostasis 30, 291, 363
Homologous pair 218, 226
Homologous structure 323
Homozygous 245
Hooke, R. 41-42
Hormone 283, 315
  animal 283-287

plant 284, 313-316
Horsetail 174-175
Human immunodeficiency
  virus (HIV) 397-399
Hydra 179, 196, 224-225
Hydrogen bond 52-53, 64
Hydrogen ion 55-56
Hydrolysis 54
Hydrophilic interaction 369
Hydrophobic interaction 369
Hydrosphere 34
Hypothalamus 277, 286, 300
Hypothesis 15-16

Immune system 394ff
Impulse 274
Incomplete dominance 249
Independent assortment 226-228, 247
Induced-fit hypothesis 418
Industrial malanism 350
Inflammatory response 393
Inorganic molecule 53-56
Insulin 64, 295
Interactions, population 121-122
Intercourse, sexual 235-237
Intestine 197, 201
Intrinsic growth rate
  (See Biotic Potential)
Intron 385
Ion 52, 433
Ionic bond 52, 433
Ionizing radiation 74
Islets of Langerhans
  (See Pancreatic Islets)
Isomer 59
Isotope 331, 428

J-shaped curve 115
Jumping gene 255

Karyotype 218
Kidney 296-301
Kind, created 161, 347
Kinetic energy 73, 372
Klinefelter syndrome 257
Krebs cycle 419-420

Lactose operon 389
Lamarck, J. 38, 324-325
Landfill 106-109, 156
Language 351
Law, scientific 16
Leaching 97, 105
Leaf 188, 413-414
Legume 101-102
Less developed country (LDC) 130

Leukocytes 202
Levels of organization 28, 33-35, 46
Lichens 125
Life cycle 40, 215
Light absorption 413ff
Limiting factor 119-120
Linked genes 252
Linnaeus, C. 36, 323
Lipase 201
Lipid 61-63
Lipoprotein 208
Lithosphere 34
Liver 200-201
Lungs 302
Lutenizing hormone (LH) 304
Lyell, C. 331
Lymph 204
Lymphocyte 394
Lysosome 194, 367

Macroevolution 339
Macrophage 394-395
Malthus, T. 326
Mammal (Mammalia) 180
Marine (oceanic) biome 148, 150
McClintock, B. 255
Mechanistic philosophy 10, 14, 49
Meiosis 215, 226-228
Melanin 251
Membrane 273, 369ff
Mendel, G. 242ff. 328
Mendelian genetics 242
Menstrual cycle 303
Meristem 188, 311
Mesoderm 306-307
Mesophyll, leaf 188
Mesozoic 333
Messenger RNA (mRNA) 386
Metabolism 29, 361ff
MHC marker 394
Microevolution 339-341
Microvillus(i) 201
Mineral cycles 92, 104-105
Mitochondrion 364, 366-368, 419-421
Mitosis 215
Molecule 28, 52
Mollusc (Mollusca) 179
Moneran (Monera) 38, 163
Monocot (monocotyledon) 174-176, 312
Monoculture 84
Monocyte (see Macrophage)
Monohybrid cross 242
Monosaccharide 59
Morgan, T.H. 329
Moss 173-174
Motor neuron 272

Multiple alleles 249, 347
Multiple genes 251, 347
Muscle 269, 280-281
Mutation 255, 336ff
Mutualism 72, 187

NAD (nicotinamide adenine
    dinucleodide) 419
NADP (nicotinamide adenine
    dinucleotide phosphate) 415
Narcotic drugs 276
Naturalism 10, 320
Natural selection,
    brief overview 121
    details 341ff
Negative feedback
    (See Feedback Control)
Nematocyst 196
Neo-Darwinism (See Evolution)
Nephron 297-299
Nerve impulse
    (See Action Potential)
Nervous system
    earthworm 269-270
    Hydra 268
    human 269ff
    planaria 269-270
Net primary productivity 79
Neuron (nerve cell) 269-271
Neurotransmitter
    (Transmitter substance) 274
Neutron 427
Neutrophil 394-395
Niche 76, 113, 325
Nitrate 100-103
Nitrogen base 65, 216
Nitrogen cycle 100
Nitrogen fixation 56, 101
Nucleic acid 65-66, 216
Nucleolus 218
Nucleotides 63, 216
Nucleus, atomic 427
Nucleus, cell 28, 218, 366-367
Nutrient 93, 206
Nutrient cycles 93
Nutrition 185, 204ff

Objective truth 6, 126
Oceanic biome 148, 150
Oncogene 257
Oocyte 234-235
Operation science 320
Operator gene 389
Operon 389
Organelle 28, 365-368
Organic molecule 53, 56

Organ/organ system 28, 198
Origin science 320
Osmoregulation 297
Osmosis 297, 372
Ovary
    human 234, 303-304
    plant 231, 310
Oviduct 234
Ovulation 234, 304
Ovule 310
Ovum 40, 213, 228
Ozone layer 99-100

Paleozoic 333
Palisade cell 188-189
Pancreas 200-201
Pancreatic islets 284-285, 295
Paradigm 16
Paramecium 170, 193-194
Parasite 39
Parasympathetic 271, 281
Pasteur, L. 42
Pathogen 393
Peppered moth 350
Pepsin 201
Peptide bond 64
Perennial plant 190
Peripheral nervous system 270-271
Peristalsis 199
Permeability 371-375
Pesticide 105
pH (scale) 55-56
Phagocyte 394
Phenotype 121, 242
Phloem 174, 188
Phospholipid 63, 369-370
Photosynthesis,
    community 76ff, 93ff
    cellular 412-415
Phototropism 314
Phylogeny 162
Phylum 37
Pinocytosis 377
Pith 190
Pituitary gland 277, 285-286
Placenta,
    mammalian 306
    plant 310
Planarian 197
Plantae 172, 187
Plasma 202
Plasmid 400
Plasmodesmata 367
Platyhelminthes 179
Pleasure center 276
Polymer 60, 63-65

Polymerase chain reaction (PCR) 402
Polysaccharide 60
Popper, K. 319
Population 28, 113
  genetics 250, 260
  growth 114ff
  human 126-132
Porifera 179
Postsynaptic neuron 274
Potential energy 73
Pregnancy 235-236, 306
Presynaptic neuron 274
Producer (see Autotroph)
Progesterone 305
Prokaryote 42, 163
Prostaglandins 309
Prostate gland 233
Protein 63, 208
Protein synthesis 64, 384, 386
Proterozoic 333
Protista 38, 166
Proton 427
Protozoan 168
Punctuated equilibrium 334
Pure science 18
Pyramids, ecological 81-82
Pyruvic acid (pyruvate) 419-420

Quaternary period 333

Races, human 260, 351
Rain forest
  biome 142, 143, 145-146
  deforestation 98-99
Receptor, sensory 271
Recessive allele 244
Recombinant DNA 400
Recycling/Reuse 106-109, 153-154
Red blood cell 202
Reductionism 10
Reflex arc 279
Replacement level fertility (RLF) 127
Repressor gene 389
Repressor protein 389
Reproduction 40, 213
  algae 229
  asexual 223
  earthworm 225
  flowering plants 231-232
  fungi 224
  human 232
  Hydra 223-225
  mosses/ferns 230
  protozoa 223
  sexual 224

Reptile (Reptilia) 180
Reservoir 94
Respiration
  cellular 416-422
  community 76ff, 93ff
Responsiveness 267, 347
Rh blood type 250
Ribonucleic acid (RNA) 384ff
Ribose sugar 59-60
Ribosome 368, 386
RNA (see Ribonucleic acid) 384ff
RNA polymerase 385
Root 192

Salivary,
  amylase 199
  gland 199
Schlieden, M.J. 42
Schwann, T. 42
Science 2
Scientific creationists 12, 321
Scientific method 15
Scientism 10
Seed 174, 229, 232-233, 310, 313
Selection (see Natural selection)
Self, biblical view of 261
Sensory neuron 271
Sex chromosome 254
Sex-linked trait 254
Sexual intercourse (human) 235-237
Sexuality, biblical view 236-237
Sexual organs 233-235
Sexual reproduction (see Reproduction)
Sickle-cell anemia 345-346
Sieve tube 191
Smith, W. ["strata"] 331
Sociobiology 10
Solute 369
Solvent 54, 369
Speciation 348-349
Species 36, 114, 161
Sperm 40, 213
Spinal cord 270, 279-282
Spindle 220-222
Sponge (Porifera) 179, 195
Spongy mesophyll 188-189
Spore 229
Sporophyte 230
Starch 60-61
Stem 190
Stem cell 272
Steno, N. 330-331
Steroid 63
Stewardship

environmental 151-154
  human nutrition 204-209
Stimulus 267, 278
Stoma (stomate) 188-189
Stomach 200
Structural gene 389
Succession, ecological 123-125
Sulfur cycle 103
Supernaturalism 321
Survivorship curve 117-118
Symbiosis 101, 187
Synapse 274
Szent-Gyorgyi, A. 49

Taiga 144, 150
Taxonomy (See Classification)
T cell 394
Teleology 10, 14
Temperature regulation 291
Territoriality 122
Tertiary period 333
Testes 225, 228, 233
Testosterone 233
Theism 10
Theology 5
Theory, scientific 16, 319-320
Thermodynamic laws 75
Thermoregulation 291
Thymine 66, 216
Thyroid gland 284-285, 296
Tissue culture 315-316
Tolerance range 138-139
Total fertility rate (TFR) 127
Toxin 275
Transcription, 384-386
  reverse, 405
Transfer RNA (tRNA) 386
Transitional form 332
Translation 384, 386
Transmitter substance 274
Transposon 255, 400
Trophic structure 39, 76
Tropical rain forest (See Rain Forest)
Tundra 143, 150
Turner syndrome 258

Ultraviolet rays 74, 256-257
Uniformitarianism 320, 331
Unsaturated fat 62, 208
Uracil 385
Urinary system 297
Uterus 233-236, 303

Vacuole 366-367
Van Helmont, J. 187
Variation, genetic 251

Vascular bundle 174, 190
Vein 188, 202
Vertebrate 180
Vessel, xylem 191
Vestigial structure 324
Villus 201
Virus 397
Vitalism 49
Vitamin 206-207

Water 54, 368
Water cycle 96-97
Water pollution 97, 105
Watson, J.D. 216, 329, 383
Weathering 97
Weismann, A. 329
White blood cell
  (see B cell, T cell)
Woody plant 190
Worldview 5, 150-151, 320

X chromosome 254
X-linked genes 254
Xylem 174, 190

Y chromosome 254

Zero population growth 126
zygote 213